T5-DHI-896

# Elementary Matrix Algebra With Applications

## second edition

**Richard J. Painter**
*Colorado State University*

**Richard P. Yantis**
*Otterbein College*

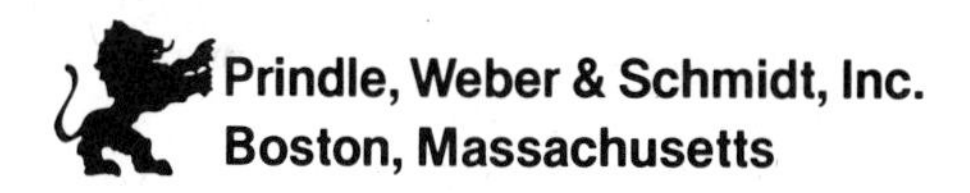

**Prindle, Weber & Schmidt, Inc.**
**Boston, Massachusetts**

**To Jan and Jane**



Printed in the United States of America
Second printing: September 1977

**Library of Congress Cataloging in Publication Data:**

Painter, Richard J.
Elementary matrix algebra with applications.

First published in 1971 under title: Elementary matrix algebra with linear programming.

Includes index.
1. Matrices. 2. Linear programming. I. Yantis, Richard P., joint author. II. Title.
QA188.P34 1977 512'.943 77-2128

ISBN 0-87150-227-5

Cover design by John J. Sorbie. Text design by Michael Michaud and the production staff of Prindle, Weber & Schmidt, Inc. Text composition by Syntax International in Times Roman and Helvetica. Drawings by Publications, Illustrations and Presentations, Inc. Cover printing by New England Book Components. Text printing and binding by Halliday Lithograph Corp.

# Preface

Today's college students are vitally interested in applications of their mathematical training. They prefer not to wait until the "next course" to see how the concepts and calculations can be used in real-world affairs. Thus, we have structured the contents to allow the college student with only a background of high school algebra, first to learn some matrix algebra, and then to see it used in a realistic setting. Applications are taken from the fields of actuarial science, business, economics, engineering, manpower planning, operations research, psychology, statistics, and transportation science. They are spread throughout the theoretical chapters, and are the heart of applied chapters 3, 6, 7, 9, and 10.

The pedagogical development of the text is based on four main themes: 1) Students learn new concepts faster and retain them longer when they are able to compare the new ideas with familiar ones; 2) It is important that the student understand and perform the computational algorithms for matrix algebra; 3) Numerous examples and exercises are a significant part of the learning process; 4) Unnecessary symbolism and unfamiliar notation often inhibit the student's ability to learn the basic facts of matrix algebra.

Chapter 1 reviews the elementary aspects of sets, functions, and real number properties presented in high school algebra. Matrices are then introduced, and the basic rules for matrix algebra are presented as analogues of the appropriate real number properties.

The development of matrix theory continues in Chapter 2 with the introduction of systems of linear equations, their representation by matrices, and their solution by the Gauss-Jordan reduction technique. This technique provides continuity for many of the examples and exercises. To simplify the arithmetic, many of the examples and exercises are designed so that all of the coefficients remain integers.

Chapter 3 discusses an important application of matrix algebra to economics — input-output analysis. It emphasizes that matrix products are particularly well-suited to handle practical situations where one requirement leads to another requirement, and it to another, and so on. This chapter is independent of the remaining text.

Chapter 4 considers both the classical definition of a vector as an arrow and the modern definition as an element of a vector space. Only vector spaces consisting of sets of row or column matrices of a certain size are dealt with in the body of the text. The results of Chapter 2 concerning linear equations are restated in vector space terminology. The dot product is introduced and used to define length, distance, and angle in a vector space.

Chapter 5 is concerned with determinants, rank, and change of basis, as well as certain aspects of linear transformations. Sections 5.6-5.10 introduce the concept of a linear transformation from a vector space into itself and find its matrix representation with respect to a fixed basis. The chapter concludes with a study of the eigenvector problem and a short study of the terminology and matrix representation of quadratic forms. Sections 5.6-5.10 are *also* independent, although they are of prime importance in many fields of application.

Linear Programming (LP) is developed in Chapter 6. Several real-world decision situations are modeled as LP problems. Elementary geometry and matrix algebra are then used to foster understanding of the well-known simplex algorithm. The algebraic solutions of the simplex algorithm are developed in coordination with a geometric solution of the same problem. As in earlier chapters, many of the exercises have been designed in a semiprogrammed fashion.

Chapter 7 studies the Transportation Problem (TP) as a specialized case of linear programming. The special algebraic structure of the TP is examined and then used to create an individually-tailored solution algorithm. The "transportation algorithm" enables its many users to solve very large problems faster and with less computer space than would otherwise be required if the standard simplex algorithm were used. Problems arising from practical situations other than those dealing with transportation *per se* are also formulated and shown to have the same special structure.

Chapter 8 provides an introduction to probability, which is defined in terms of relative frequency. Trees, conditional probability, and the differences between dependent and independent events are dealt with fully. Games of chance provide the vehicle for studying many of the topics. These sections (8.1-8.8) are also independent of the other material and can be studied any time after Chapter 1. Chapter 8 continues with a short introduction to stochastic matrices and Markov chains, and ends with an example of the use of such matrices in the insurance industry.

In Chapter 9, Manpower Planning, matrices are used first to predict the long-range structure of an organization and then to predict the effects of different recruitment, attrition, and promotion policies. In Chapter 10 we use the regression line as an example of one of two seemingly different "least squares" problems that arise in many applied fields. We conclude by demonstrating that one of the problems is merely a special case of the other.

We would like to express our appreciation to John Martindale, Locke Macdonald, Michael Michaud, Dana Andrus, and the rest of the publisher's staff for their support, advice, and extremely competent preparation of the manuscript. In addition, we would like to thank the following for their comments, corrections, and constructive criticisms:

K.F. Anderson, University of Alberta
Joseph G. Ecker, Rennselaer Polytechnic Institute
William E. Coppage, Wright State University
Richard L. Burden, Youngstown State University
John J. Hutchinson, Washington State University
Burt C. Horne, Jr., Virginia Polytechnic Institute & State University
Dennis R. Dunninger, Michigan State University
John J. Dinkel, Pennsylvania State University
Robert Glassey, Indiana University



Richard J. Painter
Richard P. Yantis

# Contents

## Matrices with Real Elements 1

## Systems of Linear Equations and Gauss-Jordan Reduction 46

## Economic Planning by Input-Output Analysis 86

## Vector Spaces and Systems of Linear Equations 98

## Determinants, Rank, and Change of Basis 137

## Linear Programming 192

## The Transportation Problem 268

8

## Probability 311

## Manpower Planning 353

## Applications in the Social and Behavioral Sciences — "Two" Least Squares Problems 363

## Appendices

The following chart illustrates the relationship of the various parts of the text.

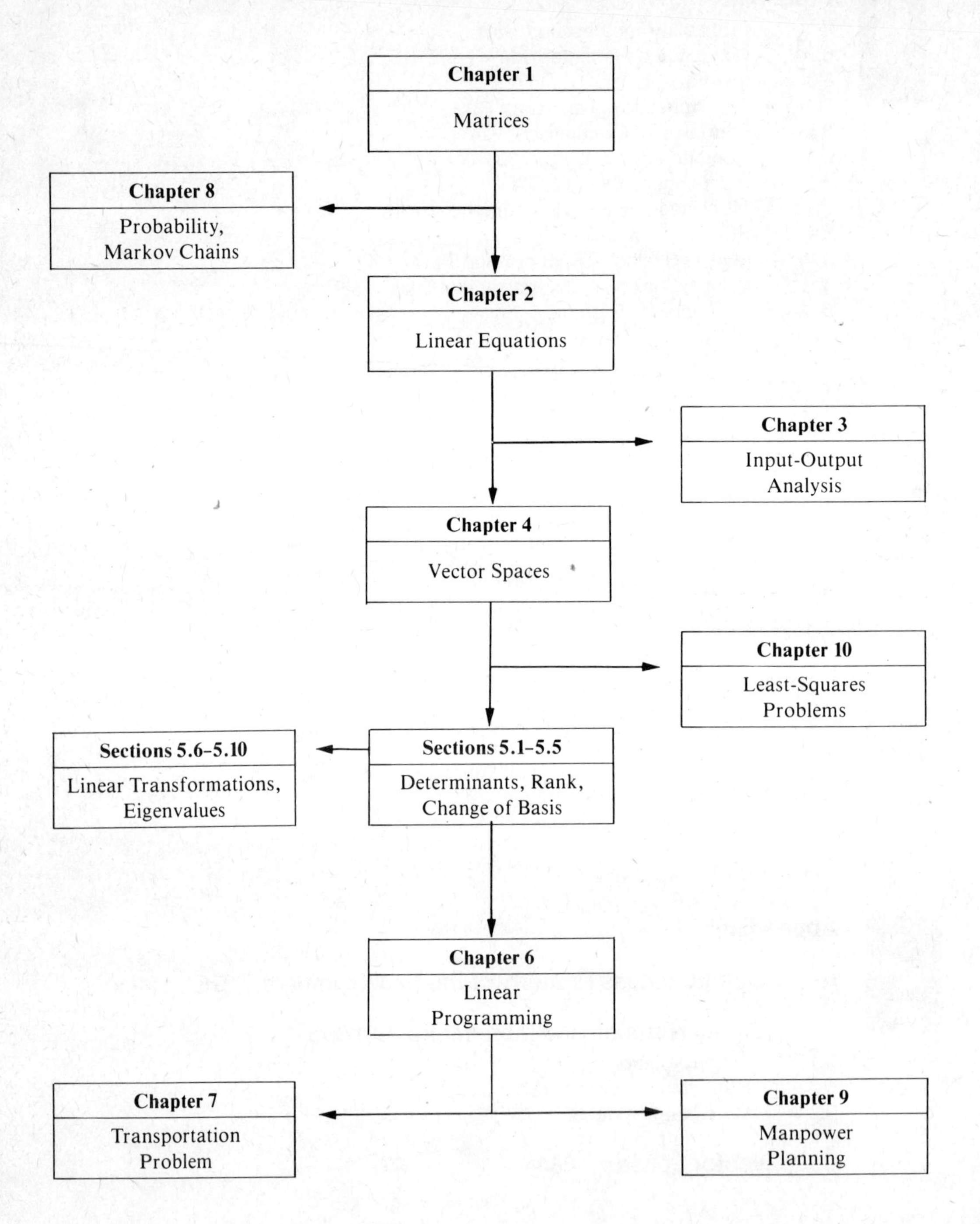

# 1 Matrices with Real Elements

Rectangular arrays of numbers known as *matrices* are the central subject of this book. In this beginning chapter we will first develop some preliminary ideas concerning the elementary concepts of *set* and *function* and will then outline the algebraic structure of the familiar real number system. Next we will proceed to define precisely what we mean by the term *matrix* and to introduce concepts of "addition" and "multiplication" for these new entities. We will conclude this chapter with an examination of the basic algebraic properties of matrices and their analogies with real number properties discussed earlier.

In the later chapters we will introduce other fundamental matrix concepts and also look at several recent practical applications of matrix methods in economics, business, transportation, manpower planning, insurance, and decision theory.* We will also provide an introduction to elementary probability so that the set of potential matrix applications might be more fully and realistically comprehended.

## 1.1 Sets and Functions

The words *set*, *element*, *belongs to* are, for us, undefined terms; and we must rely upon you to use your intuition in understanding these words. When we speak of the *set* of, say, all pages in this book, we are confident that you recognize page 1 as an *element* in that set, which is to say that page 1 *belongs to* that set.

* The Preface provides a roadmap for the interested reader.

A particular set is considered to be properly defined or specified when it is described in such a way that one can distinguish between objects which belong to the set and objects which do not belong to the set. For example, consider the set of all coins you have with you at this moment. It is clear that, if an object is given to you for consideration, you can decide whether or not it is one of your coins. On the other hand, there are some descriptions of "sets" that are so vague, or ambiguous, or so dependent upon personal opinion, that it is not possible to distinguish elements from non-elements. For instance, consider the "set" of the ten greatest American presidents. We are a little uncertain when we try to answer "What are the elements of this 'set'?" Perhaps we could all agree to include Lincoln, and we probably could also agree to exclude Andrew Johnson. But a motion to include Lyndon Johnson, Truman, Wilson, or Nixon would be likely to arouse heated debate. Thus the phrase "the ten greatest American presidents" does not properly define a set, because it does not enable one to unambiguously determine which objects are elements and which are not elements.

There are several symbols which we use to denote or to describe sets: First, we will most commonly use a capital letter such as $A$ or $T$ to stand for a given set, and, second, we will frequently describe a set by writing down a list of its elements enclosed in braces—for example, the set consisting of the first three counting numbers (positive integers) may be denoted as

$$\{1, 2, 3\}.$$

In certain instances, however, these listings could be awkward, exhausting, or even impossible to make. If we consider, even for a moment, the set consisting of the first 1000 counting numbers, the prospect of writing them all down in a list is not very appealing. Instead, we write

$$\{1, 2, 3, \ldots, 1000\}$$

and hope that whoever sees this notation will understand that the dots stand for all of the counting numbers between 4 and 999 inclusive. An extension of this device occurs when the set $\mathbb{Z}^+$ of *all* counting numbers is denoted as

$$\{1, 2, 3, \ldots\},$$

or when the set of even counting numbers is denoted as

$$\{2, 4, 6, \ldots\}.$$

Even though something is being left to the imagination in each of these cases, the first three numbers should make clear the proper interpretation of the three dots (. . .).

A third type of notation for sets, which is not ambiguous, is illustrated by the following example. Consider the set of all counting numbers which we showed above as $\{1, 2, 3, \ldots\}$. An alternative description is provided by the expression

$$\{n \mid n \text{ is a counting number}\},$$

which may be read as follows: "the set of *all* objects $n$ such that $n$ is a counting number." Or consider the set $T$ denoted by

$$\{t \mid t \text{ is an equilateral triangle}\}.$$

It says, "$T$ is the set of *all* objects $t$ such that $t$ is an equilateral triangle." In general, then, a notational shorthand

$$\{x \mid p(x)\},$$

called the "set-builder" notation, will be considered to stand for the set of *all* objects $x$ such that $x$ satisfies the properties described in the sentence $p(x)$. The braces $\{ \}$ denote *set* and the vertical bar $|$ is read *such that.*

We say that a set $B$ is a **subset** of a set $A$ if each element of $B$ is also an element of $A$. For instance, the set $E$ of all even counting numbers is a subset of the set $\mathbb{Z}^+$ of all counting numbers. We denote the situation that $B$ is a subset of the set $A$ by

$$B \subseteq A \quad \text{or, equivalently,} \quad A \supseteq B.$$

In words: "$B$ is contained in $A$" or "$A$ contains $B$." Note that this definition of subset implies that a set $A$ is always a subset of itself, since every element of $A$ is obviously an element of $A$. We distinguish between this trivial subset of $A$ and other subsets of $A$ by saying that $B$ is a **proper** subset of $A$ providing $B$ is a subset of $A$ and, further, that $A$ has at least one element which is not in $B$.

We say that two sets $C$ and $D$ are **equal** and can write

$$C = D$$

as long as the two sets have precisely the same elements. A moment's reflection will assure you that this is equivalent to saying, "$C$ is a subset of $D$ and $D$ is a subset of $C$."

When we speak of the **union** of two sets $A$ and $B$ we mean the set $C$ whose elements are exactly the elements of $A$, taken together with the elements of $B$. We denote the union of $A$ and $B$ as

$$A \cup B \qquad \text{(read "}A\text{ union }B\text{")}$$

and note that

$$A \cup B = \{x \mid x \text{ belongs to } A \text{ or } x \text{ belongs to } B\}.$$

By the **intersection** of two sets $A$ and $B$ we mean the set $C$ whose elements are precisely those elements which are both in $A$ and in $B$. We denote the intersection of $A$ and $B$ by

$$A \cap B \qquad \text{(read "}A\text{ intersect }B\text{")}$$

and note that

$$A \cap B = \{x \mid x \text{ belongs to } A \text{ and } x \text{ belongs to } B\}.$$

This definition of intersection leads us to consider a curious set. First, consider the set $C$ of all cards in an ordinary deck of playing cards. If $R$ denotes the set of all red cards (diamonds and hearts) and $B$ denotes the set of all black cards (clubs and spades) then clearly $R \subseteq C$ and $B \subseteq C$. Moreover, $R \cup B = C$ (Why?). But what is

$$R \cap B?$$

According to the above definition, $R \cap B$ *is* a set; and it is also easy to decide which objects belong to it and which do not, for clearly any object which you care to name does *not* belong to $R \cap B$. We call this set the

**null set** and denote it by $\varnothing$. Of course, there are many ways to describe $\varnothing$ other than in terms of playing cards. Observe, for example, that

$$\varnothing = \{1, 2, 3\} \cap \{4, 5, 6\},$$

and also that

$$\varnothing = \{x \mid x \text{ is a motorcycle with 79 wheels}\}.$$

As a matter of some interest, observe also that $\varnothing$ is a subset of every set $A$; for it is true (Isn't it?) that each object in $\varnothing$ is also in $A$. (This could only be false if there were objects in $\varnothing$ which were not in $A$; since there are no such objects in $\varnothing$, the statement cannot be false.) It is sometimes helpful to use a **Venn diagram** to discuss sets. If $A$, $B$ are sets and $A \subseteq B$ we show this diagrammatically as follows:

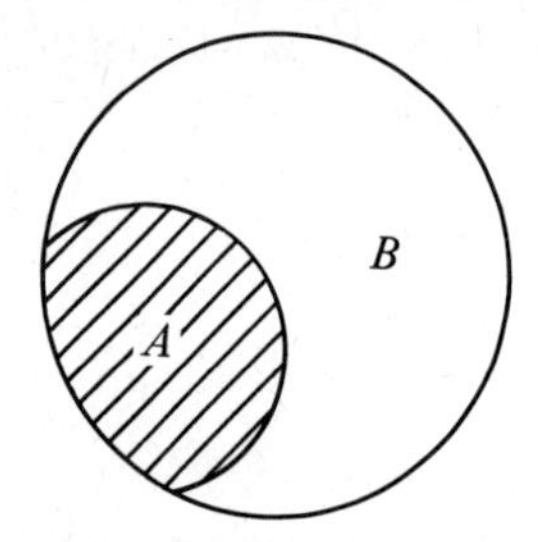


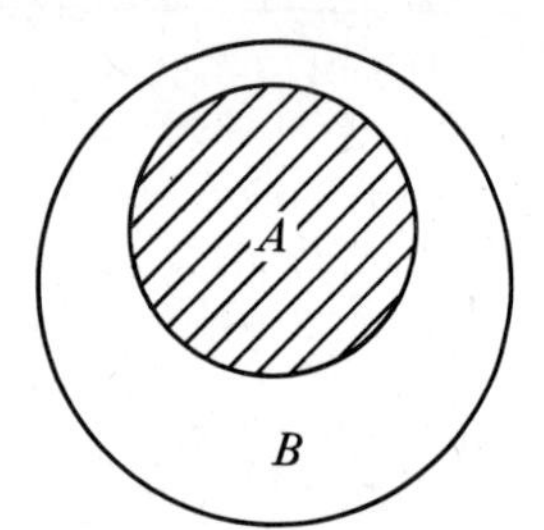


If we are interested in showing an element $x$ in $A \cap B$ we have

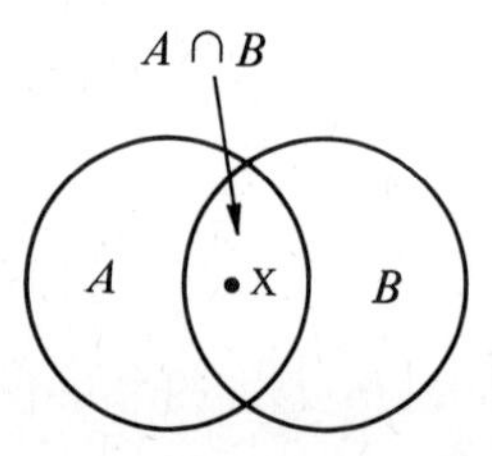


If we are interested in displaying all of $A \cap$ B, we simply shade the area common to the circles representing the sets $A$ and $B$. Thus we have

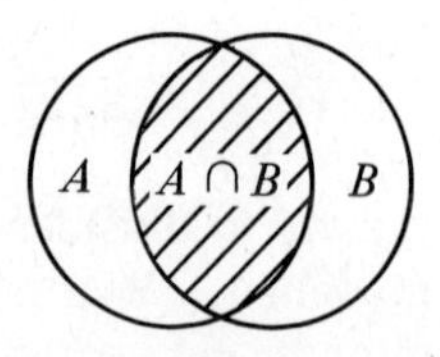


pictorially showing that $A \cap B$ consists of all elements in both $A$ and $B$.

If $A$ and $B$ are each subsets of a larger set $U$ we could then define the set $\tilde{A}$, known as the **complement** of $A$, to include all elements of $U$ *not* in $A$, and say that $\tilde{A} = U - A$. This expression is portrayed in Figure 1.1.1. If $U$ is the set of all positive integers and $A$ is the set of all even positive integers, then $\tilde{A} = U - A$ is the set of all odd positive integers. Venn diagrams can also be used to display more than two sets. (See Exercise 12.)

**Figure 1.1.1**

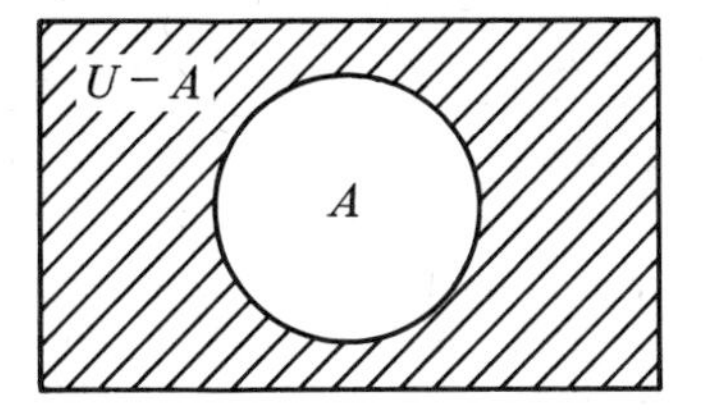


We now proceed to use our newly defined set terminology to briefly consider some of the basic vocabulary related to the idea of *function*. We will depend once more on your intuition as we introduce first the phrase **ordered pair**. By this, we mean a pair of elements $(a, b)$ from some set ($a$ and $b$ are not necessarily different elements) with the understanding that the pair $(a, b)$ is different from the pair $(b, a)$, unless $a = b$. This is why a pair is called *ordered*. Thus (2, 5) denotes an ordered pair of counting numbers, and is different from the ordered pair (5, 2). We say that the ordered pair $(a, b)$ is **equal** to the ordered pair $(c, d)$ provided that $a = c$ and $b = d$; that is, $(a, b) = (c, d)$ when the ordered pairs have the same first elements ($a$ and $c$) and have the same second elements ($b$ and $d$). Thus $(2, 5) = (4/2, 25/5)$.

Next, let us consider two sets $A$ and $B$ and define the **cartesian product** of $A$ and $B$ to be the set of *all* ordered pairs of the form $(a, b)$ where $a$ is in $A$ and $b$ is in $B$. Our symbol for this cartesian product is $A \times B$, and thus, to use our set-builder notation, we may write

$$A \times B = \{(a, b) \mid a \text{ is in } A \text{ and } b \text{ is in } B\}.$$

For example, if $A = \{i, j, k\}$ and $B = \{\#, \&\}$, then $A \times B = \{(i, \#), (i, \&), (j, \#), (j, \&), (k, \#), (k, \&)\}$. One may, of course, form the cartesian product of a set with itself. For two examples, consider that $A \times A = \{(i, i), (i, j), (i, k), (j, i), (j, j), (j, k), (k, i), (k, j), (k, k)\}$ and that $B \times B = \{(\#, \#), (\#, \&), (\&, \#), (\&, \&)\}$.

We now define a **function** as a set of ordered pairs (a subset of a cartesian product) with the property that no two pairs have the same first element. For example, the set of ordered pairs $C = (i, \#), (j, \&), (k, \#)$ is a function, while $D = (i, \#), (i, \&), (k, \&)$ is not a function. $C$ is a function because no two pairs have the same first element. $D$ fails to be a function because it contains the pairs $(i, \#)$ and $(i, \&)$ which are distinct pairs that have the same first element.

Several other technical terms may now be described. We consider a function $f$, which is, by definition, a subset of a set of ordered pairs, call the set of all first elements in the pairs of $f$ the **domain** of $f$, and call the set of all second elements in the pairs the **range** of $f$. We also say that a function $f$ is a function **from** its domain **onto** its range. Thus the function $C$, defined in the preceding paragraph, is a function from $A$ onto $B$ since $A$ is the domain of $C$ and $B$ is the range of $C$. Consider the further example of the function $E = \{(i, \&), (k, \&)\}$. This function is also a subset in $A \times B$, but it is not a function from $A$ onto $B$. Rather, it is a function from $\{i, k\}$ onto $\{\&\}$, where $\{i, k\}$ is, of course, a subset of $A$ and $\{\&\}$ is a subset of $B$. We also say that a function is **into** any set which contains its range as a subset. Hence we would say that $E$ is a function from $\{i, k\}$ **into** $B$. Finally, by a **one-to-one** function we mean a function with the additional property that different pairs have different second elements.

You may already be familiar with another definition of *function* which states "a function $f$ from $A$ onto $B$ is a correspondence which associates with each element $a$ of $A$ a unique element $b = f(a)$ of B." This is an "active" definition of function. The association of $b$ with $a$ corresponds in our "passive" definition to the pair $(a, b)$. We have merely spared ourselves the nuisance of explaining what a "correspondence" is. Note that the symbol $f$ stands for the function itself, so that $f$ is a set of pairs or $f$ is a correspondence, while $f(a)$ denotes an element of the range of $f$ which may be thought of as either the second member of the pair $(a, f(a))$ or as the *image* of $a$ under the correspondence $f$. It is common in much mathematical writing to refer to a certain function as "the function $f(x)$." It is important to understand that this is a brief way of saying "the function whose typical element is the pair $(x, f(x))$." We in no way demand that you discard the "active" definition and think only in terms of the "passive" one. On the contrary, both definitions have important connotations for our intuitive understanding of *function* and the reader should be familar with both.

To illustrate the foregoing, consider the function $f$ in $\mathbb{R} \times \mathbb{R}$ first defined as the set of all ordered pairs of the form $(x, 2x + 1)$. We may write this as

$$f = \{(x, f(x)) \mid x \in \mathbb{R} \quad \text{and} \quad f(x) = 2x + 1\}.$$

The pairs (3, 7), (−2, −3), and (0, 1) would each be elements of this function. This function's domain is the set of all real numbers, as is its range. It is also a one-to-one function since each range element corresponds to *exactly one* element in the domain.

The same function $f$ may be described as the function which associates with each real number $x$ the number $2x + 1$; and this we may symbolize as

$$f : x \to 2x + 1.$$

The phrase "the function $f(x) = 2x + 1$" is then just another way of describing $f$.

A pictorial representation of a function is helpful in understanding these definitions. If we represent the domain $D_f$ and the range $R_f$ of the function $f = \{(1, 2), (3, 4), (5, 6)\}$ as in Figure 1.1.2, then we may illustrate the correspondence between the individual elements of each pair of the set of ordered pairs.

Figure 1.1.2

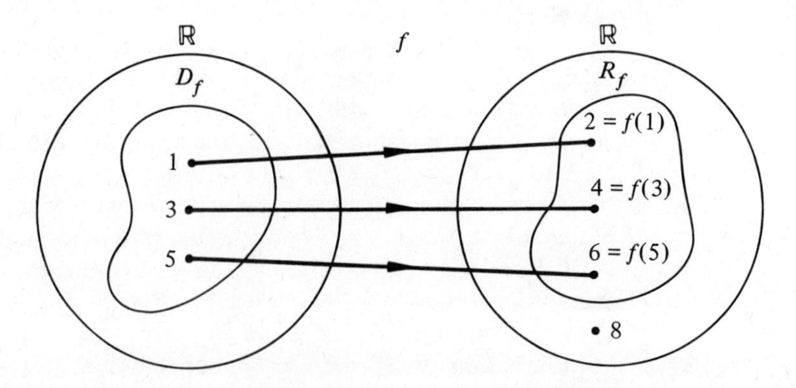


In this case $f$ is a one-to-one function from $D_f$ onto $R_f$. Note that $f$ is into $\mathbb{R}$ but not onto $\mathbb{R}$, since $\mathbb{R}$ contains $R_f$ as a proper subset.

For the function $g = \{(-2, 3), (-1, 2), (0, 2)\}$, we have the representation in Figure 1.1.3. Note in this case that $g$ is not a one-to-one function.

Figure 1.1.3

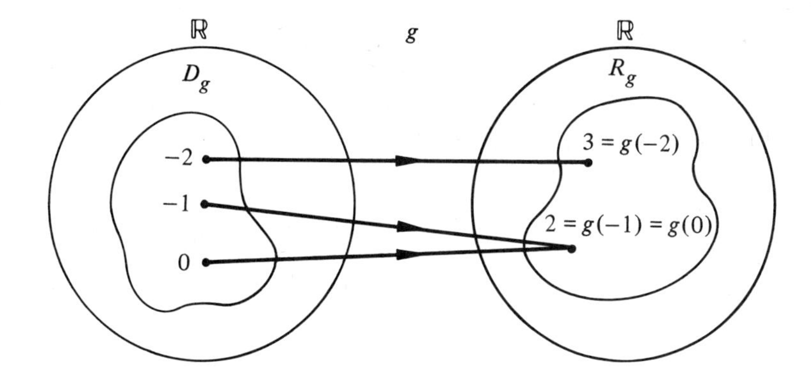


## 1.1 Exercises

1. Describe in words the following sets (where $\mathbb{Z}^+$ denotes the set of positive integers):
   a. $\{x \mid x - 5 \text{ is in } \mathbb{Z}^+\}$.
   b. $\{y \mid y \text{ is a positive integer}\}$.
   c. $\{2x - 1 \mid x \text{ is in } \mathbb{Z}^+\}$.

2. Describe the following sets using the set-builder notation:
   a. The set of all positive integers which are multiples of 5.
   b. The set of all negative integers.
   c. The set of all triangles which are also rectangles.

3. a. Is $\{1, 2, 3\}$ a subset of $\{1, 2, 3\}$? A proper subset?
   b. Is $\{1, 2, 3\}$ a proper subset of $\mathbb{Z}^+$?

4. Let $A = \{1, 2, 3, 4\}$, $B = \{-1, 2, 7\}$, $C = \{1, 4, 6, 8\}$.
   a. Find $A \cup B$ and $B \cup C$.
   b. Find $A \cap B$.
   c. Find $A \cap (B \cup C)$.
   d. Find $A \cap C$.
   e. Find $(A \cap B) \cup (A \cap C)$. Compare c and e.
   f. Find $B \cup (A \cap C)$.
   g. Find $(B \cup A) \cap (B \cup C)$. Compare f and g.

5. How many elements are in $\varnothing$? How many elements are in $\{\varnothing\}$? Would you say that $\varnothing = \{\varnothing\}$?

6. If $E$ denotes the even positive integers and $D$ denotes the odd positive integers, what is $E \cup D$?

7. If $A \subseteq B$, what is $A \cup B$? What is $A \cap B$?

8. What is $\varnothing \cup A$? What is $\varnothing \cap A$?

9. Show that if $B \subseteq C$ and $A$ is any set, then $A \cap B \subseteq A \cap C$.

10. a. Show that for the three sets $A = \{1, 3, 4, 6, 7\}$, $B = \{3, 4, 7\}$, $C = \{1, 2, 3, 7, 10\}$,
$$A \cap (B \cup C) = (A \cap B) \cup (A \cap C).$$
b. Use the definition of set equality to verify the equation for any three sets $A, B, C$.

11. How many subsets are there of $\{1\}$? Of $\{1, 2\}$? Of $\{1, 2, 3\}$? Of $\{1, 2, 3, \ldots, n\}$?

12. Draw a Venn diagram illustrating sets $A$, $B$, and $C$ in Exercise 4 above. Show $A \cup B$, $A \cap B$, $A \cap C$, and $(A \cap B) \cup (A \cap C)$.

13. Which of the following sets are functions?
a. $\{(1, 2), (1, 3), (2, 4)\}$. b. $\{(x, y), (y, z), (z, w)\}$.
c. $\{(a, b) \mid a \text{ is in } \mathbb{Z}^+ \text{ and } b = a^2\}$.
d. $\{(a, b) \mid a \text{ and } b \text{ are in the Jones family and } b \text{ is a daughter of } a\}$.
e. $\{(t, u) \mid t \text{ and } u \text{ are plane triangles and } u \text{ is congruent to } t\}$.

14. What is the domain and what is the range of each of the following functions?
a. $f = \{(a, b), (b, c), (c, d)\}$.
b. $g = \{(1, 4), (2, 5), (3, 4)\}$.
c. $h = \{(x, y) \mid x \text{ and } y \text{ are in } \mathbb{Z}^+ \text{ and } y = 2x - 2\}$.

15. In Exercise 14 above, which of the functions are one-to-one?

16. If $f$ is a function and $D_f$, $R_f$ denote its domain and range respectively, we may represent the function pictorially as illustrated in this section in Figure 1.1.2. Represent the following functions pictorially.
a. $f = \{(1, 2), (2, 3), (3, 4)\}$.
b. $g = \{(1, 1), (2, 1), (3, 4)\}$.
c. $h = \{(-1, 3), (2, 4), (4, 4)\}$.
Which of these functions are one-to-one? How does this show up in pictorial representation?

17. If $f$ and $g$ are functions it is sometimes possible to define a new function $f \circ g = h$ called the *composite* of $f$ and $g$. For each $x$ in the domain of $h$, $h(x) = (f \circ g)(x)$ is defined to be the element $f[g(x)]$. The domain of $h$ consists of the subset of the domain of $g$ upon which we can perform the indicated operations. Let $f = \{(1, 2), (3, 5), (4, 6)\}$ and let $g = \{(-1, 3), (2, 7), (4, 1)\}$. Then $h = f \circ g = \{(-1, 5), (4, 2)\}$ (Figure 1.1.4).
a. Find $g \circ f$. b. Find $h \circ f$. c. Find $g \circ h$.

**Figure 1.1.4**

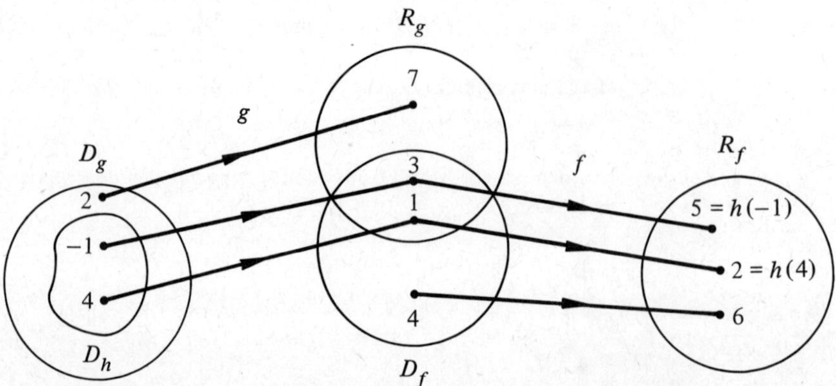

### Suggested Reading

P. Halmos. *Naive Set Theory*. New York: Van Nostrand Reinhold, 1960.

V. Larney. *Abstract Algebra*. Boston: Prindle, Weber and Schmidt, 1975.

N. Vilenkin. *Stories about Sets*. New York: Academic Press, 1968.

### New Terms*



## 1.2 The Real Numbers

In this section, we wish to show that real numbers are an example of a highly important number system and to refresh your memory with respect to several useful properties of that system. To begin, we assume that you are in some way familiar with the system $\mathbb{R}$ of real numbers. You should be aware that the real numbers contain the set of counting numbers $\mathbb{Z}^+$, that $\mathbb{R}$ contains the set of all integers $\mathbb{Z}$ (positive integers, negative integers, and 0), and that $\mathbb{R}$ contains the set $\mathbb{Q}$ of rational numbers (quotients of integers). You should know that there do exist some real numbers, such as $\pi$, $\sqrt{2}$, $\log_{10} 2$, which are *not* rational numbers, and be aware that there is a one-to-one function (correspondence) from the real numbers onto the set of points of a line (the straight line of Euclidean geometry). We do not propose to construct the real numbers from a simpler number system—such a process is described in the references for this section—but rather we simply assume the existence of such a set and proceed to describe its structure.

Perhaps the simplest part of the algebraic structure of the real numbers is that which is related to *addition*. We shall not press the matter, but we ask you to consider that addition of real numbers is a function from $\mathbb{R} \times \mathbb{R}$ into $\mathbb{R}$ in the sense that addition associates with each pair $(a, b)$ in $\mathbb{R} \times \mathbb{R}$ a sum $a + b$ in $\mathbb{R}$. That is, the operation *addition* is a collection of ordered pairs whose first elements are themselves ordered pairs of real numbers—that is, elements of $\mathbb{R} \times \mathbb{R}$—and whose second elements are real numbers. As examples, the following ordered pairs belong to the addition function for real numbers:

$$((2, 5), 7), ((1, 0), 1), ((-3, 6), 3), ((2, -2), 0), ((\pi, 3), \pi + 3)$$

and the following pairs do *not* belong to the addition function for real numbers:

$$((2, 5), 10), ((-3, 6), -18), ((1, 2), 2), ((1, 0), 0), ((\pi, 3), 3\pi).$$

* New terms are listed with page numbers in order of appearance.

Addition of reals, however, is not just *any* function from $\mathbb{R} \times \mathbb{R}$ into $\mathbb{R}$ (such a function is often called a **binary operation**, but one with special properties.) The four most basic properties we list below and label A1, A2, A3, A4.

**A1** Addition for real numbers is **associative** in the sense that

$$(a + b) + c = a + (b + c)$$

for every choice of $a, b, c$ in $\mathbb{R}$.

**A2** There is in $\mathbb{R}$ an **identity element** 0 (usually called **zero**) for addition in the sense that for each choice of $a$ in $\mathbb{R}$,

$$a + 0 = a = 0 + a.$$

**A3** For each choice of $a$ in $\mathbb{R}$, there is in $\mathbb{R}$ an element which we denote by $-a$ such that

$$a + (-a) = 0 = (-a) + a.$$

We call the element $-a$ the **additive inverse** of $a$.

**A4** Addition for real numbers is **commutative** in the sense that

$$a + b = b + a$$

for each choice of $a, b$ in $\mathbb{R}$.

Let us now consider a few simple consequences of these properties of addition for the real numbers.

(1) *Uniqueness of the zero.* There is only one identity for addition of real numbers; for if both 0 and $0'$ are identities, then

$$0 = 0 + 0' = 0'.$$

Hence two or more identities cannot exist.

(2) *Uniqueness of additive inverses.* Let $a$ be any element of $\mathbb{R}$, and consider $-a$ whose existence is postulated in A3. We claim that there is no other element $x$ in $\mathbb{R}$ which acts as an additive inverse for $a$. For if $x + a = 0$, then

$$(x + a) + (-a) = 0 + (-a) = (-a),$$

the last equality being true by virtue of A2. Now associativity (A1) tells us that

$$(x + a) + (-a) = x + [a + (-a)] = x + 0 = x.$$

Thus $(x + a) + (-a)$ is equal to both $-a$ and to $x$, and hence $x = -a$. Therefore $a$ has only one (a unique) additive inverse.

(3) *Unique solutions of equations.* Let $a$ and $b$ be real numbers. There is a unique real number $x$ such that

$$x + a = b.$$

You may be surprised if we make the following display

$$\begin{aligned} x + a &= b \\ (x + a) + (-a) &= b + (-a) \\ x + [a + (-a)] &= b + (-a) \\ x &= b + (-a) \end{aligned}$$

and then announce, "$b + (-a)$ is such an $x$." Actually, the above display shows that *if* there is a solution $x$, then the only solution possible is $b + (-a)$. The following argument verifies that $b + (-a)$ actually is a solution:

$$\begin{aligned} [b + (-a) + a &= b + [(-a) + a] \\ &= b + 0 \\ &= b. \end{aligned}$$

Hence $b + (-a)$ is a number with the required property.

We have proved these properties for real numbers under the assumption that addition for the real numbers satisfies A1, A2, A3, A4, but we now want to call your attention to the fact that these three derived properties are not peculiar to the real numbers, but are true for any "addition" on any set which satisfies the four properties A1–A4. Note here and elsewhere in the text that $a - b$ is simply a shorthand notation for $a + (-b)$.

Let us now consider a second binary operation for the real numbers—*multiplication*. It, too, is a function from $\mathbb{R} \times \mathbb{R}$ into $\mathbb{R}$. The second display of ordered pairs on page 9 belongs to the multiplication function for real numbers. Multiplication has the following properties.

**M1** Multiplication is **associative**; that is,

$$(ab)c = a(bc)$$

for any choice of $a$, $b$, $c$ in $\mathbb{R}$.

**M2** There is an **identity element** 1 for multiplication such that for each $a$ in $\mathbb{R}$,

$$1a = a1 = a.$$

**M3** For each $a$ in $R$, except $a = 0$, there is an element $a^{-1}$ in $\mathbb{R}$ such that

$$a^{-1}a = aa^{-1} = 1.$$

The element $a^{-1}$ is called the **multiplicative inverse** of $a$.

**M4** Multiplication is **commutative**; that is, for any choice of $a$, $b$ in $\mathbb{R}$,

$$ab = ba.$$

In the case of real numbers it so happens that the operations of addition and multiplication are linked in an interesting way. For these combined operations the following **distributive properties** are valid:

**D1** $a(b + c) = ab + ac$, and

**D2** $(b + c)a = ba + ca$, for every choice of $a$, $b$, $c$ in $\mathbb{R}$.

Some consequences of these further properties of the real numbers are:

1. The multiplicative identity 1 is unique.
2. The multiplicative inverse of an element $a \neq 0$ is unique.
3. The multiplicative inverse of $a^{-1}$ is $a$; that is, $(a^{-1})^{-1} = a$.

The proofs of these three properties are left to the reader as Exercise 9.

4. If $ab = 0$ then either $a = 0$ or $b = 0$ or both $a$ and $b$ are 0. For if one is not zero, say $a \neq 0$, then $b = 1b = a^{-1}ab = a^{-1}0 = 0$.
5. For any element $a \neq 0$ and any element $b$, there is one and only one solution $x$ in $\mathbb{R}$ of

$$ax = b.$$

Again, if such a solution $x$ exists, then, since $a \neq 0$, $a^{-1}$ exists by M3 and

$$\begin{aligned} a^{-1}(ax) &= a^{-1}b \\ (a^{-1}a)x &= a^{-1}b \\ 1x &= a^{-1}b \\ x &= a^{-1}b \end{aligned}$$

and hence there is at most one solution $x$. Moreover, $a(a^{-1}b) = (aa^{-1})b = 1b = b$, so that $a^{-1}b$ is actually a solution. Notice that here we had to require $a \neq 0$ in order to guarantee the existence of $a^{-1}$.
6. If $a \neq 0$ then the "linear equation"

$$ax + b = c$$

has a unique solution $x$ for each choice of $b$, $c$ in $\mathbb{R}$. In order to verify this, we merely note that the above equation has exactly the same set of solutions as the equation

$$ax = c + (-b)$$

and, as we have seen in 5. above, the latter equation has the unique solution $x = a^{-1}[c + (-b)]$.

The real numbers $\mathbb{R}$, with their usual addition and multiplication, are seen to be an example of a number system which satisfies A1–A4, M1–M4, and D1, D2. Such a system, if it has more than one element, is called a **field**. There are other familiar examples of fields—the rational numbers form one such example, the complex numbers another. And there are some, perhaps unfamiliar, simple examples of fields. (See Exercise 1.2.6 for one such example.)

We are generally familiar with the relation $\leqslant$ (read "less than or equal to") defined for the real numbers. We write $2 \leqslant 5$, and $-7 \leqslant -4$, etc. Considering $\leqslant$ as a **relation** in $\mathbb{R}$, that is, as a set of ordered pairs from $\mathbb{R} \times \mathbb{R}$, we would say that the pairs (2, 5) and (−7, −4) are elements of $\leqslant$. Thus when we write $a \leqslant b$, we are simply stating that the pair $(a, b)$ is an element of the relation $\leqslant$.

The relation $\leqslant$ is called an *order relation*, and because $\mathbb{R}$ is a field with such a relation, we call $\mathbb{R}$ an *ordered field*. The rational numbers provide another example of an ordered field, but the complex numbers do *not* form an ordered field. (We would have to decide whether or not $i = \sqrt{-1}$ is less than 1, for example.) For a complete explanation see the reference by N. McCoy.

The property which distinguishes the real numbers from all other ordered fields—the completeness property—will not be described here, but a brief discussion may be found in Appendix A.

## 1.2 Exercises

1. Consider the set $S = \{a, b, c\}$ and the binary operation "+" defined by the table

| $+$ | $a$ | $b$ | $c$ |
|---|---|---|---|
| $a$ | $a$ | $b$ | $c$ |
| $b$ | $b$ | $c$ | $a$ |
| $c$ | $c$ | $a$ | $b$ |

To find the sum $b + c$, for example, we look in the $b$-row, $c$-column and determine that $b + c = a$. Similarly, $c + a = c$.

a. Verify that
   1. $a + c = c + a$.
   2. $(a + b) + c = a + (b + c)$.
   3. $(c + c) + b = c + (c + b)$.
   4. $b + a = b$.

b. Which element of $S$ acts as an identity element for "+"?

c. Is "+" a commutative operation? What property of the arrangement of the elements of the table is important for this question?

d. Find the additive inverse of each of the elements of $S$.

e. What would be involved in a proof that the "+" operation is associative?

2. In Exercise 1 above, find the unique solution for the equation $x + b = c$.

3. Consider addition for the real numbers. What is
   a. the additive identity?
   b. the additive inverse of 5?
   c. the additive inverse of $(-5)$?

4. Show in general that in a set with addition satisfying A1–A4, the additive inverse of $(-a)$ is $a$.

5. Consider the set $S = \{a, b, c\}$, with addition "+" and multiplication "·" as defined in the tables below:

| $+$ | $a$ | $b$ | $c$ |
|---|---|---|---|
| $a$ | $a$ | $b$ | $c$ |
| $b$ | $b$ | $c$ | $a$ |
| $c$ | $c$ | $a$ | $b$ |

| $\cdot$ | $a$ | $b$ | $c$ |
|---|---|---|---|
| $a$ | $a$ | $a$ | $a$ |
| $b$ | $a$ | $b$ | $c$ |
| $c$ | $a$ | $c$ | $b$ |

a. Verify that
   1. $a \cdot c = c \cdot a$.
   2. $a \cdot (b \cdot c) = (a \cdot b) \cdot c$.
   3. $b \cdot (c \cdot a) = (b \cdot c) \cdot a$.
   4. $c \cdot b = b \cdot c$.
   5. $a \cdot b = a$.

b. Show that "·" is a commutative operation.
c. What is the identity element for "·"?
d. Verify that $b$ is its own multiplicative inverse.
e. Find $c^{-1}$ and $a^{-1}$ if they exist.

6. For the set $S$ and operations defined in Exercise 5, show that

$$(a + b) \cdot c = (a \cdot c) + (b \cdot c)$$

and

$$(b + c) \cdot a = \_\_\_\_ + \_\_\_\_ = a.$$

These are examples of property ____. Any set $S = \{a, b, c\}$ with two operations satisfying $A_1, \ldots, A_4, M_1, \ldots, M_4, D_1, D_2$, is called a ____.

7. Show that subtraction in the real numbers is neither an associative nor a commutative operation.

8. Show that division (excluding division by 0, of course) is neither associative nor commutative.

9. Prove properties (1), (2), (3) for multiplication of real numbers.

### Suggested Reading

E. Landau. *Foundations of Analysis.* Bronx, N.Y.: Chelsea Publishing Co., 1951.
N. McCoy. *Introduction to Modern Algebra*, 3d. ed., Boston: Allyn and Bacon, 1975.

### New Terms

binary operation, 10
associative, 10
identity element, 10
zero, 10
additive inverse, 10
commutative, 10
multiplicative inverse, 11
distributive properties, 11
field, 12
relation, 12

## 1.3 What Is a Matrix?

By a **matrix** we shall mean a rectangular array of real numbers, arranged in horizontal rows and vertical columns. For example

$$\begin{bmatrix} 1 & 2 \\ 2 & 3 \end{bmatrix}, \quad \begin{bmatrix} -1 & 2 & 3 \\ 0 & 1 & -1 \\ 2 & 1 & 4 \end{bmatrix}, \quad [2]$$

are matrices. In addition the array

$$\begin{bmatrix} 3 & -2 & 5 & 2\pi \\ 0 & 1/3 & \log_{10} 2 & 17 \\ -1 & e & 0.007 & 1/4 \end{bmatrix}$$

is a matrix with three rows and four columns. We number the rows from top to bottom in the array and the columns from left to right.

Thus $\quad 3 \quad -2 \quad 5 \quad 2\pi$

is the first row of our last example, and

$$\begin{matrix} 5 \\ \log_{10} 2 \\ 0.007 \end{matrix}$$

is the third column.

We shall denote matrices in several different ways. First, we shall use capital letters to stand for matrices; thus,

$$A = \begin{bmatrix} 3 & -2 & 5 & 2\pi \\ 0 & 1/3 & \log_{10} 2 & 17 \\ -1 & e & 0.007 & 1/4 \end{bmatrix}$$

means $A$ is a symbol representing this particular array of numbers. We will say that $A$ is a $3 \times 4$ matrix when we wish to call attention to the fact that $A$ has 3 rows and 4 columns. Here there should be no ambiguity in the use of capital letters to designate matrices and their use to denote sets. The context should make clear the distinction.

Second, we may write down an array of lower-case letters

$$\begin{bmatrix} a & b & c & d \\ e & f & g & h \\ i & j & k & m \end{bmatrix}$$

where each letter stands for some, perhaps unspecified, real number. But more frequently we shall write down a typical $3 \times 4$ matrix $A$ as

$$\begin{bmatrix} a_{11} & a_{12} & a_{13} & a_{14} \\ a_{21} & a_{22} & a_{23} & a_{24} \\ a_{31} & a_{32} & a_{33} & a_{34} \end{bmatrix}.$$

Here the doubly-subscripted letter $a_{12}$ refers to the element of $A$ in the first row, second column; $a_{34}$ refers to the element in the third row, fourth column; and, in general, $a_{ij}$ refers to the element of $A$ in the $i$th row, $j$th column. That is, the first subscript always refers to the *row* and the second subscript to the column in which the element appears. Hence a third way to denote our matrix $A$ is to refer to the $3 \times 4$ matrix.

$$A = [a_{ij}]$$

where we have written inside the brackets only one typical element—the one in the $i, j$ position of $A$. As $i$ and $j$ vary, $a_{ij}$ represents all particular elements of $A$.

Example 1.3.1 In Winesburg, the Police Department has three precincts. The vehicle requirement matrix for Winesburg is

| | *Cars* | *Motorcycles* |
|---|---|---|
| Precinct 1 | 3 | 6 |
| Precinct 2 | 2 | 3 |
| Precinct 3 | 4 | 8 |

from which it can be seen, for example, that Precinct 2 requires 3 motorcycles, and Precinct 3, 4 cars. ∎

Example 1.3.2 The developer of two housing projects may list his material requirements in a matrix as

$$\begin{array}{c} \\ \text{Project 1} \\ \text{Project 2} \end{array} \begin{array}{c} \begin{array}{ccc} \quad\textit{Paint} & \quad\quad\textit{Lumber} & \quad\quad\textit{Labor} \end{array} \\ \begin{bmatrix} \text{1500 gal.} & \text{30,000 bd. ft.} & \text{10,000 man hrs.} \\ \text{2000 gal.} & \text{40,000 bd. ft.} & \text{15,000 man hrs.} \end{bmatrix} \end{array}.$$ ∎

The reader should note that the row and column headings in the two previous examples are arbitrary, so that there are many different matrices displaying the same information.

We say that two matrices $A = [a_{ij}]$ and $B = [b_{ij}]$ are **equal**, and write $A = B$, provided that they have the same number of rows and the same number of columns and that corresponding elements are equal.

## 1.3 Exercises

1. Define the term "matrix."

2. A $2 \times 3$ matrix has 2 ________ and ________ ________.

3. The symbol $a_{23}$ stands for the element of the matrix $A$ in the ________ ________ and the ________ ________.

4. What do we mean by the symbol $a_{ij}$? $[a_{ij}]$?

5. Which of the following arrays are not matrices?

   a. $\begin{bmatrix} 1 & 2 & 3 \\ 4 & 5 & 6 \end{bmatrix}$ b. $\begin{bmatrix} 1 & 2 & 3 \\ 0 & 2 & \\ 1 & & \end{bmatrix}$ c. $[2]$ d. $\begin{bmatrix} \pi \\ 0 \\ \sqrt{2} \end{bmatrix}$

6. Construct an Age-Height-Weight matrix for the members of your family.

7. How would you describe the shape of the matrix and the pattern of its elements in the following examples?

$$A = \begin{bmatrix} 1 & 2 \\ 3 & 4 \end{bmatrix}, \quad B = \begin{bmatrix} 1 & 2 & 3 \\ 0 & 4 & 5 \\ 0 & 0 & 6 \end{bmatrix}, \quad C = \begin{bmatrix} 1 & 0 & 0 \\ 0 & 2 & 0 \\ 0 & 0 & 3 \end{bmatrix},$$

$$D = \begin{bmatrix} 1 & 0 \\ 0 & 1 \end{bmatrix}, \quad E = \begin{bmatrix} 1 & 2 & 3 \\ 2 & 4 & 5 \\ 3 & 5 & 6 \end{bmatrix}, \quad F = \begin{bmatrix} 1 & 2 & 3 \\ 4 & 5 & 6 \end{bmatrix}.$$

8. Your company has three divisions $A$, $B$, $C$. Division $A$ has 1000 men, 100 women, and produces 5000 units of your product per day. Division $B$ has 650 men, 48

women, and produces 2300 units. Division $C$, the clerical division, has only 50 men and 40 women. It produces none of the product. Arrange this data in matrix form.

### New Terms



## 1.4 Matrix Watching—Some Special Types of Matrices

In the course of our discussion, and particularly in the next section, we shall have occasion to use matrices of special shapes or matrices whose elements form a definite pattern. We should like to describe these various types of matrices here briefly and then proceed to our development of the algebra of matrices.

First, we shall call a matrix **square** if it has the same number of rows as columns. Thus

$$\begin{bmatrix} 1 & 2 \\ 3 & 4 \end{bmatrix}$$

is a square $2 \times 2$ matrix. We see that

$$\begin{bmatrix} 1 & 2 & 3 \\ 4 & 5 & 6 \\ 7 & 8 & 9 \end{bmatrix}$$

is a square $3 \times 3$ matrix, and that $[-0.27]$ is a square $1 \times 1$ matrix. A typical square matrix with $n$ rows and $n$ columns may be denoted by the display

$$\begin{bmatrix} a_{11} & a_{12} & \cdots & a_{1n} \\ a_{21} & a_{22} & \cdots & a_{2n} \\ \hdotsfor{4} \\ a_{n1} & a_{n2} & \cdots & a_{nn} \end{bmatrix}.$$

A matrix which is not square will be called **rectangular**, and, conversely, when we say "rectangular" we shall mean *not square.*

The elements $a_{11}, a_{22}, a_{33}, \ldots, a_{nn}$ are said to constitute the **main diagonal** of a square matrix. Thus $3, -2, \pi$ are the main diagonal elements of the $3 \times 3$ matrix

$$\begin{bmatrix} 3 & 1 & 5 \\ 2 & -2 & e \\ 0.4 & 0 & \pi \end{bmatrix}.$$

If a square matrix is such that all of its elements, except perhaps the main diagonal elements, are zero, then the matrix is called a **diagonal matrix**. Thus, all of the following are diagonal matrices:

$$\begin{bmatrix} 2 & 0 & 0 \\ 0 & -1 & 0 \\ 0 & 0 & e \end{bmatrix}, \quad \begin{bmatrix} 1 & 0 & 0 \\ 0 & 0 & 0 \\ 0 & 0 & 3 \end{bmatrix}, \quad \begin{bmatrix} 0 & 0 & 0 \\ 0 & 0 & 0 \\ 0 & 0 & 0 \end{bmatrix}.$$

As a special case of a diagonal matrix there is the diagonal matrix whose diagonal elements are all equal to each other. For example, all of the following matrices are of this type. Such a matrix is called a **scalar matrix**.

$$\begin{bmatrix} 1 & 0 & 0 \\ 0 & 1 & 0 \\ 0 & 0 & 1 \end{bmatrix}, \quad \begin{bmatrix} 0 & 0 & 0 \\ 0 & 0 & 0 \\ 0 & 0 & 0 \end{bmatrix}, \quad \begin{bmatrix} -0.3 & 0 & 0 \\ 0 & -0.3 & 0 \\ 0 & 0 & -0.3 \end{bmatrix}.$$

Another special type of square matrix is the **upper triangular** matrix, which has the property that all the elements below the main diagonal are zero (regardless of what the diagonal or above-the-diagonal elements are). Thus every diagonal matrix is also upper triangular; and so also are the following examples:

$$\begin{bmatrix} 1 & 2 & 3 \\ 0 & 4 & 5 \\ 0 & 0 & 6 \end{bmatrix}, \quad \begin{bmatrix} 1 & 2 & 0 \\ 0 & 0 & 4 \\ 0 & 0 & 5 \end{bmatrix}, \quad \begin{bmatrix} 0 & 1 & 2 \\ 0 & 0 & 3 \\ 0 & 0 & 4 \end{bmatrix}.$$

Another way of describing in general an upper triangular $n \times n$ matrix $A = [a_{ij}]$ is to require that

$$a_{ij} = 0 \quad \text{for} \quad i > j.$$

**Lower triangular** matrices are defined in a similar manner, with $a_{ij} = 0$ for $i < j$.

Two useful types of matrices are the **row matrices**, which consist of only one row, and the **column matrices** which consist of only one column; for example the $1 \times 5$ matrix

$$[1, 2, 3, 0, -7]$$

is a row matrix; and the $3 \times 1$ matrix

$$\begin{bmatrix} 0 \\ -2 \\ 1 \end{bmatrix}$$

is a column matrix. Indeed, we may think of the $m \times n$ matrix $A = [a_{ij}]$ as consisting of $m$ row matrices

$$\begin{array}{l} [a_{11} \quad a_{12} \quad \cdots \quad a_{1n}], \\ [a_{21} \quad a_{22} \quad \cdots \quad a_{2n}], \\ \cdots\cdots\cdots\cdots\cdots\cdots \\ [a_{m1} \quad a_{m2} \quad \cdots \quad a_{mn}] \end{array}$$

or as consisting of $n$ column matrices

$$\begin{bmatrix} a_{11} \\ a_{21} \\ \vdots \\ a_{m1} \end{bmatrix}, \begin{bmatrix} a_{12} \\ a_{22} \\ \vdots \\ a_{m2} \end{bmatrix}, \ldots, \begin{bmatrix} a_{1n} \\ a_{2n} \\ \vdots \\ a_{mn} \end{bmatrix},$$

For reasons which we shall see later, it is common to refer to such row matrices or column matrices as **vectors**. We shall also find it convenient in much of what follows to think of matrices as having been partitioned into smaller matrices consisting entirely of either rows or columns as shown above. For example, we may write

$$A = \begin{bmatrix} 1 & -1 & 0 \\ 2 & 2 & 1 \\ 3 & 4 & 2 \end{bmatrix}$$

as a matrix whose elements are column matrices themselves; that is, $A = [A_1, A_2, A_3]$ where

$$A_1 = \begin{bmatrix} 1 \\ 2 \\ 3 \end{bmatrix}, \quad A_2 = \begin{bmatrix} -1 \\ 2 \\ 4 \end{bmatrix}, \quad \text{and} \quad A_3 = \begin{bmatrix} 0 \\ 1 \\ 2 \end{bmatrix}.$$

Finally, we define the **transpose** $A^T$ of an $m \times n$ matrix $A$ to be the $n \times m$ matrix whose rows are the columns of $A$ (in order) and whose columns are the rows of $A$ (in order). For example, if

$$A = \begin{bmatrix} 1 & 2 & 3 & 4 \\ 5 & 6 & 7 & 8 \\ 9 & 10 & 11 & 12 \end{bmatrix},$$

then

$$A^T = \begin{bmatrix} 1 & 5 & 9 \\ 2 & 6 & 10 \\ 3 & 7 & 11 \\ 4 & 8 & 12 \end{bmatrix}.$$

For a square matrix, the transpose $A^T$ of $A$ is simply the matrix obtained by the reflection of the elements in the main diagonal. And, of course, the transpose of a row matrix is a column matrix, the transpose of an upper triangular matrix is lower triangular, and the transpose of a diagonal matrix is itself. We may use the transpose to define a special subset of matrices which are very important in applications. The square matrix $A$ is said to be a **symmetric** matrix provided that $A = A^T$. This definition is equivalent to saying that $a_{ij} = a_{ji}$ for each $i$ and $j$. The following matrices are symmetric:

$$\begin{bmatrix} 1 & 2 \\ 2 & 0 \end{bmatrix}, \quad \begin{bmatrix} -1 & 0 & -2 \\ 0 & 3 & 7 \\ -2 & 7 & 5 \end{bmatrix}, \quad \begin{bmatrix} 0 & 1 & 2 & 3 \\ 1 & 0 & 1 & 4 \\ 2 & -1 & 0 & 5 \\ 3 & 4 & 5 & 0 \end{bmatrix}.$$

## 1.4 Exercises

Consider the following special categories of matrices:
a. square b. rectangular c. diagonal d. scalar e. upper triangular
f. lower triangular g. row matrix h. column matrix i. symmetric.
After each of the following examples of matrices, write the letters corresponding to the categories in which the matrix belongs.

1. $\begin{bmatrix} 1 & 0 & 0 \\ 0 & 0 & 0 \end{bmatrix}$

2. $\begin{bmatrix} 0 & 0 & 0 \\ 0 & 1 & 0 \\ 0 & 0 & 1 \end{bmatrix}$

3. $[1]$

4. $\begin{bmatrix} 1 & 2 & 0 & 0 & 0 & 6 \end{bmatrix}$

5. $\begin{bmatrix} 1 & 0 & 0 & 0 & 0 & 0 \\ 0 & 0 & 0 & 0 & 0 & 1 \\ 0 & 0 & 1 & 0 & 0 & 0 \\ 0 & 0 & 0 & 0 & 0 & 0 \\ 0 & 0 & 0 & 0 & 0 & 0 \\ 1 & 0 & 0 & 0 & 0 & 1 \end{bmatrix}$

6. $\begin{bmatrix} 3 & 0 & 1 \\ 0 & 2 & 1 \\ 0 & 0 & 1 \end{bmatrix}$

7. $\begin{bmatrix} 2 \\ 1 \\ 3 \\ 5 \end{bmatrix}$

8. $\begin{bmatrix} 3 & 4 \\ 4 & -1 \end{bmatrix}$

9. $\begin{bmatrix} 5 & 0 & 0 & 0 \\ 2 & 1 & 0 & 0 \\ 3 & 1 & 6 & 0 \\ 6 & 4 & 3 & 2 \end{bmatrix}$

10. $\begin{bmatrix} 1 & 2 & 3 & 4 & 5 \\ 0 & 6 & 7 & 8 & 9 \\ 0 & 0 & 10 & 11 & 12 \\ 0 & 0 & 0 & 13 & 14 \end{bmatrix}$

11. Write down an example of a $2 \times 2$ matrix (if one exists) which is
    a. upper triangular and not diagonal.
    b. lower triangular, diagonal, and not scalar.
    c. symmetric, scalar, and not diagonal.
    d. symmetric.

12. Find the transposes of the matrices in Exercises 1, 7, 8, 9. Which are symmetric?

13. a. For $A = \begin{bmatrix} 1 & -1 \\ 3 & 2 \end{bmatrix}$, find $(A^T)^T$.
    b. Form a conjecture about $(A^T)^T$ for any $2 \times 2$ matrix.

### New Terms

square matrix, 17
rectangular matrix, 17
main diagonal, 17
diagonal matrix, 17
scalar matrix, 18
upper triangular, 18
lower triangular, 18
row matrix, 18
column matrix, 18
vectors, 18
transpose, 19
symmetric matrix, 19

## 1.5 Addition of Matrices

We now begin our development of the algebra of matrices with the definition of the first of three operations, namely, *addition*. We consider two $m \times n$ matrices $A = [a_{ij}]$ and $B = [b_{ij}]$ and define the **sum** of $A$ and $B$, written as $A + B$, as follows:

$$A + B = [a_{ij} + b_{ij}];$$

that is, $A + B$ is the $m \times n$ matrix obtained by simply adding corresponding elements of $A$ and $B$. For example,

$$\begin{bmatrix} 1 & 2 & 3 \\ 0 & -2 & 4 \end{bmatrix} + \begin{bmatrix} -3 & 1 & 5 \\ 2 & 1 & 3 \end{bmatrix}$$

$$= \begin{bmatrix} 1 + (-3) & 2 + 1 & 3 + 5 \\ 0 + 2 & -2 + 1 & 4 + 3 \end{bmatrix} = \begin{bmatrix} -2 & 3 & 8 \\ 2 & -1 & 7 \end{bmatrix}.$$

We ask the reader to note carefully that addition is defined for two matrices $A$ and $B$ which have the same number of rows and the same number of columns—we say in this case that $A$ and $B$ are **conformable for addition**. It should thus be clear that

$$A = \begin{bmatrix} 1 & 2 & 3 \\ 0 & -2 & 4 \end{bmatrix} \quad \text{and} \quad B = \begin{bmatrix} 3 & 2 \\ 1 & 1 \\ 5 & 3 \end{bmatrix}$$

are *not* conformable for addition, so that a sum for $A$ and $B$ is not defined.

Let us now select two positive integers $m$ and $n$ and restrict our attention for the moment to the set of all $m \times n$ matrices with real elements. We denote this set of matrices by $\mathcal{M}_{m,n}$ and proceed to show that addition in $\mathcal{M}_{m,n}$ satisfies the properties A1–A4 of Section 1.2. We will thus have shown that each of these additive real number properties has a direct analogue in the set of $m \times n$ matrices.

**A1** Addition in $\mathcal{M}_{m,n}$ is associative.

**Proof** If $A, B, C$ are in $\mathcal{M}_{m,n}$ with $A = [a_{ij}]$, $B = [b_{ij}]$, and $C = [c_{ij}]$, then

$$\begin{aligned} (A + B) + C &= ([a_{ij}] + [b_{ij}]) + [c_{ij}] \\ &= [a_{ij} + b_{ij}] + [c_{ij}] \\ * \Big\{ &= [(a_{ij} + b_{ij}) + c_{ij}] \\ &= [a_{ij} + (b_{ij} + c_{ij})] \Big. \\ &= [a_{ij}] + [b_{ij} + c_{ij}] \\ &= A + (B + C). \end{aligned}$$

The step in the above display marked* follows from the associativity of addition in $\mathbb{R}$. ∎

Example 1.5.1

$$\left(\begin{bmatrix} 1 & 2 & 3 \\ 4 & 5 & 6 \end{bmatrix} + \begin{bmatrix} -1 & 1 & 4 \\ 2 & 1 & 1 \end{bmatrix}\right) + \begin{bmatrix} 0 & 1 & 5 \\ 2 & 2 & 2 \end{bmatrix}$$

$$= \begin{bmatrix} 0 & 3 & 7 \\ 6 & 6 & 7 \end{bmatrix} + \begin{bmatrix} 0 & 1 & 5 \\ 2 & 2 & 2 \end{bmatrix} = \begin{bmatrix} 0 & 4 & 12 \\ 8 & 8 & 9 \end{bmatrix}.$$

$$\begin{bmatrix} 1 & 2 & 3 \\ 4 & 5 & 6 \end{bmatrix} + \left(\begin{bmatrix} -1 & 1 & 4 \\ 2 & 1 & 1 \end{bmatrix} + \begin{bmatrix} 0 & 1 & 5 \\ 2 & 2 & 2 \end{bmatrix}\right)$$

$$= \begin{bmatrix} 1 & 2 & 3 \\ 4 & 5 & 6 \end{bmatrix} + \begin{bmatrix} -1 & 2 & 9 \\ 4 & 3 & 3 \end{bmatrix} = \begin{bmatrix} 0 & 4 & 12 \\ 8 & 8 & 9 \end{bmatrix}.$$

**A2** There is an *additive identity* element $0 = [0]$ (usually called the **zero matrix**) in $\mathcal{M}_{m,n}$ whose entries are zero in each position and such that for each $A = [a_{ij}]$ in $\mathcal{M}_{m,n}$

$$A + 0 = A = 0 + A.$$

**Proof**

$$\begin{aligned} A + 0 &= [a_{ij}] + [0] = [a_{ij} + 0] = [a_{ij}] \\ &= A = [0 + a_{ij}] = [0] + [a_{ij}] = 0 + A. \end{aligned}$$ ∎

(*Note.* Observe that the symbol 0 is used in the above proof in two different senses. In the left-hand side of the equation $A + 0$, it means the $m \times n$ matrix of zeros. In $[a_{ij} + 0]$ it means the zero scalar. The context, however, makes it clear which is intended, since a scalar cannot be added to a matrix.)

Example 1.5.2

$$\begin{bmatrix} 1 & 2 & 3 \\ 4 & 5 & 6 \end{bmatrix} + \begin{bmatrix} 0 & 0 & 0 \\ 0 & 0 & 0 \end{bmatrix} = \begin{bmatrix} 1 & 2 & 3 \\ 4 & 5 & 6 \end{bmatrix} = \begin{bmatrix} 0 & 0 & 0 \\ 0 & 0 & 0 \end{bmatrix} + \begin{bmatrix} 1 & 2 & 3 \\ 4 & 5 & 6 \end{bmatrix}.$$ ∎

**A3** For each choice of $A = [a_{ij}]$ in $\mathcal{M}_{m,n}$ there is a matrix

$$-A = [-a_{ij}] \quad \text{such that} \quad A + (-A) = 0 = (-A) + A.$$

**Proof**

$$\begin{aligned} A + (-A) &= [a_{ij}] + [-a_{ij}] = [a_{ij} + (-a_{ij})] = [0] = 0 \\ &= [-a_{ij} + a_{ij}] = [-a_{ij}] + [a_{ij}] = -A + A. \end{aligned}$$ ∎

Example 1.5.3

$$\begin{bmatrix} 1 & 2 & 3 \\ 4 & 5 & 6 \end{bmatrix} + \begin{bmatrix} -1 & -2 & -3 \\ -4 & -5 & -6 \end{bmatrix} = \begin{bmatrix} 0 & 0 & 0 \\ 0 & 0 & 0 \end{bmatrix}.$$ ∎

We shall find it convenient henceforth to write $B - A$ when we mean $B + (-A)$.

**A4** Addition in $\mathcal{M}_{m,n}$ is *commutative.*

**Proof** For each $A = [a_{ij}]$ and $B = [b_{ij}]$ in $\mathcal{M}_{m,n}$, $A + B = [a_{ij}] + [b_{ij}] = [a_{ij} + b_{ij}] = [b_{ij} + a_{ij}] = [b_{ij}] + [a_{ij}] = B + A$. ∎

Example 1.5.4

$$\begin{aligned} \begin{bmatrix} 1 & 2 & 3 \\ 4 & 5 & 6 \end{bmatrix} + \begin{bmatrix} -1 & 1 & 4 \\ 2 & 1 & 1 \end{bmatrix} &= \begin{bmatrix} 0 & 3 & 7 \\ 6 & 6 & 7 \end{bmatrix} \\ &= \begin{bmatrix} -1 & 1 & 4 \\ 2 & 1 & 1 \end{bmatrix} + \begin{bmatrix} 1 & 2 & 3 \\ 4 & 5 & 6 \end{bmatrix}. \end{aligned}$$ ∎

Thus $\mathcal{M}_{m,n}$ enjoys properties A1–A4 just as the real numbers do. Since we used only A1–A4 to establish Property (1), *Uniqueness of the zero*, Property (2), *Uniqueness of additive inverses*, and Property (3), *Unique solution of equations* for addition of real numbers, then there must exist direct analogues of these properties for addition in $\mathcal{M}_{m,n}$. The proofs of these properties for matrix addition are left to you in Exercises 14, 15 and 17.

Example 1.5.5 Gilt State Winery does not grow grapes of its own but imports grapes from two different regions in $N$ (north) and $S$ (south) areas of the state.

The following matrices represent the purchase price of each variety and the transportation cost.

$$N = \begin{array}{c} \\ \end{array}\begin{bmatrix} \$58{,}500 & 76{,}500 & 51{,}500 \\ 2{,}070 & 3{,}250 & 2{,}200 \end{bmatrix} \begin{array}{l} \text{Purchase price} \\ \text{Transportation expense} \end{array}$$

(Columns: *Gamay*, *Cabernet Sauvignon*, *Chenin Blanc*)

$$S = \begin{bmatrix} 56{,}500 & 74{,}500 & 53{,}500 \\ 2{,}450 & 3{,}450 & 2{,}550 \end{bmatrix} \begin{array}{l} \text{Purchase price} \\ \text{Transportation expense} \end{array}$$

The matrix $N + S$, where

$$N + S = \begin{bmatrix} 115{,}000 & 151{,}000 & 105{,}000 \\ 4{,}520 & 6{,}700 & 4{,}750 \end{bmatrix} \begin{array}{l} \text{Purchase price} \\ \text{Transportation expense} \end{array}$$

represents the purchase price and transportation expense for the total grape importation to the winery. ∎

## 1.5 Exercises

In Exercises 1–8, find the indicated matrix sum $A + B$ (if it is defined).

1. $\begin{bmatrix} 1 & 2 & 3 \\ 4 & 5 & 6 \end{bmatrix} + \begin{bmatrix} -2 & -4 & -6 \\ -8 & -10 & -12 \end{bmatrix}$

2. $\begin{bmatrix} 1 & 2 & 3 & 4 \end{bmatrix} + \begin{bmatrix} 5 & 2 & 1 & 3 \\ 1 & 2 & 3 & 4 \end{bmatrix}$

3. $\begin{bmatrix} 1 & 2 & 3 & 4 \end{bmatrix} + \begin{bmatrix} 5 & 2 & 1 & 3 \end{bmatrix}$

4. a. $\begin{bmatrix} 1 \\ 2 \\ -5 \\ 0.3 \end{bmatrix} + \begin{bmatrix} 0.3 \\ -0.2 \\ 0.4 \\ -0.1 \end{bmatrix}$ b. $\begin{bmatrix} 1 & 5 \\ 2 & 6 \\ 3 & 7 \\ 4 & 8 \end{bmatrix} + \begin{bmatrix} 1 \\ 1 \\ 2 \\ -3 \end{bmatrix}$

5. Suppose that the Gilt State Winery (Example 1.5.5) expands its grape purchases from the eastern ($E$) and western ($W$) portions of the state. If the matrix

$$E + W = \begin{bmatrix} 50{,}000 & 40{,}000 & 65{,}000 \\ 2{,}000 & 2{,}000 & 2{,}500 \end{bmatrix}$$

represents the new purchase prices and transportation expenses as shown in the original example, find the sum of these expenditures for north, south, east and west.

6. $\begin{bmatrix} 1 & 2 & 3 \\ 4 & 5 & 6 \end{bmatrix} + \begin{bmatrix} 0 & 0 & 0 \\ 0 & 0 & 0 \end{bmatrix}$

7. $\begin{bmatrix} 0 & 0 \\ 0 & 0 \\ 0 & 0 \end{bmatrix} + \begin{bmatrix} 1 & -1 \\ 2 & -3 \\ 3 & -4 \end{bmatrix}$

8. $\begin{bmatrix} 1 & 2 & 3 \\ 4 & 5 & 6 \end{bmatrix} + \begin{bmatrix} -1 & -2 & -3 \\ -4 & -5 & -6 \end{bmatrix}$

9. Let $A = \begin{bmatrix} 1 & -1 & 2 \\ -2 & 3 & 1 \end{bmatrix}$ and $B = \begin{bmatrix} 2 & 1 & 3 \\ 5 & 6 & -1 \end{bmatrix}$. Is there a matrix $X$ such that

$$X + A = B?$$

10. Is the sum of two lower triangular $n \times n$ matrices also lower triangular? What about the sum of two diagonal matrices?

11. a. Find the transpose of each of the matrices $A$, $B$ in Exercise 1.
    b. Compare $(A + B)^T$ with $A^T + B^T$.
    c. Verify for any $2 \times 2$ matrices $A$, $B$, that $(A + B)^T = A^T + B^T$.

12. Find the additive inverses of the following matrices.

a. $[1 \ \ 2 \ \ 3 \ \ 4]$ b. $\begin{bmatrix} -1 \\ 2 \\ -3 \\ 4 \end{bmatrix}$ c. $\begin{bmatrix} -2.5 & -3.1 \\ 4.6 & 7.4 \end{bmatrix}$

13. What is the additive identity for $\mathscr{M}_{1,3}$? For $\mathscr{M}_{4,1}$? For $\mathscr{M}_{3,6}$?

14. Prove that for fixed positive integers $m$, $n$ the matrix 0 (additive identity) for $\mathscr{M}_{m,n}$ is unique.

15. Prove that for fixed positive integers $m$, $n$ the additive inverse $-A$ of $A = [a_{ij}]$ is unique.

16. For Exercise 9, show that there is one and only one (a unique) solution.

17. Show in general that if $A$, $B$ are in the set $\mathscr{M}_{m,n}$ then there is a unique solution $X$ for the equation

$$X + A = B.$$

Does this solution belong to $\mathscr{M}_{m,n}$?

**New Terms**

sum (of matrices), 20 conformable for addition, 21 zero matrix, 22

## 1.6 Matrix Multiplication

In this section we define the operation of multiplication for certain matrices and develop some elementary properties of this multiplication. Several more elaborate properties are stated and illustrated by example, but the proofs of these latter properties are relegated to Appendix B, "Sigma Notation and More Matrix Algebra."

We first define multiplication of a $1 \times n$ row matrix $A$ by an $n \times 1$ column matrix $B$ as follows:

$$AB = [a_1, a_2, \ldots, a_n]\begin{bmatrix} b_1 \\ b_2 \\ \vdots \\ b_n \end{bmatrix} = [a_1 b_1 + a_2 b_2 + \cdots + a_n b_n].$$

Here a $\mathbf{1} \times n$ matrix multiplied on the right by an $n \times \mathbf{1}$ matrix yields a $1 \times 1$ matrix product (the dimensions in boldface give us the dimensions of the product). The reader should note carefully that this definition only makes sense when the two matrices involved have the same number of elements. For example,

$$[2 \quad 3 \quad 4]\begin{bmatrix}5\\6\\7\end{bmatrix} = [2 \cdot 5 + 3 \cdot 6 + 4 \cdot 7] = [56],$$

and
$$[-1, 0.2, 2.1]\begin{bmatrix}0.2\\-3\\1\end{bmatrix} = [-0.2 - 0.6 + 2.1] = [1.3],$$

but $[1 \quad 2 \quad 3 \quad 4]\begin{bmatrix}1\\2\\3\end{bmatrix}$ is simply not defined.

Since it turns out that the set of $1 \times 1$ matrices has exactly the same algebraic structure as the real numbers, we will usually not distinguish between them. Thus we will consider the matrix $[5]$ and the real number 5 to be algebraically interchangeable.

## 1.6 Exercises

In Exercises 1–11 find the indicated matrix product (if it is defined).

1. $[1 \quad 2 \quad 3]\begin{bmatrix}4\\5\\6\end{bmatrix}$

2. $[1 \quad 2 \quad 3]\begin{bmatrix}4\\5\end{bmatrix}$

3. $[1 \quad 2 \quad 3 \quad 4]\begin{bmatrix}5\\6\\7\end{bmatrix}$

4. $[-2.1 \quad 3.7]\begin{bmatrix}2.2\\-1.5\end{bmatrix}$

5. $[a \quad b \quad c]\begin{bmatrix}2\\3\\4\end{bmatrix}$

6. $[-x \quad 2y \quad -z]\begin{bmatrix}h\\k\\-3p\end{bmatrix}$

7. $[-1 \quad 2 \quad 3 \quad -4 \quad 5]\begin{bmatrix}-2j\\3k\\-4k\\6j\\-k\end{bmatrix}$

8. $[a \quad b \quad c]\begin{bmatrix}a\\b\\c\end{bmatrix}$

9. $[0 \quad 1 \quad 0]\begin{bmatrix}a\\b\\c\end{bmatrix}$

10. $[0 \quad 0 \quad 0]\begin{bmatrix} 1 \\ 3 \\ -2 \end{bmatrix}$

11. $[a \quad b \quad c]\begin{bmatrix} 0 \\ 0 \\ 1 \end{bmatrix}$

12. Jim carries 250 papers per week and receives 5 cents per paper. He also babysits 6 hours per week for 75 cents per hour. Verify that the matrix product $[250, 6]\begin{bmatrix} 5 \\ 75 \end{bmatrix}$ correctly computes his weekly income (in cents).

With our first type of multiplication clearly in mind, we now proceed to a slightly more elaborate type; namely, we define the multiplication of a $1 \times n$ row matrix $A$ by an $n \times k$ matrix $B$ as

$$AB = [a_1, a_2, \ldots, a_n]\overbrace{\begin{bmatrix} b_1 & c_1 & \cdots & e_1 \\ b_2 & c_2 & \cdots & e_2 \\ \cdot & \cdot & \cdots & \cdot \\ b_n & c_n & \cdots & e_n \end{bmatrix}}^{k \text{ columns}} = [p_1, p_2, \ldots, p_k],$$

where, as in the simpler case when $B$ consisted of just one column,

$$\begin{aligned} p_1 &= a_1b_1 + a_2b_2 + \cdots + a_nb_n \\ p_2 &= a_1c_1 + a_2c_2 + \cdots + a_nc_n \\ &\cdots\cdots\cdots\cdots\cdots\cdots \\ p_k &= a_1e_1 + a_2e_2 + \cdots + a_ne_n. \end{aligned}$$

That is, we simply think of the $n \times k$ matrix $B$ as consisting of $k$ column matrices (each $n \times 1$) and form the individual products as we did earlier. For example,

$$\begin{aligned} [1 \quad 2 \quad 3]\begin{bmatrix} 2 & -1 & 0 \\ 1 & 2 & 1 \\ 3 & 1 & 2 \end{bmatrix} &= [2 + 2 + 9, -1 + 4 + 3, 0 + 2 + 6] \\ &= [13 \quad 6 \quad 8]. \end{aligned}$$

We draw your attention to the following feature of this more general type of multiplication; namely, that a $\mathbf{1} \times n$ matrix multiplied on the right by an $n \times \mathbf{k}$ matrix yields a $1 \times k$ product, and that each element of the product matrix is a sum of $n$ terms.

## 1.6 Exercises (continued)

Find the indicated product (if it is defined).

13. $[1 \quad 2]\begin{bmatrix} 3 & 6 \\ 4 & 7 \\ 5 & 8 \end{bmatrix}$

14. $[1 \quad 2]\begin{bmatrix} 2 & 3 \\ 1 & 4 \end{bmatrix}$

15. $[a \quad b \quad c]\begin{bmatrix} 1 & 0 & 0 \\ 0 & 1 & 0 \\ 0 & 0 & 1 \end{bmatrix}$

16. $[1 \quad 2 \quad 3 \quad 4]\begin{bmatrix} 2 & 1 & 2 & 1 \\ 1 & 3 & 1 & 1 \\ 3 & 2 & 3 & 0 \end{bmatrix}$

17. $[1 \quad 2 \quad 3 \quad 4]\begin{bmatrix} 5 & -1 & 0 & 1 & 1 \\ 6 & -3 & 1 & 4 & 1 \\ 7 & 2 & 3 & 3 & 2 \\ 8 & 1 & 2 & -2 & -3 \end{bmatrix}$

18. $[1 \quad 2 \quad 3 \quad 4]\begin{bmatrix} 1 & 2 \\ 2 & 1 \\ 3 & 3 \\ 6 & 7 \\ 8 & 9 \end{bmatrix}$

We are now able to define general matrix multiplication in terms of the special multiplication previously explained. To illustrate the procedure, we let

$$A = \begin{bmatrix} 1 & 2 & 3 \\ 5 & 6 & 7 \end{bmatrix} \quad \text{and} \quad B = \begin{bmatrix} 1 & 3 & 1 \\ 2 & 2 & 1 \\ 3 & 1 & 1 \end{bmatrix}$$

and partition $A$ into two row matrices $A_1$ and $A_2$ as illustrated below:

$$\begin{bmatrix} A_1 \\ \hline A_2 \end{bmatrix} = \begin{bmatrix} [1 \quad 2 \quad 3] \\ \hline [5 \quad 6 \quad 7] \end{bmatrix}.$$

The product $AB$ may now be easily described as the matrix whose rows are the row matrices $A_1B$ and $A_2B$; that is,

$$AB = \begin{bmatrix} A_1B \\ \hline A_2B \end{bmatrix}$$

where $$A_1B = [1 \quad 2 \quad 3]\begin{bmatrix} 1 & 3 & 1 \\ 2 & 2 & 1 \\ 3 & 1 & 1 \end{bmatrix} = [14, 10, 6]$$

and $$A_2B = [5 \quad 6 \quad 7]\begin{bmatrix} 1 & 3 & 1 \\ 2 & 2 & 1 \\ 3 & 1 & 1 \end{bmatrix} = [38, 34, 18].$$

Thus $$\begin{bmatrix} 1 & 2 & 3 \\ \hline 5 & 6 & 7 \end{bmatrix}\begin{bmatrix} 1 & 3 & 1 \\ 2 & 2 & 1 \\ 3 & 1 & 1 \end{bmatrix} = \begin{bmatrix} 14 & 10 & 6 \\ \hline 38 & 34 & 18 \end{bmatrix}.$$

Here $A$ is a $\mathbf{2} \times 3$ matrix, $B$ is a $3 \times \mathbf{3}$ matrix and the product $AB$ is a $2 \times 3$ matrix.

In our most general situation, then, if $A$ is an $\mathbf{m} \times p$ matrix

$$A = \begin{bmatrix} a_{11} & \cdots & a_{1p} \\ a_{21} & \cdots & a_{2p} \\ \cdot & \cdots & \cdot \\ a_{m1} & \cdots & a_{mp} \end{bmatrix}$$

and $B$ is a $p \times \mathbf{n}$ matrix

$$B = \begin{bmatrix} b_{11} & \cdots & b_{1n} \\ b_{21} & \cdots & b_{2n} \\ \cdot & \cdots & \cdot \\ b_{p1} & \cdots & b_{pn} \end{bmatrix}$$

then the product $AB$ is defined as follows. We first partition $A$ as before into $m$ row matrices $A_1, A_2, \ldots, A_m$ each of which is a $1 \times p$ matrix; thus

$$A = \begin{bmatrix} A_1 \\ \hline A_2 \\ \hline \cdots \\ \hline A_m \end{bmatrix} = \begin{bmatrix} a_{11} & \cdots & a_{1p} \\ \hline a_{21} & \cdots & a_{2p} \\ \hline \cdot & \cdots & \cdot \\ \hline a_{m1} & \cdots & a_{mp} \end{bmatrix}.$$

The product $AB$ is the $m \times n$ matrix whose rows are $A_1B, A_2B, \ldots, A_mB$; that is,

$$AB = \begin{bmatrix} A_1B \\ \hline A_2B \\ \hline \cdots \\ \hline A_mB \end{bmatrix},$$

where, for example,

$$A_1B = [a_{11} \quad a_{12} \quad \cdots \quad a_{1p}] \begin{bmatrix} b_{11} & \cdots & b_{1n} \\ b_{21} & \cdots & b_{2n} \\ \cdot & \cdots & \cdot \\ b_{p1} & \cdots & b_{pn} \end{bmatrix},$$

and in general

$$A_iB = [a_{i1} \quad a_{i2} \quad \cdots \quad a_{ip}] \begin{bmatrix} b_{11} & \cdots & b_{1n} \\ b_{21} & \cdots & b_{2n} \\ \cdot & \cdots & \cdot \\ b_{p1} & \cdots & b_{pn} \end{bmatrix}$$

for each $i = 1, 2, \ldots, m$.

Another way of thinking of the product $C = AB$ of two matrices $A$ and $B$ is to think of a typical element $c_{ij}$ (lying in the $i$th row, $j$th column of $C$) as being the product $A_iB_j$, where $A_i$ is the $i$th row of $A$ and $B_j$ is the $j$th column of $B$. For example, the element $c_{23}$ in the second row, third column of the product

$$C = \begin{bmatrix} 1 & 2 & 3 \\ 4 & 5 & 6 \end{bmatrix} \begin{bmatrix} 1 & 2 & 2 \\ 0 & 1 & -2 \\ -2 & 3 & 1 \end{bmatrix}$$

is

$$c_{23} = A_2B_3 = [4 \quad 5 \quad 6] \begin{bmatrix} 2 \\ -2 \\ 1 \end{bmatrix} = 4.$$

We call your attention to the fact that in the general description above, $A$ is $\mathbf{m} \times p$, $B$ is $p \times \mathbf{n}$, and the product $AB$ is $m \times n$, with each element of the product consisting of a sum of $p$ terms. Moreover, the matrix product $AB$ is defined only when the *column* dimension of the left factor $A$ is the same as the *row* dimension of the right factor $B$. Such matrices are said to be **conformable for multiplication in the order** $AB$.

The following examples should be studied carefully. They not only provide practice in computation, but also provide some instances which illustrate that matrix multiplication is not completely analogous to multiplication of real numbers.

Example 1.6.1 Let

$$A = \begin{bmatrix} 1 & 1 & 2 \\ 0 & 1 & 3 \end{bmatrix} \quad \text{and} \quad B = \begin{bmatrix} 1 & 0 & 2 \\ 2 & 1 & 0 \\ 1 & 2 & 1 \end{bmatrix}.$$

Then the product $AB$ is defined and

$$AB = \begin{bmatrix} 1 & 1 & 2 \\ 0 & 1 & 3 \end{bmatrix} \begin{bmatrix} 1 & 0 & 2 \\ 2 & 1 & 0 \\ 1 & 2 & 1 \end{bmatrix}$$

$$= \begin{bmatrix} 1\cdot 1 + 1\cdot 2 + 2\cdot 1 & 1\cdot 0 + 1\cdot 1 + 2\cdot 2 & 1\cdot 2 + 1\cdot 0 + 2\cdot 1 \\ 0\cdot 1 + 1\cdot 2 + 3\cdot 1 & 0\cdot 0 + 1\cdot 1 + 3\cdot 2 & 0\cdot 2 + 1\cdot 0 + 3\cdot 1 \end{bmatrix}$$

$$= \begin{bmatrix} 5 & 5 & 4 \\ 5 & 7 & 3 \end{bmatrix}.$$

Note that while the product $AB$ *is* defined, the product $BA$ is not defined. Since the column dimension of $B$ is not the same as the row dimension of $A$, our original row matrix by column matrix multiplication cannot be performed. ∎

Example 1.6.2 Let

$$C = \begin{bmatrix} 1 & 1 & 2 \\ 0 & 1 & 3 \end{bmatrix} \quad \text{and} \quad D = \begin{bmatrix} 1 & 0 \\ 2 & 1 \\ 1 & 2 \end{bmatrix}.$$

Then

$$CD = \begin{bmatrix} 1 & 1 & 2 \\ 0 & 1 & 3 \end{bmatrix} \begin{bmatrix} 1 & 0 \\ 2 & 1 \\ 1 & 2 \end{bmatrix} = \begin{bmatrix} 5 & 5 \\ 5 & 7 \end{bmatrix}.$$

In this instance the product $DC$ is also defined and

$$DC = \begin{bmatrix} 1 & 0 \\ 2 & 1 \\ 1 & 2 \end{bmatrix} \begin{bmatrix} 1 & 1 & 2 \\ 0 & 1 & 3 \end{bmatrix} = \begin{bmatrix} 1 & 1 & 2 \\ 2 & 3 & 7 \\ 1 & 3 & 8 \end{bmatrix}.$$

Here the products $CD$ and $DC$ are each defined but are certainly not equal. They are both square (as must be the case when products are defined in both orders). Thus $CD$ is $2 \times 2$ while $DC$ is $3 \times 3$. ∎

Example 1.6.3 Let

$$A = \begin{bmatrix} 1 & 2 \\ 3 & 4 \end{bmatrix} \quad \text{and} \quad B = \begin{bmatrix} 1 & 2 \\ 0 & -1 \end{bmatrix}.$$

Then, using the fact that $c_{ij} = A_i B_j$,

$$AB = [A_i B_j] = \begin{bmatrix} A_1 \\ A_2 \end{bmatrix} [B_1 \quad B_2] = \begin{bmatrix} 1 & 2 \\ \hline 3 & 4 \end{bmatrix} \left[\begin{array}{c|c} 1 & 2 \\ 0 & -1 \end{array}\right] = \begin{bmatrix} 1 & 0 \\ 3 & 2 \end{bmatrix}.$$

and $$BA = \begin{bmatrix} 1 & 2 \\ 0 & -1 \end{bmatrix}\begin{bmatrix} 1 & 2 \\ 3 & 4 \end{bmatrix} = \begin{bmatrix} 7 & 10 \\ -3 & -4 \end{bmatrix}.$$

Both of the above products are defined; each is $2 \times 2$, but $AB \neq BA$. We shall discover by experience that it is only in very special situations that matrix products commute, that is, that $AB = BA$. ▮

Example 1.6.4 Let $$A = [1 \quad 2 \quad 3] \text{ and } B = \begin{bmatrix} 3 \\ 2 \\ 1 \end{bmatrix}.$$

Then $$AB = [1 \quad 2 \quad 3]\begin{bmatrix} 3 \\ 2 \\ 1 \end{bmatrix} = [10],$$

while $$BA = \begin{bmatrix} B_1A \\ B_2A \\ B_3A \end{bmatrix} = \begin{bmatrix} 3 \\ \hdashline 2 \\ \hdashline 1 \end{bmatrix}[1 \quad 2 \quad 3] = \begin{bmatrix} 3 & 6 & 9 \\ 2 & 4 & 6 \\ 1 & 2 & 3 \end{bmatrix}.$$ ▮

Example 1.6.5 Consider the two products

$$\begin{bmatrix} 1 & 0 & 0 \\ 0 & 1 & 0 \\ 0 & 0 & 1 \end{bmatrix}\begin{bmatrix} a_{11} & a_{12} & a_{13} \\ a_{21} & a_{22} & a_{23} \\ a_{31} & a_{32} & a_{33} \end{bmatrix} = \begin{bmatrix} a_{11} & a_{12} & a_{13} \\ a_{21} & a_{22} & a_{23} \\ a_{31} & a_{32} & a_{33} \end{bmatrix},$$

and $$\begin{bmatrix} a_{11} & a_{12} & a_{13} \\ a_{21} & a_{22} & a_{23} \\ a_{31} & a_{32} & a_{33} \end{bmatrix}\begin{bmatrix} 1 & 0 & 0 \\ 0 & 1 & 0 \\ 0 & 0 & 1 \end{bmatrix} = \begin{bmatrix} a_{11} & a_{12} & a_{13} \\ a_{21} & a_{22} & a_{23} \\ a_{31} & a_{32} & a_{33} \end{bmatrix}.$$ ▮

It is easy to verify the above products (you should check this) and thus we see that

$$I_3 = \begin{bmatrix} 1 & 0 & 0 \\ 0 & 1 & 0 \\ 0 & 0 & 1 \end{bmatrix}$$

acts as a multiplicative identity for $3 \times 3$ matrices in the sense that, for any $A$ in $\mathcal{M}_{3,3}$,

$$AI_3 = A = I_3A.$$

We will have more to say about this situation in a later section, and will in general use the symbol $I_n$ to denote the $n \times n$ multiplicative identity.

In Section 1.4 we explained what is meant by the *transpose* of a matrix $A$; to repeat, it is the matrix whose rows (in order) are the columns of $A$ and whose columns are the rows of $A$. We present a result concerning matrix products and transposes as a conclusion to this section on matrix multiplication.

If $A$ and $B$ are matrices conformable for multiplication in the order $AB$, then

$$(AB)^T = B^TA^T.$$

Thus, the transpose of a product is the product of the transposes of the factors *in reverse order*. We will give two examples, and will prove this result in a case in which both $A$ and $B$ are $2 \times 2$ matrices. The general argument is placed in Appendix B.

Example 1.6.6 Let $A = \begin{bmatrix} 1 & 2 & 3 \\ 4 & 5 & 6 \end{bmatrix}$ and $B = \begin{bmatrix} 2 & 1 & 0 \\ -1 & 2 & 1 \\ 0 & -1 & 1 \end{bmatrix}$.

Then

$$AB = \begin{bmatrix} 1 & 2 & 3 \\ 4 & 5 & 6 \end{bmatrix} \begin{bmatrix} 2 & 1 & 0 \\ -1 & 2 & 1 \\ 0 & -1 & 1 \end{bmatrix} = \begin{bmatrix} 0 & 2 & 5 \\ 3 & 8 & 11 \end{bmatrix};$$

thus

$$(AB)^T = \begin{bmatrix} 0 & 3 \\ 2 & 8 \\ 5 & 11 \end{bmatrix}$$

while

$$B^T A^T = \begin{bmatrix} 2 & -1 & 0 \\ 1 & 2 & -1 \\ 0 & 1 & 1 \end{bmatrix} \begin{bmatrix} 1 & 4 \\ 2 & 5 \\ 3 & 6 \end{bmatrix} = \begin{bmatrix} 0 & 3 \\ 2 & 8 \\ 5 & 11 \end{bmatrix},$$

so that $(AB)^T = B^T A^T$. ∎

Example 1.6.7 Let $C = [1 \quad 2 \quad 3]$ and $D = \begin{bmatrix} 4 \\ 5 \\ 6 \end{bmatrix}$. Then $CD = [1 \quad 2 \quad 3] \begin{bmatrix} 4 \\ 5 \\ 6 \end{bmatrix} = [32]$,

so that $(CD)^T = [32]$ also. $D^T C^T = [4 \quad 5 \quad 6] \begin{bmatrix} 1 \\ 2 \\ 3 \end{bmatrix} = [32]$, so that in this case, as above, $(CD)^T = D^T C^T$. ∎

Now, if $A$ and $B$ are arbitrary $2 \times 2$ matrices with

$$A = \begin{bmatrix} a_{11} & a_{12} \\ a_{21} & a_{22} \end{bmatrix} \quad \text{and} \quad B = \begin{bmatrix} b_{11} & b_{12} \\ b_{21} & b_{22} \end{bmatrix},$$

then

$$AB = \begin{bmatrix} a_{11}b_{11} + a_{12}b_{21} & a_{11}b_{12} + a_{12}b_{22} \\ a_{21}b_{11} + a_{22}b_{21} & a_{21}b_{12} + a_{22}b_{22} \end{bmatrix},$$

so that

$$(AB)^T = \begin{bmatrix} a_{11}b_{11} + a_{12}b_{21} & a_{21}b_{11} + a_{22}b_{21} \\ a_{11}b_{12} + a_{12}b_{22} & a_{21}b_{12} + a_{22}b_{22} \end{bmatrix},$$

while

$$B^T A^T = \begin{bmatrix} b_{11} & b_{21} \\ b_{12} & b_{22} \end{bmatrix} \begin{bmatrix} a_{11} & a_{21} \\ a_{12} & a_{22} \end{bmatrix} = \begin{bmatrix} b_{11}a_{11} + b_{21}a_{12} & b_{11}a_{21} + b_{21}a_{22} \\ b_{12}a_{11} + b_{22}a_{12} & b_{12}a_{21} + b_{22}a_{22} \end{bmatrix}$$

$$= \begin{bmatrix} a_{11}b_{11} + a_{12}b_{21} & a_{21}b_{11} + a_{22}b_{21} \\ a_{11}b_{12} + a_{12}b_{22} & a_{21}b_{12} + a_{22}b_{22} \end{bmatrix}.$$

The last step follows from the commutativity of multiplication of the real numbers. Thus we have shown for $2 \times 2$ matrices that $B^T A^T = (AB)^T$.

Example 1.6.8 To show a practical application of matrix multiplication we continue with Example 1.3.2 (see page 16). Let $A$ be the requirement matrix for the two projects

$$A = \begin{array}{c} \\ \end{array}\begin{bmatrix} \textit{Paint} & \textit{Lumber} & \textit{Labor} \\ 1500 \text{ gals.} & 30{,}000 \text{ bd. ft.} & 10{,}000 \text{ man hrs.} \\ 2000 \text{ gals.} & 40{,}000 \text{ bd. ft.} & 15{,}000 \text{ man hrs.} \end{bmatrix} \begin{array}{l} \\ \text{Project 1} \\ \text{Project 2} \end{array}$$

and let $B$ be the unit cost matrix

$$B = \begin{bmatrix} \$7.00 \\ 0.25 \\ 5.00 \end{bmatrix} \begin{array}{l} \text{Paint/gal.} \\ \text{Lumber/bd. ft.} \\ \text{Labor/man hr.} \end{array}$$

Then the project cost matrix is the matrix product $AB$ where (you should be ready to check this)

$$AB = \begin{bmatrix} 1500 & 30{,}000 & 10{,}000 \\ 2000 & 40{,}000 & 15{,}000 \end{bmatrix} \begin{bmatrix} 7.00 \\ 0.25 \\ 5.00 \end{bmatrix} = \begin{bmatrix} 68{,}000 \\ 99{,}000 \end{bmatrix} \begin{array}{l} \text{Project 1} \\ \text{Project 2} \end{array}$$ ∎

Example 1.6.9 As a second practical example, let us suppose that Jim, Jack, and Jerry work on a piecework basis, i.e., they are paid a specific amount for each unit of product that they make. On a particular day Jim made 4 units of product $P$, 3 units of product $Q$, and 2 units of product $S$; Jack made 5, 4, and 1 units of these products, respectively; while Jerry produced 3, 1, and 5 units. They are each paid \$1 for each unit of product $P$, \$3 for each unit of $Q$, and \$2 for each unit of $S$.

The total wage that each boy is to receive for his day's work can be found by computing the matrix product $AB$ shown below.

$$AB = \begin{array}{r} \\ \text{Jim} \\ \text{Jack} \\ \text{Jerry} \end{array} \begin{array}{c} \begin{array}{ccc} P & Q & S \end{array} \\ \begin{bmatrix} 4 & 3 & 2 \\ 5 & 4 & 1 \\ 3 & 1 & 5 \end{bmatrix} \end{array} \begin{array}{c} \textit{Wage/unit} \\ \begin{bmatrix} 1 \\ 3 \\ 2 \end{bmatrix} \end{array} = \begin{bmatrix} 17 \\ 19 \\ 16 \end{bmatrix}$$ ∎

## 1.6 Exercises (continued)

In Exercises 19–26, calculate the indicated matrix product.

19. $\begin{bmatrix} 1 & 0 & 0 \\ 0 & 0 & 0 \\ 0 & 0 & 0 \end{bmatrix} \begin{bmatrix} a & b & c \\ d & e & f \\ g & h & i \end{bmatrix}$

20. $\begin{bmatrix} 0 & 0 & 0 \\ 0 & 1 & 0 \\ 0 & 0 & 0 \end{bmatrix} \begin{bmatrix} a & b & c \\ d & e & f \\ g & h & i \end{bmatrix}$

21. $\begin{bmatrix} 0 & 0 & 0 \\ 0 & 0 & 0 \\ 0 & 0 & 1 \end{bmatrix} \begin{bmatrix} a & b & c \\ d & e & f \\ g & h & i \end{bmatrix}$

22. $\begin{bmatrix} a & b & c \\ d & e & f \\ g & h & i \end{bmatrix} \begin{bmatrix} 1 & 0 & 0 \\ 0 & 0 & 0 \\ 0 & 0 & 0 \end{bmatrix}$

23. $\begin{bmatrix} a & b & c \\ d & e & f \\ g & h & i \end{bmatrix} \begin{bmatrix} 0 & 0 & 0 \\ 0 & 1 & 0 \\ 0 & 0 & 0 \end{bmatrix}$

24. $\begin{bmatrix} a & b & c \\ d & e & f \\ g & h & i \end{bmatrix} \begin{bmatrix} 0 & 0 & 0 \\ 0 & 0 & 0 \\ 0 & 0 & 1 \end{bmatrix}$

25. $\begin{bmatrix} 1 & 1 & 0 \\ 0 & 1 & 0 \\ 0 & 0 & 1 \end{bmatrix}\begin{bmatrix} a & b & c \\ d & e & f \\ g & h & i \end{bmatrix}$

26. $\begin{bmatrix} 2 & 3 & 0 \\ 0 & 1 & 0 \\ 0 & 0 & 1 \end{bmatrix}\begin{bmatrix} a & b & c \\ d & e & f \\ g & h & i \end{bmatrix}$

27. Describe in words the matrix product

$$\begin{bmatrix} x & y & 0 \\ 0 & 1 & 0 \\ 0 & 0 & 1 \end{bmatrix}\begin{bmatrix} a & b & c \\ d & e & f \\ g & h & i \end{bmatrix}.$$

28. Find $\begin{bmatrix} a & b & c \\ d & e & f \\ g & h & i \end{bmatrix}\begin{bmatrix} x & y & 0 \\ 0 & 1 & 0 \\ 0 & 0 & 1 \end{bmatrix}.$

In Exercises 29–31, calculate the indicated matrix product.

29. $\begin{bmatrix} 1 & 2 & 1 \\ 0 & 1 & 3 \end{bmatrix}\begin{bmatrix} 1 & 2 & 1 & 1 \\ 0 & 1 & 1 & 4 \\ 1 & 0 & 3 & 2 \end{bmatrix}$

30. $\begin{bmatrix} 1.3 & 2.1 \\ 4.3 & -1.2 \end{bmatrix}\begin{bmatrix} -1.1 & 2.1 & -0.1 \\ -0.3 & 3.1 & -1.1 \end{bmatrix}$

31. $\begin{bmatrix} 5 & 1 & 2 & 3 \\ 2 & 1 & 4 & 1 \\ -1 & -1 & 1 & 1 \\ 0 & 1 & 0 & 1 \end{bmatrix}\begin{bmatrix} 2 & 1 \\ 1 & 0 \\ 3 & 0 \\ 1 & 2 \end{bmatrix}$

32. As in Example 1.6.8, suppose that 4 projects are undertaken in which the paint, lumber, and labor requirements are represented by columns 1–3, respectively. Matrix

$$A = \begin{bmatrix} 1000 & 25{,}000 & 9{,}000 \\ 1500 & 30{,}000 & 10{,}000 \\ 2500 & 40{,}000 & 15{,}000 \\ 3000 & 45{,}000 & 20{,}000 \end{bmatrix}.$$

If the unit cost matrix

$$B = \begin{bmatrix} 10.00 \\ 0.50 \\ 6.00 \end{bmatrix},$$

find the project cost matrix $P$ and the total cost of all 4 projects.

33. Let $A = \begin{bmatrix} 1 & 2 \\ 3 & 4 \end{bmatrix}$, $B = \begin{bmatrix} 1 & 2 \\ -1 & 1 \end{bmatrix}$, and $C = \begin{bmatrix} 1 & 0 \\ 3 & -2 \end{bmatrix}$.

Is $(A + B)C = AC + BC$?

34. Compute $\begin{bmatrix} 1 & 2 \\ 0 & 1 \\ 1 & 1 \end{bmatrix}\begin{bmatrix} 1 & 2 & 0 & 1 & 3 & 4 & 6 \\ -1 & -1 & 1 & 0 & 1 & -2 & 1 \end{bmatrix}$

**New Terms**

conformable for multiplication, 29

## 1.7 Distributive Properties

In Exercise 33 of the preceding set, a numerical example of the property

$$(A + B)C = AC + BC$$

was presented. It is analogous to the distributive property **D2** for real numbers (see Section 1.2), which states

**D2** *Right Distributive Property.* If $A$ and $B$ are conformable for addition and are also conformable for multiplication with $C$ on the right then

$$(A + B)C = AC + BC.$$

Similarly, we also have

**D1** *Left Distributive Property.*

$$A(B + C) = AB + AC,$$

providing the individual sums and products are defined.

We have placed the general proofs of these important properties in Appendix B for the interested reader to study separately. Here we merely illustrate these properties by further example and give the proof of the right distributive property in a simple case where $A$, $B$, and $C$ are each $2 \times 2$ matrices.

Example 1.7.1 $(A + B)C$

$$= \left(\begin{bmatrix} 2 & 1 & 3 & 1 \\ -1 & 0 & 2 & -1 \end{bmatrix} + \begin{bmatrix} -1 & 1 & -2 & 0 \\ 1 & 1 & -1 & 2 \end{bmatrix}\right) \begin{bmatrix} 1 & -1 & 0 \\ 1 & 1 & 1 \\ 2 & -1 & 1 \\ 1 & 1 & 0 \end{bmatrix}$$

$$= \begin{bmatrix} 1 & 2 & 1 & 1 \\ 0 & 1 & 1 & 1 \end{bmatrix} \begin{bmatrix} 1 & -1 & 0 \\ 1 & 1 & 1 \\ 2 & -1 & 1 \\ 1 & 1 & 0 \end{bmatrix} = \begin{bmatrix} 6 & 1 & 3 \\ 4 & 1 & 2 \end{bmatrix},$$

while

$$AC = \begin{bmatrix} 2 & 1 & 3 & 1 \\ -1 & 0 & 2 & -1 \end{bmatrix} \begin{bmatrix} 1 & -1 & 0 \\ 1 & 1 & 1 \\ 2 & -1 & 1 \\ 1 & 1 & 0 \end{bmatrix} = \begin{bmatrix} 10 & -3 & 4 \\ 2 & -2 & 2 \end{bmatrix}$$

and

$$BC = \begin{bmatrix} -1 & 1 & -2 & 0 \\ 1 & 1 & -1 & 2 \end{bmatrix} \begin{bmatrix} 1 & -1 & 0 \\ 1 & 1 & 1 \\ 2 & -1 & 1 \\ 1 & 1 & 0 \end{bmatrix} = \begin{bmatrix} -4 & 4 & -1 \\ 2 & 3 & 0 \end{bmatrix}$$

so that

$$AC + BC = \begin{bmatrix} 10 & -3 & 4 \\ 2 & -2 & 2 \end{bmatrix} + \begin{bmatrix} -4 & 4 & -1 \\ 2 & 3 & 0 \end{bmatrix}$$

$$= \begin{bmatrix} 6 & 1 & 3 \\ 4 & 1 & 2 \end{bmatrix} = (A + B)C. \quad \blacksquare$$

Example 1.7.2 $D(E + F)$

$$= [1 \quad 2 \quad 3]\left(\begin{bmatrix} 1 & 2 & 1 & 2 \\ -1 & 1 & 4 & -1 \\ 0 & -3 & 2 & 3 \end{bmatrix} + \begin{bmatrix} 2 & -1 & 1 & -1 \\ 0 & 1 & -2 & 3 \\ 1 & 2 & 0 & -1 \end{bmatrix}\right)$$

$$= [1 \quad 2 \quad 3]\begin{bmatrix} 3 & 1 & 2 & 1 \\ -1 & 2 & 2 & 2 \\ 1 & -1 & 2 & 2 \end{bmatrix} = [4 \quad 2 \quad 12 \quad 11].$$

$$DE + DF = [1 \quad 2 \quad 3]\begin{bmatrix} 1 & 2 & 1 & 2 \\ -1 & 1 & 4 & -1 \\ 0 & -3 & 2 & 3 \end{bmatrix}$$

$$+ [1 \quad 2 \quad 3]\begin{bmatrix} 2 & -1 & 1 & -1 \\ 0 & 1 & -2 & 3 \\ 1 & 2 & 0 & -1 \end{bmatrix}$$

$$= [-1 \quad -5 \quad 15 \quad 9] + [5 \quad 7 \quad -3 \quad 2]$$

$$= [4 \quad 2 \quad 12 \quad 11].$$

Thus $$D(E + F) = DE + DF. \quad \blacksquare$$

Now let us prove the right distributive property in the case where $A$, $B$, and $C$ are general $2 \times 2$ matrices. Let

$$A = \begin{bmatrix} a_{11} & a_{12} \\ a_{21} & a_{22} \end{bmatrix}, \quad B = \begin{bmatrix} b_{11} & b_{12} \\ b_{21} & b_{22} \end{bmatrix}, \quad \text{and} \quad C = \begin{bmatrix} c_{11} & c_{12} \\ c_{21} & c_{22} \end{bmatrix}.$$

Then

$$(A+B)C = \begin{bmatrix} a_{11}+b_{11} & a_{12}+b_{12} \\ a_{21}+b_{21} & a_{22}+b_{22} \end{bmatrix}\begin{bmatrix} c_{11} & c_{12} \\ c_{21} & c_{22} \end{bmatrix}$$

$$= \begin{bmatrix} (a_{11} + b_{11})c_{11} + (a_{12} + b_{12})c_{21} & (a_{11} + b_{11})c_{12} + (a_{12} + b_{12})c_{22} \\ (a_{21} + b_{21})c_{11} + (a_{22} + b_{22})c_{21} & (a_{21} + b_{21})c_{12} + (a_{22} + b_{22})c_{22} \end{bmatrix}$$

$$= \begin{bmatrix} (a_{11}c_{11}+a_{12}c_{21})+(b_{11}c_{11}+b_{12}c_{21}) & (a_{11}c_{12}+a_{12}c_{22})+(b_{11}c_{12}+b_{12}c_{22}) \\ (a_{21}c_{11}+a_{22}c_{21})+(b_{21}c_{11}+b_{22}c_{21}) & (a_{21}c_{12}+a_{22}c_{22})+(b_{21}c_{12}+b_{22}c_{22}) \end{bmatrix}$$

$$= \begin{bmatrix} a_{11}c_{11}+a_{12}c_{21} & a_{11}c_{12}+a_{12}c_{22} \\ a_{21}c_{11}+a_{22}c_{21} & a_{21}c_{12}+a_{22}c_{22} \end{bmatrix} + \begin{bmatrix} b_{11}c_{11}+b_{12}c_{21} & b_{11}c_{12}+b_{12}c_{22} \\ b_{21}c_{11}+b_{22}c_{21} & b_{21}c_{12}+b_{22}c_{22} \end{bmatrix}$$

$$= AC+BC. \quad \blacksquare$$

## 1.7 Exercises

Verify the left or right distributive property in the following exercises.

1. $[2 \quad 1 \quad 5 \quad 3]\left(\begin{bmatrix} 1 \\ -1 \\ -1 \\ 2 \end{bmatrix} + \begin{bmatrix} 1 \\ 1 \\ 1 \\ -1 \end{bmatrix}\right)$

2. $\begin{bmatrix} 2 & 1 & 2 \\ 1 & 3 & 1 \\ 4 & 1 & 4 \end{bmatrix}\left(\begin{bmatrix} -1 & 1 & -1 \\ 0 & -1 & 1 \\ -2 & 1 & 2 \end{bmatrix} + \begin{bmatrix} 1 & 2 & 1 \\ 3 & 1 & 5 \\ 4 & 2 & 0 \end{bmatrix}\right)$

3. $([-1 \quad 2 \quad 1] + [1 \quad -1 \quad 2])\begin{bmatrix} 1 \\ 2 \\ 1 \end{bmatrix}$

4. $\left(\begin{bmatrix} 2 & 0 & 2 \\ 1 & 4 & 3 \end{bmatrix} + \begin{bmatrix} 3 & 1 & -2 \\ 1 & 0 & 2 \end{bmatrix}\right)\begin{bmatrix} 1 & 0 & 1 & 0 & 0 \\ 1 & 1 & 0 & 1 & 0 \\ 2 & 1 & 0 & 0 & 1 \end{bmatrix}$.

5. $\begin{bmatrix} 3 & 1 & 2 \\ 1 & 1 & 1 \end{bmatrix}\left(\begin{bmatrix} 2 & 1 \\ -1 & 2 \\ 1 & 1 \end{bmatrix} + \begin{bmatrix} -1 & -1 \\ 1 & -1 \\ 0 & 1 \end{bmatrix}\right)$

6. $\left(\begin{bmatrix} 1 & 2 \\ 3 & 4 \\ 5 & 6 \end{bmatrix} + \begin{bmatrix} -1 & -1 \\ 1 & -2 \\ -3 & 2 \end{bmatrix}\right)\begin{bmatrix} 1 & 2 & 3 & 4 \\ -1 & 1 & -2 & 3 \end{bmatrix}$

7. $\left(\begin{bmatrix} 1 \\ -1 \\ -1 \\ 2 \end{bmatrix} + \begin{bmatrix} 1 \\ 1 \\ 1 \\ -1 \end{bmatrix}\right)[2 \quad 1 \quad 5 \quad 3]$ (Compare with Exercise 1 above.)

8. $\begin{bmatrix} 1 & 0 & 1 & 0 & 0 \\ 1 & 1 & 0 & 1 & 0 \\ 2 & 1 & 0 & 0 & 1 \end{bmatrix}\left(\begin{bmatrix} 1 & 3 & 1 \\ 2 & 0 & 2 \\ 1 & 1 & 1 \\ 1 & 4 & 3 \\ 0 & 0 & 1 \end{bmatrix} + \begin{bmatrix} 2 & 1 & -4 \\ 3 & 1 & -2 \\ 1 & 2 & 3 \\ 1 & 0 & 2 \\ 1 & 1 & 0 \end{bmatrix}\right)$

9. In Example 1.6.9 suppose that the matrix

$$C = \begin{bmatrix} 2 & 1 & 7 \\ 3 & 3 & 4 \\ 1 & 6 & 2 \end{bmatrix}$$

represents the production of Jim, Jack, and Jerry on the following day. Show that the distributive law may be used to compute the total wage that each boy will receive for the two days' work. Find the "total wage" matrix by two methods.

## 1.8 Some Algebraic Properties of Square Matrices

We have already seen in Section 1.5 of this chapter that the set $\mathscr{M}_{m,n}$ of all $m \times n$ real matrices satisfies the matrix analogues of properties A1–A4 of addition for real numbers. In this section we shall discover that certain matrices satisfy further properties with analogues in the real number system. We shall see also that while there are striking similarities, there are also significant differences between these two algebraic structures.

We now wish to explore further multiplicative properties of matrices, and thus develop other analogies with the real numbers. In order to do this we consider a set of matrices in which any two matrices are conformable for addition and for multiplication. Thus we must restrict our attention to the set $\mathscr{M}_{n,n}$ of all real $n \times n$ (square) matrices. As in the earlier sections of this chapter, we have placed the more difficult proofs in Appendix B. We shall illustrate our properties with examples from $\mathscr{M}_{2,2}$ and $\mathscr{M}_{3,3}$.

For convenience of reference, we re-list the properties for real numbers whose analogues will be discussed in this section:

**M1** Associativity of Multiplication
**M2** Multiplicative Identity
**M3** Multiplicative Inverses
**M4** Commutativity of Multiplication
**Property 1** Uniqueness of Identity
**Property 2** Uniqueness of Multiplicative Inverse
**Property 3** $(a^{-1})^{-1} = a$.
**Property 4** If $ab = 0$ then $a = 0$ or $b = 0$.
**Property 5** Unique solution of $ax = b$.
**Property 6** Unique solution of $ax + b = c$.

Our first matrix analogue of real number multiplication is

**M1** *Matrix multiplication is associative. Thus, if $A$, $B$, $C$ are in $\mathscr{M}_{n,n}$, then*

$$A(BC) = (AB)C.$$

In other words, the product $BC$ multiplied on the left by $A$ is precisely the same as $AB$ multiplied on the right by $C$. You should note here and in Exercises 3 and 4 of the next set, that this property is not peculiar simply to square matrices, but is valid whenever the matrices are conformable for multiplication in the order indicated. We consider an example from $\mathscr{M}_{2,2}$ and let

$$A = \begin{bmatrix} 1 & -1 \\ 2 & 3 \end{bmatrix}, \quad B = \begin{bmatrix} 2 & 1 \\ -1 & 0 \end{bmatrix}, \quad \text{and} \quad C = \begin{bmatrix} 1 & 1 \\ -1 & 2 \end{bmatrix}.$$

Then
$$AB = \begin{bmatrix} 1 & -1 \\ 2 & 3 \end{bmatrix}\begin{bmatrix} 2 & 1 \\ -1 & 0 \end{bmatrix} = \begin{bmatrix} 3 & 1 \\ 1 & 2 \end{bmatrix},$$

so that
$$(AB)C = \begin{bmatrix} 3 & 1 \\ 1 & 2 \end{bmatrix}\begin{bmatrix} 1 & 1 \\ -1 & 2 \end{bmatrix} = \begin{bmatrix} 2 & 5 \\ -1 & 5 \end{bmatrix}.$$

On the other hand,

$$BC = \begin{bmatrix} 2 & 1 \\ -1 & 0 \end{bmatrix}\begin{bmatrix} 1 & 1 \\ -1 & 2 \end{bmatrix} = \begin{bmatrix} 1 & 4 \\ -1 & -1 \end{bmatrix},$$

and hence $A(BC) = \begin{bmatrix} 1 & -1 \\ 2 & 3 \end{bmatrix}\begin{bmatrix} 1 & 4 \\ -1 & -1 \end{bmatrix} = \begin{bmatrix} 2 & 5 \\ -1 & 5 \end{bmatrix}.$

Thus $(AB)C = A(BC)$.

## 1.8 Exercises

1. Verify the associative property for matrix multiplication in the case that

   a. $A = \begin{bmatrix} 1 & 2 \\ 3 & 4 \end{bmatrix}, B = \begin{bmatrix} -2 & 1 \\ -1 & 0 \end{bmatrix}, C = \begin{bmatrix} 1 & 3 \\ 0 & 2 \end{bmatrix}.$

   b. $A = \begin{bmatrix} 3 & 1 & -2 \\ 2 & 1 & 3 \\ 1 & 0 & 1 \end{bmatrix}, B = \begin{bmatrix} -1 & 2 & -1 \\ 0 & 1 & 2 \\ 1 & -1 & 0 \end{bmatrix}, C = \begin{bmatrix} 1 & 1 & 1 \\ 2 & 0 & 1 \\ 0 & -1 & 0 \end{bmatrix}.$

2. Prove the associative property in the case of $2 \times 2$ matrices; that is, let

$$A = \begin{bmatrix} a_{11} & a_{12} \\ a_{21} & a_{22} \end{bmatrix}, \quad B = \begin{bmatrix} b_{11} & b_{12} \\ b_{21} & b_{22} \end{bmatrix}, \quad C = \begin{bmatrix} c_{11} & c_{12} \\ c_{21} & c_{22} \end{bmatrix};$$

   compute $A(BC)$ and $(AB)C$ separately and compare the results.

3. Let

$$A = \begin{bmatrix} 2 \\ 1 \end{bmatrix}, \quad B = [1 \quad -1 \quad 2], \quad C = \begin{bmatrix} 1 & 0 & 1 & -1 \\ -1 & 1 & 2 & 0 \\ 2 & -1 & -1 & 1 \end{bmatrix},$$

   and show that $A(BC) = (AB)C$, even in this case of non-square matrices.

4. Verify that, with $A = [1 \quad 2 \quad 1]$, $B = \begin{bmatrix} 3 & 1 \\ -1 & 0 \\ 2 & 1 \end{bmatrix}$, $C = \begin{bmatrix} 1 \\ -1 \end{bmatrix}$, we again have $A(BC) = (AB)C$.

5. Let $A = \begin{bmatrix} 1 & 2 \\ 3 & 4 \end{bmatrix}$ and $B = \begin{bmatrix} -1 & 6 \\ 2 & 8 \end{bmatrix}$. Verify that

   a. $(A^T)^T = A$.
   b. $(AB)^T = B^T A^T$.
   c. $(A + B)^T = A^T + B^T$.

6. Verify the statement of Exercise 5 for arbitrary $2 \times 2$ matrices.

7. Show that if each of $A$ and $B$ is a $2 \times 2$ symmetric matrix, then their sum $A + B$ is symmetric.

8. Is the product $AB$ in Exercise 7 symmetric?

Our next property is

**M2** *There is a multiplicative identity* $I_n$ *in* $\mathscr{M}_{n,n}$ *such that for each* $A$ *in* $\mathscr{M}_{n,n}$

$$I_nA = AI_n = A.$$

Indeed, such a matrix $I_n$ is obtained by setting

$$I_n = \begin{bmatrix} 1 & 0 & 0 & \cdots & 0 & 0 \\ 0 & 1 & 0 & \cdots & 0 & 0 \\ 0 & 0 & 1 & \cdots & 0 & 0 \\ \cdot & \cdot & \cdot & \cdots & \cdot & \cdot \\ 0 & 0 & 0 & \cdots & 1 & 0 \\ 0 & 0 & 0 & \cdots & 0 & 1 \end{bmatrix}.$$

That is, $I_n$ is the $n \times n$ diagonal matrix whose main diagonal elements are all 1's. That $I_n$ as defined above acts as an identity may be verified by inspection. It is readily computed in the following example from $\mathscr{M}_{2,2}$.

$$\begin{bmatrix} 1 & 0 \\ 0 & 1 \end{bmatrix}\begin{bmatrix} a_{11} & a_{12} \\ a_{21} & a_{22} \end{bmatrix} = \begin{bmatrix} a_{11} & a_{12} \\ a_{21} & a_{22} \end{bmatrix} = \begin{bmatrix} a_{11} & a_{12} \\ a_{21} & a_{22} \end{bmatrix}\begin{bmatrix} 1 & 0 \\ 0 & 1 \end{bmatrix}.$$

## 1.8 Exercises (continued)

9. Calculate the product $\begin{bmatrix} 1 & 0 \\ 0 & 1 \end{bmatrix}\begin{bmatrix} a_{11} & a_{12} & a_{13} & a_{14} \\ a_{21} & a_{22} & a_{23} & a_{24} \end{bmatrix}$.

   Here $I_2$ acts as a left identity for all $2 \times 4$ matrices (indeed for all $2 \times n$ matrices, where $n$ is any positive integer).

10. Is there a right identity for $\mathscr{M}_{2,4}$, and, if so, what is it?

Now that we have established the existence of a multiplicative identity $I_n$ for $\mathscr{M}_{n,n}$, it is reasonable to ask whether or not the matrix analogue of M3 is true in $\mathscr{M}_{n,n}$: Does each non-zero matrix $A$ in $\mathscr{M}_{n,n}$ have a multiplicative inverse (denoted by $A^{-1}$) such that

$$AA^{-1} = A^{-1}A = I_n?$$

The answer to this question is: No; *some square matrices do have inverses and some do not.* (Those matrices which have inverses will be called **non-singular** and those which do not have inverses will be called **singular**.) The question "How can one determine whether or not a given matrix has an inverse?" and some other closely related questions will only be mentioned briefly in this section and will be dealt with more fully in our next chapter.

Let us note that at least one non-zero member of $\mathscr{M}_{n,n}$ has a multiplicative inverse—namely, $I_n$, which is its own inverse, since

$$I_nI_n = I_n.$$

Our old familiar matrix $A = \begin{bmatrix} 1 & 2 \\ 3 & 4 \end{bmatrix}$ in $\mathscr{M}_{n,n}$ also has a multiplicative inverse—namely,

$$A^{-1} = \begin{bmatrix} 1 & 2 \\ 3 & 4 \end{bmatrix}^{-1} = \begin{bmatrix} -2 & 1 \\ 3/2 & -1/2 \end{bmatrix}.$$

(You should verify this by direct calculation of the products $AA^{-1}$ and $A^{-1}A$.)

In order to see that some non-zero matrices do not have inverses, consider the non-zero matrix

$$A = \begin{bmatrix} 1 & 0 \\ 0 & 0 \end{bmatrix}.$$

Were there an inverse $A^{-1}$ for $A$ with, say

$$A^{-1} = \begin{bmatrix} x & y \\ z & w \end{bmatrix},$$

then $I_2 = AA^{-1}$, which means that

$$I_2 = \begin{bmatrix} 1 & 0 \\ 0 & 1 \end{bmatrix} = \begin{bmatrix} 1 & 0 \\ 0 & 0 \end{bmatrix}\begin{bmatrix} x & y \\ z & w \end{bmatrix} = \begin{bmatrix} x & y \\ 0 & 0 \end{bmatrix}.$$

Since this equality is impossible (compare the elements in the 2, 2 position), $A$ cannot have an inverse.

It is not purely the number of zero elements in a matrix, however, that causes it to have no inverse. Consider the matrix

$$B = \begin{bmatrix} 1 & 2 \\ 2 & 4 \end{bmatrix}$$

which has no zero elements at all. $B$ also has no inverse. Suppose $B$ did have an inverse

$$B^{-1} = \begin{bmatrix} x & y \\ z & w \end{bmatrix}.$$

Then, in particular, $I_2 = BB^{-1}$, so that

$$\begin{bmatrix} 1 & 0 \\ 0 & 1 \end{bmatrix} = \begin{bmatrix} 1 & 2 \\ 2 & 4 \end{bmatrix}\begin{bmatrix} x & y \\ z & w \end{bmatrix} = \begin{bmatrix} x + 2z & y + 2w \\ 2x + 4z & 2y + 4w \end{bmatrix}.$$

By equating elements in the 1, 1 position we find that we must have

$$1 = x + 2z,$$

while if we equate elements in the 2, 1 position we find that at the same time we must have

$$0 = 2x + 4z.$$

That such numbers $x$ and $z$ cannot exist is easily seen by subtracting twice the first equation from the second to obtain the absurd result that

$$-2 = 0.$$

From these examples, then, we see that a direct analogue of M3 (multiplicative inverses) does not hold in $\mathscr{M}_{n,n}$.

The matrix analogue of M4 (commutativity of multiplication) does not hold in $\mathscr{M}_{n,n}$ either, as we have seen in Example 1.6.3. We have seen, however, that *some* matrices do commute (a matrix and its inverse, for instance). We shall not probe the question of commutativity in our discussions here, but let the reader be warned to assume that products are not guilty of commuting until they are proven otherwise.

Of course, a member $A$ of $\mathscr{M}_{n,n}$ does commute with itself, and we are able to define through associativity the positive integral powers of $A$ inductively; thus

$$A^1 = A, \qquad A^2 = AA, \qquad A^3 = AA^2, \ldots, \qquad A^{k+1} = AA^k.$$

In the special case that $A$ has an inverse, we remain consistent if we define, for positive integers $k$,

$$A^{-k} = (A^{-1})^k \quad \text{and} \quad A^0 = I_n.$$

For example, $A^{-2} = (A^{-1})^2 = (A^{-1})(A^{-1})$. Thus,

$$\begin{aligned} A^{-2}(A^2) &= (A^{-1}A^{-1})(AA) \\ &= A^{-1}(A^{-1}A)A \\ &= A^{-1}I_nA \\ &= A^{-1}A \\ &= I_n. \end{aligned}$$

## 1.8 Exercises (continued)

In Exercises 11–16, discover whether or not the given matrix has a multiplicative inverse; and find the inverse if it exists.

11. $\begin{bmatrix} 3 & 0 & 0 \\ 0 & 3 & 0 \\ 0 & 0 & 3 \end{bmatrix}$

12. $\begin{bmatrix} 2 & 0 \\ 0 & 3 \end{bmatrix}$

13. $\begin{bmatrix} 0 & 2 \\ 3 & 0 \end{bmatrix}$

14. $\begin{bmatrix} 3 & 0 & 0 \\ 0 & 0 & 0 \\ 0 & 0 & 3 \end{bmatrix}$

15. $\begin{bmatrix} 2 & 1 \\ -4 & -2 \end{bmatrix}$

16. $\begin{bmatrix} 1 & 0 & 1 \\ 0 & 1 & 0 \\ 0 & 0 & 1 \end{bmatrix}$

17. Let $A$ and $B$ belong to $\mathscr{M}_{n,n}$. Is $(A+B)^2 = A^2 + 2AB + B^2$? Is $(A+B)(A-B) = A^2 - B^2$? (*Hint*: Is there anything else that we need to know? Consider an example from $\mathscr{M}_{2,2}$.)

18. Expand $(A + I_n)^2$. What is $(A + I_n)(A - I_n)$?

19. Let

$$A = \begin{bmatrix} 0 & 1 & 1 \\ 0 & 0 & 1 \\ 0 & 0 & 0 \end{bmatrix}.$$

Calculate $A^2$ and $A^3$. Notice that $A$ satisfies the equation $X^3 = 0$ but does not satisfy $X^2 = 0$.

20. Use the fact that $(AB)^T = B^T A^T$ to prove that $(A^T)^{-1} = (A^{-1})^T$. Verify this statement in Exercises 13 and 16 above.

Let us now examine which of Properties 1–6 for the real numbers have analogues in $\mathcal{M}_{n,n}$.

**Property 1** *Uniqueness of $I_n$.* As with real numbers, if there were an identity $E$ other than $I_n$, then, necessarily,

$$I_n = I_n E = E.$$

Hence there is only one identity for multiplication in $\mathcal{M}_{n,n}$.

**Property 2** *Uniqueness of Inverses.* Not every non-zero member of $\mathcal{M}_{n,n}$ has an inverse. However, if a certain member $A$ does have an inverse, then the inverse is unique; for if both $A^{-1}$ and $X$ are multiplicative inverses of $A$, then $AX = I_n$. Multiplying both sides of this equation on the left by $A^{-1}$ we have

$$\begin{aligned} A^{-1}(AX) &= A^{-1}I_n \\ (A^{-1}A)X &= A^{-1} \\ I_n X &= A^{-1} \\ X &= A^{-1}. \end{aligned}$$

Thus there can be only one multiplicative inverse for $A$.

**Property 3** *If $A^{-1}$ exists, then $(A^{-1})^{-1} = A$.* This property is true because $A$ does act as an inverse for $A^{-1}$ in the sense that $AA^{-1} = A^{-1}A = I_n$. Since inverses are unique by Property 2, it follows that $(A^{-1})^{-1} = A$.

Thus Property 1 has a direct analogue for matrices while Properties 2 and 3, although not valid for every non-zero matrix, are valid for those matrices which possess inverses.

**Property 4** "$ab = 0$ implies $a = 0$ or $b = 0$". The situation degenerates further when we calculate the product

$$AB = \begin{bmatrix} 1 & 2 \\ 2 & 4 \end{bmatrix} \begin{bmatrix} 2 & 4 \\ -1 & -2 \end{bmatrix}$$

and find that it is
$$\begin{bmatrix} 0 & 0 \\ 0 & 0 \end{bmatrix} = 0.$$

Thus $AB = 0$, and yet neither factor is zero, in direct contradiction to the matrix analogue of Property 4. We shall not explore the question of *divisors of zero* in this book, but let the reader be warned again that the familiar argument

$$ax = ay \quad \text{implies} \quad x = y \text{ (or } a = 0\text{)},$$

which is so frequently employed in solving equations involving real numbers can *not* be used with matrices. For if, with $A$, $X$, $Y$ members of $\mathcal{M}_{n,n}$ we have

$$AX = AY,$$

then it *is* true that $A(X - Y) = 0.$

But, at this point, we cannot conclude that either $X - Y$ is zero or that $A$ is zero. Hence we cannot further conclude that $X = Y$ (or $A = 0$).

**Property 5** "$ax = b$ has the unique solution $a^{-1}b$ for $a \neq 0$" does have a matrix analogue for matrices to the following extent. For each member $A$ of $\mathscr{M}_{n,n}$ which has an inverse (is non-singular) and each member $B$ of $\mathscr{M}_{n,n}$, there is one and only one solution $X$ in $\mathscr{M}_{n,n}$ of the matrix equation

$$AX = B.$$

To verify this, we note that $A^{-1}B$ is one solution, since $A(A^{-1}B) = (AA^{-1})B = I_nB = B$. Now, if $X$ is any solution whatever, so that $AX = B$, then, multiplying on the left by $A^{-1}$ (which we assume exists) we obtain

$$\begin{aligned} A^{-1}(AX) &= A^{-1}B \\ (A^{-1}A)X &= A^{-1}B \\ I_nX &= A^{-1}B \\ X &= A^{-1}B. \end{aligned}$$

Thus, the only solution is $A^{-1}B$.

**Property 6** We have a similar matrix analogue for Property 6 in that if $A$ has an inverse and $B$ and $C$ are any members of $\mathscr{M}_{n,n}$, then the equation

$$AX + B = C$$

has the unique solution

$$X = A^{-1}(C - B).$$

The restricted analogue of Property 5 gives us valuable information about the equation $AX = B$ when $A$ is non-singular, but says nothing about the situation when $A$ is singular. Nor does it touch on the question of solutions of

$$AX = B$$

when $A$ is not square. These and the related question, "How do we find the inverse of a non-singular matrix?" will be answered in the next chapter.

In summary, then, the set $\mathscr{M}_{n,n}$ satisfies A1–A4, M1, and D1, D2. Such an algebraic structure is called a *ring*. Moreover $\mathscr{M}_{n,n}$ satisfies M2 (identity) and does not satisfy M4 (commutativity) so that $\mathscr{M}_{n,n}$ may properly be termed a *non-commutative ring with identity*.

## 1.8 Exercises (continued)

21. Show for

$$A = \begin{bmatrix} 2 & -1 \\ 4 & -2 \end{bmatrix}, \quad X = \begin{bmatrix} 0 & 1 \\ 3 & 2 \end{bmatrix}, \quad Y = \begin{bmatrix} -1 & -1 \\ 1 & -2 \end{bmatrix}$$

that $AX = AY$, even though $X \neq Y$.

22. Solve the matrix equation $AX = B$, where

$$A = \begin{bmatrix} 1 & 2 \\ 3 & 4 \end{bmatrix}, \quad B = \begin{bmatrix} 1 & -2 \\ 1 & 3 \end{bmatrix}.$$

23. Can you solve the matrix equation $AX = B$ with

$$A = \begin{bmatrix} 1 & 2 \\ 2 & 4 \end{bmatrix}, \quad B = \begin{bmatrix} 3 & 1 \\ 6 & 2 \end{bmatrix}?$$

**New Terms**

non-singular matrix, 39 singular matrix, 39

## 1.9 Scalar Product

In this section we define the multiplication of a matrix $A$ by a scalar (real number) $s$, and develop a few elementary properties of this "mixed" multiplication.

Thus if $s$ is a scalar and $A = [a_{ij}]$ is any $m \times n$ matrix, we define the product of $A$ by $s$, as

$$sA = [sa_{ij}].$$

To multiply the matrix $A$ by the scalar $s$, we simply multiply each element of $A$ by $s$. Thus

$$2\begin{bmatrix} 1 & 2 \\ 3 & 4 \end{bmatrix} = \begin{bmatrix} 2 \cdot 1 & 2 \cdot 2 \\ 2 \cdot 3 & 2 \cdot 4 \end{bmatrix} = \begin{bmatrix} 2 & 4 \\ 6 & 8 \end{bmatrix}.$$

You should now verify that the additive inverse $-A$ of $A$ is simply $(-1)A$. Moreover, our *scalar matrices*, which we defined in Section 1.4 are now seen to be simply scalar multiples of the identity matrix $I_n$. Thus

$$\begin{bmatrix} 3 & 0 & 0 \\ 0 & 3 & 0 \\ 0 & 0 & 3 \end{bmatrix} = 3\begin{bmatrix} 1 & 0 & 0 \\ 0 & 1 & 0 \\ 0 & 0 & 1 \end{bmatrix} = 3I_3.$$

The following three properties of this scalar multiplication are easily proved. Further numerical examples illustrating their use are found in the exercises.

**S1** $1A = A$ for each $A$ in $\mathcal{M}_{n,n}$.

**Proof**

$$1A = [1a_{ij}] = [a_{ij}] = A. \quad \blacksquare$$

Example 1.9.1

$$1\begin{bmatrix} 2 & 3 \\ 4 & 5 \end{bmatrix} = \begin{bmatrix} 1 \cdot 2 & 1 \cdot 3 \\ 1 \cdot 4 & 1 \cdot 5 \end{bmatrix} = \begin{bmatrix} 2 & 3 \\ 4 & 5 \end{bmatrix}. \quad \blacksquare$$

**S2** For each pair of scalars $s$, $t$ in $R$, $(s + t)A = sA + tA$ (a distributive property).

**Proof**

$$\begin{aligned}(s + t)A = (s + t)[a_{ij}] &= [(s + t)a_{ij}] \\ &= [sa_{ij} + ta_{ij}] = [sa_{ij}] + [ta_{ij}] \\ &= sA + tA. \quad \blacksquare\end{aligned}$$

Example 1.9.2

$$(2+3)\begin{bmatrix}2 & 3\\ 4 & 5\end{bmatrix} = \begin{bmatrix}(2+3)2 & (2+3)3\\ (2+3)4 & (2+3)5\end{bmatrix} = \begin{bmatrix}2\cdot 2+3\cdot 2 & 2\cdot 3+3\cdot 3\\ 2\cdot 4+3\cdot 4 & 2\cdot 5+3\cdot 5\end{bmatrix}$$
$$= \begin{bmatrix}2\cdot 2 & 2\cdot 3\\ 2\cdot 4 & 2\cdot 5\end{bmatrix} + \begin{bmatrix}3\cdot 2 & 3\cdot 3\\ 3\cdot 4 & 3\cdot 5\end{bmatrix}$$
$$= 2\begin{bmatrix}2 & 3\\ 4 & 5\end{bmatrix} + 3\begin{bmatrix}2 & 3\\ 4 & 5\end{bmatrix}.$$ ∎

**S3** For each pair of scalars $s, t$ in $\mathbb{R}$, $(st)A = s(tA)$ (an associative property).

**Proof** $(st)A = (st)[a_{ij}] = [(st)a_{ij}] = [s(ta_{ij})] = s[ta_{ij}] = s(tA).$ ∎

Example 1.9.3

$$(2\cdot 3)\begin{bmatrix}4 & 5\\ 6 & 7\end{bmatrix} = \begin{bmatrix}(2\cdot 3)4 & (2\cdot 3)5\\ (2\cdot 3)6 & (2\cdot 3)7\end{bmatrix} = \begin{bmatrix}2(3\cdot 4) & 2(3\cdot 5)\\ 2(3\cdot 6) & 2(3\cdot 7)\end{bmatrix}$$
$$= 2\begin{bmatrix}3\cdot 4 & 3\cdot 5\\ 3\cdot 6 & 3\cdot 7\end{bmatrix} = 2\left(3\begin{bmatrix}4 & 5\\ 6 & 7\end{bmatrix}\right).$$ ∎

## 1.9 Exercises

1. Show that $(-2)\begin{bmatrix}-1 & 2 & 1\\ 3 & 1 & 5\end{bmatrix} = 3\begin{bmatrix}-1 & 2 & 1\\ 3 & 1 & 5\end{bmatrix} - 5\begin{bmatrix}-1 & 2 & 1\\ 3 & 1 & 5\end{bmatrix}.$

2. Show that $(-12)\begin{bmatrix}1 & 1 & -1 & 6\\ 2 & 1 & 3 & 1\end{bmatrix} = (-4)\left(3\begin{bmatrix}1 & 1 & -1 & 6\\ 2 & 1 & 3 & 1\end{bmatrix}\right).$

3. Show that for $A = \begin{bmatrix}1 & -2\\ 3 & 4\end{bmatrix}$, $B = \begin{bmatrix}4 & 6\\ 5 & 8\end{bmatrix}$, $a = 2$,
$$a(A+B) = aA + aB.$$

4. Prove that for any $2 \times 2$ matrices $A$, $B$ and the scalar $a$,
   a. $a(A+B) = aA + aB$.
   b. $a(AB) = (aA)B = A(aB)$.

5. Verify that $(aA)^T = aA^T$ for $a$ and $A$ as in Exercise 3.

6. Prove the statement in Exercise 5 for $2 \times 2$ matrices.

7. For a given semester, Jim received a grade of $A$ in two courses, $B$ in one course, and $C$ in two courses, so that his grade matrix $G = [A, B, C, D, F] = [2, 1, 2, 0, 0]$. At his school the quality point matrix $Q = [4, 3, 2, 1, 0]^T$, where an $A$ is valued at 4 quality points, etc. Show that the matrix $1/5\ GQ$ correctly represents Jim's grade point average of 3.0. Construct matrices $G$, $Q$ for your own case, and compute your own most recent GPA.

# 2 Systems of Linear Equations and Gauss-Jordan Reduction

## 2.1 Matrix Representation of Systems of Linear Equations

One of the most important uses of matrices and matrix notation occurs in the study of systems of linear equations. Such equations arise in many different contexts, some of which we shall see later. In this chapter we shall observe how matrices can be used to represent such systems. We shall learn how to solve systems geometrically, when possible, and then proceed to develop a method, known as Gauss-Jordan reduction, for solving any system of linear equations. In order to introduce some terminology and to familiarize the reader with two equivalent matrix formulations we shall now consider two small systems of linear equations.

Example 2.1.1

$$\begin{aligned} 2x_1 + 3x_2 &= 0 \\ 4x_1 + 2x_2 &= 0. \end{aligned}$$ ∎

In the equations of Example 2.1.1 we say that we have two homogeneous linear equations in the two unknowns $x_1$ and $x_2$. Each equation is said to be a **homogeneous equation** because its right-hand side is zero. If each of a set of equations is homogeneous then the set of equations is called **homogeneous**. It is possible to express the set of equations of Example 2.1.1 very compactly in matrix notation. If we let

$$A = \begin{bmatrix} 2 & 3 \\ 4 & 2 \end{bmatrix}, \quad X = \begin{bmatrix} x_1 \\ x_2 \end{bmatrix}, \quad \text{and} \quad 0 = \begin{bmatrix} 0 \\ 0 \end{bmatrix},$$

then we see that the matrix equation $AX = 0$, where

$$AX = \begin{bmatrix} 2 & 3 \\ 4 & 2 \end{bmatrix} \begin{bmatrix} x_1 \\ x_2 \end{bmatrix} = \begin{bmatrix} 2x_1 + 3x_2 \\ 4x_1 + 2x_2 \end{bmatrix} = \begin{bmatrix} 0 \\ 0 \end{bmatrix} = 0$$

must be satisfied by a pair $x_1$, $x_2$ that happens to satisfy the original set of equations. On the other hand, any $X = \begin{bmatrix} x_1 \\ x_2 \end{bmatrix}$ that satisfies the matrix equation $AX = 0$ also provides values for $x_1$, $x_2$ that satisfy the original set of equations. Thus we say that $AX = 0$ is an equivalent representation of the original set of equations in the sense that a solution of either problem provides a solution to the other. We wish now to find a matrix $X$, if one exists, which will satisfy the matrix equation. Such a matrix $X$ is called a **solution** of the matrix equation, and the set of all such matrices is called the **complete** or **general solution set** for the matrix equation.

There is an alternative matrix representation of this set of linear equations which will be of considerable assistance in our later work. If we let $A_1 = \begin{bmatrix} 2 \\ 4 \end{bmatrix}$ denote the first column of $A$ and let $A_2 = \begin{bmatrix} 3 \\ 2 \end{bmatrix}$ denote the second column, then we could say that our problem is to find $x_1$, $x_2$, if they exist, such that

$$x_1 \begin{bmatrix} 2 \\ 4 \end{bmatrix} + x_2 \begin{bmatrix} 3 \\ 2 \end{bmatrix} = \begin{bmatrix} 0 \\ 0 \end{bmatrix},$$

or such that

$$x_1 A_1 + x_2 A_2 = 0.$$

We shall now consider a similar set of equations in which the right-hand constants are not all zeros. Such a set of equations is said to be **nonhomogeneous**.

Example 2.1.2

$$\begin{aligned} 2x_1 + 3x_2 &= 5 \\ 4x_1 + 2x_2 &= 6. \end{aligned}$$

If we let $A = \begin{bmatrix} 2 & 3 \\ 4 & 2 \end{bmatrix}$ and $X = \begin{bmatrix} x_1 \\ x_2 \end{bmatrix}$ as before, and denote $\begin{bmatrix} 5 \\ 6 \end{bmatrix}$ by $B$, then we have the equivalent matrix equation $AX = B$. In the second matrix format, we wish to find numbers $x_1$ and $x_2$ such that

$$x_1 \begin{bmatrix} 2 \\ 4 \end{bmatrix} + x_2 \begin{bmatrix} 3 \\ 2 \end{bmatrix} = \begin{bmatrix} 5 \\ 6 \end{bmatrix},$$

or such that

$$x_1 A_1 + x_2 A_2 = B.$$ ∎

Example 2.1.3

A certain grandfather left an estate worth $330,000 to be divided among his four grandchildren—Dick, Jane, Jim, and Mary—from oldest to youngest. Construct a set of 4 linear equations to divide the estate according to his 4 wishes:

Wish 1. All of the money should be divided.

Wish 2. The youngest 2 children should receive, in total, 1/2 the amount that the 2 oldest receive in total.

Wish 3. The boys should receive, in total, 6/5 of the amount that the girls receive in total.

Wish 4. Jim should receive $10,000 more than Mary receives.

Letting $x_1$ denote in thousands of dollars, the amount that Dick is to receive, $x_2$, the amount that Jane is to receive, etc., we can write one equation for each of the 4 wishes as follows:

Wish 1: $x_1 + x_2 + x_3 + x_4 = 330$

Wish 2: $x_3 + x_4 = 1/2(x_1 + x_2)$

Wish 3: $x_1 + x_3 = (6/5)(x_2 + x_4)$

Wish 4: $x_3 = x_4 + 10$

Rearranging the order of the variables and multiplying appropriately to get rid of fractions, we obtain

$$\begin{aligned} x_1 + \phantom{6}x_2 + \phantom{5}x_3 + \phantom{6}x_4 &= 330 \\ x_1 + \phantom{6}x_2 - 2x_3 - 2x_4 &= 0 \\ 5x_1 - 6x_2 + 5x_3 - 6x_4 &= 0 \\ x_3 - \phantom{6}x_4 &= 10 \end{aligned}$$

with

$$A = \begin{bmatrix} 1 & 1 & 1 & 1 \\ 1 & 1 & -2 & -2 \\ 5 & -6 & 5 & -6 \\ 0 & 0 & 1 & -1 \end{bmatrix} \quad \text{and} \quad B = \begin{bmatrix} 330 \\ 0 \\ 0 \\ 10 \end{bmatrix}.$$ ∎

## 2.1 Exercises

1. Consider the set of equations

$$\begin{aligned} x_1 - 4x_2 &= -5 \\ -2x_1 + 2x_2 &= 4. \end{aligned}$$

a. Find $A$, $B$, $X$ which represent this system as $AX = B$.
b. Find $A_1$, $A_2$, and $B$ such that $x_1A_1 + x_2A_2 = B$. We say that $B$ is a **linear combination** of $A_1$ and $A_2$.
c. Is the system a homogeneous system?
d. Verify that $X = \begin{bmatrix} -1 \\ 1 \end{bmatrix}$ is a solution for the system.

2. a. Write the system of equations corresponding to the matrix equation $AX = B$, if

$$A = \begin{bmatrix} 1 & 2 & 3 \\ -1 & 4 & 5 \end{bmatrix}, \quad X = \begin{bmatrix} x_1 \\ x_2 \\ x_3 \end{bmatrix}, \quad B = \begin{bmatrix} 0 \\ -1 \end{bmatrix}.$$

b. Express the set of equations in a. in the alternative matrix format.
c. Verify that $X = [1/3, (-1/6), 0]^T$ is a solution.

3. Consider the system of equations

$$\begin{aligned} x_1 + 2x_2 - \phantom{3}x_3 &= 0 \\ 2x_1 - \phantom{2}x_2 + 3x_3 &= 0. \end{aligned}$$

a. Find $A$, $X$, $B$ such that $AX = B$.
b. Is the system homogeneous?
c. Can you think of any solution matrix $X$ that will satisfy the equation?
d. Verify that $[-1, 1, 1]^T$ is a solution.

4. a. Describe verbally the following set of points in the $x$, $y$-plane:

$$S = \{(x, y) \mid y = 2x + 1\}$$

b. Represent $S$ geometrically by drawing its graph.
c. Consider $T = \{(x, y) \mid y = -x + 1\}$. Draw the graph of $T$ on the same set of axes as you did $S$.
d. Geometrically, what is $S \cap T$?
e. What is the relationship between your solution in d. and the solution of the set of equations

$$\begin{aligned} y - 2x &= 1 \\ y + \phantom{2}x &= 1? \end{aligned}$$

5. Let $X_1$ and $X_2$ be solutions of a homogeneous equation $AX = 0$. Use the properties of matrix addition and multiplication to show that:
a. $X_1 + X_2$ is a solution to $AX = 0$.
b. $cX_1$ is a solution to $AX = 0$, for any scalar $c$ (you need to use the properties of scalar multiplication here).
c. For any scalars $c_1$, $c_2$,

$$Y = c_1X_1 + c_2X_2 \quad \text{is a solution to} \quad AX = 0.$$

6. Verify the statements of Exercise 5 above for the case where

$$A = \begin{bmatrix} 1 & -1 \\ 2 & -2 \end{bmatrix}, \quad \text{and} \quad X_1 = \begin{bmatrix} 1 \\ 1 \end{bmatrix}, \quad X_2 = \begin{bmatrix} -3 \\ -3 \end{bmatrix}.$$

7. A college student has two part-time jobs and works at them 40 hours per week in total. His hourly wages are \$3/hr on Job 1 and \$2/hr on Job 2. Formulate a set of linear equations which will tell how many hours per week he works at each job if his total weekly wage is \$90.

8. A certain builder builds garages and boats. To build a garage he needs $a$ board-feet of lumber and $c$ man-hours. To build a boat he needs $b$ board-feet of lumber and $d$ man-hours. Formulate a set of two equations in two unknowns which will tell the builder how many of each to build to exactly use up his available $r_1$ board-feet of lumber and his $r_2$ man-hours.

**New Terms**

homogeneous equation, 46
homogeneous set of equations, 46
solution of a matrix equation, 47
complete or general solution set, 47
non-homogeneous set of equations, 47
linear combination, 48

## 2.2 Geometrical Solutions

It is possible to find solutions of each of the sets of equations in the previous section geometrically, since the graph of each equation is a straight line in the $x_1$, $x_2$-plane.

Example 2.2.1

$$2x_1 + 3x_2 = 0$$
$$4x_1 + 2x_2 = 0.$$

By setting one of the variables equal to some constant we can solve for the value of the remaining variable and obtain a point on the graph of the straight line represented by the equation. For example, letting $x_2 = 2$ in the first equation we find that $x_1 = (-3/2)x_2 = (-3/2)(2) = -3$, so that $(x_1, x_2) = (-3, 2)$ is on the first line. If we let $x_2 = -2$, we find that $x_1 = (-3/2)(-2) = 3$, so that $(3, -2)$ is also on the line. Drawing an $x_1$, $x_2$-axis system as shown in Figure 2.2.1, we can then plot the points $(-3, 2)$ and $(3, -2)$ and use them to obtain the graph of the straight line. For equation 2, we find that $x_1 = (-1/2)x_2$. If $x_2 = 2$, then $x_1 = -1$ so that $(-1, 2)$ is on the second line. When $x_2 = -2$, then $x_1 = 1$, and $(1, -2)$ is also on the line. We plot and connect these two points to obtain the graph of the second line as shown in Figure 2.2.1. ▮

**Figure 2.2.1**

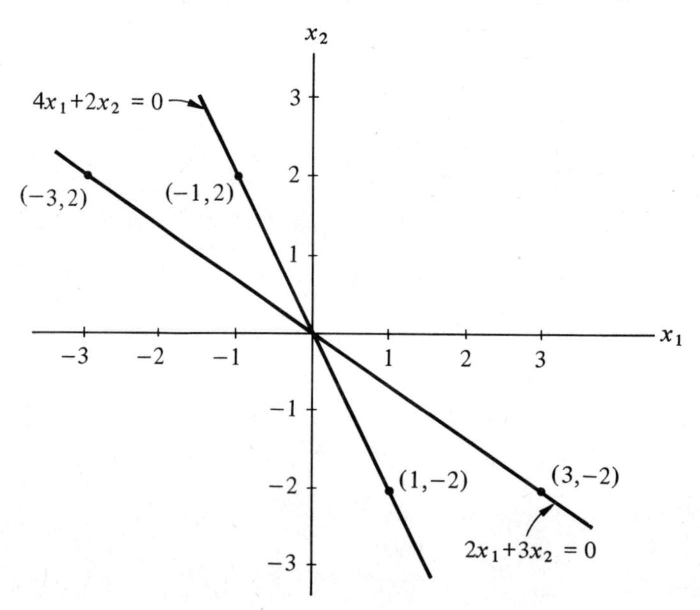


In Example 2.2.1, the two straight lines, whose equations are given above, intersect only at the point (0, 0) in the plane. We would thus say that $0 = \begin{bmatrix} 0 \\ 0 \end{bmatrix}$ is the *unique* (only) solution. For such a homogeneous set of equations we say that 0 is the **trivial solution**, since it is always true that a homogeneous set of linear equations will be satisfied if we set all of the $x_i$ equal to zero. Therefore, we will be concerned with finding non-trivial solutions for such a set of equations.

The geometrical solution for the second example is shown in Figure 2.2.2.

Example 2.2.2

$$2x_1 + 3x_2 = 5$$
$$4x_1 + 2x_2 = 6.$$

Figure 2.2.2

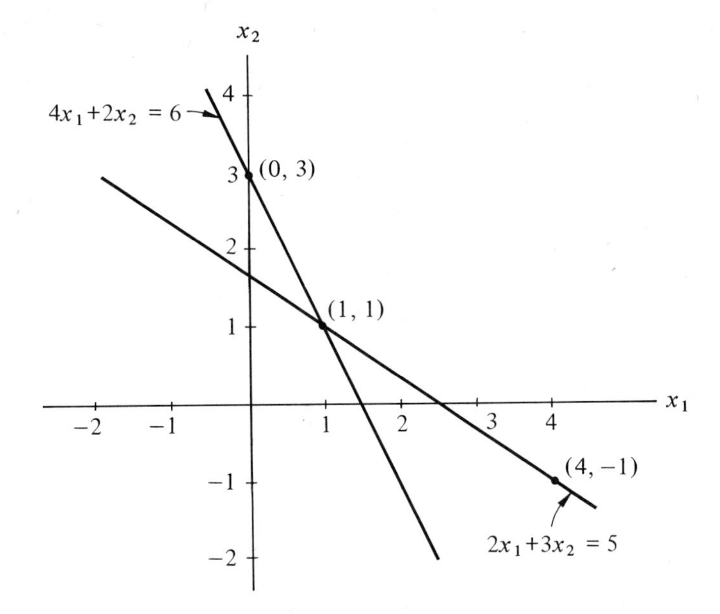


We see from the graphical solution of Example 2.2.2 that we again have a unique point of intersection, namely the point (1, 1). Thus again, the matrix $X = \begin{bmatrix} 1 \\ 1 \end{bmatrix}$ is our desired solution matrix for the matrix equation $AX = B$. We can verify this by substitution in the original set of linear equations or by checking to see that

$$AX = \begin{bmatrix} 2 & 3 \\ 4 & 2 \end{bmatrix} \begin{bmatrix} 1 \\ 1 \end{bmatrix}$$

is indeed equal to the matrix $\begin{bmatrix} 5 \\ 6 \end{bmatrix} = B$. ∎

Unfortunately, this geometrical approach fails us when we have more than three unknowns or variables; and we must defer to a strictly algebraic procedure for our solutions. It will be very helpful, however, even when we are dealing with larger systems, for the reader to refer to the geometry of two or three dimensions and to reflect upon the problems of intersecting lines or of intersecting planes, and the possible solutions that result.

Example 2.2.3 The set of points $(x_1, x_2, x_3)$ which satisfy the equation

$$x_1 + x_2 + x_3 = 1$$

all lie on a plane (a surface of two dimensions) in 3-dimensional space. Letting two of the variables, say $x_1$ and $x_2$, be 0, we find that $x_3 = 1$, or that the point (0, 0, 1) lies on the plane in question. Letting $x_2 = x_3 = 0$, we have $x_1 = 1$, so that (1, 0, 0) is on the plane. Similarly, (0, 1, 0) is also on the plane. Connecting these points as shown in Figure 2.2.3, we have a pictorial representation of part of the plane. ∎

Figure 2.2.3

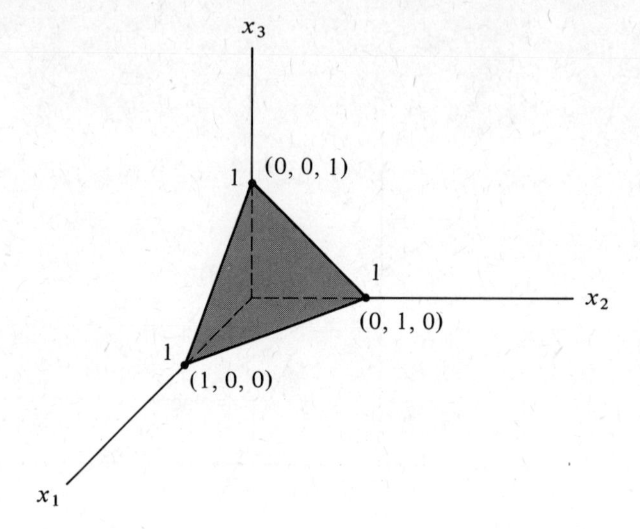


In general, two such planes which intersect will intersect in a line as shown in Figure 2.2.4. You should now consider the various intersection possibilities which arise when a third plane is introduced.

There is yet another geometrical way of looking at sets of equations and their solutions. Consider the equations of Example 2.2.1 and refer to Figure 2.2.1.

$$\begin{aligned} 2x_1 + 3x_2 &= 0 \\ 4x_1 + 2x_2 &= 0. \end{aligned}$$

Let $A$ be the set of points in the plane which satisfy the first equation (the points of $A$ form a straight line), and let $B$ denote the set of points in the plane which satisfy the second equation (also a straight line). Then the set of points which satisfy both equations is the set of points that are common to both $A$ and $B$; that is, $A \cap B$. In this case, $A \cap B = \{(0, 0)\}$. If there were no points in the plane that satisfied both equations then the intersection would be the null set $\varnothing$. When one thinks of a set of equations in this fashion, it is easy to comprehend that we start with a certain set of potential solution

Figure 2.2.4

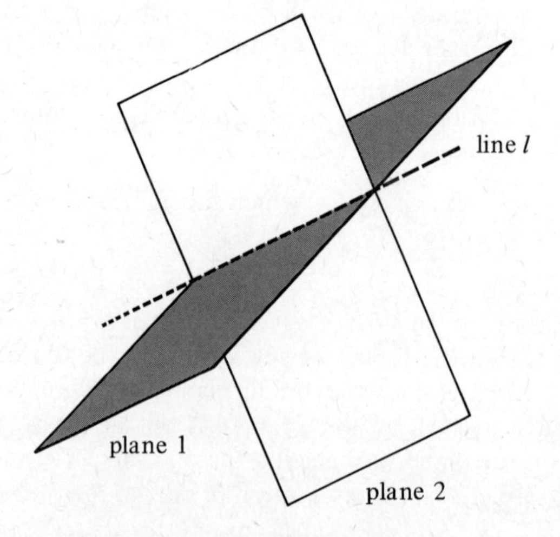

points and then successively eliminate certain subsets as we add restrictive equations. Thus, in Example 2.2.1, we start with the set of all points in the plane. The first equation, alone, removes from consideration all points of the plane except those on a certain straight line. The second equation, alone, removes all points except those on another straight line. The two equations, together, thus succeed in deleting from consideration every point in the plane except one, namely (0, 0).

Before we proceed to a description of the general problem, let us pause to reiterate some of the important terms and methods so far discussed in this chapter.

At this point you should be able to recognize given sets of equations as being *homogeneous* or *non-homogeneous.* You should know what we mean by a *solution* to such a set of equations. You should be able to express a set of equations in either of the two matrix formats covered, and also be able to solve small sets of equations graphically. We recall that a homogeneous set of equations always has at least the *trivial* solution. Can you think of any small homogeneous set of equations which has more than this one solution?

With these introductory ideas in mind, we now turn to the general problem of finding solution matrices $X$ (if any exist) for arbitrarily large systems of linear equations, that is, systems with any number of equations and any number of unknowns.

## 2.2 Exercises

1. The solution which always exists for a homogeneous set of equations is called the __________ solution.

2. A system of three homogeneous equations in two unknowns $x_1$, $x_2$ is represented graphically by three __________ which intersect at __________.

3. When we say that a solution to a system of linear equations is *unique*, what do we mean?

4. Solve the following systems of equations graphically:

   a. $x_1 - x_2 = 0$
   $2x_1 + x_2 = 0$

   b. $x_1 + 2x_2 = 3$
   $-3x_1 + x_2 = -2$

   c. $2x_1 - x_2 = 3$
   $x_1 + 4x_2 = 6$
   $4x_1 + 7x_2 = 15$

   d. $x_1 + x_2 = 4$
   $-x_1 - x_2 = -4.$

5. Describe the geometric counterpart of the following:
   a. one linear equation in two unknowns
   b. two linear equations in two unknowns
   c. one homogeneous linear equation in three unknowns
   d. two linear equations in three unknowns
   e. three homogeneous linear equations in three unknowns.

6. Use your intuitive knowledge of geometry in 2 and 3 dimensions to answer the following:
   a. In which of the cases of Exercise 5 above must solutions exist?
   b. In which of the cases of Exercise 5 above must more than one solution exist?

7. In Exercise 4a. above, what set of points is excluded by the first equation? By the second equation? What set of points is excluded by the pair of equations?

8. Graphically find the solution for Exercise 2.1.8 for $a = 2000$, $b = 1000$, $c = 200$, $d = 300$, $r = 10{,}000$, and $r_2 = 1800$.

9. Graphically solve Exercise 2.1.7.

### New Terms

trivial solution, 50

## 2.3 The General Problem

If we let $m$ and $n$ be any positive integers, then we can formulate our general problem as that of finding values for the unknowns $x_1, x_2, \ldots, x_n$ such that

$$\begin{aligned} a_{11}x_1 + a_{12}x_2 + \cdots + a_{1n}x_n &= b_1 \\ a_{21}x_1 + a_{22}x_2 + \cdots + a_{2n}x_n &= b_2 \\ &\cdots \\ a_{m1}x_1 + a_{m2}x_2 + \cdots + a_{mn}x_n &= b_m, \end{aligned}$$

where we assume that the constants $a_{ij}$ and $b_i$ are known real numbers for all $i$ and $j$. To express this problem in the first matrix format, we merely let

$$A = [a_{ij}] = \begin{bmatrix} a_{11} & a_{12} & \cdots & a_{1n} \\ a_{21} & a_{22} & \cdots & a_{2n} \\ & & \cdots & \\ a_{m1} & a_{m2} & \cdots & a_{mn} \end{bmatrix}, \quad X = \begin{bmatrix} x_1 \\ x_2 \\ \vdots \\ x_n \end{bmatrix}, \quad B = \begin{bmatrix} b_1 \\ b_2 \\ \vdots \\ b_m \end{bmatrix},$$

from which we obtain the matrix representation

$$AX = B$$

as before. In this general case, as in our previous examples, we will say that the set of equations is **homogeneous** if $B = 0$. Otherwise, the set will be called **non-homogeneous**.

To obtain the second matrix formulation we let $A_i$ denote column $i$ of $A$, so that

$$A_1 = \begin{bmatrix} a_{11} \\ a_{21} \\ \vdots \\ a_{m1} \end{bmatrix}, \quad A_2 = \begin{bmatrix} a_{12} \\ a_{22} \\ \vdots \\ a_{m2} \end{bmatrix}, \quad \ldots, \quad A_n = \begin{bmatrix} a_{1n} \\ a_{2n} \\ \vdots \\ a_{mn} \end{bmatrix},$$

We then wish to find constants $x_1, x_2, \ldots, x_n$, if they exist, such that

$$x_1A_1 + x_2A_2 + \cdots + x_nA_n = B.$$

Example 2.3.1

$$\begin{aligned} x_1 + 2x_2 + 3x_3 + x_4 &= 7 \\ 2x_1 + x_2 + x_3 + 2x_4 &= 6 \\ -x_1 + x_2 + 2x_3 - x_4 &= 1. \end{aligned}$$

In this non-homogeneous set of three equations in four unknowns we see that $m = 3$ and $n = 4$, while

$$A = \begin{bmatrix} 1 & 2 & 3 & 1 \\ 2 & 1 & 1 & 2 \\ -1 & 1 & 2 & -1 \end{bmatrix}, \quad X = \begin{bmatrix} x_1 \\ x_2 \\ x_3 \\ x_4 \end{bmatrix}, \quad B = \begin{bmatrix} 7 \\ 6 \\ 1 \end{bmatrix}.$$ ∎

The reader may have observed that, in reality, all of the known information is contained in the **augmented matrix** denoted by $[A \mid B]$, in which we merely write the column of constants, $B$, as an additional column to obtain a slightly larger matrix. In the general case the augmented matrix is the partitioned matrix

$$[A \mid B] = \left[\begin{array}{cccc|c} a_{11} & a_{12} & \cdots & a_{1n} & b_1 \\ a_{21} & a_{22} & \cdots & a_{2n} & b_2 \\ \cdot & \cdot & \cdots & \cdot & \vdots \\ a_{m1} & a_{m2} & \cdots & a_{mn} & b_n \end{array}\right],$$

while in Example 2.3.1 the augmented matrix is

$$[A \mid B] = \left[\begin{array}{cccc|c} 1 & 2 & 3 & 1 & 7 \\ 2 & 1 & 1 & 2 & 6 \\ -1 & 1 & 2 & -1 & 1 \end{array}\right].$$

We observe that in both cases we have essentially removed the unknowns from the sets of equations to obtain the augmented matrices.

It may prove helpful to the reader to consider the general problem in the same manner that we considered the two-dimensional example earlier. When we have $m$ equations in $n$ unknowns we are looking for the set of elements in $n$-dimensional space (a concept which may be rather vague at the moment—you may think of these elements as merely column matrices with $n$ elements just as we may think of 2-space as consisting of column matrices with two elements) which satisfy each of the $m$ equations. Letting $S_1$ denote the subset of $n$-space which satisfies equation 1, and $S_2$ the subset which satisfies equation 2, etc., then the solution set for the system of $m$ equations may be considered to be the intersection of the $m$ sets $S_1$, $S_2, \ldots, S_m$; that is,

$$S_1 \cap S_2 \cap \cdots \cap S_m.$$

If there are no points in the intersection then there are no solutions to the set of equations. If it should be the case that equation 3, for example, excludes no points from consideration not already excluded by equation 1 and equation 2, then we say that equation 3 is a **redundant equation,** with respect to equation 1 and equation 2. Equivalently, we may say that equation 3 is redundant if

$$S_1 \cap S_2 = S_1 \cap S_2 \cap S_3.$$

For example, if we were to add a third equation to the two of Example 2.2.1, say

$$x_1 - x_2 = 0$$

whose solution points also form a straight line through the origin as shown in Figure 2.3.1, then it is seen that the new equation has really excluded no new points, and is thus redundant. It could just as well have been omitted as far as its effect upon the ultimate solution is concerned.

**Figure 2.3.1**

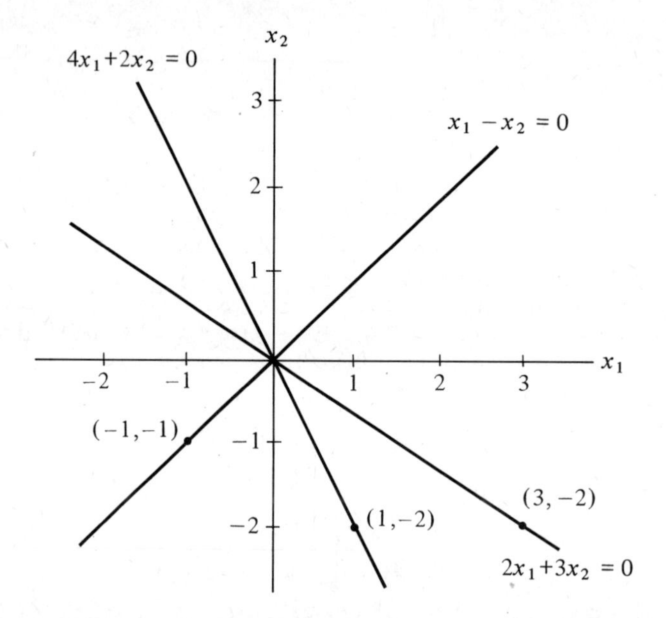


Several questions arise about the existence and number of solutions to our general problem. All of these questions have geometrical counterparts in two- and three-dimensional problems concerning the existence and number of points of intersection of lines and planes. Thus, for a given set of equations with matrix format $AX = B$, one may ask:

1. Are there any solutions? (Do the lines or planes intersect?)
2. If so, is the solution unique?
3. If there is at least one solution and it is not unique, how many solutions are there?
4. What is the complete set of solutions?

These are the types of questions to which we will develop answers in the following sections. After a slight diversion we shall develop a simple but systematic and easily-computerized technique for solving the general problem when $A$ is square and non-singular. We will thus gain experience with the essential ideas involved so that the extension of our technique to the singular and non-square cases will not be difficult.

## 2.3 Exercises

1. Let

$$A = \begin{bmatrix} 1 & -1 & 2 & 4 \\ 0 & 2 & 3 & 7 \\ 6 & 3 & 4 & 5 \end{bmatrix}, \quad B = \begin{bmatrix} 2 & 1 \\ 3 & 3 \\ 4 & 7 \end{bmatrix}, \quad C = \begin{bmatrix} 1 & 2 \\ 3 & 4 \end{bmatrix}, \quad D = \begin{bmatrix} 1 \\ 2 \\ 3 \end{bmatrix}, \quad E = \begin{bmatrix} 3 \\ 7 \end{bmatrix}.$$

Write down the augmented matrices for the following equations for which the multiplication makes sense, and suppose that $X$, $Y$, $Z$ are column matrices of appropriate length.

a. $AX = D$ b. $AY = E$ c. $BZ = D$ d. $CX = E$.

2. Consider the system of equations:

$$\begin{aligned} x_1 - 2x_2 &= -1 \\ 2x_1 + 3x_2 &= 5 \\ -x_1 + 5x_2 &= 4 \end{aligned}$$

a. Describe $S_1$, $S_2$, $S_3$ as they apply to this problem.
b. Find $S_1 \cap S_2$ and $S_2 \cap S_3$.
c. In order that any solution exist, it is necessary that $S_1 \cap S_2 =$ ______.
d. Here we observe that when we have two unknowns we can have no more than two non-redundant equations. If we have $n$ unknowns, how many non-redundant equations do you think we can have and still have a solution?

3. Let $A$ be any $2 \times 2$ matrix, $B = \begin{bmatrix} 1 \\ 0 \end{bmatrix}$, $C = \begin{bmatrix} 0 \\ 1 \end{bmatrix}$. Suppose that we solve the equations $AX = B$ and $AY = C$ for $X$ and $Y$, and solutions exist.

a. Consider $A[X \mid Y] = [AX \mid AY] = \begin{bmatrix} 1 & 0 \\ 0 & 1 \end{bmatrix}$. What can you conclude about $A^{-1}$?
b. Let $A = \begin{bmatrix} 0 & -1 \\ -1 & 0 \end{bmatrix}$. Use the technique described above to find $A^{-1}$.

### New Terms

augmented matrix, 55 redundant equation, 55

## 2.4 A Special Case—Solution by Analogy

Because it fits with our philosophy of presenting new ideas by comparison with previously well-understood concepts, we shall digress from our general solution procedure in order to investigate the solution of certain problems in which $A$ is a square matrix. In this case it is evident that the number of equations is the same as the number of unknowns. Examples 2.2.1 and 2.2.2 are of this type. There is a special solution technique which is sometimes available for the solution of equations of this form.

The method which we wish to illustrate arises as an analogue of the solution of the single equation

$$ax = b,$$

when $a$, $x$, and $b$ are each real numbers and $a$ is not zero. To be specific, suppose that we wish to find a number $x$ such that

$$2x = 3.$$

As we indicated in our discussion of the real numbers, we merely multiply both sides of the equation by 1/2 to obtain

$$(1/2)(2x) = (1/2 \cdot 2)x = 1x = x$$

and
$$1/2(3) = 3/2$$

so that
$$x = 3/2.$$

Thus the only possible solution is 3/2. The solution above depends upon our being able to find the multiplicative inverse of the real number $a = 2$; that is, we found a number $a^{-1}$ such that $a^{-1}a = 1$. In general, we can always find the multiplicative inverse of a non-zero real number $a$ (namely, $a^{-1} = 1/a$), and hence we can always solve such an equation by multiplying both sides of the equation by $a^{-1}$. Thus to solve $ax = b$, we multiply both sides by $1/a$ to obtain

$$\left(\frac{1}{a}\right)(ax) = \left(\frac{1}{a} \cdot a\right)x = 1x = x$$

and
$$\left(\frac{1}{a}\right)b = a^{-1}b,$$

so that
$$x = a^{-1}b,$$

from which we note that $x = a^{-1}b$ is the only possible solution.

Let us now recall Example 2.2.2 in its matrix formulation $AX = B$, where

$$A = \begin{bmatrix} 2 & 3 \\ 4 & 2 \end{bmatrix}, \quad X = \begin{bmatrix} x_1 \\ x_2 \end{bmatrix}, \quad \text{and} \quad B = \begin{bmatrix} 5 \\ 6 \end{bmatrix},$$

and describe the analogy which exists with the above procedure. We suppose that we know a matrix which we have called $A^{-1}$ such that $AA^{-1} = A^{-1}A = I_2$. Then $A(A^{-1}B) = (AA^{-1})B = I_2B = B$ so that $A^{-1}B$ is a solution. Moreover, multiplying each side of the equation $AX = B$ by $A^{-1}$ on the left, we obtain

$$A^{-1}(AX) = (A^{-1}A)X = I_2X = X = A^{-1}B$$

implying that $X = A^{-1}B$ is the unique solution. The same argument, of course, is valid for any size non-singular matrix $A$. In the specific example above, $A^{-1}$ does exist and we check that

$$A^{-1} = \begin{bmatrix} -2/8 & 3/8 \\ 4/8 & -2/8 \end{bmatrix}$$

by multiplication. Thus

$$A^{-1}A = \begin{bmatrix} -2/8 & 3/8 \\ 4/8 & -2/8 \end{bmatrix}\begin{bmatrix} 2 & 3 \\ 4 & 2 \end{bmatrix} = \begin{bmatrix} 1 & 0 \\ 0 & 1 \end{bmatrix}$$
$$= AA^{-1} = \begin{bmatrix} 2 & 3 \\ 4 & 2 \end{bmatrix}\begin{bmatrix} -2/8 & 3/8 \\ 4/8 & -2/8 \end{bmatrix}.$$

We now calculate $X = A^{-1}B$ as

$$X = A^{-1}B = \begin{bmatrix} -2/8 & 3/8 \\ 4/8 & -2/8 \end{bmatrix}\begin{bmatrix} 5 \\ 6 \end{bmatrix} = \begin{bmatrix} 1 \\ 1 \end{bmatrix},$$

the same solution which we found earlier by geometrical considerations.

To further illustrate this method, recall Example 2.2.1 in which $A$ is the same as above, but $B = 0$. Since the same inverse suffices, we have

$$X = A^{-1}B = A^{-1}0 = \begin{bmatrix} 0 \\ 0 \end{bmatrix},$$

which is the same solution we obtained earlier. Although this method appears very easy at first glance, we must point out two important drawbacks as far as our present situation is concerned. First, even if the method were applicable, we do not at present know how to calculate $A^{-1}$. Secondly, the method does not work for all square matrices $A$ because, as was shown in the previous chapter, not all square matrices have a multiplicative inverse.

We thus observe that this method does have limitations, but in those cases in which $A^{-1}$ does exist and you happen to know what it is (for example, if we tell you), you should be able to find the solution $X$ very easily.

In the next section we shall begin to describe a solution technique which will always work, in the sense that if there are any solutions, we will find all of them, and if there are none, then we will discover that fact also. The procedure is called *Gauss-Jordan reduction* after the men who received credit for its discovery. As we indicated earlier, we shall first consider the case where $A$ is a non-singular square matrix and then proceed to the more general case in which $A$ may not even be square.

## 2.4 Exercises

1. Consider the set of equations $AX = B$ where

$$A = \begin{bmatrix} 1 & -1 \\ -1 & 2 \end{bmatrix}, \quad X = \begin{bmatrix} x_1 \\ x_2 \end{bmatrix}, \quad B = \begin{bmatrix} -1 \\ 4 \end{bmatrix}.$$

a. Verify that $A^{-1} = \begin{bmatrix} 2 & 1 \\ 1 & 1 \end{bmatrix}$.

b. Use $A^{-1}$ to find $X$. Verify that $X$ is a solution.

c. Solve the equation $A^{-1}\,Y = 0$, where $Y = \begin{bmatrix} y_1 \\ y_2 \end{bmatrix}$.

2. Let $A = \begin{bmatrix} 1 & 0 & 1 \\ -1 & 1 & 0 \\ 0 & 1 & 2 \end{bmatrix}$, $X = \begin{bmatrix} x_1 \\ x_2 \\ x_3 \end{bmatrix}$, $B = \begin{bmatrix} -2 \\ 0 \\ -1 \end{bmatrix}$, and consider $AX = B$.

a. Verify that $A^{-1} = \begin{bmatrix} 2 & 1 & -1 \\ 2 & 2 & -1 \\ -1 & -1 & 1 \end{bmatrix}$.

b. Use $A^{-1}$ to find $X$. Verify that $X$ is a solution.

c. Solve the equation $A^{-1}Y = \begin{bmatrix} -1 \\ 1 \\ -1 \end{bmatrix}$.

3. Let $A = \begin{bmatrix} 1 & -1 \\ 0 & 1 \end{bmatrix}$, $X = \begin{bmatrix} x_1 \\ x_2 \end{bmatrix}$, and $B = \begin{bmatrix} 0 \\ 1 \end{bmatrix}$.

a. Verify that $A^{-1} = \begin{bmatrix} 1 & 1 \\ 0 & 1 \end{bmatrix}$.

b. Solve the equation $AX = B$. Verify that $X$ is a solution.

c. Solve the equation $A^{-1}Y = B$.

## 2.5 Gauss-Jordan Reduction for Non-singular Square Matrices

The basic principle underlying the algorithm (computational procedure) known as **Gauss-Jordan reduction** is the following:

*One may exchange the augmented matrix representation $[A \mid B]$ of the original set of equations for a new but equivalent augmented matrix representation, in the sense that any solution matrix for either represented problem is also a solution for the other.*

We wish to systematically exchange augmented matrix representations until we obtain a form so simple that the solution matrix $X$ is obvious. For example, from the augmented matrix representation

$$\left[\begin{array}{cc|c} 1 & 0 & 5 \\ 0 & 1 & -3 \end{array}\right]$$

for the set of equations

$$\begin{aligned} 1x_1 + 0x_2 &= 5 \\ 0x_1 + 1x_2 &= -3 \end{aligned}$$

we readily observe that $X = \begin{bmatrix} 5 \\ -3 \end{bmatrix}$ is the desired solution. Our goal, then, is to be able to start with augmented matrices of a more complicated form, and to systematically "reduce" them to equivalent augmented matrices in successively simpler form, terminating when we have reached the simplest possible augmented matrix representation. Using $A^{(1)}$ to denote $A$ and $B^{(1)}$ to denote $B$ at the start of our problem, and then exchanging $A^{(1)}$ for $A^{(2)}$ and $B^{(1)}$ for $B^{(2)}$, we will generate a sequence $[A|B] = [A^{(1)}|B^{(1)}]$, $[A^{(2)}|B^{(2)}]$, $[A^{(3)}|B^{(3)}], \ldots, [A^{(s)}|B^{(s)}]$ of augmented matrix representations of $s$ different sets of linear equations—all equivalent in that they have exactly the same solutions.

A problem that immediately arises is the following one. What reduction operations may we perform on a given augmented matrix and insure that the set of solution matrices does not change? The allowable operations are simply the analogues of the operations that one may perform on a set of equations without changing the solution set. Namely, one may perform operations of the following three types without affecting the solutions:

**Type 1** Multiply any equation by a non-zero constant.

Example 2.5.1

$$x_1 + x_2 = 2 \qquad \text{and} \qquad 2x_1 + 2x_2 = 4$$

have exactly the same solution set. ∎

**Type 2** Interchange any two equations.

Example 2.5.2

$$\begin{aligned} x_1 + x_2 &= 2 \\ x_1 - x_2 &= 4 \end{aligned} \qquad \text{and} \qquad \begin{aligned} x_1 - x_2 &= 4 \\ x_1 + x_2 &= 2 \end{aligned}$$

have exactly the same solution set. ∎

**Type 3** Replace any equation by the *new* equation obtained by multiplying any other equation by a constant and adding the result to the original equation.

Example 2.5.3

$$\begin{aligned} x_1 + x_2 &= 2 \\ x_1 - x_2 &= 4 \end{aligned} \qquad \text{and} \qquad \begin{aligned} x_1 + x_2 &= 2 \\ 3x_1 + x_2 &= 8 \end{aligned}$$

have the same solution set; namely $X = \begin{bmatrix} x_1 \\ x_2 \end{bmatrix} = \begin{bmatrix} 3 \\ -1 \end{bmatrix}$.

The original second equation has been replaced by a new equation obtained by multiplying equation 1 by the constant 2 and adding the resulting equation to equation 2. If the constant of multiplication in the operation of Type 3 happens to be 1, then we have merely added equations; while if the constant is $-1$, then we have subtracted. ∎

*In order to obtain the allowable operations on the augmented matrix, we merely replace the word "equation(s)" by "row(s)" in the above description of Type 1, Type 2, and Type 3 operations.*

Let us now consider the following set of equations from Example 2.1.2 and its equivalent augmented matrix form. We will systematically exchange the sets of equations and their respective augmented matrix forms, using operations of the three types just described, until we reach sets from which the solution is obvious.

| *Equations* | *Augmented Matrix Form* |
|---|---|
| $\begin{aligned} 2x_1 + 3x_2 &= 5 \\ 4x_1 + 2x_2 &= 6 \end{aligned}$ | $\left[\begin{array}{cc\|c} 2 & 3 & 5 \\ 4 & 2 & 6 \end{array}\right]$ |

We replace the first set of equations with an equivalent set in which the coefficient of $x_1$ in equation 1 is a 1 (In some sets of equations this will require an interchange of equations. Why?). We accomplish this by multiplying the original equation 1 by 1/2, an operation of Type 1.

$$\begin{aligned} 1 \cdot x_1 + (3/2)x_2 &= 5/2 \\ 4x_1 + \quad 2x_2 &= 6 \end{aligned} \qquad \left[\begin{array}{cc|c} 1 & 3/2 & 5/2 \\ 4 & 2 & 6 \end{array}\right]$$

We proceed now to eliminate $x_1$ from all of the equations except equation 1 (namely, equation 2 in this case) by means of operations of Type 3. To do so, we replace the present second equation by the new equation obtained by multiplying equation 1 by $(-4)$ and adding it to equation 2 to obtain

$$\begin{aligned} 1 \cdot x_1 + (3/2)x_2 &= 5/2 \\ 0 \cdot x_1 - \quad 4x_2 &= -4 \end{aligned} \qquad \left[\begin{array}{cc|c} 1 & 3/2 & 5/2 \\ 0 & -4 & -4 \end{array}\right].$$

We now replace the last set of equations by a set in which the coefficient of $x_2$ in equation 2 is the constant 1, by multiplying equation 2 by $(-1/4)$, another operation of Type 1.

$$\begin{aligned} 1 \cdot x_1 + (3/2)x_2 &= 5/2 \\ 0 \cdot x_1 + \quad 1 \cdot x_2 &= 1 \end{aligned} \qquad \left[\begin{array}{cc|c} 1 & 3/2 & 5/2 \\ 0 & 1 & 1 \end{array}\right].$$

We conclude by eliminating $x_2$ from equation 1 by another Type 3 operation in which we replace equation 1 by the new equation obtained from multiplying equation 2 by $(-3/2)$ and adding the result to equation 1. Thus we have

$$\begin{aligned} 1 \cdot x_1 + 0 \cdot x_2 &= 1 \\ 0 \cdot x_1 + 1 \cdot x_2 &= 1 \end{aligned} \qquad \left[\begin{array}{cc|c} 1 & 0 & 1 \\ 0 & 1 & 1 \end{array}\right].$$

We see that in this case we have been able, by means of operations of Type 1 and Type 3 to obtain an equivalent augmented matrix representation from which we can readily find the solution to the original set of equations. What is it?

Although not used in the foregoing example, operations of Type 2 are sometimes required. For instance, if the original set of equations had been

$$\begin{aligned} 0 \cdot x_1 + 3x_2 &= 1 \\ 2x_1 - \quad x_2 &= 3 \end{aligned}$$

then our algorithm fails at the first step (since the coefficient of $x_1$ in equation 1 is 0, and hence has no multiplicative inverse). At this point we merely interchange the two equations and proceed as before. The reader should note that this situation might occur at later steps in the algorithm, and, if so, that we remedy the situation by interchanging the faulty equation with any following equation which has a non-zero coefficient for the unknown in question. (In the non-singular case there will always be such an equation.) We shall illustrate this further after the next example.

Example 2.5.4

$$\begin{aligned} x_1 - \quad x_2 + \quad x_3 &= 3 \\ 4x_1 - 3x_2 - \quad x_3 &= 6 \\ 3x_1 + \quad x_2 + 2x_3 &= 4 \end{aligned}$$

In this case the original augmented matrix representation is

$$\left[\begin{array}{ccc|c} 1 & -1 & 1 & 3 \\ 4 & -3 & -1 & 6 \\ 3 & 1 & 2 & 4 \end{array}\right].$$

In order to proceed to an equivalent augmented matrix representation in a simpler form, we wish to find one in which the only non-zero element in column 1 is the constant 1 in row 1. In other words, we are exchanging the original set of equations for a new set of equations with the same solution set, but with the helpful property that $x_1$ has been eliminated from all of the equations except equation 1. In order to accomplish this, we merely perform two operations of Type 3, replacing row 2 and row 3 as follows:

For row 2, we multiply row one by $(-4)$ and add the result to the original row 2 to obtain the equivalent augmented matrix

$$\left[\begin{array}{ccc|c} 1 & -1 & 1 & 3 \\ 0 & 1 & -5 & -6 \\ 3 & 1 & 2 & 4 \end{array}\right];$$

and we denote this change by writing $R_2 \leftarrow -4R_1 + R_2$, which means that the contents of row 2 have been replaced as indicated. For row 3, we multiply row 1 by $(-3)$ and add the result to the original row 3 to obtain

$$\left[\begin{array}{ccc|c} 1 & -1 & 1 & 3 \\ 0 & 1 & -5 & -6 \\ 0 & 4 & -1 & -5 \end{array}\right],$$

and note that $R_3 \leftarrow -3R_1 + R_3$.

We now proceed to essentially eliminate $x_2$ from all equations except the second, obtaining an equivalent augmented matrix representation in which the only non-zero element in column 2 is the number 1 in row 2. We first obtain

$$\left[\begin{array}{ccc|c} 1 & 0 & -4 & -3 \\ 0 & 1 & -5 & -6 \\ 0 & 4 & -1 & -5 \end{array}\right],$$

in which $R_1 \leftarrow R_2 + R_1$, in order to have the desired zero in the 1, 2 position, and then let $R_3 \leftarrow -4R_2 + R_3$ to arrive at

$$\left[\begin{array}{ccc|c} 1 & 0 & -4 & -3 \\ 0 & 1 & -5 & -6 \\ 0 & 0 & 19 & 19 \end{array}\right],$$

with the desired zero in the 3, 2 position.

To achieve an even simpler form, we perform an operation of Type 1, multiplying row 3 (equation 3) by 1/9 to obtain

$$\left[\begin{array}{ccc|c} 1 & 0 & -4 & -3 \\ 0 & 1 & -5 & -6 \\ 0 & 0 & 1 & 1 \end{array}\right],$$

where $R_3 \leftarrow (1/19)R_3$, and then obtain zeros in positions 1, 3 and 2, 3 by essentially eliminating $x_3$ from all of the corresponding equations except for equation 3. First, we change row 2 to get

$$\left[\begin{array}{ccc|c} 1 & 0 & -4 & -3 \\ 0 & 1 & 0 & -1 \\ 0 & 0 & 1 & 1 \end{array}\right],$$

in which we let $R_2 \leftarrow 5R_3 + R_2$, and then change row 1 to finally arrive at

$$\left[\begin{array}{ccc|c} 1 & 0 & 0 & 1 \\ 0 & 1 & 0 & -1 \\ 0 & 0 & 1 & 1 \end{array}\right],$$

by letting $R_1 \leftarrow 4R_3 + R_1$. We can then see immediately that our solution is

$$X = \begin{bmatrix} x_1 \\ x_2 \\ x_3 \end{bmatrix} = \begin{bmatrix} 1 \\ -1 \\ 1 \end{bmatrix},$$

as we may verify by direct substitution in the original set of equations. ∎

The technique illustrated above is a straightforward method of solution that is sufficient to solve systems of linear equations in which the coefficient matrix $A$ is square and non-singular. The procedure may be very easily programmed for use on a digital computer. *By the use of certain allowable row operations which do not affect the solution set, we have merely exchanged* $[A \mid B]$ *for a sequence of equivalent representations—first eliminating* $x_1$ *from all equations except the first, then eliminating* $x_2$ *from all equations except the second, and proceeding, in general, until* $x_n$ *has been removed from all equations except the nth.* We should re-emphasize here that we refer to the final numbering of the equations after they have been juggled (perhaps more than once) to insure that the (perhaps revised) coefficient of $x_i$ is not 0 in equation $i$. The following example illustrates the procedure in a case where such juggling is required.

Example 2.5.5

$$\begin{aligned} x_1 + x_2 + x_3 &= 3 \\ 2x_1 + 2x_2 + x_3 &= 5 \\ x_1 + 2x_2 + 3x_3 &= 6. \end{aligned}$$ ∎

We shall use the symbol $\sim$ to mean "is equivalent to." Proceeding as before, we find that

$$\left[\begin{array}{ccc|c} 1 & 1 & 1 & 3 \\ 2 & 2 & 1 & 5 \\ 1 & 2 & 3 & 6 \end{array}\right] \sim \left[\begin{array}{ccc|c} 1 & 1 & 1 & 3 \\ 0 & 0 & -1 & -1 \\ 0 & 1 & 2 & 3 \end{array}\right].$$

$$R_2 \leftarrow -2R_1 + R_2$$
$$R_3 \leftarrow -1R_1 + R_3$$

At this point we notice that the 2, 2 element is 0. Since the 3, 2 element is non-zero, namely 1, we interchange the last two rows (indicated by $R_2 \leftrightarrow R_3$) to obtain

$$\left[\begin{array}{ccc|c} 1 & 1 & 1 & 3 \\ 0 & 1 & 2 & 3 \\ 0 & 0 & -1 & -1 \end{array}\right],$$
$$R_2 \leftrightarrow R_3$$

from which we would proceed as before:

$$\left[\begin{array}{ccc|c} 1 & 1 & 1 & 3 \\ 0 & 1 & 2 & 3 \\ 0 & 0 & -1 & -1 \end{array}\right] \sim \left[\begin{array}{ccc|c} 1 & 0 & -1 & 0 \\ 0 & 1 & 2 & 3 \\ 0 & 0 & -1 & -1 \end{array}\right]$$
$$R_1 \leftarrow -R_2 + R_1$$

$$\sim \left[\begin{array}{ccc|c} 1 & 0 & -1 & 0 \\ 0 & 1 & 2 & 3 \\ 0 & 0 & 1 & 1 \end{array}\right]$$
$$R_3 \leftarrow -R_3$$

$$\sim \left[\begin{array}{ccc|c} 1 & 0 & 0 & 1 \\ 0 & 1 & 0 & 1 \\ 0 & 0 & 1 & 1 \end{array}\right].$$
$$R_1 \leftarrow R_3 + R_1$$
$$R_2 \leftarrow -2R_3 + R_2$$

The unique solution is the matrix $X = \begin{bmatrix} 1 \\ 1 \\ 1 \end{bmatrix}$.

We thus have a system for solving $AX = B$ when $A$ is square and non-singular. We will first use this procedure to develop a method to find $A^{-1}$ and then extend the Gauss-Jordan algorithm to cover the case when $A$ is singular. Perhaps you can design the method for computing $A^{-1} = \begin{bmatrix} a & b \\ c & d \end{bmatrix}$ yourself. Consider, in the $2 \times 2$ case, the problem of finding $A^{-1}$ when $A = \begin{bmatrix} 4 & 1 \\ 3 & 2 \end{bmatrix}$. What two matrix equations are to be solved? Could they be solved concurrently?

## 2.5 Exercises

1. What are the operations which one may perform on a system of equations without changing the solution set?

2. Consider the set of equations

$$2x_1 - x_2 = 3$$
$$4x_1 + x_2 = 7.$$

a. Find $[A^{(1)} \mid B^{(1)}] = [A \mid B]$.
b. Let $R_1 \leftarrow (1/2)R_1$, and find $[A^{(2)} \mid B^{(2)}]$.

3. Consider the equation $AX = B$ with $A = \begin{bmatrix} 1 & 1 \\ -1 & 2 \end{bmatrix}$, $B = \begin{bmatrix} -1 \\ 4 \end{bmatrix}$.

a. Find $[A^{(1)} \mid B^{(1)}]$.
b. Let $R_2 \leftarrow R_1 + R_2$ and find $[A^{(2)} \mid B^{(2)}]$.
c. Let $R_2 \leftarrow (1/3)R_2$ and find $[A^{(3)} \mid B^{(3)}]$.
d. Let $R_1 \leftarrow -R_2 + R_1$ and find $[A^{(4)} \mid B^{(4)}]$.
e. What is $X$?

4. Use the technique of this section to solve the following sets of equations.

a. $\begin{aligned} x_1 + 2x_2 &= 1 \\ 2x_1 + 5x_2 &= 3. \end{aligned}$

b. $\begin{aligned} 2x_1 - 3x_2 &= -2 \\ -x_1 + 4x_2 &= 6. \end{aligned}$

c. $\begin{aligned} x_1 + x_2 &= 1 \\ -x_1 + x_2 &= 0. \end{aligned}$

d. $\begin{aligned} x_1 + x_2 &= 0 \\ -x_1 + x_2 &= 1. \end{aligned}$

5. Use the results of Exercise 4c. and 4d. above to find the inverse of the matrix $A = \begin{bmatrix} 1 & 1 \\ -1 & 1 \end{bmatrix}$.

6. Solve the equation $AX = B$ with

a. $A = \begin{bmatrix} 2 & -1 & 2 \\ 4 & -1 & 3 \\ 1 & 2 & 1 \end{bmatrix}$, $B = \begin{bmatrix} 2 \\ 5 \\ 4 \end{bmatrix}$.

b. $A = \begin{bmatrix} 2 & -1 & 3 \\ 3 & 1 & -1 \\ 1 & 2 & 3 \end{bmatrix}$, $B = \begin{bmatrix} 1 \\ 2 \\ -6 \end{bmatrix}$.

c. $A = \begin{bmatrix} 1 & 1 & 1 \\ 2 & 3 & 5 \\ -1 & 2 & 2 \end{bmatrix}$, $B = \begin{bmatrix} 0 \\ -4 \\ -6 \end{bmatrix}$.

7. Solve the equation $AX = B$, with

$$A = \begin{bmatrix} 1 & -1 & 1 & -1 \\ 0 & 1 & 0 & 1 \\ 1 & 0 & -1 & 0 \\ 0 & 1 & 0 & -1 \end{bmatrix}, \quad B = \begin{bmatrix} -2 \\ 4 \\ 0 \\ 0 \end{bmatrix}.$$

8. Solve the equation $AX = B$, with

$$A = \begin{bmatrix} 1 & 0 & 1 & -1 & 1 \\ 2 & 1 & 1 & -1 & 3 \\ 0 & 0 & 1 & -1 & 2 \\ 3 & 1 & 4 & -4 & 4 \\ 0 & -1 & 2 & 3 & 0 \end{bmatrix}, \quad B = \begin{bmatrix} -3 \\ -2 \\ -3 \\ -11 \\ 3 \end{bmatrix}$$

9. Find a solution for the equations of Example 2.1.3.

The next two exercises are for calculator enthusiasts.

10. Solve the system $AX = R$ with $A = \begin{bmatrix} \pi & \sqrt{2} \\ 22/7 & 1.414 \end{bmatrix}$ and $R = \begin{bmatrix} 6 \\ -8 \end{bmatrix}$.

    Note that the rows of $A$ are very nearly the same, and that the solution values are relatively large. This is not accidental.

11. Solve the following system of linear equations:

    a. $$\begin{aligned} \pi x_1 + 2.5x_2 &= 6.8 \\ -\sqrt{2}x_1 + ex_2 &= 10.7 \end{aligned}$$

    b. $$\begin{aligned} x_1 + x_2 + x_3 &= 6 \\ \pi x_1 + \pi^2 x_2 + \pi^3 x_3 &= \sqrt{7} \\ -\sqrt{3}x_1 + 5x_2 + 4x_3 &= \sqrt{10} \end{aligned}$$

### New Terms

Gauss-Jordan reduction, 60

## 2.6 The Use of Gauss-Jordan Reduction to Find $A^{-1}$

We have noted previously that if we knew $A^{-1}$ then we could find the unique $X$ which satisfies

$$AX = B$$

by multiplication of both sides of the equation on the left by $A^{-1}$ to obtain

$$X = A^{-1}B.$$

In Example 2.5.5 we started with an augmented matrix of the form $[A \mid B]$ and obtained an equivalent augmented matrix of the form $[I_3 \mid X]$. The matrix $[A \mid B]$ is an example of a **partitioned matrix**. If $B_1$ and $B_2$ are matrices with the same number of rows then the matrix $[B_1 \mid B_2]$ can always be formed. If $A$ is a matrix that is conformable with each of $B_1$ and $B_2$, then the product $A[B_1 \mid B_2]$ is defined and

$$A[B_1 \mid B_2] = [AB_1 \mid AB_2].$$

For example,

$$\begin{bmatrix} 1 & 2 \\ 3 & 4 \end{bmatrix} \left[\begin{array}{cc|cc} 1 & 2 & 1 & 2 \\ 1 & 2 & 2 & 1 \end{array}\right] = \left[\begin{array}{cc|cc} 3 & 6 & 5 & 4 \\ 7 & 14 & 11 & 10 \end{array}\right].$$

We use this type of product in the following examination of the partitioned matrix $[I_3 \mid X]$. We see that

$$[I_3 \mid X] = [I_3 \mid A^{-1}B]$$

and note that the same result would have been obtained by multiplying the augmented matrix $[A \mid B]$ on the left by $A^{-1}$ to obtain

$$A^{-1}[A \mid B] = [A^{-1}A \mid A^{-1}B] = [I_3 \mid A^{-1}B] = [I_3 \mid X].$$

Thus we observe that the sequence of row operations which we performed on $[A \mid B]$ to reduce it to $[I_3 \mid X]$ had the same effect as if we had multiplied $[A \mid B]$ on the left by $A^{-1}$.

What if we had further augmented $A$ with the appropriate identity matrix $I_3$ to form the matrix $[A \mid B \mid I_3]$? It can be proved that the effect on $I_3$ would then have been as if $I_3$ were multiplied by $A^{-1}$ to obtain

$$A^{-1}[A \mid B \mid I_3] = [A^{-1}A \mid A^{-1}B \mid A^{-1}I_3] = [I_3 \mid X \mid A^{-1}]$$

at the completion of our sequence of row operations. *By performing the same operations on the rows of $I_3$ that we perform on the rows of $A$, we obtain $A^{-1}$ as a by-product of our solution process.* Let us verify this in the case of Example 2.1.2, where we had the following system of equations:

$$\begin{aligned} 2x_1 + 3x_2 &= 5 \\ 4x_1 + 2x_2 &= 6. \end{aligned}$$

$$[A|B|I_2] = \left[\begin{array}{cc|c|cc} 2 & 3 & 5 & 1 & 0 \\ 4 & 2 & 6 & 0 & 1 \end{array}\right] \sim \left[\begin{array}{cc|c|cc} 1 & 3/2 & 5/2 & 1/2 & 0 \\ 4 & 2 & 6 & 0 & 1 \end{array}\right]$$
$$R_1 \leftarrow (1/2)R_1$$

$$\sim \left[\begin{array}{cc|c|cc} 1 & 3/2 & 5/2 & 1/2 & 0 \\ 0 & -4 & -4 & -2 & 1 \end{array}\right]$$
$$R_2 \leftarrow -4R_1 + R_2$$

$$\sim \left[\begin{array}{cc|c|cc} 1 & 3/2 & 5/2 & 1/2 & 0 \\ 0 & 1 & 1 & 1/2 & -1/4 \end{array}\right]$$
$$R_2 \leftarrow (-1/4)R_2$$

$$\sim \left[\begin{array}{cc|c|cc} 1 & 0 & 1 & -1/4 & 3/8 \\ 0 & 1 & 1 & 1/2 & -1/4 \end{array}\right].$$
$$R_1 \leftarrow (-3/2)R_2 + R_1$$

We verify by direct multiplication that

$$\begin{bmatrix} 2 & 3 \\ 4 & 2 \end{bmatrix}\begin{bmatrix} -1/4 & 3/8 \\ 1/2 & -1/4 \end{bmatrix} = \begin{bmatrix} 1 & 0 \\ 0 & 1 \end{bmatrix} = \begin{bmatrix} -1/4 & 3/8 \\ 1/2 & -1/4 \end{bmatrix}\begin{bmatrix} 2 & 3 \\ 4 & 2 \end{bmatrix},$$

and thus that we have found $A^{-1}$ in this case. It should be clear that there is no need to carry the $B$ matrix through the above manipulations if $A^{-1}$ is the only output desired.

## 2.6 Exercises

1. Find $A^{-1}$ if $A$ is

   a. $\begin{bmatrix} 1 & 2 \\ 2 & 1 \end{bmatrix}$ b. $\begin{bmatrix} 1 & -1 \\ 0 & 1 \end{bmatrix}$ c. $\begin{bmatrix} 2 & 3 \\ 4 & 5 \end{bmatrix}$

2. Use the inverses found in Exercise 1 above to solve the equations $AX = B$ where $B = \begin{bmatrix} 1 \\ 1 \end{bmatrix}$.

3. Find $A^{-1}$ if $A$ is

   a. $\begin{bmatrix} 1 & 1 & 1 \\ 0 & 1 & 0 \\ 0 & 0 & 1 \end{bmatrix}$ b. $\begin{bmatrix} 1 & -1 & 1 \\ -1 & 1 & 0 \\ 0 & -1 & 1 \end{bmatrix}$ c. $\begin{bmatrix} 2 & -1 & 3 \\ 1 & 0 & 2 \\ -1 & 2 & 1 \end{bmatrix}$

4. Find $A^{-1}$ if $A = \begin{bmatrix} 1 & -1 & 1 & -1 \\ 0 & 1 & 0 & 1 \\ 1 & 0 & -1 & 0 \\ 0 & 1 & 0 & -1 \end{bmatrix}$.

5. Use your calculator to find $A^{-1}$ if $A = \begin{bmatrix} \pi & \sqrt{5} \\ -2 & e \end{bmatrix}$.

**New Terms**

partitioned matrix, 67

## 2.7 Gauss-Jordan Reduction. The General Case—Redundant Equations

Now that we have a reasonably good idea of how we can work with row operations and understand that we are really only eliminating the variables in a systematic fashion, we shall study the situations that arise when $A$ is singular, or indeed when $A$ is rectangular rather than square. In either case the previously described procedure needs some revision because of the new circumstances which will arise. One of those situations occurs when one or more of the original equations is redundant (with respect to those preceding it).

Consider the singular matrix $A = \begin{bmatrix} 1 & 2 \\ 2 & 4 \end{bmatrix}$ that we encountered in Chapter 1. Suppose that we attempted to use our technique to solve a set of equations, say

$$\begin{aligned} x_1 + 2x_2 &= 3 \\ 2x_1 + 4x_2 &= 6 \end{aligned}$$

which has $A$ for its coefficient matrix. Proceeding as before,

$$\left[\begin{array}{cc|c} 1 & 2 & 3 \\ 2 & 4 & 6 \end{array}\right] \sim \left[\begin{array}{cc|c} 1 & 2 & 3 \\ 0 & 0 & 0 \end{array}\right].$$
$$R_2 \leftarrow -2R_1 + R_2$$

We now observe that it is impossible to eliminate $x_2$ from equation 1 by any row operation because the element in the 2, 2 position of the last matrix is 0.

It will be worthwhile to consider the geometry of the situation in an attempt to describe what has taken place. We first note that the original equation 2 is merely a multiple of equation 1 and that in reality we started with only one independent equation, namely

$$x_1 + 2x_2 = 3,$$

in that no new points were excluded from the potential solution set by the additional equation. Geometrically, the set of points in the $x_1$, $x_2$-plane which satisfy the first equation form a straight line, and there are an infinite number of them. The set of points which satisfy the second equation also form a straight line and it has turned out that the two straight lines are really the same line. *Whenever, in the course of our application of the Gauss-Jordan reduction technique, we create an entire row of zeros, we shall have discovered the fact that the original equation whose coefficients appeared in that row was really redundant and that its presence excluded no points as solutions which were not excluded by the previous equations.* In this particular case, we may find all of the possible solutions by letting $x_2$ take on some arbitrary value, say $a$, and then solving for $x_1$ to find that

$$x_1 = 3 - 2a.$$

A solution matrix $X$ is thus any matrix of the form

$$X = \begin{bmatrix} 3 - 2a \\ a \end{bmatrix},$$

where $a$ may be any real number whatsoever. For example, if $a = 0$, we have $X_1 = \begin{bmatrix} 3 \\ 0 \end{bmatrix}$; or if $a = 1$, we have $X_2 = \begin{bmatrix} 1 \\ 1 \end{bmatrix}$, each of which is a solution to the original set of equations. *We shall use this technique of arbitrary assignment of values to a particular subset of the variables a great deal.* We note that $X$ as used above really denotes a set of matrices—a different one for each value of $a$.

Let us now investigate what would have happened if we had tried to use our standard technique to find $A^{-1}$. We would have proceeded to obtain

$$\left[\begin{array}{cc|c|cc} 1 & 2 & 3 & 1 & 0 \\ 2 & 4 & 6 & 0 & 1 \end{array}\right] \sim \left[\begin{array}{cc|c|cc} 1 & 2 & 3 & 1 & 0 \\ 0 & 0 & 0 & -2 & 1 \end{array}\right].$$

Since there are no non-zero elements remaining in row 2 of the final left-hand submatrix we can proceed no further in our reduction to $I_2$ and may legitimately conclude that $A^{-1}$ does not exist. If you happen to be familiar with $2 \times 2$ determinants, you will observe that the determinant of $A$, namely $\begin{bmatrix} 1 & 2 \\ 2 & 4 \end{bmatrix}$, is zero. *This is always a sure sign that $A^{-1}$ does not exist.* We shall discuss determinants of square matrices of all sizes in Chapter 5.

Example 2.7.1

$$\begin{aligned} x_1 + x_2 + x_3 &= 3 \\ 2x_1 - x_2 + 2x_3 &= 3 \\ 4x_1 + x_2 + 4x_3 &= 9. \end{aligned}$$

$$\left[\begin{array}{ccc|c} 1 & 1 & 1 & 3 \\ 2 & -1 & 2 & 3 \\ 4 & 1 & 4 & 9 \end{array}\right] \sim \left[\begin{array}{ccc|c} 1 & 1 & 1 & 3 \\ 0 & -3 & 0 & -3 \\ 0 & -3 & 0 & -3 \end{array}\right]$$

$$R_2 \leftarrow -2R_1 + R_2$$
$$R_3 \leftarrow -4R_1 + R_3$$

$$\sim \left[\begin{array}{ccc|c} 1 & 1 & 1 & 3 \\ 0 & 1 & 0 & 1 \\ 0 & -3 & 0 & -3 \end{array}\right] \sim \left[\begin{array}{ccc|c} 1 & 0 & 1 & 2 \\ 0 & 1 & 0 & 1 \\ 0 & 0 & 0 & 0 \end{array}\right].$$

$$R_2 \leftarrow (-1/3)R_2 \qquad R_1 \leftarrow -R_2 + R_1$$
$$R_3 \leftarrow 3R_2 + R_3$$

We see that the original equation 3 was redundant. Geometrically, the plane represented by the third equation passes through the line of intersection of the first two planes, a situation illustrated in Figure 2.7.1. ∎

Figure 2.7.1

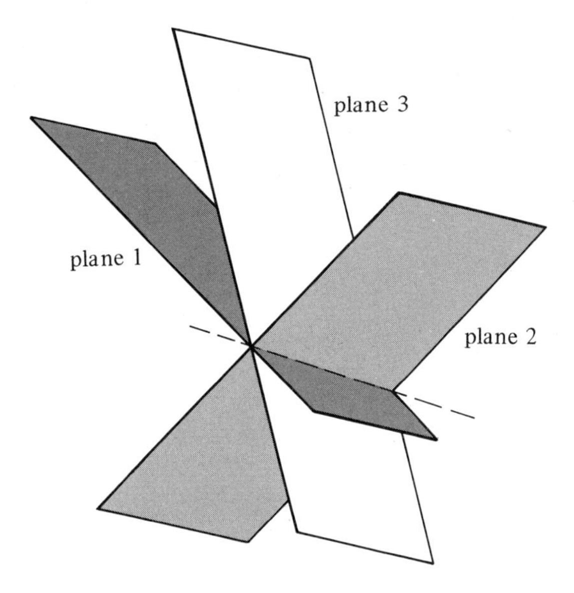


Observing the final matrix of Example 2.7.1, namely,

$$\left[\begin{array}{ccc|c} 1 & 0 & 1 & 2 \\ 0 & 1 & 0 & 1 \\ 0 & 0 & 0 & 0 \end{array}\right],$$

we note that the value of $x_3$ is not fixed by the third equation (the element in the 3, 3 position is 0). Thus we are permitted to arbitrarily assign any value, say $a$, to $x_3$. We call such a variable a **free variable**. (We shall discuss free variables more fully in the next section.) However, $x_2$ is fixed by equation 2, and is clearly equal to 1. Also, $x_1$ is fixed by equation 1, but only in terms of the parameter $a$. For any $a$ that we have assigned to $x_3$, we have $x_1 = 2 - a$. Thus the general solution set consists of matrices $X$ of the form

$$X = \begin{bmatrix} 2 - a \\ 1 \\ a \end{bmatrix},$$

for any value of $a$. We could have split the solution matrix into two parts and written

$$X = \begin{bmatrix} 2 \\ 1 \\ 0 \end{bmatrix} + a \begin{bmatrix} -1 \\ 0 \\ 1 \end{bmatrix} = X_P + X_H.$$

We use the notation $X_P$ to denote a **particular solution** to our equation $AX = B$—that is, a single matrix $X_P$ such that $AX_P = B$. You can verify that $X_P$ is a particular solution to the original set of equations and that $X_H$ is a solution to the homogeneous set of equations obtained when we let 0 replace the original constants on the right-hand side. (This will have more meaning to those of you who have studied differential equations. We shall discuss it in more detail later.)

Although we started with a square coefficient matrix in both of our examples, the phenomenon of redundancy could also have occurred had $A$ been rectangular rather than square.

Example 2.7.2

$$\begin{aligned} 2x_1 + 3x_2 &= 5 \\ 4x_1 + 2x_2 &= 6 \\ 2x_1 - x_2 &= 1. \end{aligned}$$

$$\left[\begin{array}{cc|c} 2 & 3 & 5 \\ 4 & 2 & 6 \\ 2 & -1 & 1 \end{array}\right] \sim \left[\begin{array}{cc|c} 1 & 3/2 & 5/2 \\ 0 & -4 & -4 \\ 0 & -4 & -4 \end{array}\right] \sim \left[\begin{array}{cc|c} 1 & 0 & 1 \\ 0 & 1 & 1 \\ 0 & 0 & 0 \end{array}\right].$$

$$\begin{array}{ll} R_1 \leftarrow (1/2)R_1 & R_2 \leftarrow (-1/4)R_2 \\ R_2 \leftarrow -4R_1 + R_2 & R_1 \leftarrow (-3/2)R_2 + R_1 \\ R_3 \leftarrow -2R_1 + R_3 & R_3 \leftarrow 4R_2 + R_3 \end{array}$$

Thus $X = \begin{bmatrix} 1 \\ 1 \end{bmatrix}$ is the unique solution. ∎

The reader may have noticed a startling similarity between Examples 2.2.2 and 2.7.2. Geometrically, we merely added another straight line and it just happened to go through the point of intersection of the original two straight lines. If it had not, then there would have been no solution at all. We shall point out a valid general conclusion from this particular example in which we had more equations (3) than variables (2). *If, in our general problem, with m equations in n unknowns,*

$$m > n,$$

*then there will be no solution unless at least*

$$m - n$$

*of the equations are redundant*. Geometrically, we may illustrate this statement by pointing out that two intersecting straight lines in the plane (not three or more) determine a point, as do three properly intersecting planes in 3-space (not four or more). In general, $n$ non-redundant equations is the maximum number that we may have in $n$-space if a solution is to exist. We shall suppose henceforth that $m \leqslant n$ since any other problem for which a solution exists can be reduced to this case by using Gauss-Jordan reduction to determine the non-redundant equations.

## 2.7 Exercises

1. Use Gauss-Jordan reduction to solve the following systems of linear equations:

a. $$\begin{aligned} x_1 + x_2 &= 2 \\ -2x_1 - 2x_2 &= -4. \end{aligned}$$

b. $$\begin{aligned} x_1 - x_2 + 2x_3 &= 3 \\ 2x_1 - x_2 + x_3 &= 3 \\ -3x_1 + 2x_2 - 3x_3 &= -6. \end{aligned}$$

c. $x_1 + x_2 = 0.$

d.
$$\begin{aligned} x_1 + x_2 + x_3 - x_4 &= 1 \\ -x_1 \quad\quad + 3x_3 + x_4 &= -1 \\ x_1 + 2x_2 + 5x_3 - x_4 &= 1 \\ x_1 - x_2 - 4x_3 + 2x_4 &= 1. \end{aligned}$$

2. There is a method shown in the text of G. Hadley (see Suggested Reading) of controlling arithmetic errors during the Gauss-Jordan reduction process. It consists of merely adjoining an additional column to the right of the augmented matrix and *performing the same operations on it* that we do on the $B$ column. If we call this new column $S$, then, to start with, the element in row $i$ of $S$ is merely the sum of the elements of row $i$ of the augmented matrix; that is,

$$s_i = a_{i1} + a_{i2} + \cdots + a_{in} + b_i.$$

As we perform the row operations on *all* the columns and go from augmented matrix to augmented matrix, this *new* column should always record the present sum of the elements in each row. If not, an arithmetic mistake has been made and we should correct it before proceeding. We verify that this procedure is valid for

$$A = \begin{bmatrix} 1 & 2 \\ 3 & 4 \end{bmatrix}, \quad B = \begin{bmatrix} -1 \\ -1 \end{bmatrix}.$$

$$\left[\begin{array}{cc|c|c} 1 & 2 & -1 & 2 \\ 3 & 4 & -1 & 6 \end{array}\right] \sim \underset{R_2 \leftarrow -3R_1 + R_2}{\left[\begin{array}{cc|c|c} 1 & 2 & -1 & 2 \\ 0 & -2 & 2 & 0 \end{array}\right]} \sim \underset{\substack{R_2 \leftarrow (-1/2)R_2 \\ R_1 \leftarrow -2R_2 + R_1}}{\left[\begin{array}{cc|c|c} 1 & 0 & 1 & 2 \\ 0 & 1 & -1 & 0 \end{array}\right]}.$$

Use the extra column and solve $AX = B$ where

a. $A = \begin{bmatrix} 3 & 6 \\ -1 & 2 \end{bmatrix}, B = \begin{bmatrix} -6 \\ 2 \end{bmatrix}$

b. $A = \begin{bmatrix} 1 & 1 & 0 \\ 3 & 4 & 1 \\ -1 & 2 & 3 \end{bmatrix}, B = \begin{bmatrix} 4 \\ 15 \\ 5 \end{bmatrix}.$

3. a. Solve the equation $AX = 0$ where

$$A = \begin{bmatrix} -1 & 2 & 3 \\ 3 & -5 & 2 \\ 5 & -8 & 7 \end{bmatrix}.$$

b. If we let the free variable $x_3 = a$, say, what particular solution do you get when $a = 1, -2, 3$?

c. Verify by direct multiplication by $A$ that your solution for $a = 1$ is valid.

4. Let

$$\left\{X \middle| X = \begin{bmatrix} 1 - 2a + 3b \\ 2 + 4a - b \end{bmatrix}\right\}$$

be the set of all solutions of the matrix equation $AX = B$.

a. Show that $X$ may be expressed as a sum of two matrices, one a constant matrix $X_P$ and the other a matrix $X_H$ with the parameters (arbitrary constants) $a$ and $b$.

b. Show that $X_H$ may be further decomposed as a sum of two matrices, one containing $a$'s only, and the other containing only $b$'s.

5. Suppose that the following augmented matrices are the result of the Gauss-Jordan reduction algorithm. In each case find the number of redundant original equations and the solution. Break the solution matrix up into its constant part $X_P$ and the remainder $X_H$. Find the particular solution obtained when each of the free variables is set equal to 1 simultaneously.

a. $\left[\begin{array}{cccc|c} 1 & 0 & -1 & 2 & 3 \\ 0 & 1 & 2 & 3 & 4 \\ 0 & 0 & 0 & 0 & 0 \\ 0 & 0 & 0 & 0 & 0 \end{array}\right]$.
b. $\left[\begin{array}{ccc|c} 1 & 0 & 0 & 4 \\ 0 & 1 & 0 & -1 \\ 0 & 0 & 1 & 0 \\ 0 & 0 & 0 & 0 \\ 0 & 0 & 0 & 0 \end{array}\right]$.

c. $\left[\begin{array}{ccc|c} 1 & 0 & 2 & -1 \\ 0 & 1 & 0 & 3 \\ 0 & 0 & 0 & 0 \\ 0 & 0 & 0 & 0 \end{array}\right]$.
d. $\left[\begin{array}{cccccc|c} 1 & 0 & 0 & 0 & 5 & -1 & 2 \\ 0 & 1 & 0 & 1 & 4 & 3 & 3 \\ 0 & 0 & 1 & 0 & 0 & 1 & -1 \\ 0 & 0 & 0 & 0 & 0 & 0 & 0 \end{array}\right]$.

### Suggested Reading

G. Hadley. *Linear Programming*. Reading, Mass: Addison-Wesley, 1962.

### New Terms

free variable, 71    particular solution of a matrix equation, 71

## 2.8 Gauss-Jordan Reduction. The General Case—Free Variables

Consider the following example in which we have fewer non-redundant equations than unknowns.

Example 2.8.1

$$\begin{aligned} x_1 + x_2 + \phantom{2}x_3 &= 3 \\ 2x_1 - x_2 + 2x_3 &= 3. \end{aligned}$$

The Gauss-Jordan reduction process can be used to solve this system much as it was used in the square, non-singular case. The discriminating student may have already noticed that these are merely the non-redundant equations from Example 2.7.1. The iterative procedure for this system is exactly the same as it was for the first two rows of that example.

$$\left[\begin{array}{ccc|c} 1 & 1 & 1 & 3 \\ 2 & -1 & 2 & 3 \end{array}\right] \sim \underset{R_2 \leftarrow -2R_1 + R_2}{\left[\begin{array}{ccc|c} 1 & 1 & 1 & 3 \\ 0 & -3 & 0 & -3 \end{array}\right]} \sim \underset{R_2 \leftarrow (-1/3)R_2}{\left[\begin{array}{ccc|c} 1 & 1 & 1 & 3 \\ 0 & 1 & 0 & 1 \end{array}\right]}$$

$$\sim \underset{R_1 \leftarrow -R_2 + R_1}{\left[\begin{array}{ccc|c} 1 & 0 & 1 & 2 \\ 0 & 1 & 0 & 1 \end{array}\right]}.$$

The solution is then obtained in the same manner as previously. Nothing of an extraordinary nature has arisen. The *last* $n - m = 3 - 2 = 1$ variable(s) could be arbitrarily assigned to obtain the solution. Let us illustrate that such is not always the case by the following example. ∎

Example 2.8.2

$$\begin{aligned} x_1 + x_2 + x_3 + x_4 &= 4 \\ x_1 + x_2 + 2x_3 + 3x_4 &= 7 \\ 3x_1 + 3x_2 + 5x_3 + 4x_4 &= 15. \end{aligned}$$

At the first step in the Gauss-Jordan reduction to obtain a solution to Example 2.8.2 we obtain

$$\left[\begin{array}{cccc|c} 1 & 1 & 1 & 1 & 4 \\ 1 & 1 & 2 & 3 & 7 \\ 3 & 3 & 5 & 4 & 15 \end{array}\right] \sim \left[\begin{array}{cccc|c} 1 & 1 & 1 & 1 & 4 \\ 0 & 0 & 1 & 2 & 3 \\ 0 & 0 & 2 & 1 & 3 \end{array}\right]$$

$$\begin{aligned} R_2 &\leftarrow -R_1 + R_2 \\ R_3 &\leftarrow -3R_1 + R_2 \end{aligned}$$

when we eliminate $x_1$ from all equations except the first. The problem now arises when we attempt to eliminate $x_2$ from all equations except the second. We find that the coefficient of $x_2$ in equation 2 is 0. When this has taken place in our previous examples we have merely scanned the elements of column 2 below the 2, 2 position until we found a non-zero element and then interchanged rows. There was always such a non-zero element unless the remaining equations were redundant and all elements in the succeeding rows were zeros. Such is not the case here and we are unable to eliminate $x_2$ from equation 1. In such a situation we skip $x_2$ and attempt to eliminate $x_3$ from all equations except *equation* 2. We can do this unless the element in the 2, 3 position is also 0. If it were 0, we would scan column 3 below the 2, 3 position for non-zero elements. If we found one, we would interchange rows and proceed as before. If not, we would repeat the above procedure for the remaining variables until a non-zero element was found. In this example the 2, 3 element is 1, so we use it to delete $x_3$ from all equations except the second, as follows:

$$\left[\begin{array}{cccc|c} 1 & 1 & 1 & 1 & 4 \\ 0 & 0 & 1 & 2 & 3 \\ 0 & 0 & 2 & 1 & 3 \end{array}\right] \sim \left[\begin{array}{cccc|c} 1 & 1 & 0 & -1 & 1 \\ 0 & 0 & 1 & 2 & 3 \\ 0 & 0 & 0 & -3 & -3 \end{array}\right].$$

$$\begin{aligned} R_1 &\leftarrow -R_2 + R_1 \\ R_3 &\leftarrow -2R_2 + R_3 \end{aligned}$$

We now delete $x_4$ from all equations except the third.

$$\left[\begin{array}{cccc|c} 1 & 1 & 0 & -1 & 1 \\ 0 & 0 & 1 & 2 & 3 \\ 0 & 0 & 0 & -3 & -3 \end{array}\right] \sim \left[\begin{array}{cccc|c} 1 & 1 & 0 & 0 & 2 \\ 0 & 0 & 1 & 0 & 1 \\ 0 & 0 & 0 & 1 & 1 \end{array}\right].$$

$$\begin{aligned} R_3 &\leftarrow (-1/3)R_3 \\ R_2 &\leftarrow -2R_3 + R_2 \\ R_1 &\leftarrow R_3 + R_1 \end{aligned}$$

Instead of assigning the last $n - m = 4 - 3 = 1$ variable(s) arbitrarily, we assign $x_2$ arbitrarily in this case to obtain

$$X = \begin{bmatrix} 2 - a \\ a \\ 1 \\ 1 \end{bmatrix}.$$

For emphasis, we again point out that we could express $X$ as a sum

$$X = X_P + X_H = \begin{bmatrix} 2 \\ 0 \\ 1 \\ 1 \end{bmatrix} + a \begin{bmatrix} -1 \\ 1 \\ 0 \\ 0 \end{bmatrix},$$

where $X_p$ is one particular solution to the original set of equations and $X_H$ is the solution to the homogeneous set of equations with coefficient matrix $A$. ∎

In Example 2.8.1 we had a variable $x_3$ for which there was no (non-redundant) equation 3. Thus we obviously could not use equation 3 to eliminate $x_3$ from all other equations as we did for $x_1$ and $x_2$. (We used equation 1 to eliminate $x_1$, equation 2 to eliminate $x_2$, etc.). Such a variable as $x_3$ we have called a free variable and we know that if there is any solution at all, this variable may be arbitrarily assigned.

In Example 2.8.2 we had a variable, namely $x_2$, which we were unable to eliminate from one of the previous equations, equation 1. In this case, $x_2$ would also be called a free variable and be freely assigned any value. If there is any solution at all, such a variable may receive an arbitrary assignment.

*If we are dealing originally with $m$ equations in $n$ variables and it turns out that there are $k$ redundant equations, as indicated in the final matrix by the number of rows consisting entirely of zeros, then the number of free variables is $n - (m - k)$. Each of these free variables may be arbitrarily assigned values.* We sometimes say that we have an $[n - (m - k)]$-dimensional infinity of solutions or that we have $n - (m - k)$ degrees of freedom.

For those readers who are familiar with vectors as arrows, and with the elementary ideas of 3-dimensional geometry, the following example may help to explain the workings of a free variable. It will not impede your progress to skip this example until after you have studied Section 4.7.

Example 2.8.3

$$1x_1 + 0x_2 + 0x_3 = 2$$
$$0x_1 + 1x_2 + 0x_3 = 3.$$

Using Gauss-Jordan reduction, we obtain $\left[\begin{array}{ccc|c} 1 & 0 & 0 & 2 \\ 0 & 1 & 0 & 3 \end{array}\right]$, from which we see that

$$\left\{ X \mid X = \begin{bmatrix} 2 \\ 3 \\ a \end{bmatrix} = \begin{bmatrix} 2 \\ 3 \\ 0 \end{bmatrix} + a \begin{bmatrix} 0 \\ 0 \\ 1 \end{bmatrix} \right\} = \{X_P + X_H\}$$

is the complete solution. Now, $x_1 = 2$ and $x_2 = 3$ each represent planes in 3-space as shown in Figure 2.8.1.

Figure 2.8.1

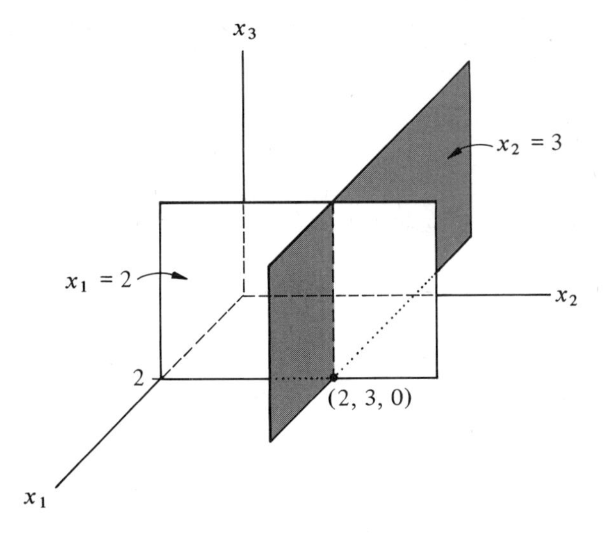


The dotted line of intersection is as shown. The entire line represents the solution set of the set of equations. When we split $X$ into $\begin{bmatrix} 2 \\ 3 \\ 0 \end{bmatrix}$ and $a\begin{bmatrix} 0 \\ 0 \\ 1 \end{bmatrix}$ we are saying that any solution vector $X$ may be written as a sum of two vectors $X_P$ and $X_H$. Drawing $X_P = \begin{bmatrix} 2 \\ 3 \\ 0 \end{bmatrix}$ and $X_H = \begin{bmatrix} 0 \\ 0 \\ 2 \end{bmatrix}$, with $a = 2$, in Figure 2.8.2 we see that the solution vector $X = \begin{bmatrix} 2 \\ 3 \\ 2 \end{bmatrix}$ has been located.

Figure 2.8.2

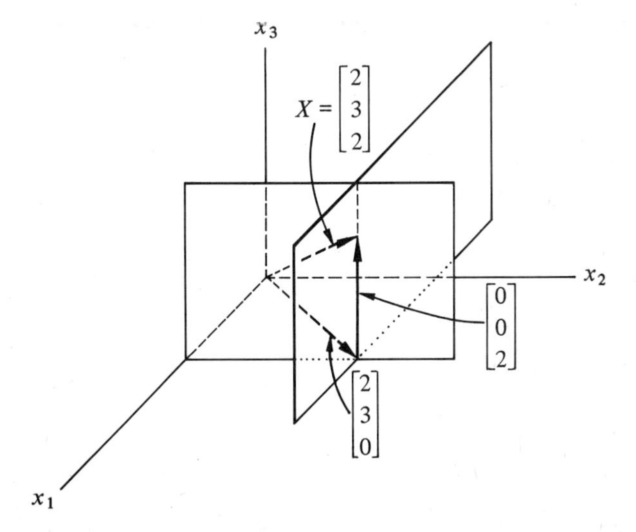

## 2.8 Exercises

1. a. Solve the following set of linear equations:

$$\begin{aligned} x_1 - x_2 + x_3 - x_4 &= 0 \\ x_3 + x_4 &= 0. \end{aligned}$$

b. Find the particular solution obtained when both free variables are set equal to 1 simultaneously.

c. Find the particular solution obtained when both free variables are set equal to $-3$ simultaneously.

d. Let $Y$ be the sum of the solutions in b. and c. Show that $Y$ is a solution of the original problem.

2. a. Solve the equation $AX = B$ by Gauss-Jordan reduction:

$$A = \begin{bmatrix} 2 & 3 & 1 & 0 \\ 3 & 2 & 0 & 1 \end{bmatrix}, \qquad B = \begin{bmatrix} 5 \\ 5 \end{bmatrix}.$$

b. Find the particular solution that results when both free variables are set equal to 0.

c. We have ______ non-redundant equations in ______ unknowns, and a ______-dimensional infinity of solutions.

d. What matrix appears in columns 3 and 4 of the final augmented matrix?

e. Decompose $X$ into $X_P$ and $X_H$. Verify that $AX_P = B$ and that $AX_H = 0$.

3. a. Solve the equation $AX = B$ by Gauss-Jordan reduction, where

$$A = \begin{bmatrix} 1 & 2 & -5 \\ 2 & 4 & -3 \end{bmatrix}, \qquad B = \begin{bmatrix} 0 \\ 0 \end{bmatrix}.$$

b. In this case we have ______ non-redundant equations in ______ unknowns, and a ______-dimensional infinity of solutions, as evidenced by the ______ free variable(s) in the solution.

c. Find and verify the particular solution that results from setting the free variable equal to 2. Is any scalar multiple of this solution also a solution?

4. a. Solve the equation $AX = B$ where

$$A = \begin{bmatrix} 1 & 2 & -1 & 2 & 4 \\ 3 & -1 & 4 & 6 & -5 \\ 4 & 2 & -3 & -1 & -2 \end{bmatrix}, \qquad B = \begin{bmatrix} 5 \\ 5 \\ -2 \end{bmatrix}.$$

b. We have ______ non-redundant equations in ______ unknowns. There is a ______-dimensional infinity of solutions, as evidenced by the ______ free variables.

c. Find and verify the particular solution that results when each of the free variables is set equal to 1.

5. a. Consider the equation $AX = B$ where

$$A = \begin{bmatrix} 1 & 1 & -2 & 1 & 3 \\ 2 & -1 & 2 & 2 & 6 \\ 3 & 2 & -4 & -3 & -9 \end{bmatrix}, \qquad B = \begin{bmatrix} 2 \\ 4 \\ 6 \end{bmatrix}.$$

If there is 1 redundant equation, then there should be a ________-dimensional infinity of solutions; while if none of the equations are redundant, there should be ________ free variables and a ________-dimensional infinity of solutions.

b. Solve the equation.

c. Decompose $X$ into $X_P$ and $X_H$, and verify that $AX_P = B$ and $AX_H = 0$.

6. a. Solve the set of equations $AX = B$ for

$$A = \begin{bmatrix} e^2 & e^3 & e^4 \\ -\cos 30^\circ & \sin 30^\circ & 1 \end{bmatrix}, \quad \text{and} \quad B = \begin{bmatrix} \pi \\ e \end{bmatrix}.$$

b. Find $X_P$ for $a = 1$.

## 2.9 Gauss-Jordan Reduction. The General Case—A Synopsis

We have now covered all of the situations that may arise when $A$ is not non-singular, and there is at least one solution matrix $X$. To illustrate that not all sets of equations have solutions, consider the following example.

Example 2.9.1

$$x_1 + x_2 = 1$$
$$x_1 + x_2 = 2.$$

Clearly there are no matrices $X = \begin{bmatrix} x_1 \\ x_2 \end{bmatrix}$, the sum of whose elements is both 1 and 2. Thus there can be no solution. ∎

How does this fact that the set of equations above has no solution become apparent in the solution algorithm? We answer the question by attempting to solve the system:

$$\left[\begin{array}{cc|c} 1 & 1 & 1 \\ 1 & 1 & 2 \end{array}\right] \sim \left[\begin{array}{cc|c} 1 & 1 & 1 \\ 0 & 0 & 1 \end{array}\right]$$
$$R_2 \leftarrow -R_1 + R_2.$$

The last row of the second matrix provides the clue, since it is a representation of the equation

$$0 \cdot x_1 + 0 \cdot x_2 = 1,$$

which obviously has no solution. At some point in the algorithm, *such a row will always be created in a problem which has no solution.* The algorithm should be terminated when this occurs. The geometrical situation is shown in Figure 2.9.1. The two equations represent non-intersecting parallel lines.

*How do we then proceed, in general, to find an $X$ such that $AX = B$? Using the ad hoc remedies of*

1. *juggling the order of the rows (equations) as required, and*
2. *skipping columns (variables that become free) as required,*

*we start with the augmented matrix $[A \mid B] = [A^{(1)} \mid B^{(1)}]$ and proceed*

Figure 2.9.1

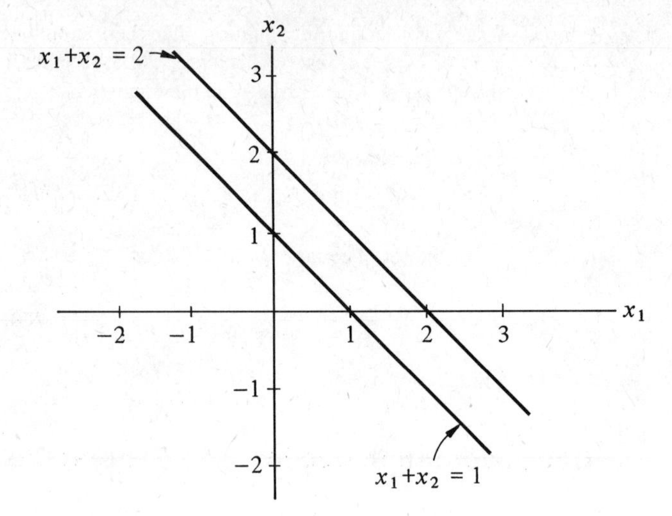


*through a sequence of s equivalent augmented matrices until we reach the matrix*

$$[A^{(s)} \mid B^{(s)}]$$

*in which $A^{(s)}$ is of a very special form.* To be specific, in the case where a solution exists, suppose that we have $m$ equations in $n$ unknowns and $k$ of the original equations are redundant, with $k \leqslant m$. Then

1. The last $k \geqslant 0$ rows of $A^{(s)}$ consist entirely of zeros.
2. The first non-zero element in each of the first $m - k$ rows of $A^{(s)}$ is a 1.
3. Each such "first 1" in a row is the only non-zero element in its column.
4. These "first 1's" are in an echelon form; that is, if the first 1 in row 1 appears in column $a_1$, the first 1 in row 2 appears in column $a_2$, and so forth, then

$$a_1 < a_2 < \cdots < a_{m-k}$$

(the $a_j$ here should not be confused with the matrix elements $a_{ij}$).

Such a matrix is said to be in **row-reduced echelon form**. *The variables $x_{a_1}, x_{a_2}, \ldots, x_{a_{m-k}}$ corresponding to the "first 1's" are then determined in terms of the constants of $B^{(s)}$ and the remaining variables (if any) which may be freely assigned arbitrary values.* We emphasize that one does not usually know the number $k$ at the beginning of the solution process.

Let us now consider several augmented matrices in row-reduced echelon form and use them to write down the general solutions to the corresponding systems of equations, if any exist.

Example 2.9.2

$$[A^{(s)} \mid B^{(s)}] = \left[\begin{array}{ccc|c} 1 & 0 & 2 & 3 \\ 0 & 1 & -1 & 4 \\ 0 & 0 & 0 & 0 \end{array}\right].$$

In this case, $m = 3$, $n = 3$, and $k = 1$ as indicated by the one row consisting entirely of zeros. Here $x_3$ is a free variable that may be arbitrarily assigned. Suppose we let $x_3 = a$. Then from row 2, we find that

$$x_2 = 4 + a,$$

and from row 1, then, we find that

$$x_1 = 3 - 2a,$$

so that

$$\left\{X \mid X = \begin{bmatrix} 3 - 2a \\ 4 + a \\ a \end{bmatrix} = \begin{bmatrix} 3 \\ 4 \\ 0 \end{bmatrix} + a\begin{bmatrix} -2 \\ 1 \\ 1 \end{bmatrix}\right\} = \{X_P + X_H\}$$

is the general solution to the original matrix equation $AX = B$. (What is $A^{(s)}X_H$?) To check the above solution, we verify that

$$\begin{bmatrix} 1 & 0 & 2 \\ 0 & 1 & -1 \\ 0 & 0 & 0 \end{bmatrix}\begin{bmatrix} 3 - 2a \\ 4 + a \\ a \end{bmatrix} = \begin{bmatrix} 3 \\ 4 \\ 0 \end{bmatrix} = B^{(s)},$$

and recall that the matrix equation $A^{(s)}X = B^{(s)}$ is equivalent to the original matrix equation $AX = B$, whatever it was. (Can we tell what it was?) Geometrically, two of the three planes (represented by the original three equations) intersected in a straight line. The third plane contained this same straight line and was thus redundant. We thus have an $n - (m - k)$ or $3 - (3 - 1) = 1$ dimensional infinity of solutions (a straight line in 3-dimensional space as shown in Figure 2.9.2). Before you continue reading, you should be sure you understand what the situation would be, had the 3, 4 element of $[A^{(s)} \mid B^{(s)}]$ been 1 rather than 0. ∎

Figure 2.9.2

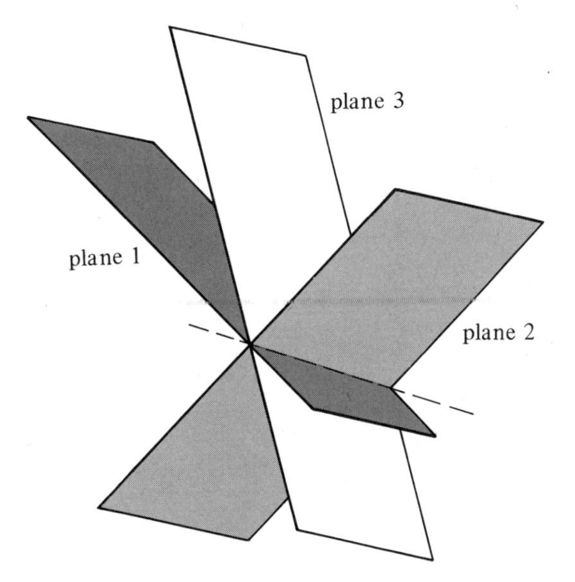


Example 2.9.3 $m = 3, n = 5$.

$$[A^{(s)} \mid B^{(s)}] = \left[\begin{array}{ccccc|c} 1 & 0 & 0 & 2 & -5 & 2 \\ 0 & 1 & 0 & -1 & 2 & 4 \\ 0 & 0 & 1 & 4 & 3 & -3 \end{array}\right].$$

Here, $k = 0$, indicating that none of the original equations was redundant. The variables $x_4$ and $x_5$ may be freely assigned. Suppose that we let $x_4 = a$ and $x_5 = b$. Then, from the last row,

$$x_3 = -3 - 4a - 3b;$$

and proceeding upward through the matrix we obtain

$$\begin{aligned} x_2 &= 4 + a - 2b \\ x_1 &= 2 - 2a + 5b, \end{aligned}$$

so that our parametric solution vector (matrix) $X$ in 5-space (the set of all column matrices with five elements) is

$$X = \begin{bmatrix} 2 & - & 2a & + & 5b \\ 4 & + & 1a & - & 2b \\ -3 & - & 4a & - & 3b \\ 0 & + & 1a & + & 0b \\ 0 & + & 0a & + & 1b \end{bmatrix}.$$

Again we note that $X$ may be split into two parts, $X = X_P + X_H$, where

$$X_P = \begin{bmatrix} 2 \\ 4 \\ -3 \\ 0 \\ 0 \end{bmatrix} \quad \text{and} \quad X_H = a\begin{bmatrix} -2 \\ 1 \\ -4 \\ 1 \\ 0 \end{bmatrix} + b\begin{bmatrix} 5 \\ -2 \\ -3 \\ 0 \\ 1 \end{bmatrix}.$$

You should verify that

1. $A^{(s)}X_P = B^{(s)}$, and
2. $A^{(s)}X_H = 0$.

Here we have an $n - (m - k) = 5 - (3 - 0) = 2$-dimensional infinity of solutions, one for each specific choice of $a$ and $b$. For instance, if we set $a = b = 1$, we find that

$$\begin{bmatrix} 5 \\ 3 \\ -10 \\ 1 \\ 1 \end{bmatrix}$$

is a solution. ∎

Example 2.9.4 $m = 4, n = 6$.

$$[A^{(s)} \mid B^{(s)}] = \left[\begin{array}{cccccc|c} 1 & 2 & 0 & 0 & 1 & -2 & -1 \\ 0 & 0 & 1 & 0 & -1 & 3 & 2 \\ 0 & 0 & 0 & 1 & 2 & -4 & 0 \\ 0 & 0 & 0 & 0 & 0 & 0 & 0 \end{array}\right].$$

In this example $k = 1$. The 0 in the 2, 2 position indicates that column 2 was skipped and that $x_2$ is thus a free variable. The variables $x_5$ and

$x_6$ are also free so we expect a 3-dimensional infinity of solutions, and find, letting $x_2 = a, x_5 = b, x_6 = c$, that

$$\left\{ X \mid X = \begin{bmatrix} -1-2a-1b+2c \\ 0+1a+0b+0c \\ 2+0a+1b-3c \\ 0+0a-2b+4c \\ 0+0a+1b+0c \\ 0+0a+0b+1c \end{bmatrix} = \begin{bmatrix} -1 \\ 0 \\ 2 \\ 0 \\ 0 \\ 0 \end{bmatrix} + a\begin{bmatrix} -2 \\ 1 \\ 0 \\ 0 \\ 0 \\ 0 \end{bmatrix} + b\begin{bmatrix} -1 \\ 0 \\ 1 \\ -2 \\ 1 \\ 0 \end{bmatrix} + c\begin{bmatrix} 2 \\ 0 \\ -3 \\ 4 \\ 0 \\ 1 \end{bmatrix} \right.$$

is the general solution. It is left to the reader to consider the product of $A^{(s)}$ and the individual matrices shown in the above display of the general solution. By this time you should have some idea of what to expect. ∎

Example 2.9.5 $m = 3, n = 4.$

$$\left[\begin{array}{cccc|c} 1 & 0 & 0 & 0 & 3 \\ 0 & 1 & 0 & 2 & 4 \\ 0 & 0 & 1 & 3 & -1 \\ 0 & 0 & 0 & 0 & 2 \end{array}\right]$$

The last row indicates that no solution exists. ∎

*We note that in all those cases in which any solution exists, we found an $n - (m - k)$-dimensional infinity of solutions. Thus there were $n - (m - k)$ variables which could be freely assigned values.* If the solution is to be unique, it must be true that $n - (m - k) = 0$ or that $n = m - k$. (What happens if $n < m - k$?)

At this point we now have a computational algorithm for solving any system of linear equations for which a solution exists (called a **consistent system**), and for discovering any system which has no solution (**inconsistent system**).

The reader with some experience in the use of digital computers should appreciate the fact that the Gauss-Jordan reduction procedure is rather easily programmed, and that large-scale systems of equations may be solved in this manner in an amazingly short time. As an exercise, construct a flow chart for the reduction process (or write a full program) including the input-output instructions (assuming that everything can be stored in fast-access memory).

A project of additional interest for the reader with access to a computing center would be to investigate the various "canned" subroutines available for solution of systems of equations or for computing the inverse of a square matrix. If computer time is easily available, you might invert a 10 × 10 matrix on the machine and then attempt to do so with pencil and paper. Hurrah for the computer!

We will pause now to allow you to gain some practical experience in actually solving systems of linear equations by using the Gauss-Jordan reduction process. The exercises have been chosen so that they exhibit all of the idiosyncrasies that may arise.

In the next chapter we will introduce some new terminology pertaining to certain sets of matrices or vectors called *vector spaces*. We shall use it to formulate some of the standard results in the study of linear equations to which we have alluded from time to time in this chapter. If the reader works carefully through the following exercises, these results will appear natural and easy to understand.

## 2.9 Exercises

1. Use Gauss-Jordan reduction to solve the matrix equation $AX = B$. Before you start each problem, determine the minimum number of free variables that must be present if any solution exists.

a. $A = \begin{bmatrix} 2 & 0 & 1 & 3 & -2 \\ -1 & 3 & 4 & 2 & -1 \\ 5 & -3 & -1 & 7 & -5 \end{bmatrix}, B = \begin{bmatrix} 0 \\ -2 \\ 0 \end{bmatrix}.$

b. $A = \begin{bmatrix} 0 & 1 & 2 & 0 & 4 \\ -2 & 0 & 0 & 1 & 1 \end{bmatrix}, B = \begin{bmatrix} 6 \\ 1 \end{bmatrix}.$

c. $A = \begin{bmatrix} 1 & 1 & 1 & 1 \\ -1 & -1 & -1 & 1 \\ -1 & -1 & 1 & 1 \\ -1 & 1 & 1 & 1 \end{bmatrix}, B = \begin{bmatrix} 8 \\ -4 \\ 0 \\ 6 \end{bmatrix}.$

d. $A = \begin{bmatrix} 1 & -1 & 2 & -2 \\ 3 & -3 & -4 & 5 \end{bmatrix}, B = \begin{bmatrix} 3 \\ 7 \end{bmatrix}.$

e. $A = \begin{bmatrix} 1 & 2 & 0 \\ -1 & 2 & 3 \\ 3 & 6 & 3 \\ 0 & 1 & 1 \end{bmatrix}, B = \begin{bmatrix} 3 \\ -2 \\ 6 \\ 0 \end{bmatrix}.$

f. $A = \begin{bmatrix} 1 & -1 & 1 \\ 0 & 1 & 1 \\ -2 & 3 & -1 \end{bmatrix}, B = \begin{bmatrix} -1 \\ -2 \\ 1 \end{bmatrix}.$

2. What occurs in the Gauss-Jordan reduction process which causes "juggling" the order of the rows as we proceed from one of the equivalent augmented matrices to the next?

3. Let $A = \begin{bmatrix} 1 & 2 \\ 3 & 4 \end{bmatrix}$ and suppose that we form a new matrix $C$ from $A$ by letting $R_2 \leftarrow -3R_1 + R_2$, so that $C = \begin{bmatrix} 1 & 2 \\ 0 & -2 \end{bmatrix}$. We can perform this operation by using matrix multiplication appropriately. For example, if we let $E_1 = \begin{bmatrix} 1 & 0 \\ -3 & 1 \end{bmatrix}$ and compute $E_1A = \begin{bmatrix} 1 & 0 \\ -3 & 1 \end{bmatrix}\begin{bmatrix} 1 & 2 \\ 3 & 4 \end{bmatrix} = \begin{bmatrix} 1 & 2 \\ 0 & -2 \end{bmatrix}$, we have $C$. How did we form $E_1$? Very simply, we performed upon the identity matrix $I_2$ the same operation we performed upon $A$ to yield $C$. Matrices like $E_1$ are called **elementary**

matrices. Now consider the matrix equation $AX = B$, where $A = \begin{bmatrix} 3 & 4 \\ 5 & 7 \end{bmatrix}$, $B = \begin{bmatrix} -2 \\ -4 \end{bmatrix}$.

a. Use Gauss-Jordan reduction to solve the problem, and keep an account of the elementary matrices $E_1, E_2, \ldots, E_s$ that would perform steps $1, 2, \ldots, s$.

c. Find the product $(E_s E_{s-1} \cdots E_2 E_1)A$.

d. From the results of c. what is the matrix $E_s E_{s-1} \cdots E_2 E_1$ in relation to $A$?

4. Find $A^{-1}$ by finding the product $E_s E_{s-1} \cdots E_2 E_1$, where the $E_i$ are the elementary matrices that will reduce $A$ to a row-reduced echelon matrix $I$ if

a. $A = \begin{bmatrix} 2 & 7 \\ 1 & 4 \end{bmatrix}$ b. $A = \begin{bmatrix} -1 & 0 & 1 \\ 2 & 1 & 0 \\ 0 & 2 & 3 \end{bmatrix}$.

5. How do we determine the number of free variables that may be arbitrarily assigned? To which variables do we choose to assign values freely?

6. What do we mean when we say that a variable has been "skipped" as the Gauss-Jordan reduction algorithm proceeded? Give an example of an augmented matrix which has a variable that has been skipped.

7. Let $A_1 = \begin{bmatrix} 1 \\ 3 \end{bmatrix}$, $A_2 = \begin{bmatrix} 1 \\ 4 \end{bmatrix}$, $B = \begin{bmatrix} -1 \\ -5 \end{bmatrix}$. Find numbers $x_1, x_2$, if they exist, such that $x_1A_1 + x_2A_2 = B$. If such $x$'s exist, then we say that $B$ is a **linear combination** of $A_1$ and $A_2$.

8. Show that $\begin{bmatrix} -1 \\ 2 \\ 3 \end{bmatrix}$ is a linear combination of $\begin{bmatrix} 1 \\ 0 \\ 0 \end{bmatrix}$, $\begin{bmatrix} 0 \\ 1 \\ 0 \end{bmatrix}$, and $\begin{bmatrix} 0 \\ 0 \\ 1 \end{bmatrix}$. Is every column vector with three elements such a linear combination?

9. Let $A_1 = \begin{bmatrix} 1 \\ 0 \\ 1 \end{bmatrix}$, $A_2 = \begin{bmatrix} 1 \\ 1 \\ 0 \end{bmatrix}$, $A_3 = \begin{bmatrix} 0 \\ 1 \\ 1 \end{bmatrix}$. Decide whether $B = \begin{bmatrix} 0 \\ 0 \\ 2 \end{bmatrix}$ is a linear combination of the $A_i$.

### New Terms

row-reduced echelon form, 80
consistent system, 83
inconsistent system, 83
elementary matrix, 84
linear combination, 85

# 3 Economic Planning by Input-Output Analysis

## 3.1 Linear Equations in Input-Output Analysis

The main purpose of this chapter is to introduce some of the elementary aspects of input-output analysis as they were developed by the economist Wassily Leontief in the 1930's. The significance of this work has been emphasized by Leontief's selection to receive the Nobel Prize for Economics in 1974. The chapter will also give the reader some insight into how matrix operations can be used effectively to model interactions of complex real-world systems.

As we shall see, the mathematical model uses the elements of a square **input-output matrix** to portray the interdependencies of the various industries within an economy. Many nations, including the United States, have constructed such matrices to describe various aspects of their own economies.* The size of such national matrices is dependent upon the level of aggregation of the various industries into so-called "sectors". One could, for example, describe a "commercial transportation sector" including the commercial aircraft industry, the commercial bus industry, AMTRAC, etc. In such a case where several industries have been "aggregated" into one composite industry, the size of the national matrix is greatly reduced—at the expense of considerable detail about the individual industries. Theil, *et al.* list the 35 sectors that were utilized by the Dutch to create a $35 \times 35$ matrix for their economy.† Much larger input-output matrices exist, however, and are regularly used for economic forecasting and planning.

This extremely useful model of an economy is designed to assist planners who have the responsibility to predict future demands upon the capacities

---

* For a discussion of the assumptions and some of the problems associated with input-output modeling, see any of the excellent references cited in the Suggested Reading.

† H. Theil, J. C. G. Boot, and T. Kloek, *Operations Research and Quantitative Economics* (New York: McGraw-Hill, 1965), pp. 56–59.

of the various industries. In some countries, of course, it is additionally true that decisions to increase or to decrease capacity can be made by the government who can then unilaterally implement its decision with sufficient lead-time to guarantee success. In this country, however, implementation is disseminated among many corporate Boards of Directors, who act independently upon demand predictions of such federal agencies as the Department of Commerce and the Federal Energy Agency.

For simplicity we shall describe the situation for a greatly reduced economy consisting of 4 industries—auto, coal, electric power, and steel. In such a case, the input-output matrix $A$, as shown in Table 3.1.1, has 4 rows and 4 columns, one for each of the industries. As we shall see, the precise values of the elements of $A$ (here known as **input-output coefficients**) are functions of the unique corporate production processes under which the several industries operate; $a_{ij}$ generally reflects the *amount* of product of industry $i$ that is consumed in the process of producing one unit of industry $j$. Once these processes have been established and consumer demands for the individual industry products become known, one can begin to ask about their hidden demand implications for the other producers. In Figure 3.1.1 we have partially portrayed the industry interdependencies to show that the total outputs of the industries may have to be much greater than that required just to satisfy final consumer demand. The steel industry, for example, must supply not only its own final consumer demand, but also the steel requirements of the production processes of the auto, coal and power industries, as indicated by the arrows. It is the precise production implications of the inter-industry and intra-industry transfers represented in the figure that the matrix $A$ portrays. The other notation will be explained shortly.

One way to begin to establish the numerical values of the $a_{ij}$ would be to consider the various industries in terms of their product units (namely, numbers of cars and trucks, tons of coal, kilowatt-hours of electrical power, and tons of steel) and to formulate questions about the relationships between these units, such as,

How many kilowatt-hours of power are required to produce a ton of coal?

How many tons of coal are required to produce a ton of steel?

**Table 3.1.1 Input-Output Coefficients**

| *To industry* / *From industry* | *auto* 1 | *coal* 2 | *electricity* 3 | *steel* 4 |
|---|---|---|---|---|
| *auto 1* | $a_{11}$ | $a_{12}$ | $a_{13}$ | $a_{14}$ |
| *coal 2* | $a_{21}$ | $a_{22}$ | $a_{23}$ | $a_{24}$ |
| *electricity 3* | $a_{31}$ | $a_{32}$ | $a_{33}$ | $a_{34}$ |
| *steel 4* | $a_{41}$ | $a_{42}$ | $a_{43}$ | $a_{44}$ |

In the former case we would have established the value of $a_{32}$. What matrix element would we have found in the latter case?

**Figure 3.1.1** **Industry Interdependencies: Industry i Input into the Production Process of Industry j**

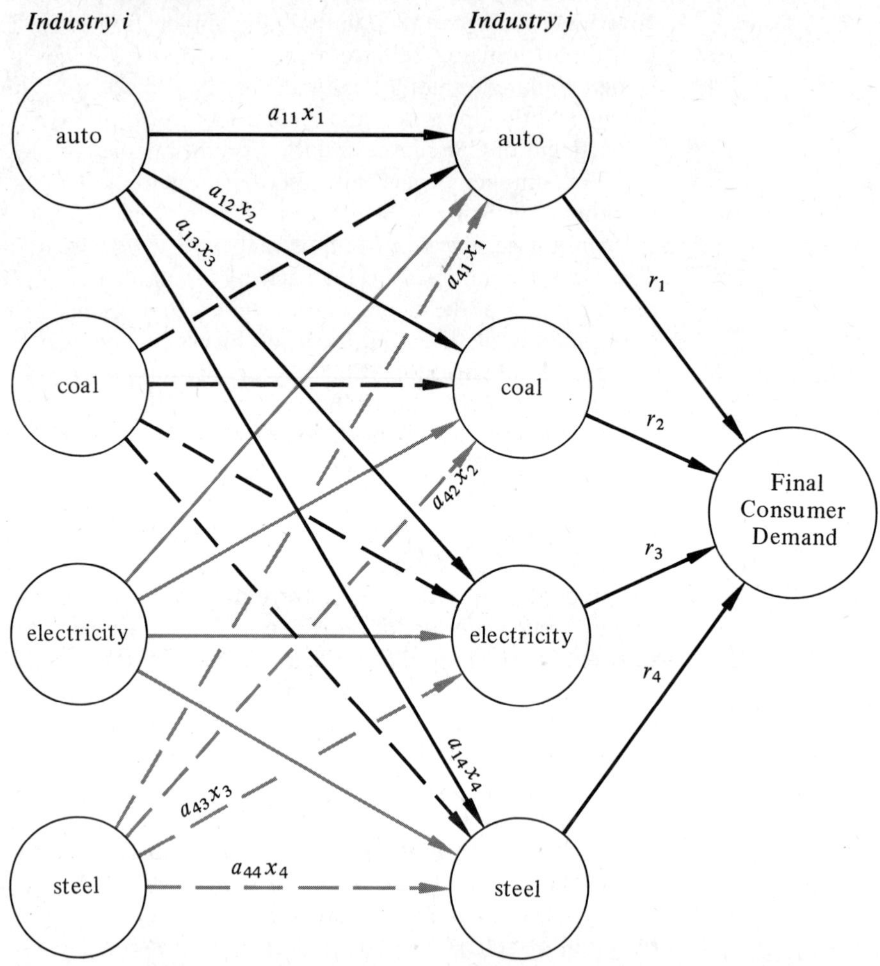


For simplicity, however, and because historical data are usually available in dollar values spent by the various industries for products of the other industries, we shall proceed by using dollar equivalents of the several product units. In this case we are interested in finding answers to a different set of questions—ones like

> What is the value of the electricity that is required to produce a dollar's worth of coal?
>
> What is the value of the coal required to produce a dollar's worth of steel?

The former value would be called $a_{32}$ in the Leontief matrix $A$ for our simplified economy.

Let us suppose for the present that we have been provided with an input-output matrix $A$ in which $a_{ij}$ represents the dollar value of production from industry $i$ that is required to produce a dollar's worth of industry $j$.*

* We will see later how the matrix $A$ can be obtained from past sales data.

If $x_j$ represents the total production dollar value of industry $j$ for the upcoming period (an amount to be determined) then $x_j$ must be sufficient not only to supply consumer demand for product $j$, denoted by $r_j$, but also to supply product $j$ to each of the four production processes. For example, consider the total value of electrical power production $x_3$:

$$\begin{aligned} x_3 = a_{31}&\left(\frac{\text{\$ worth of electricity}}{\text{\$ auto production}}\right) x_1 \text{ (\$ worth of auto production)} \\ + a_{32}&\left(\frac{\text{\$ worth of electricity}}{\text{\$ coal production}}\right) x_2 \text{ (\$ worth of coal production)} \\ + a_{33}&\left(\frac{\text{\$ worth of electricity}}{\text{\$ electricity production}}\right) x_3 \text{ (\$ worth of electricity production)} \\ + a_{34}&\left(\frac{\text{\$ worth of electricity}}{\text{\$ steel production}}\right) x_4 \text{ (\$ worth of steel production)} \\ + r_3& \text{ (\$ worth of electricity).} \end{aligned}$$

Similarly,

$$x_2 = a_{21}x_1 + a_{22}x_2 + a_{23}x_3 + a_{24}x_4 + r_2, \qquad \text{(coal)}$$

where the total value of coal production, $x_2$, is expressed as a sum of the dollar amounts needed to supply the coal demands of the auto industry, $a_{21}x_1$; the coal industry, $a_{22}x_2$; the electric power industry, $a_{23}x_3$; the steel industry, $a_{24}x_4$; and the final consumer demand for coal, namely $r_2$.

In its entirety, then, the complete set of equations would be

$$\begin{aligned} x_1 &= a_{11}x_1 + a_{12}x_2 + a_{13}x_3 + a_{14}x_4 + r_1 & &\text{(auto)} \\ x_2 &= a_{21}x_1 + a_{22}x_2 + a_{23}x_3 + a_{24}x_4 + r_2 & &\text{(coal)} \\ x_3 &= a_{31}x_1 + a_{32}x_2 + a_{33}x_3 + a_{34}x_4 + r_3 & &\text{(electrical power)} \\ x_4 &= a_{41}x_1 + a_{42}x_2 + a_{43}x_3 + a_{44}x_4 + r_4 & &\text{(steel).} \end{aligned}$$

In matrix format, we let $X = [x_1, x_2, x_3, x_4]^T$ denote the **total output matrix** and $R = [r_1, r_2, r_3, r_4]^T$ the **final consumer demand matrix**. Thus we have $X = AX + R$. $AX = X - R$ represents the amounts of the various products consumed in the production processes.

Since $X = IX$, we write the above in the more useful forms

$$\begin{aligned} IX &= AX + R \\ IX - AX &= R \\ (I - A)X &= R. \end{aligned}$$

Whenever $(I - A)^{-1}$ exists* we may solve for $X$ uniquely as

$$X = (I - A)^{-1}R.$$

Since it may well be the case that we would want to find $X$ vectors for many different $R$ vectors, it might be beneficial to solve for $(I - A)^{-1}$ rather than use Gauss-Jordan reduction each time. Once the columns $C_1$, $C_2$, $C_3$, $C_4$

* It can be shown that such an $X$ will be non-negative for appropriate $A$ matrices. See J. C. G. Boot, *Mathematical Readings in Economics and Management Science* (Englewood Cliffs, N.J.: Prentice-Hall, 1967), pp. 169–170.

of $(I - A)^{-1}$ are known then we may express $X$ as a linear combination of them. Thus

$$X = r_1C_1 + r_2C_2 + r_3C_3 + r_4C_4.$$

Example 3.1.1 Consider a 3-product economy with input-output matrix $A$ given as

$$A = \begin{bmatrix} 0.4 & 0.1 & 0.1 \\ 0.2 & 0.3 & 0.1 \\ 0.1 & 0.2 & 0.3 \end{bmatrix}.$$

Then

$$I - A = \begin{bmatrix} 0.6 & -0.1 & -0.1 \\ -0.2 & 0.7 & -0.1 \\ -0.1 & -0.2 & 0.7 \end{bmatrix}$$

and

$$(I - A)^{-1} = 1/384 \begin{bmatrix} 705 & 135 & 120 \\ 225 & 615 & 120 \\ 165 & 195 & 600 \end{bmatrix}.$$

Thus, for any $R = [r_1, r_2, r_3]^T$,

$$X = (I - A)^{-1}R = r_1 \begin{bmatrix} 705/384 \\ 225/384 \\ 165/384 \end{bmatrix} + r_2 \begin{bmatrix} 135/384 \\ 615/384 \\ 195/384 \end{bmatrix} + r_3 \begin{bmatrix} 120/384 \\ 120/384 \\ 600/384 \end{bmatrix}.$$

Suppose that we have a consumer demand matrix $R$ such that

$$R = \begin{bmatrix} 3840 \\ 3840 \\ 7680 \end{bmatrix}.$$

Then the required production matrix $X$ is given by

$$X = \begin{bmatrix} 7050 \\ 2250 \\ 1650 \end{bmatrix} + \begin{bmatrix} 1350 \\ 6150 \\ 1950 \end{bmatrix} + \begin{bmatrix} 2400 \\ 2400 \\ 12000 \end{bmatrix} = \begin{bmatrix} 10800 \\ 10800 \\ 15600 \end{bmatrix}.$$

Thus we must produce \$10,800 worth of Products 1 and 2 and \$15,600 worth of Product 3. ∎

## 3.1 Exercises

1. Consider a 2-product economy with industries $I_1$ and $I_2$ producing the two products. Suppose that to make a dollar's worth of $I_1$'s product, 20 cents' worth of that product and 10 cents' worth of $I_2$'s product are required. Similarly, to make a dollar's worth of $I_2$'s product, 30 cents' worth of $I_1$'s product and 40 cents' worth of $I_2$'s product are consumed.
   a. Construct the input-output matrix $A$.
   b. How much of the individual products would be consumed in the various production processes if a *total* of \$50 worth of $I_1$'s product and \$100 worth of $I_2$'s product were produced? (First use a direct computational process and then compute the appropriate matrix product.)

c. How much of each product would be left for consumer use?

d. Suppose that the industries' economists decided to satisfy a final consumer demand matrix $R = [20, 110]^T$. What total output matrix $X$ would be required?

e. Consider $(I - A)^{-1}$ as it was used in d. above. Find the matrix $D$ such that $D + I = (I - A)^{-1}$. Show that $DR = AX$.

2. Verify that if $A$ is a square matrix and $A = bB$, where $b$ is a scalar, then $A^{-1} = (1/b)B^{-1}$. (This result can be used to simplify the calculation of the appropriate matrix in Exercise 3 below.)

3. Let $A = \begin{bmatrix} 0.2 & 0 & 0.2 \\ 0 & 0.1 & 0.3 \\ 0 & 0.2 & 0.3 \end{bmatrix}$

a. Interpreting $A$ as an input-output matrix, what is the meaning of the 0 elements?
b. Use the result of Exercise 2 above, with $b = 1/10$, to find $(I - A)^{-1}$ by Gauss-Jordan reduction.
c. Find the required final output matrix $X$ that would supply a final demand matrix $R^T = [4560, 4560, 4560]$.
d. Verify that $AX = X - R$.

### Suggested Reading

J. C. G. Boot. *Mathematical Readings in Economics and Management Science.* Englewood Cliffs, N.J.: Prentice-Hall, 1967.

S. I. Gass. *Linear Programming.* 3d. ed. New York: McGraw-Hill, 1969.

G. Hadley, and M. C. Kemp. *Finite Mathematics in Business and Economics.* New York: American Elsevier, 1972.

H. Theil, J. C. G. Boot, and T. Kloek. *Operations Research and Quantitative Economics.* New York: McGraw-Hill, 1965.

### New Terms

input-output matrix, 86
input-output coefficients, 87
total output matrix, 89
final consumer demand matrix, 89

## 3.2 Powers of the Matrix *A* and the "Ripple Effect"

Before we proceed to comment further on the general aspects of the Leontief model, let us consider another small example in order to illustrate the ability of powers of the matrix $A$ to reflect the *ripple effects* that characterize the production process. We shall also see that we could eliminate the necessity to invert $I - A$ to find a satisfactory solution matrix in certain cases.

Example 3.2.1 For this purpose a 2-product economy will suffice.

$$A = \begin{bmatrix} 0.2 & 0.1 \\ 0.3 & 0.4 \end{bmatrix}$$

so that $$I - A = \begin{bmatrix} 0.8 & -0.1 \\ -0.3 & 0.6 \end{bmatrix}$$

and $$(I - A)^{-1} = \begin{bmatrix} 6/4.5 & 1/4.5 \\ 3/4.5 & 8/4.5 \end{bmatrix}$$

so that $$X = (I - A)^{-1}R = r_1\begin{bmatrix} 6/4.5 \\ 3/4.5 \end{bmatrix} + r_2\begin{bmatrix} 1/4.5 \\ 8/4.5 \end{bmatrix}.$$

If, for simplicity of calculation, we require $R = \begin{bmatrix} 90 \\ 180 \end{bmatrix}$, we find that $X = \begin{bmatrix} 160 \\ 380 \end{bmatrix}$. Thus, of the 160 units of Product 1 made, 70 were produced for internal use and, of the 380 of Product 2, 380 − 180 = 200 were for internal use. We shall examine the situation to see exactly where the additional products were utilized and also what additional production they imply. This succession of *additional* requirements is the ripple effect mentioned above.

To produce \$1 of Product 1 requires an additional \$0.2 of Product 1 and \$0.3 of Product 2. Thus \$90 of Product 1 requires \$18 of Product 1 and \$27 of Product 2, in addition to the original requirement. But the extra \$18 of Product 1 also requires 18(\$0.2) or \$3.60 additional Product 1 and 18(\$0.3) or \$5.40 additional Product 2. To make matters worse, the additional \$27 worth of Product 2 requires 27(\$0.1) or \$2.70 additional Product 1 and 27(\$0.4) or \$10.80 of Product 2; and each of these has further implications—and so it goes.

Perhaps the situation can be more succinctly illustrated by a branching process as in Figure 3.2.1. Here we classify the implied requirements as *2nd order* requirements, *3rd order* requirements, etc., calling the original $R$ prescription the *1st order* requirement. Thus we have 2nd order requirements for $P_1$ of 18 from the 1st order $P_1$ requirement and for $P_1$ of 18 from the 1st order $P_2$ requirement, for a total 2nd order requirement for $P_1$ of 36. Similarly, for $P_2$, we have a 2nd order requirement of 27 + 72 = 99.

Noting that the matrix $R$ provides the 1st order requirements directly, let us compute $AR$:

$$AR = \begin{bmatrix} 0.2 & 0.1 \\ 0.3 & 0.4 \end{bmatrix}\begin{bmatrix} 90 \\ 180 \end{bmatrix} = \begin{bmatrix} 18 + 18 \\ 27 + 72 \end{bmatrix} = \begin{bmatrix} 36 \\ 99 \end{bmatrix}.$$

Conveniently, $AR$ contains the 2nd order requirements.

Referring to Figure 3.2.1, let us find the 3rd order requirement for $P_1$. Thus we find a 3rd order requirement of 3.6 + 2.7 = 6.3 from the original $P_1$ need, and a requirement for 3.6 + 7.2 = 10.8 from the original $P_2$, for a total 3rd order requirement for $P_1$ of 17.1. For the 3rd order $P_2$ requirement, we find 5.4 + 10.8 = 16.2 from $P_1$ and 5.4 + 28.8 = 34.2 from $P_2$. Thus we have a total of 50.4 for $P_2$ in 3rd order requirements.

**Figure 3.2.1 Production Requirements of a 2-Product Economy**

| 1st Order | 2nd Order | 3rd Order | 4th Order |
|---|---|---|---|
| $R = \begin{bmatrix} 90 \\ 180 \end{bmatrix}$, | $AR = \begin{bmatrix} 18+18 \\ 27+72 \end{bmatrix}$, | $A^2R = \begin{bmatrix} (3.6+2.7)+(3.6+7.2) \\ (5.4+10.8)+(5.4+28.8) \end{bmatrix} = \begin{bmatrix} 17.1 \\ 50.4 \end{bmatrix}$, | $A^3R = \begin{bmatrix} 8.46 \\ 25.29 \end{bmatrix}$ |

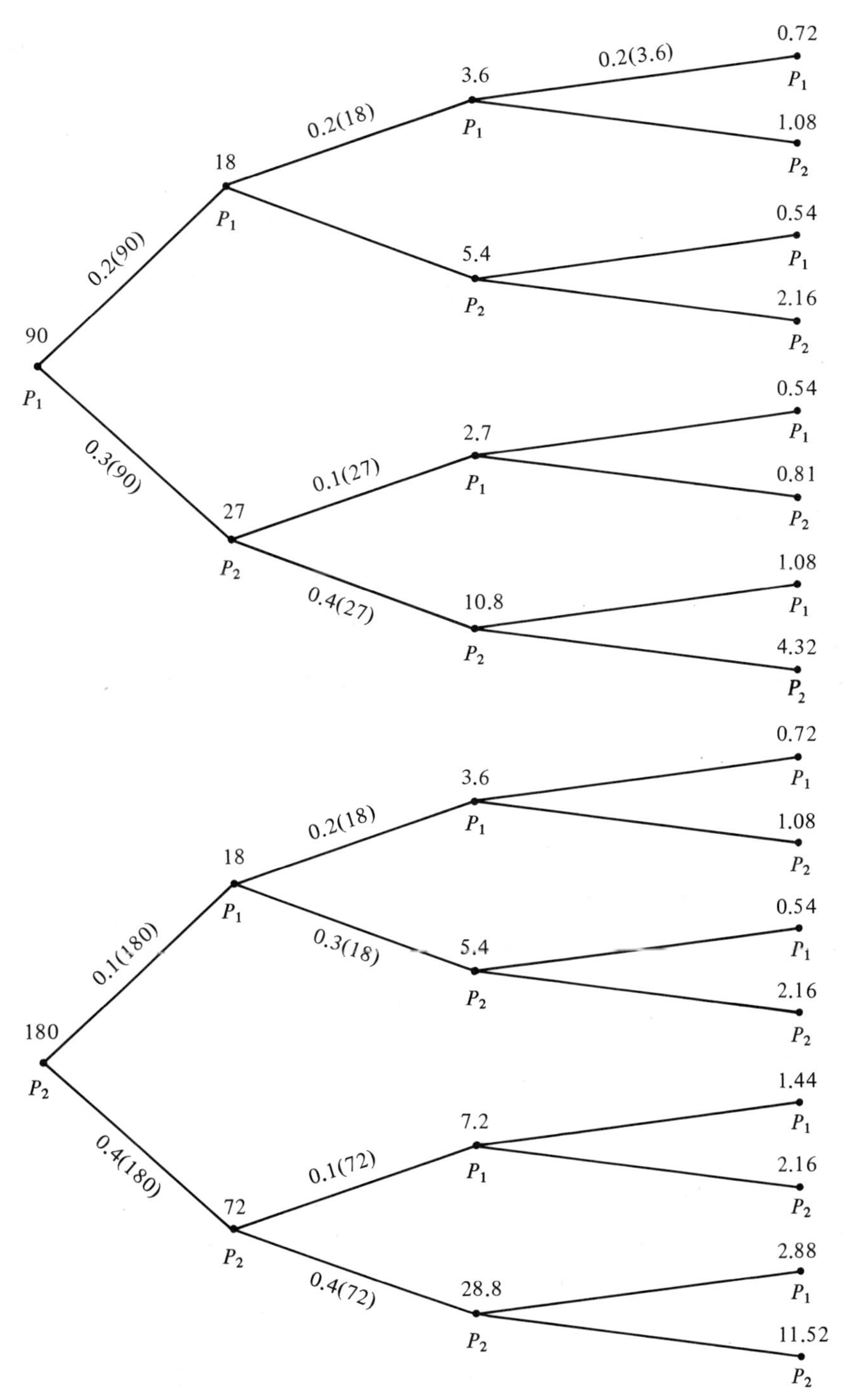

Computing

$$A^2R = \begin{bmatrix} 0.07 & 0.06 \\ 0.18 & 0.19 \end{bmatrix} \begin{bmatrix} 90 \\ 180 \end{bmatrix} = \begin{bmatrix} 6.3 + 10.8 \\ 16.2 + 34.2 \end{bmatrix} = \begin{bmatrix} 17.1 \\ 50.4 \end{bmatrix}$$

we find that $A^2R$ provides the solution for the 3rd order requirements.

As you may verify for $n = 4$, the $n$th order requirements are found by computing $A^{n-1}R$, and the total requirements may be found by computing

$$X = R + AR + A^2R + A^3R + \cdots + A^nR + \cdots$$

where, by this notation, we mean that the process is theoretically repeated *ad infinitum*. In practice, however, we will find that for relatively small values of $n$, $A^nR$ is close to the 0 matrix, so that, as a practical matter, further multiplication would be useless. ∎

Let us take advantage of what we have seen and make a general observation. We have seen that

$$X = (I - A)^{-1}R$$

and that

$$X = R + AR + A^2R + \cdots + A^nR + \cdots$$
$$= (I + A + A^2 + \cdots + A^n + \cdots)R.$$

We should not be surprised then if it turned out that

$$(I - A)^{-1} = I + A + A^2 + \cdots + A^n + \cdots.$$

This is indeed the case for economically reasonable input-output matrices $A$ in which the column sums are less than 1*.

We could have perhaps proceeded to establish this relationship in a more straightforward manner by noting that for any positive integer $n$,

$$(I - A)(I + A + A^2 + \cdots + A^n) = I - A^{n+1}$$

as can be easily seen by expansion of the product on the left. If, as $n$ increases, each of the elements of $A^{n+1}$ gets closer and closer to zero, then

$$(I - A)(I + A + A^2 + \cdots + A^n)$$

is approximately $I$ so that $(I - A)^{-1}$ is approximated by

$$I + A + A^2 + \cdots + A^n.$$

In such a case an approximate solution for $X$ may be obtained by letting

$$X = (I + A + A^2 + \cdots + A^n)R$$
$$= R + AR + A^2R + \cdots + A^nR.$$

In so doing, as we have seen, $X$ is built up as a sum of 1st order, 2nd order, and higher order terms, until the amounts added at some stage become negligible.

Let us now turn to the problem of establishing a matrix $A$ for a 4-industry simplified economy. We shall assume that the economy remains

---

* This intuitively plausible result actually requires a proof more sophisticated than the level of this book.

stable over the period with which we are concerned. This will certainly be the case if the production process requirements and the product prices remain the same.

We shall refer to Table 3.2.1 which contains a variable representation of the dollar sales data for each of our industries for the previous period. In the table, $s_{ij}$ denotes sales by industry $i$ to industry $j$, with $j = 5$ denoting the final consumer demand.* $S_i$ represents the total sales by industry $i$, and $S_i = s_{i1} + s_{i2} + \cdots + s_{i5}$.

To find $a_{ij}$, the dollar value of industry $i$'s product required to make one dollar's worth of product $j$, we divide $s_{ij}$ by $S_j = s_{j1} + s_{j2} + \cdots + s_{j5}$. Thus $a_{11} = s_{11}/S_1$, $a_{12} = s_{12}/S_2$, etc.

**Table 3.2.1 Previous Period Sales (in Dollars) for a 4-Industry Economy.**

| *From industry* \ *To industry* | *1* | *2* | *3* | *4* | *Final Consumer Demand* | *Total Sales* |
|---|---|---|---|---|---|---|
| *1* | $s_{11}$ | $s_{12}$ | $s_{13}$ | $s_{14}$ | $s_{15}$ | $S_1$ |
| *2* | $s_{21}$ | $s_{22}$ | $s_{23}$ | $s_{24}$ | $s_{25}$ | $S_2$ |
| *3* | $s_{31}$ | $s_{32}$ | $s_{33}$ | $s_{34}$ | $s_{35}$ | $S_3$ |
| *4* | $s_{41}$ | $s_{42}$ | $s_{43}$ | $s_{44}$ | $s_{45}$ | $S_4$ |

Example 3.2.2 If, for the last year reported, sales (in millions of dollars) were as indicated in Table 3.2.2 for the 4-industry economy, we could create the appropriate matrix $A$.

**Table 3.2.2 Previous Period Sales (in Millions of Dollars) for a 4-Industry Economy.**

| *From* \ *To* | *1* | *2* | *3* | *4* | *Final Consumer Demand* | *Total Sales* |
|---|---|---|---|---|---|---|
| *1* | 2 | 1 | 3 | 2 | 12 | 20 |
| *2* | 1 | 1 | 8 | 4 | 1 | 15 |
| *3* | 5 | 1 | 2 | 4 | 13 | 25 |
| *4* | 8 | 2 | 2 | 3 | 5 | 20 |

Using the formulas above, we obtain

* Clearly $s_{ij} \geqslant 0$, and $s_{ij} > 0$ if industry $j$ requires industry $i$'s product in its total production process.

$$A = \begin{bmatrix} 2/20 & 1/15 & 3/25 & 2/20 \\ 1/20 & 1/15 & 8/25 & 4/20 \\ 5/20 & 1/15 & 2/25 & 4/20 \\ 8/20 & 2/15 & 2/25 & 3/20 \end{bmatrix}.$$ ∎

In conclusion, we should like to point out that the Leontief model as it is developed here is a static model in that it considers only one period. (See Hadley for a consideration of a "dynamic" multi-period case.)*

## 3.2 Exercises

1. Consider the matrix $A$ of Exercise 3.1.1.
   a. Approximate $(I - A)^{-1}$ by computing $A^2$, $A^3$, $A^4$, and $A^5$, using the appropriate formula. Compare this result with that you achieved in Exercise 3.1.1d.
   b. Compute $AR$, $A^2R$, $A^3R$, $A^4R$, and $A^5R$, for $R^T = [20, 110]$. Use the appropriate formula to obtain an approximation for $X$.

2. Construct a tree similar to that of Figure 3.2.1 by using a common-sense approach to the requirements of each order. Verify that $AR$, $A^2R$, and $A^3R$ correctly reflect the 2nd order, 3rd order, and 4th order requirements.

3. Consider Exercise 3.1.3.
   a. Approximate $(I - A)^{-1}$ by computing $A^2$, $A^3$, and $A^4$ and using the appropriate formula. Compare this result with that of Exercise 3.1b.
   b. Compute $AR$, $A^2R$, $A^3R$, and $A^4R$, for $R^T = [4560, 4560, 4560]$. Use the results to approximate $X$.

4. Consider the $I - A$ matrices in Example 3.2.1 and Exercise 3.1.1.
   a. How are they related?
   b. Compare their inverses. How are they related?
   c. What possible general theorem can you conjecture?

5. Consider the matrices

$$A = \begin{bmatrix} 0.4 & 0.1 & 0.1 \\ 0.2 & 0.3 & 0.1 \\ 0.1 & 0.2 & 0.3 \end{bmatrix} \quad \text{and} \quad R = \begin{bmatrix} 3840 \\ 3840 \\ 7680 \end{bmatrix}$$

   of Example 3.1.1.
   a. Use the fact that $A^2R = A(AR)$, $A^3R = A(A^2R)$, etc., to compute $A^6R$. Thus, it is not necessary to compute explicitly the various powers of $A$.
   b. Approximate $X$ by using the results of a.

---

* G. Hadley, *Linear Programming* (Reading, Mass.: Addison-Wesley, 1962), pp. 504ff.

6. Consider Example 3.1.1. Suppose that a new demand vector

$$R' = R + \begin{bmatrix} 0 \\ 3840 \\ 0 \end{bmatrix} = R + 3840 \begin{bmatrix} 0 \\ 1 \\ 0 \end{bmatrix}$$

is prescribed. Show that

$$X' = X + 3840C_2,$$

where $C_2$ is column 2 of $(I - A)^{-1}$. Can you conjecture a general result?

# 4 Vector Spaces and Systems of Linear Equations

## 4.1 Linear Combinations and Vector Spaces

Many of the standard results concerning linear equations may be succinctly phrased in terminology arising from a study of so-called "vector spaces." In order to do this we shall introduce the elementary ideas involved and use them in Sections 4.5 and 4.6 in a discussion regarding our earlier questions about the existence and uniqueness of solutions of linear equations. We shall then relate our newly-defined algebraic vectors to the geometric *arrow* vectors that you may be more familiar with from past experience.

We have previously noted that any solution $X$ of the matrix equation $AX = B$ may be decomposed into $X_P$ and $X_H$ in such a manner that $AX_P = B$ and $AX_H = 0$. You may have noticed that $X_H$ is always of a particular form. In Example 2.9.3, where

$$X = \begin{bmatrix} 2 \\ 4 \\ -3 \\ 0 \\ 0 \end{bmatrix} + a\begin{bmatrix} -2 \\ 1 \\ -4 \\ 1 \\ 0 \end{bmatrix} + b\begin{bmatrix} 5 \\ -2 \\ -3 \\ 0 \\ 1 \end{bmatrix},$$

we have

$$X_P = \begin{bmatrix} 2 \\ 4 \\ -3 \\ 0 \\ 0 \end{bmatrix} \quad \text{and} \quad X_H = a\begin{bmatrix} -2 \\ 1 \\ -4 \\ 1 \\ 0 \end{bmatrix} + b\begin{bmatrix} 5 \\ -2 \\ -3 \\ 0 \\ 1 \end{bmatrix}.$$

In Example 2.9.4 in which

$$X = \begin{bmatrix} -1 \\ 0 \\ 2 \\ 0 \\ 0 \\ 0 \end{bmatrix} + a \begin{bmatrix} -2 \\ 1 \\ 0 \\ 0 \\ 0 \\ 0 \end{bmatrix} + b \begin{bmatrix} -1 \\ 0 \\ 1 \\ -2 \\ 1 \\ 0 \end{bmatrix} + c \begin{bmatrix} 2 \\ 0 \\ -3 \\ 4 \\ 0 \\ 1 \end{bmatrix},$$

we have

$$X_P = \begin{bmatrix} -1 \\ 0 \\ 2 \\ 0 \\ 0 \\ 0 \end{bmatrix} \quad \text{and} \quad X_H = a \begin{bmatrix} -2 \\ 1 \\ 0 \\ 0 \\ 0 \\ 0 \end{bmatrix} + b \begin{bmatrix} -1 \\ 0 \\ 1 \\ -2 \\ 1 \\ 0 \end{bmatrix} + c \begin{bmatrix} 2 \\ 0 \\ -3 \\ 4 \\ 0 \\ 1 \end{bmatrix}.$$

For a particular problem, we now consider the set of matrices $H = \{X_H\}$, where the constants $a$, $b$, $c$, etc. take on all real values. To begin with, we shall call any $n \times 1$ matrix a **vector of order n**. (Although we will not always say so explicitly, whenever we state a definition, theorem, or property that applies to a set of $n \times 1$ column matrices, we will suppose that an analogous statement holds for $1 \times n$ row matrices.) Any single matrix constructed above by particular choices of the scalars involved in $X_H$ is called a linear combination of the defining matrices involved. To make our definition specific, we will say that if $A_1, A_2, \ldots, A_p$ are each $n \times 1$ column matrices and $x_1, x_2, \ldots, x_p$ are any scalars, then the column matrix

$$B = x_1 A_1 + x_2 A_2 + \cdots + x_p A_p$$

is a **linear combination** of the $A_i$.

For $a = b = 2$ in Example 2.9.3,

$$X_H = \begin{bmatrix} 6 \\ -2 \\ -14 \\ 2 \\ 2 \end{bmatrix} = 2 \begin{bmatrix} -2 \\ 1 \\ -4 \\ 1 \\ 0 \end{bmatrix} + 2 \begin{bmatrix} 5 \\ -2 \\ -3 \\ 0 \\ 1 \end{bmatrix},$$

and we say that $X_H$ is a linear combination of $[-2, 1, -4, 1, 0]^T$ and $[5, -2, -3, 0, 1]^T$. Similarly, in Example 2.9.4, for $a = b = c = 1$, we see that

$$X_H = \begin{bmatrix} -1 \\ 1 \\ -2 \\ 2 \\ 1 \\ 1 \end{bmatrix} = 1 \begin{bmatrix} -2 \\ 1 \\ 0 \\ 0 \\ 0 \\ 0 \end{bmatrix} + 1 \begin{bmatrix} -1 \\ 0 \\ 1 \\ -2 \\ 1 \\ 0 \end{bmatrix} + 1 \begin{bmatrix} 2 \\ 0 \\ -3 \\ 4 \\ 0 \\ 1 \end{bmatrix}$$

is a linear combination of the vectors $[-2, 1, 0, 0, 0, 0]^T$, $[-1, 0, 1, -2, 1, 0]^T$ and $[2, 0, -3, 4, 0, 1]^T$.

The problem of finding scalars $x_i$ which will express $B$ as a linear combination of the $A_i$ is just the general linear equation problem which we have examined in Chapter 2. On the other hand, if we are given a set of vectors $A_i$ and a set of scalars $x_i$, then it is a simple matter to find $B$.

If $S$ is a set of $n \times 1$ column matrices, then we shall call $S$ a **vector space** provided that

1. $S$ is closed under the operation of addition; that is, if $Y_1$ and $Y_2$ are in $S$, then $Y_1 + Y_2$ is in $S$.
2. $S$ is closed under scalar multiplication; that is, if $Y$ is in $S$ and $c$ is any scalar, then the vector $cY$ is also in $S$.

*Under this definition the set of all $n \times 1$ column matrices is a vector space.** There are many subsets of $\mathcal{M}_{n,1}$ which are also vector spaces. In a moment we shall show that if $S$ is any subset of $\mathcal{M}_{n,1}$ which consists of all possible linear combinations of a finite set of vectors in $\mathcal{M}_{n,1}$ then $S$ is a vector space.

To illustrate the method of proof numerically, consider the set

$$S = \left\{ Y \mid Y = x_1A_1 + x_2A_2,\ A_1 = \begin{bmatrix} 1 \\ 2 \\ 3 \end{bmatrix},\ A_2 = \begin{bmatrix} -1 \\ 3 \\ -1 \end{bmatrix} \right\}.$$

Thus $S$ is the set of all possible linear combinations of $[1, 2, 3]^T$ and $[-1, 3, -1]^T$. For $x_1 = 2$, $x_2 = 3$, we have

$$Y_1 = 2A_1 + 3A_2 = \begin{bmatrix} -1 \\ 13 \\ 3 \end{bmatrix}$$

which is an element of $S$; and for $x_1 = -1$, $x_2 = -2$, we compute that

$$Y_2 = -A_1 - 2A_2 = \begin{bmatrix} 1 \\ -8 \\ -1 \end{bmatrix}$$

is also in $S$. If $S$ is to be closed under addition and scalar multiplication then it must be true, for example, that

$$Y_3 = Y_1 + Y_2 = \begin{bmatrix} 0 \\ 5 \\ 2 \end{bmatrix}$$

and

$$Y_4 = cY_1 = \begin{bmatrix} -1c \\ 13c \\ 3c \end{bmatrix}$$

---

* For the definition of an abstract vector space, see Appendix C. The vector space we defined above is an important special case.

are in $S$. Since

$$Y_3 = \begin{bmatrix} 0 \\ 5 \\ 2 \end{bmatrix} = 1\begin{bmatrix} 1 \\ 2 \\ 3 \end{bmatrix} + 1\begin{bmatrix} -1 \\ 3 \\ -1 \end{bmatrix} = 1A_1 + 1A_2,$$

we see that $Y_3$ is a linear combination of $A_1$ and $A_2$, and is thus an element of $S$. Similarly,

$$Y_4 = \begin{bmatrix} -c \\ 13c \\ 3c \end{bmatrix} = (2c)\begin{bmatrix} 1 \\ 2 \\ 3 \end{bmatrix} + (3c)\begin{bmatrix} -1 \\ 3 \\ -1 \end{bmatrix}$$

is also in $S$.

*We now claim that if $S$ consists of all possible linear combinations of a given finite set of $p$ vectors, then $S$ is a vector space.* Since $H$, as defined above, is such a set, we will then know that $H$ is a vector space. We present the proof for $p = 2$, and show that $S$ is closed under addition and scalar multiplication.

**Proof** Let $S = \{Y \mid Y = x_1A_1 + x_2A_2\}$. For addition, suppose that $Y_1 = a_1A_1 + a_2A_2$ and $Y_2 = b_1A_1 + b_2A_2$ are in $S$. Then

$$\begin{aligned} Y_3 = Y_1 + Y_2 &= (a_1A_1 + a_2A_2) + (b_1A_1 + b_2A_2) \\ &= a_1A_1 + (a_2A_2 + b_1A_1) + b_2A_2 && \text{(Why?)} \\ &= a_1A_1 + (b_1A_1 + a_2A_2) + b_2A_2 && \text{(Why?)} \\ &= (a_1A_1 + b_1A_1) + (a_2A_2 + b_2A_2) && \text{(Why?)} \\ &= (a_1 + b_1)A_1 + (a_2 + b_2)A_2 && \text{(Why?)} \\ &= c_1A_1 + c_2A_2. \end{aligned}$$

Therefore $Y_3$ is a linear combination of the $A_i$ and is in $S$, so that $S$ is closed under addition.

For scalar multiplication, suppose that $Y_1 = a_1A_1 + a_2A_2$ as before, and let $c$ be any scalar. Then

$$\begin{aligned} Y_4 = cY_1 &= c(a_1A_1 + a_2A_2) = c(a_1A_1) + c(a_2A_2) && \text{(Why?)} \\ &= (ca_1)A_1 + (ca_2)A_2 = c_1A_1 + c_2A_2. \end{aligned}$$

Thus $Y_4$ is in $S$, implying that $S$ is closed under scalar multiplication. ∎

To further illustrate the last property demonstrated in the proof, let

$$A_1 = \begin{bmatrix} 1 \\ 2 \\ 3 \end{bmatrix}, \qquad A_2 = \begin{bmatrix} -1 \\ 3 \\ -1 \end{bmatrix}$$

as previously, and let

$$Y = 2A_1 + 3A_2 = \begin{bmatrix} -1 \\ 13 \\ 3 \end{bmatrix}$$

be in $S$. For $c = 4$, for example, we want to show that

$$4Y = 4\begin{bmatrix} -1 \\ 13 \\ 3 \end{bmatrix} = \begin{bmatrix} -4 \\ 52 \\ 12 \end{bmatrix}$$

is in $S$. Since

$$\begin{bmatrix} -4 \\ 52 \\ 12 \end{bmatrix} = 8\begin{bmatrix} 1 \\ 2 \\ 3 \end{bmatrix} + 12\begin{bmatrix} -1 \\ 3 \\ -1 \end{bmatrix} = 8A_1 + 12A_2,$$

it is also in $S$.

We should mention that not every set of vectors has the two closure properties, so that there do exist vector sets which are not vector spaces. First, the scalar multiplication property makes it evident that a vector space is an infinite set of vectors unless it contains the zero vector alone (see Exercise 7). However, there are many infinite vector sets which are not vector spaces. Consider the set of all positive multiples of $\begin{bmatrix} 1 \\ 0 \end{bmatrix}$. This set is closed under addition but not under scalar multiplication since, for example, $(-1)\begin{bmatrix} 1 \\ 0 \end{bmatrix} = \begin{bmatrix} -1 \\ 0 \end{bmatrix}$ is not in the set. As a do-it-yourself second illustration, let

$$T = \left\{Y \mid Y = a\begin{bmatrix} 1 \\ 0 \end{bmatrix}\right\} \cup \left\{Y \mid Y = b\begin{bmatrix} 0 \\ 1 \end{bmatrix}\right\},$$

where, as usual, $a$ and $b$ are arbitrary scalars. Is $T$ closed under addition and scalar multiplication and thus a vector space (see Exercise 9)?

Any subset of $S$ which is also a vector space will be called a **subspace** of $S$. Thus

$$S_1 = \left\{Y \mid Y = x_1\begin{bmatrix} 1 \\ 0 \\ 1 \end{bmatrix} + x_2\begin{bmatrix} 0 \\ 1 \\ 2 \end{bmatrix}\right\}$$

is a subspace of

$$S = \left\{Y \mid Y = x_1\begin{bmatrix} 1 \\ 0 \\ 1 \end{bmatrix} + x_2\begin{bmatrix} 0 \\ 1 \\ 2 \end{bmatrix} + x_3\begin{bmatrix} 1 \\ 1 \\ 1 \end{bmatrix}\right\}.$$

Note in this example that the defining vectors of $S_1$ are a proper subset of the defining vectors of $S$. However, there are many subspaces of $S$ other than those constructed strictly as the set of all linear combinations of a proper subset of the defining vectors of $S$. For example,

$$Y_1 = 2\begin{bmatrix} 1 \\ 0 \\ 1 \end{bmatrix} + 3\begin{bmatrix} 0 \\ 1 \\ 2 \end{bmatrix} + 0\begin{bmatrix} 1 \\ 1 \\ 1 \end{bmatrix} = \begin{bmatrix} 2 \\ 3 \\ 8 \end{bmatrix}$$

and

$$Y_2 = 1\begin{bmatrix} 1 \\ 0 \\ 1 \end{bmatrix} + 1\begin{bmatrix} 0 \\ 1 \\ 2 \end{bmatrix} + (-1)\begin{bmatrix} 1 \\ 1 \\ 1 \end{bmatrix} = \begin{bmatrix} 0 \\ 0 \\ 2 \end{bmatrix}$$

are in $S$, and $\{Y \mid Y = a_1Y_1 + a_2Y_2\}$ is another subspace of $S$.

In Section 4.7 we shall relate the elements of our vector spaces—which we have called vectors—to the geometrical vectors with which you may already be familiar.

## 4.1 Exercises

1. Show that the following vectors are linear combinations of the vectors $E_1 = \begin{bmatrix} 1 \\ 0 \end{bmatrix}$ and $E_2 = \begin{bmatrix} 0 \\ 1 \end{bmatrix}$ by finding the appropriate scalars:

   a. $\begin{bmatrix} 2 \\ 0 \end{bmatrix}$ b. $\begin{bmatrix} 3 \\ 5 \end{bmatrix}$ c. $\begin{bmatrix} \pi \\ e \end{bmatrix}$ d. $\begin{bmatrix} 1/2 \\ 3/8 \end{bmatrix}$.

2. Use Gauss-Jordan reduction to show that the following vectors are linear combinations of the vectors $\begin{bmatrix} 1 \\ 2 \end{bmatrix}, \begin{bmatrix} -1 \\ 3 \end{bmatrix}$ by finding the appropriate scalars.

   a. $\begin{bmatrix} 0 \\ 5 \end{bmatrix}$ b. $\begin{bmatrix} 4 \\ 3 \end{bmatrix}$ c. $\begin{bmatrix} 2 \\ 0 \end{bmatrix}$ d. $\begin{bmatrix} -3 \\ -1 \end{bmatrix}$.

3. Solve simultaneously the four sets of equations $AX = B_1$, $AX = B_2$, $AX = B_3$, $AX = B_4$ represented by the augmented matrix

$$[A \mid B_1 B_2 B_3 B_4] = \left[\begin{array}{cc|cccc} 1 & -1 & 0 & 4 & 2 & -3 \\ 2 & 3 & 5 & 3 & 0 & -1 \end{array}\right].$$

   Relate your solution to Exercise 2.

4. By searching for the appropriate scalars, decide whether or not the following vectors are linear combinations of $\begin{bmatrix} 1 \\ 0 \\ 1 \end{bmatrix}$ and $\begin{bmatrix} 0 \\ 1 \\ 1 \end{bmatrix}$:

   a. $\begin{bmatrix} 0 \\ 0 \\ 2 \end{bmatrix}$ b. $\begin{bmatrix} -2 \\ 1 \\ -1 \end{bmatrix}$ c. $\begin{bmatrix} 2 \\ 3 \\ 4 \end{bmatrix}$.

5. Consider the set $S = \left\{ Y \mid Y = x_1 \begin{bmatrix} 0 \\ 1 \\ 1 \end{bmatrix} + x_2 \begin{bmatrix} 1 \\ 0 \\ 0 \end{bmatrix} \right\}$.

   a. Is $S$ a vector space?
   b. Find two different vectors $A$ and $B$ in $S$.
   c. Find the scalars which prove that $A + B$ is in $S$.
   d. Find a proper subset $T$ of $S$ which is a subspace of $S$.

6. Consider the set

$$S = \left\{ Y \mid Y = x_1 \begin{bmatrix} 1 \\ 0 \\ 1 \\ 0 \end{bmatrix} + x_2 \begin{bmatrix} 0 \\ -1 \\ 0 \\ 0 \end{bmatrix} + x_3 \begin{bmatrix} 1 \\ 1 \\ 1 \\ 1 \end{bmatrix} \right\}.$$

   a. Is $S$ a vector space?
   b. Find a vector $Y_1$ in $S$ other than the three listed.
   c. Find scalars which verify that $3Y_1$ is in $S$.
   d. Find scalars which verify that 0 is in $S$.

7. Is $\{0\}$ a vector space? Why?

8. Consider the set $S$ from Exercise 5. Let

$$T = \left\{ Y \mid Y = y_1 \begin{bmatrix} 0 \\ 2 \\ 2 \end{bmatrix} + y_2 \begin{bmatrix} 2 \\ 0 \\ 0 \end{bmatrix} \right\}.$$

Show that $S = T$ by showing that $T \subseteq S$ and $S \subseteq T$.

9. Consider the set $T = \left\{ Y \mid Y = a \begin{bmatrix} 1 \\ 0 \end{bmatrix} \right\} \cup \left\{ Y \mid Y = b \begin{bmatrix} 0 \\ 1 \end{bmatrix} \right\}$ in the text. Show that this set is not closed under addition, but that it is closed under scalar multiplication.

10. Show that 0 is in every vector space.

11. Prove that the non-empty intersection of two vector spaces is a vector space.

12. Verify that $T = \left\{ Y \mid Y = a \begin{bmatrix} 1 \\ 0 \\ 1 \end{bmatrix} + b \begin{bmatrix} 0 \\ 1 \\ 1 \end{bmatrix} \right\}$ is a subspace of

$$S = \left\{ Y \mid Y = a \begin{bmatrix} 1 \\ -1 \\ 0 \end{bmatrix} + b \begin{bmatrix} 2 \\ 1 \\ 3 \end{bmatrix} + c \begin{bmatrix} 3 \\ 0 \\ 0 \end{bmatrix} \right\}.$$

### New Terms



## 4.2 Spanning Sets and Linear Dependence

We shall say that a finite set of vectors $T \subseteq S$ **spans** a vector space $S$ provided that any vector in $S$ may be written as a linear combination of the vectors in $T$.

Example 4.2.1 Let

$$S = \left\{ Y \mid Y = a_1 \begin{bmatrix} 1 \\ 0 \\ 0 \end{bmatrix} + a_2 \begin{bmatrix} 0 \\ 1 \\ 0 \end{bmatrix} \right\}.$$

Then

$$T_1 = \left\{ \begin{bmatrix} 1 \\ 0 \\ 0 \end{bmatrix}, \begin{bmatrix} 0 \\ 1 \\ 0 \end{bmatrix} \right\}$$

spans $S$ as does

$$T_2 = \left\{ \begin{bmatrix} 2 \\ 0 \\ 0 \end{bmatrix}, \begin{bmatrix} 0 \\ 2 \\ 0 \end{bmatrix} \right\}.$$

However,
$$T_3 = \left\{ \begin{bmatrix} 2 \\ 0 \\ 0 \end{bmatrix}, \begin{bmatrix} 0 \\ 0 \\ 2 \end{bmatrix} \right\}$$
does not span $S$ since, for example,
$$\begin{bmatrix} 1 \\ 1 \\ 0 \end{bmatrix}$$
is in $S$ and is not in the set of all linear combinations of vectors in $T_3$. ▮

With the definition of a spanning set in mind you might wonder what the knowledge of such a subset $T$ of $S$ tells us about $S$. You could attempt, for example, to define a measure of the size of $S$ as the number of vectors in $T$. For this integer to serve as a definition of a unique size for $S$, however, it is apparent that each spanning set for $S$ would have to contain the same number of vectors.

Example 4.2.1 (continued)
$$T_4 = \left\{ \begin{bmatrix} 1 \\ 0 \\ 0 \end{bmatrix}, \begin{bmatrix} 0 \\ 1 \\ 0 \end{bmatrix}, \begin{bmatrix} 0 \\ 2 \\ 0 \end{bmatrix} \right\}$$
is a spanning set for $S$ which contains 3 vectors as opposed to the 2 vectors in each of $T_1$ and $T_2$. ▮

From this example we are forced to conclude that there may be different spanning sets for a given vector space and that the spanning sets may contain different numbers of vectors. Such a measure of size for $S$ would not be unique, so another approach to the idea of size is required. We shall examine this question later.

The next example portrays what must be done to show that a given set of vectors spans a vector space when the latter has been constructed as a set of linear combinations.

Example 4.2.2 Let $S = \{Y \mid Y = x_1A_1 + x_2A_2 + x_3A_3\}$, and consider the three sets
$$T_1 = \{A_1, A_2, A_3\},$$
$$T_2 = \{2A_1, 2A_2, 2A_3\},$$
and
$$T_3 = \{A_1, A_2, A_3, A_1 + A_2, A_1 + A_3, A_2 + A_3\}.$$
$T_1$ spans $S$ by definition. We may show that $T_2$ spans $S$ by showing that we may write any $Y$ in $S$ as a linear combination of the vectors in $T_2$, namely, $2A_1$, $2A_2$, $2A_3$. Since $Y = x_1A_1 + x_2A_2 + x_3A_3 = (\frac{1}{2}x_1)(2A_1) + (\frac{1}{2}x_2)(2A_2) + (\frac{1}{2}x_3)(2A_3)$ is such a linear combination, $T_2$ does span $S$. Can you now show that $T_3$ spans $S$? ▮

Realizing that the number of vectors in different spanning sets may differ, we pursue the following question: *Does there exist a spanning set for S which contains a minimum number of vectors?* The answer to this question is "Yes." Such sets are of pre-eminent importance as we shall see later. To find out how we can determine which spanning sets contain "surplus" vectors which could be discarded to obtain such a minimal spanning set, we consider the concept of linear dependence.

In general, we shall say that the set $\{A_1, A_2, \ldots, A_p\}$ of $n \times 1$ column vectors is a **linearly dependent set of vectors** provided that there exist scalars $c_1, c_2, \ldots, c_p$, *not all* 0, such that

$$c_1A_1 + c_2A_2 + \cdots + c_pA_p = 0.$$

You may already be familiar with this idea in its simplest form, that of a set of 2 vectors in the plane which happen to lie in the same line, as shown in Figure 4.2.1.

**Figure 4.2.1**

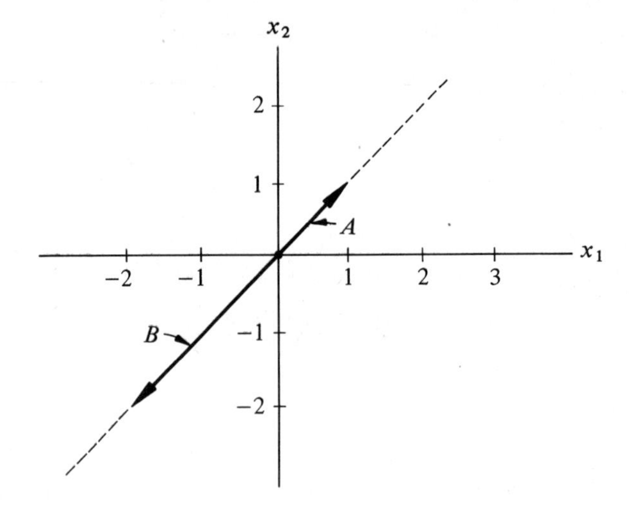


In the figure, $B$ is merely a scalar multiple of $A$; and it happens here that $B = -2A$. To put this in a form which is more useful when more than 2 vectors are being considered, we note that

$$2A + 1B = 0.$$

We have found constants $c_1 = 2$ and $c_2 = 1$, not both zero, such that

$$c_1A + c_2B = 0.$$

We have satisfied our general definition in this case. In reality, we have found a fancy way of saying that $B$ is a non-zero multiple of $A$.

Example 4.2.3 Let $A, B, C$ be the vectors shown in Figure 4.2.2. In this case $C = 2A + B$, so that

$$2A + B + (-1)C = 0.$$

Although none of the 3 vectors is a scalar multiple of any other one, there do exist scalars $c_1 = 2, c_2 = 1, c_3 = -1$, such that

$$c_1A + c_2B + c_3C = 0.$$

Thus $\{A, B, C\}$ is a linearly dependent set. ∎

Figure 4.2.2

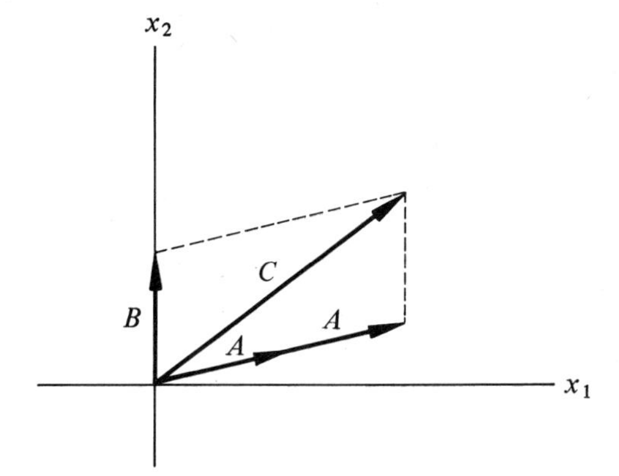


If a set of vectors is not linearly dependent, then it is said to be **linearly independent**. If we let $C$ and $D$ denote the vectors in Figure 4.2.3, we see that neither is a scalar multiple of the other and it should be intuitively clear that the only values of $c_1$ and $c_2$ such that

$$c_1C + c_2D = 0$$

are $c_1 = c_2 = 0$. Thus $\{C, D\}$ is a linearly independent set. Rather than saying the latter, we may sometimes say that $C$ and $D$ are linearly independent vectors.

Figure 4.2.3

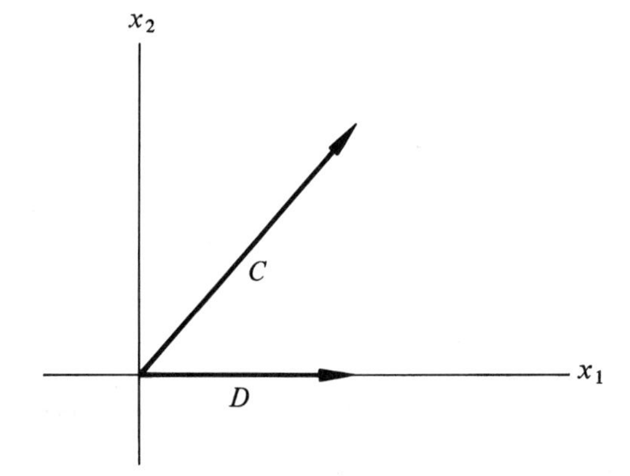


We now reconsider the general definition of linear dependence. Suppose that we have a set of vectors and a set of scalars such that

$$c_1A_1 + c_2A_2 + \cdots + c_pA_p = 0,$$

and that $c_1 \neq 0$. Then $c_1{}^{-1} = 1/c_1$ exists, and multiplying both sides of the equation by $1/c_1$, we have

$$A_1 + \frac{c_2}{c_1}A_2 + \cdots + \frac{c_p}{c_1}A_p = 0,$$

or

$$A_1 = -\frac{c_2}{c_1}A_2 - \cdots - \frac{c_p}{c_1}A_p.$$

We have just shown that if $c_1 \neq 0$ then $A_1$ is a linear combination of the remaining $p - 1$ vectors $A_2, \ldots, A_p$—an immediate generalization of the idea of being a scalar multiple of one vector. Since, in general, at least one of the $c_i \neq 0$, we know that in a linearly dependent set we could solve for at least one of the $A_i$ in a similar fashion, and that *at least one of the $p$ vectors is a linear combination of the remaining ones.* Thus the concept of a linearly dependent set of vectors is intimately associated with the idea of scalar multiple and linear combination.

In order to build our confidence in our knowledge of these fundamental definitions, let us consider some further examples.

Example 4.2.4 Consider

$$T_4 = \left\{ \begin{bmatrix} 1 \\ 0 \\ 0 \end{bmatrix}, \begin{bmatrix} 0 \\ 1 \\ 0 \end{bmatrix}, \begin{bmatrix} 0 \\ 2 \\ 0 \end{bmatrix} \right\}$$

of Example 4.2.1. Since

$$0 \begin{bmatrix} 1 \\ 0 \\ 0 \end{bmatrix} + 2 \begin{bmatrix} 0 \\ 1 \\ 0 \end{bmatrix} + (-1) \begin{bmatrix} 0 \\ 2 \\ 0 \end{bmatrix} = \begin{bmatrix} 0 \\ 0 \\ 0 \end{bmatrix},$$

we conclude that $T_4$ is a linearly dependent set. ∎

Example 4.2.5 Let

$$T = \left\{ \begin{bmatrix} 1 \\ 0 \\ 1 \end{bmatrix}, \begin{bmatrix} 0 \\ 1 \\ -1 \end{bmatrix}, \begin{bmatrix} 1 \\ 1 \\ 0 \end{bmatrix} \right\}.$$

Are there constants $c_1, c_2, c_3$, not all zero, such that

$$c_1 \begin{bmatrix} 1 \\ 0 \\ 1 \end{bmatrix} + c_2 \begin{bmatrix} 0 \\ 1 \\ -1 \end{bmatrix} + c_3 \begin{bmatrix} 1 \\ 1 \\ 0 \end{bmatrix} = \begin{bmatrix} 0 \\ 0 \\ 0 \end{bmatrix}?$$

We recognize this as a homogeneous linear equation problem with the $c_i$ as unknowns. Solving, we find that

$$1 \begin{bmatrix} 1 \\ 0 \\ 1 \end{bmatrix} + 1 \begin{bmatrix} 0 \\ 1 \\ -1 \end{bmatrix} + (-1) \begin{bmatrix} 1 \\ 1 \\ 0 \end{bmatrix} = \begin{bmatrix} 0 \\ 0 \\ 0 \end{bmatrix},$$

so that $T$ is another linearly dependent set. ∎

Example 4.2.6 Consider

$$T_2 = \left\{ \begin{bmatrix} 2 \\ 0 \\ 0 \end{bmatrix}, \begin{bmatrix} 0 \\ 2 \\ 0 \end{bmatrix} \right\}$$

of Example 4.2.1. We wish to find scalars $c_1, c_2$, not both zero, such that

$$c_1 \begin{bmatrix} 2 \\ 0 \\ 0 \end{bmatrix} + c_2 \begin{bmatrix} 0 \\ 2 \\ 0 \end{bmatrix} = \begin{bmatrix} 0 \\ 0 \\ 0 \end{bmatrix}.$$

By observation, however, we see that both $c_1$ and $c_2$ *must be* 0 in order to solve this equation. Thus, $T_2$ is a linearly independent set. It is apparent that neither of the vectors is a linear combination (multiple) of the other one. ∎

We conclude this section by pointing out that *if we have a set of nonzero vectors* $\{A_1, A_2, \ldots, A_p\}$, *and one of them is a linear combination of the others, then the entire set is linearly dependent.* Suppose, for example, that

$$A_3 = c_1A_1 + c_2A_2 + c_4A_4 + \cdots + c_pA_p.$$

Then we know that

$$c_1A_1 + c_2A_2 + (-1)A_3 + c_4A_4 + \cdots + c_pA_p = 0.$$

Since $(-1) \neq 0$, the set is linearly dependent.

Example 4.2.7 Consider

$$A_1 = \begin{bmatrix} 1 \\ 2 \\ 3 \end{bmatrix}, \quad A_2 = \begin{bmatrix} -1 \\ 3 \\ 4 \end{bmatrix}, \quad A_3 = 2A_1 - A_2 = \begin{bmatrix} 3 \\ 1 \\ 2 \end{bmatrix}.$$

Since $2A_1 + (-1)A_2 + (-1)A_3 = 0$, we have that $\{A_1, A_2, A_3\}$ is linearly dependent. ∎

## 4.2 Exercises

1. What do we mean when we say that a set $T$ spans a vector space $S$?

2. Let $S = \left\{ Y \mid Y = a\begin{bmatrix} 1 \\ 0 \\ 2 \end{bmatrix} + b\begin{bmatrix} 0 \\ 1 \\ -1 \end{bmatrix} \right\}$. Find a spanning set $T$ for $S$.

3. The set $T = \{A_1, A_2, A_3\}$ is said to be a linearly dependent set provided that there exist scalars $c_i$, ________, such that ________. On the other hand, $T$ is said to be a linearly independent set if ________________.

4. If $c_1A_1 + c_2A_2 = 0$ holds only in case $c_1 = c_2 = 0$, we say that $\{A_1, A_2\}$ is ________.

5. Use the definition of linear dependence to decide whether or not the following sets are linearly dependent. If dependent, find scalars required by the definition.

   a. $T = \left\{ \begin{bmatrix} 1 \\ 2 \end{bmatrix}, \begin{bmatrix} -1 \\ 3 \end{bmatrix} \right\}$

   b. $T = \left\{ \begin{bmatrix} 1 \\ 0 \\ 1 \end{bmatrix}, \begin{bmatrix} 2 \\ -4 \\ -4 \end{bmatrix}, \begin{bmatrix} 0 \\ 2 \\ 3 \end{bmatrix} \right\}$

   c. $T = \left\{ \begin{bmatrix} 1 \\ 2 \end{bmatrix}, \begin{bmatrix} -1 \\ 3 \end{bmatrix}, \begin{bmatrix} 1 \\ 0 \end{bmatrix} \right\}$

   d. $T = \left\{ \begin{bmatrix} 1 \\ 2 \end{bmatrix}, \begin{bmatrix} 1 \\ 0 \end{bmatrix} \right\}$.

6. Is $B = [3, -2, 1]^T$ in the space spanned by $[2, 0, 2]^T$, $[-1, 1, 1]^T$, and $[3, 0, 1]^T$?

7. Use the definition of linear dependence to decide which of the following sets of vectors are linearly dependent. For those which are, express one of the vectors as a linear combination of the others.

a. $T_1 = \left\{ \begin{bmatrix} 1 \\ 1 \\ 1 \\ 1 \end{bmatrix}, \begin{bmatrix} 1 \\ 0 \\ 1 \\ 0 \end{bmatrix}, \begin{bmatrix} 1 \\ 0 \\ 3 \\ 1 \end{bmatrix}, \begin{bmatrix} -1 \\ -1 \\ 1 \\ 0 \end{bmatrix} \right\}$ b. $T_2 = \left\{ \begin{bmatrix} 1 \\ 1 \\ 1 \\ 1 \end{bmatrix}, \begin{bmatrix} -1 \\ 2 \\ 1 \\ 2 \end{bmatrix}, \begin{bmatrix} 0 \\ 0 \\ 0 \\ 1 \end{bmatrix} \right\}$

c. $T_3 = \left\{ \begin{bmatrix} 1 \\ 0 \end{bmatrix}, \begin{bmatrix} 0 \\ 3 \end{bmatrix} \right\}$ d. $T_4 = \left\{ \begin{bmatrix} 1 \\ 0 \\ 0 \end{bmatrix}, \begin{bmatrix} 0 \\ 0 \\ 3 \end{bmatrix}, \begin{bmatrix} 0 \\ 2 \\ 0 \end{bmatrix} \right\}$

e. $T_5 = \left\{ \begin{bmatrix} 1 \\ 0 \\ 0 \end{bmatrix}, \begin{bmatrix} 0 \\ 0 \\ 3 \end{bmatrix}, \begin{bmatrix} 0 \\ 2 \\ 0 \end{bmatrix}, \begin{bmatrix} 2 \\ -4 \\ 6 \end{bmatrix} \right\}$.

8. Which of the sets in Exercise 7 could have at least one vector deleted and still span the same space?

9. Show that

$$T_1 = \left\{ \begin{bmatrix} 1 \\ 0 \\ 2 \end{bmatrix}, \begin{bmatrix} 0 \\ 2 \\ 1 \end{bmatrix} \right\} \quad \text{and} \quad T_2 = \left\{ \begin{bmatrix} 2 \\ -2 \\ 3 \end{bmatrix}, \begin{bmatrix} 3 \\ -4 \\ 4 \end{bmatrix} \right\}$$

span the same subspace of $3 \times 1$ column vectors.

10. Let $\{A_1, A_2, A_3\}$ be a linearly independent set. Show that if $A_4$ is not a linear combination of $A_1, A_2, A_3$, then $\{A_1, A_2, A_3, A_4\}$ is linearly independent also.

11. Let $\{A_1, A_2, A_3\}$ be a linearly dependent set. Show that at least one of the $A_i$ is a linear combination of the others.

12. Show that if 0 is an element of any finite set of vectors, $T$, then the set $T$ is linearly dependent.

13. Let $A$ and $B$ be finite sets of vectors such that $A \subseteq B$. Show that a sufficient condition that $B$ be linearly dependent is that $A$ be linearly dependent. Is this condition also necessary?

14. Refer to sets $A$ and $B$ in Exercise 13. Show that a necessary condition that $B$ be linearly independent is that $A$ be linearly independent. Is this condition also sufficient?

15. If $\{A_1, A_2, A_3\}$ is a linearly independent set, show that $\{A_1, A_1 + A_2, A_1 + A_2 + A_3\}$ is also a linearly independent set.

16. Is the following statement true? If the set $T$ spans the vector space $S$, then $T$ is linearly independent?

### New Terms

spans, 104
linearly dependent set of vectors, 106
linearly independent set of vectors, 107

## 4.3 Basis and Dimension

If $T \subseteq S$ is a linearly independent spanning set for the vector space $S$, we say that $T$ is a **basis** for $S$.

Example 4.3.1 Since

$$T_1 = \left\{ \begin{bmatrix} 1 \\ 0 \\ 0 \end{bmatrix}, \begin{bmatrix} 0 \\ 1 \\ 0 \end{bmatrix} \right\}$$

is a linearly independent spanning set for

$$S = \left\{ Y \mid Y = a_1 \begin{bmatrix} 1 \\ 0 \\ 0 \end{bmatrix} + a_2 \begin{bmatrix} 0 \\ 1 \\ 0 \end{bmatrix} \right\},$$

$T$ is a basis for $S$. Similarly,

$$T_2 = \left\{ \begin{bmatrix} 2 \\ 0 \\ 0 \end{bmatrix}, \begin{bmatrix} 0 \\ 2 \\ 0 \end{bmatrix} \right\}$$

is also a basis for $S$. However, neither

$$T_3 = \left\{ \begin{bmatrix} 2 \\ 0 \\ 0 \end{bmatrix}, \begin{bmatrix} 0 \\ 0 \\ 2 \end{bmatrix} \right\} \quad \text{nor} \quad T_4 = \left\{ \begin{bmatrix} 1 \\ 0 \\ 0 \end{bmatrix}, \begin{bmatrix} 0 \\ 1 \\ 0 \end{bmatrix}, \begin{bmatrix} 0 \\ 2 \\ 0 \end{bmatrix} \right\}$$

is a basis for $S$ (Why?). ∎

Example 4.3.2 If

$$S = \left\{ Y \mid Y = a_1 \begin{bmatrix} 1 \\ 0 \end{bmatrix} + a_2 \begin{bmatrix} 0 \\ 1 \end{bmatrix} \right\},$$

then

$$T_1 = \left\{ \begin{bmatrix} 1 \\ 0 \end{bmatrix}, \begin{bmatrix} 0 \\ 1 \end{bmatrix} \right\}$$

is a basis for $S$, while

$$\left\{ \begin{bmatrix} 1 \\ 0 \end{bmatrix}, \begin{bmatrix} 0 \\ 1 \end{bmatrix}, \begin{bmatrix} 0 \\ 0 \end{bmatrix} \right\}$$

is not (Why?). ∎

We conclude from Example 4.3.2 that the zero vector cannot be an element of any basis (or, indeed, of any linearly independent set).

Suppose that we have a vector space $S$ and a basis $B \subseteq S$. We wish to show that it is the number of vectors in $B$ which should serve as the measure of the size of $S$. In order that this be a meaningful definition, it must be true that each basis for $S$ has the same number of vectors. We shall show that this is the case in two steps.

**LEMMA 4.3.1** *If $\{A_1, A_2, \ldots, A_n\}$ is a linearly dependent set of non-zero vectors, then there is some one of the vectors which is a linear combination of those preceding it in the list.*

**Proof** The set $\{A_1\}$ is linearly independent, since $A_1 \neq 0$. If $\{A_1, A_2\}$ is a linearly dependent set, in which case $A_2$ is a linear combination (scalar multiple) of $A_1$, the proof is complete. If not, then $\{A_1, A_2\}$ is an independent set; and we then consider $\{A_1, A_2, A_3\}$. If this set is linearly dependent, then there exist constants $c_1, c_2, c_3$, not all zero, such that

$$c_1A_1 + c_2A_2 + c_3A_3 = 0.$$

Now $c_3 \neq 0$, because that would imply that $\{A_1, A_2\}$ was linearly dependent (How?). Thus, $c_3^{-1}$ exists; and we can solve for $A_3$ as a linear combination of $A_1, A_2$. If $\{A_1, A_2, A_3\}$ is linearly independent, we consider $\{A_1, A_2, A_3, A_4\}$, and so forth. In general, we reason as follows. Consider the set $T_k = \{A_1, A_2, \ldots, A_k\}$. If $T_k$ is linearly independent, then we consider the set $\{A_1, A_2, \ldots, A_k, A_{k+1}\}$. If $T_k$ is the first such set that is linearly dependent, then there are scalars $c_i$, not all zero, such that

$$c_1A_1 + c_2A_2 + \cdots + c_kA_k = 0.$$

Further, $c_k \neq 0$, since this would imply that $\{A_1, A_2, \ldots, A_{k-1}\}$ was linearly dependent, contrary to our assumption that we are considering the first dependent set. Since $c_k \neq 0$, then $c_k^{-1} = 1/c_k$ exists; and

$$A_k = -\frac{c_1}{c_k}A_1 - \frac{c_2}{c_k}A_2 - \cdots - \frac{c_{k-1}}{c_k}A_{k-1}.$$

We thus have expressed $A_k$ as a linear combination of the $A$'s preceding it. Since there were only $n$ vectors to start with and the original set was dependent, in no more than $n$ steps we must find a vector which is a linear combination of those preceding it. This completes the proof of the lemma. ∎

Example 4.3.3 Let

$$\left\{A_1 = \begin{bmatrix}1\\0\\0\end{bmatrix}, A_2 = \begin{bmatrix}1\\2\\1\end{bmatrix}, A_3 = \begin{bmatrix}0\\2\\1\end{bmatrix}, A_4 = \begin{bmatrix}-1\\2\\3\end{bmatrix}\right\}$$

be a spanning set for a vector space $S$. $\{A_1\}$ is a linearly independent set. $\{A_1, A_2\}$ is also independent since $A_2$ is not a scalar multiple of $A_1$. Consider $\{A_1, A_2, A_3\}$. Since $(-1)A_1 + 1A_2 + (-1)A_3 = 0$, as we see by solving the system of equations

$$\begin{bmatrix}1 & 1 & 0\\0 & 2 & 2\\0 & 1 & 1\end{bmatrix}\begin{bmatrix}c_1\\c_2\\c_3\end{bmatrix} = \begin{bmatrix}0\\0\\0\end{bmatrix},$$

we conclude that $A_3 = -A_1 + A_2$ is a linear combination of the preceding vectors $A_1, A_2$. We note that $A_4$ is not a linear combination of $A_1, A_2, A_3$, so the conjecture that the last vector must be a linear combination of the preceding is false. ∎

We next show in Theorem 4.3.1 that the number of vectors in a spanning set for $S$ cannot be less than the number of vectors in any given linearly independent subset of $S$. We shall use this fact in the proof of Theorem 4.3.2, the main result of this section.

**THEOREM 4.3.1** *Let $S$ be the vector space spanned by $\{A_1, A_2, \ldots, A_n\}$ and let $\{B_1, B_2, \ldots, B_m\}$ be a linearly independent subset of $S$. Then $m \leqslant n$.*

**Proof** Consider the set $T_1 = \{B_1, A_1, A_2, \ldots, A_n\}$. This set is linearly dependent since $B_1$ is a linear combination of the $A_i$ (the $A_i$ span $S$). Thus, by Lemma 4.3.1, some vector $X$ of the set $T_1$ is a linear combination of those preceding it. Since $X$ cannot be $B_1$, suppose it is $A_k$. We discard $A_k$ from the set to form

$$\{B_1, A_1, \ldots, A_{k-1}, A_{k+1}, \ldots, A_n\}.$$

This new set also spans $S$ (Why?). Thus, in particular, the vector $B_2$ is a linear combination of its elements; and

$$T_2 = \{B_2, B_1, A_1, \ldots, A_{k-1}, A_{k+1}, \ldots, A_n\}$$

is a linearly dependent set. Again, some vector of the set is a linear combination of the preceding. This vector can not be one of the $B_j$ since the $B_j$ are independent. Thus it must be one of the $A_i$. We call this vector $A_r$ and discard it from $T_2$. The resulting set also spans $S$. Continuing in this fashion, we delete one of the $A_i$ vectors and add one of the $B_j$ vectors, each time creating a new spanning set for $S$ consisting of $n$ vectors. We want to show that $n \geqslant m$. We suppose to the contrary that $n < m$ and see that this leads us to conclude that the set of $B_j$ vectors is dependent, contradicting our assumption that the $B_j$ are independent. If $n < m$ we will eventually run out of $A_i$ to delete before we have all $m$ of the $B_j$ in the set. We might, for example, have

$$T_k = \{B_k, B_{k-1}, \ldots, B_1\}$$

with $k < m$, so that, in particular, $B_m$ is not in $T_k$. Since $T_k$ is a spanning set for $S$, $B_m$ would have to be a linear combination of the vectors in $T_k$. Thus the set of all the $B_j$ would be linearly dependent, in contradiction to our original assumption. Thus $m \leqslant n$, as we wished to show.

**THEOREM 4.3.2** *Every basis for $S$ contains the same number of vectors.*

**Proof** Suppose that we have two bases for $S$—both linearly independent spanning sets for $S$. Call them $B_1$ and $B_2$, and let $n_1$ and $n_2$ denote the number of vectors in $B_1$, $B_2$ respectively. We wish to show that $n_1 = n_2$. Since $B_1$ spans $S$ and $B_2$ is linearly independent, $n_1 \geqslant n_2$ by Theorem 4.3.1. However, $B_2$ spans $S$ and $B_1$ is also linearly independent, so that $n_2 \geqslant n_1$. Thus $n_1 = n_2$ and the proof is complete. ▮

Given a spanning set $T = \{A_1, A_2, \ldots, A_n\}$ for a vector space $S$, we can create a basis for $S$ by starting with $A_1$. If $A_1 \neq 0$, we consider $\{A_1, A_2\}$. If this set is independent, we consider $\{A_1, A_2, A_3\}$ and so forth, until we find a vector which is a linear combination of the preceding ones. We discard this vector and continue until we check all of the vectors in $T$, and discard those which are linear combinations of the preceding.

We have shown in Theorem 4.3.2 that the number of vectors in a basis for $S$ is unique. This non-negative integer, which represents both the maximum number of linearly independent vectors in $S$ and the minimum number of vectors required to span $S$, is called the **dimension** of $S$. If we happen to know the dimension of $S$, then the process of finding a basis for $S$ must terminate when we have found the correct number of linearly independent vectors. We thus need not continue and check every element of the spanning set being examined.

**Example 4.3.4** Suppose that

$$S = \left\{ Y \mid Y = a_1 \begin{bmatrix} 1 \\ 0 \\ 0 \\ 0 \end{bmatrix} + a_2 \begin{bmatrix} 1 \\ 2 \\ 1 \\ 0 \end{bmatrix} + a_3 \begin{bmatrix} 0 \\ 2 \\ 1 \\ 0 \end{bmatrix} + a_4 \begin{bmatrix} 0 \\ 0 \\ 1 \\ 1 \end{bmatrix} + a_5 \begin{bmatrix} 1 \\ 4 \\ 3 \\ 1 \end{bmatrix} \right\}.$$

$S$ has dimension 3, and we wish to pick an independent subset of the defining vectors as a basis. Now $\{A_1, A_2\}$ is clearly independent, but $A_3 = A_2 - A_1$, so that $\{A_1, A_2, A_3\}$ is dependent. However, we can observe that $A_4$ is not a linear combination of $A_1$ and $A_2$. Thus $\{A_1, A_2, A_4\}$ is independent and will serve as a basis for $S$. Suppose we had started with $A_2$. What would the basis have been? ∎

When we speak of "2-space" we shall mean

$$\left\{ Y \mid Y = a \begin{bmatrix} 1 \\ 0 \end{bmatrix} + b \begin{bmatrix} 0 \\ 1 \end{bmatrix} \right\},$$

that is, the set of all $2 \times 1$ column matrices with real elements. This space has dimension 2. The obvious extensions to 3-space and to $n$-space will also be exploited. However, when we speak of "a 2-space," we merely mean some subset of a vector space which is a vector space of dimension 2. The vectors themselves may have 5 components, for example. Thus, we will say that

$$\left\{ Y \mid Y = a \begin{bmatrix} 1 \\ 0 \\ 0 \end{bmatrix} + b \begin{bmatrix} 0 \\ 1 \\ 0 \end{bmatrix} \right\}$$

is a subspace of 3-space of dimension 2, or simply a 2-space.

## 4.3 Exercises

1. We say that the set $T \subseteq S$ is a basis for the vector space $S$ provided that $T$ is ____________ and that $T$ ____________.

2. If the dimension of a vector space $S$ is 4, then any basis for $S$ contains ________.

3. Decide which of the following statements are true:
   a. A subset of a basis consists of linearly independent vectors.
   b. A basis for a vector space is unique.
   c. If $T$ is a basis for $S$, then no proper subset of $T$ spans $S$.

4. Suppose that we know that a given vector space $S$ is spanned by a set $T$ which contains 6 vectors. What, if anything, can be said about the dimension of $S$?

5. Which of the following sets are not a basis for the vector space

$$S = \left\{ Y \mid Y = x_1 \begin{bmatrix} 1 \\ 0 \\ 0 \end{bmatrix} + x_2 \begin{bmatrix} 0 \\ 1 \\ 0 \end{bmatrix} + x_3 \begin{bmatrix} 0 \\ 0 \\ 1 \end{bmatrix} \right\}?$$

a. $T_1 = \left\{ \begin{bmatrix} 1 \\ 0 \\ 0 \end{bmatrix}, \begin{bmatrix} 0 \\ 1 \\ 1 \end{bmatrix}, \begin{bmatrix} 1 \\ -1 \\ -1 \end{bmatrix} \right\}$ b. $T_2 = \left\{ \begin{bmatrix} 0 \\ 0 \\ 1 \end{bmatrix}, \begin{bmatrix} 1 \\ 0 \\ 1 \end{bmatrix}, \begin{bmatrix} 2 \\ 3 \\ 4 \end{bmatrix}, \begin{bmatrix} -1 \\ 0 \\ 0 \end{bmatrix} \right\}.$

6. If

$$H = \left\{ X_H \mid X_H = a \begin{bmatrix} 1 \\ 0 \\ 1 \\ 0 \end{bmatrix} + b \begin{bmatrix} 0 \\ 1 \\ 1 \\ 0 \end{bmatrix} + c \begin{bmatrix} 0 \\ 0 \\ 3 \\ 1 \end{bmatrix} \right\}$$

what is the dimension of $H$?

7. Let a basis for 2-space consist of $\{A_1, A_2\} = \left\{ \begin{bmatrix} 2 \\ 1 \end{bmatrix}, \begin{bmatrix} 1 \\ 2 \end{bmatrix} \right\}$. Which, if any, of these vectors could be replaced by $\begin{bmatrix} 6 \\ 3 \end{bmatrix}$ to form a new basis?

8. Suppose that $A_1, A_2, A_3, A_4$ are 4 vectors in 3-space such that $\{A_1, A_2, A_3\}$ is a basis for the space spanned by all 4 $A_i$. If $A_4 = A_1 + A_2$, which of the 3 vectors, if any, could be replaced by $A_4$ to obtain a new basis?

9. Suppose that $Y = A_1 + 2A_2 + 3A_3$ is a representation of $Y$ in terms of the basis vectors in Exercise 8. Find the new representation of $Y$ if $A_2$ is replaced by $A_4$ to form a new basis $\{A_1, A_4, A_3\}$.

10. Let $\{B_1, B_2, \ldots, B_m\}$ be a basis for $m$-space, and suppose that $A = y_1B_1 + y_2B_2 + \cdots + y_mB_m$ with $y_1 \neq 0$. Prove that $\{A, B_2, \ldots, B_m\}$ is also a basis for $m$-space.

11. Let $S$ be the space spanned by

$$T = \left\{ \begin{bmatrix} 1 \\ 0 \\ 1 \\ 1 \end{bmatrix}, \begin{bmatrix} -1 \\ -3 \\ 1 \\ 0 \end{bmatrix}, \begin{bmatrix} 2 \\ 3 \\ 0 \\ 1 \end{bmatrix}, \begin{bmatrix} 2 \\ 0 \\ 2 \\ 2 \end{bmatrix} \right\}.$$

Find the dimension of $S$ and a subset of $T$ which could serve as a basis for $S$.

**New Terms**

basis of a vector space, 111
dimension of a vector space, 113

## 4.4 The Coordinates of a Vector

A basis for a vector space $S$ provides another advantage over a spanning set which happens to be linearly dependent. *A basis may be used to define a coordinate system for $S$ which provides a unique set of coordinates for each* *vector in S*. A merely linearly dependent spanning set will not suffice. Let

$$S = \left\{ Y \mid Y = a_1 \begin{bmatrix} 1 \\ 0 \\ 0 \end{bmatrix} + a_2 \begin{bmatrix} 0 \\ 1 \\ 0 \end{bmatrix} + a_3 \begin{bmatrix} 0 \\ 2 \\ 0 \end{bmatrix} \right\}.$$

Some $3 \times 1$ column matrices are elements of $S$ and some are not (Can you name one which is not?). For $a_1 = a_2 = a_3 = 1$, we see that

$$Y_1 = \begin{bmatrix} 1 \\ 3 \\ 0 \end{bmatrix} = 1 \begin{bmatrix} 1 \\ 0 \\ 0 \end{bmatrix} + 1 \begin{bmatrix} 0 \\ 1 \\ 0 \end{bmatrix} + 1 \begin{bmatrix} 0 \\ 2 \\ 0 \end{bmatrix}$$

is in $S$. With respect to this particular spanning set, we could say that $Y_1$ is represented by the coordinate set $\{1, 1, 1\}$ since, knowing the $A_i$ and this triple of constants, we could re-create $Y_1$. However, it is also the case that

$$Y_1 = \begin{bmatrix} 1 \\ 3 \\ 0 \end{bmatrix} = 1 \begin{bmatrix} 1 \\ 0 \\ 0 \end{bmatrix} + 3 \begin{bmatrix} 0 \\ 1 \\ 0 \end{bmatrix} + 0 \begin{bmatrix} 0 \\ 2 \\ 0 \end{bmatrix},$$

so that $\{1, 3, 0\}$ would also be a candidate for a coordinate set for $Y_1$. Here we have an example of non-uniqueness that is undesirable. We prefer to use a set of vectors with the property that such a **coordinate set for a given vector** is unique (much as you prefer to be known by *one* name). A basis for $S$ provides such a set. We show that this is true for a 3-dimensional vector space.

**THEOREM 4.4.1** *If $\{A_1, A_2, A_3\}$ is a basis for 3-space and $Y = a_1A_1 + a_2A_2 + a_3A_3 = b_1A_1 + b_2A_2 + b_3A_3$ then $a_1 = b_1, a_2 = b_2, a_3 = b_3$.*

**Proof** We demonstrate that the contrapositive is true; that is, we shall show that if $a_i \neq b_i$ for some $i$, then $\{A_1, A_2, A_3\}$ is not a basis for $S$. We see that, from the hypothesis of the theorem,

$$(a_1 - b_1)A_1 + (a_2 - b_2)A_2 + (a_3 - b_3)A_3 = 0.$$

Since $a_i \neq b_i$ for some $i$, we have a set of scalars $c_i = a_i - b_i$, not all zero, such that $c_1A_1 + c_2A_2 + c_3A_3 = 0$, and in which the $A_i$ cannot be linearly independent. Thus $\{A_1, A_2, A_3\}$ is not a basis for $S$. Recalling that establishing the contrapositive is equivalent to establishing the original theorem, we are finished. ∎

This uniqueness of coordinate sets, or "names" of vectors, which is provided by a basis for a vector space is one which we will exploit often.

Example 4.4.1 Let $S = \left\{ Y \mid Y = a \begin{bmatrix} 1 \\ 0 \end{bmatrix} + b \begin{bmatrix} 0 \\ 1 \end{bmatrix} \right\}$. In this basis it is a simple matter to find coordinates. Let $Y_1 = \begin{bmatrix} 3 \\ 2 \end{bmatrix} = 3 \begin{bmatrix} 1 \\ 0 \end{bmatrix} + 2 \begin{bmatrix} 0 \\ 1 \end{bmatrix}$. Thus $\{3, 2\}$ is the coordinate set denoting $\begin{bmatrix} 3 \\ 2 \end{bmatrix}$. ∎

Example 4.4.2 Let

$$S = \left\{ Y \mid Y = a\begin{bmatrix}1\\0\\1\end{bmatrix} + b\begin{bmatrix}0\\1\\0\end{bmatrix} + c\begin{bmatrix}-1\\2\\3\end{bmatrix} \right\}.$$

If we know that $Y_1 = \begin{bmatrix}0\\3\\4\end{bmatrix}$ is in $S$, we may wish to find its coordinates with respect to this basis. Thus we wish to solve the set of equations

$$x_1\begin{bmatrix}1\\0\\1\end{bmatrix} + x_2\begin{bmatrix}0\\1\\0\end{bmatrix} + x_3\begin{bmatrix}-1\\2\\3\end{bmatrix} = \begin{bmatrix}0\\3\\4\end{bmatrix}.$$

By Gauss-Jordan reduction we find that $x_1 = x_2 = x_3 = 1$. The coordinate set is thus $\{1, 1, 1\}$. ∎

Because of the notational advantages which it provides, we now consider the concept of a coordinate matrix of a given vector $X$ with respect to a fixed basis. Let $\{U_1, U_2, \ldots, U_m\}$ be a fixed basis for $\mathcal{M}_{m,1}$. If $X$ is in $\mathcal{M}_{m,1}$ such that

$$X = u_1U_1 + u_2U_2 + \cdots + u_mU_m,$$

then the column matrix

$$X_U = \begin{bmatrix}u_1\\u_2\\\vdots\\u_m\end{bmatrix}$$

is called the **coordinate matrix** of $X$ with respect to the $U$ basis. We can then represent $X$ as the matrix product

$$[U_1, U_2, \ldots, U_m]\begin{bmatrix}u_1\\u_2\\\vdots\\u_m\end{bmatrix} = UX_U = X,$$

where the square matrix $U = [U_1, U_2, \ldots, U_m]$ is the **basis matrix** whose columns are the basis vectors $U_i$, and $X_U$ is the column matrix of coordinates of $X$ with respect to the $U$ basis.

Example 4.4.3 In $\mathcal{M}_{2,1}$, let $\{U_1, U_2\} = \left\{\begin{bmatrix}1\\2\end{bmatrix}, \begin{bmatrix}1\\3\end{bmatrix}\right\}$ be a basis, so that $U = [U_1, U_2] = \begin{bmatrix}1 & 1\\2 & 3\end{bmatrix}$ is the corresponding basis matrix. Since $X = \begin{bmatrix}2\\7\end{bmatrix}$ is in $\mathcal{M}_{2,1}$, there are scalars $u_1, u_2$ such that $\begin{bmatrix}2\\7\end{bmatrix} = u_1\begin{bmatrix}1\\2\end{bmatrix} + u_2\begin{bmatrix}1\\3\end{bmatrix} = \begin{bmatrix}1 & 1\\2 & 3\end{bmatrix}\begin{bmatrix}u_1\\u_2\end{bmatrix}$, or such that $X = UX_U$. Noting that this is nothing more than a linear equation problem to solve, and that $U^{-1} = \begin{bmatrix}3 & -1\\-2 & 1\end{bmatrix}$, we can find

$$X_U = U^{-1}X = \begin{bmatrix}3 & -1\\-2 & 1\end{bmatrix}\begin{bmatrix}2\\7\end{bmatrix} = \begin{bmatrix}-1\\3\end{bmatrix}.$$

We verify that $X_U$ is valid by computing

$$X = -1\begin{bmatrix}1\\2\end{bmatrix} + 3\begin{bmatrix}1\\3\end{bmatrix} = \begin{bmatrix}2\\7\end{bmatrix}.$$ ∎

Let us now consider a very special case in Example 4.4.4.

Example 4.4.4 Let $\{U_1, U_2\} = \left\{\begin{bmatrix}1\\0\end{bmatrix}, \begin{bmatrix}0\\1\end{bmatrix}\right\}$, so that $U = \begin{bmatrix}1 & 0\\0 & 1\end{bmatrix}$. If $X = \begin{bmatrix}2\\7\end{bmatrix}$, then, since

$$X = 2U_1 + 7U_2 = \begin{bmatrix}1 & 0\\0 & 1\end{bmatrix}\begin{bmatrix}2\\7\end{bmatrix},$$

we see that $X = UX_U = I_2X_U$, so that $X = X_U$. ∎

This last example demonstrates that if we have the special case in which our basis matrix is the identity matrix, then the vector $X$ and its coordinate matrix $X_U$ are indistinguishable. In a particular application, the context of the discussion must make it clear whether a matrix is being used to represent a vector or the coordinates of a vector. We shall return to our discussion of basis matrices in Section 5.5 where we examine how the coordinates of a vector change when the basis is changed.

The previous discussion in this section should make it very clear that it is important to know when a given set of vectors is a basis. Certain sets of vectors can be quickly identified as linearly independent sets and thus as a basis for the vector space which they span. Consider $\left\{\begin{bmatrix}2\\0\end{bmatrix}, \begin{bmatrix}0\\1\end{bmatrix}\right\}$. Because of the 0 in the second vector, it is clear that the second vector could not be a non-zero multiple of the first, and thus that the two vectors are linearly independent.

Now let us look at a set containing three vectors

$$\left\{\begin{bmatrix}1\\0\\0\end{bmatrix}, \begin{bmatrix}0\\2\\0\end{bmatrix}, \begin{bmatrix}0\\0\\4\end{bmatrix}\right\}.$$

The placement of the zeros again makes it clear that these three vectors are also linearly independent. Of course, we could always use the definition and look for scalars $c_1, c_2, c_3$ such that

$$c_1\begin{bmatrix}1\\0\\0\end{bmatrix} + c_2\begin{bmatrix}0\\2\\0\end{bmatrix} + c_3\begin{bmatrix}0\\0\\4\end{bmatrix} = \begin{bmatrix}0\\0\\0\end{bmatrix}.$$

Since this requires that $c_1 = 0, 2c_2 = 0$, and $4c_3 = 0$—that is, each $c_i = 0$—the set of vectors is linearly independent. In general, then, we note that if we have a set of $p$ vectors $\{A_1, A_2, \ldots, A_p\}$ with the property that each $A_i$ has at least $p - 1$ zeros and each $A_i$ has a non-zero element in a row in which all of the other vectors have a 0, then the set must be linearly independent. We recall that the set $H$ of Example 2.9.3 was constructed as the set of all linear combinations of the vectors in the set

$$\left\{\begin{bmatrix}-2\\1\\-4\\1\\0\end{bmatrix}, \begin{bmatrix}5\\-2\\-3\\0\\1\end{bmatrix}\right\},$$

which has this property, as does the corresponding set of vectors

$$\left\{\begin{bmatrix}-2\\1\\0\\0\\0\\0\end{bmatrix}, \begin{bmatrix}-1\\0\\1\\-2\\1\\0\end{bmatrix}, \begin{bmatrix}2\\0\\-3\\4\\0\\1\end{bmatrix}\right\}$$

of Example 2.9.4. A reconsideration of the manner in which $X_H$ was constructed should convince you that this will always be the case. *Thus, our $H$ vector spaces are constructed as sets of linear combinations of linearly independent vectors to start with.* These defining vectors form a basis for $H$ in each case, and the dimension of $H$ is readily obtainable.

## 4.4 Exercises

1. Let $\left\{\begin{bmatrix}1\\0\end{bmatrix}, \begin{bmatrix}0\\1\end{bmatrix}\right\}$ be a basis for 2-space. Find the coordinates of $\begin{bmatrix}1\\-1\end{bmatrix}$ and $\begin{bmatrix}3\\4\end{bmatrix}$ with respect to this basis.

2. Determine by observation which of the following sets are linearly independent.

   a. $\left\{\begin{bmatrix}1\\0\\0\end{bmatrix}, \begin{bmatrix}0\\0\\3\end{bmatrix}, \begin{bmatrix}0\\2\\0\end{bmatrix}\right\}$

   b. $\left\{\begin{bmatrix}2\\0\\0\\3\end{bmatrix}, \begin{bmatrix}0\\2\\0\\1\end{bmatrix}, \begin{bmatrix}0\\0\\5\\4\end{bmatrix}\right\}$

   c. $\left\{\begin{bmatrix}1\\0\\1\end{bmatrix}, \begin{bmatrix}2\\4\\0\end{bmatrix}, \begin{bmatrix}-2\\0\\-2\end{bmatrix}\right\}$

   d. $\left\{\begin{bmatrix}3\\4\\1\\0\\0\end{bmatrix}, \begin{bmatrix}2\\3\\0\\1\\0\end{bmatrix}, \begin{bmatrix}-1\\2\\0\\0\\1\end{bmatrix}\right\}$

3. Let $\left\{\begin{bmatrix}1\\2\end{bmatrix}, \begin{bmatrix}2\\1\end{bmatrix}\right\}$ be a basis for 2-space. Find the coordinates of $\begin{bmatrix}1\\2\end{bmatrix}, \begin{bmatrix}3\\3\end{bmatrix}$, and $\begin{bmatrix}-3\\0\end{bmatrix}$ with respect to this basis.

4. Let $\left\{\begin{bmatrix}1\\0\\0\end{bmatrix}, \begin{bmatrix}0\\1\\0\end{bmatrix}, \begin{bmatrix}0\\0\\1\end{bmatrix}\right\}$ and $\left\{\begin{bmatrix}1\\0\\0\end{bmatrix}, \begin{bmatrix}1\\1\\0\end{bmatrix}, \begin{bmatrix}1\\1\\1\end{bmatrix}\right\}$ be bases for 3-space. Find the coordinates of $\begin{bmatrix}3\\2\\1\end{bmatrix}$ and $\begin{bmatrix}2\\2\\1\end{bmatrix}$ with respect to each basis.

5. The results of this problem will be of special interest for those who plan to study linear programming in Chapter 6.

   Let $A = \begin{bmatrix}2 & 3 & 1 & 0\\3 & 2 & 0 & 1\end{bmatrix} = [A_1, A_2, A_3, A_4]$ and let $S$ be the space spanned by the columns $A_i$ of $A$.
   a. Which possible pairs $\{A_i, A_j\}$ could serve as a basis for $S$?
   b. If $\{A_3, A_4\}$ is chosen, find the coordinate sets which represent each of the $A_i$ with respect to this basis.
   c. If $\{A_3, A_1\}$ is chosen, find the coordinate sets.
   d. If $\{A_2, A_1\}$ is chosen, find the coordinate sets.
   e. Express $\begin{bmatrix}6\\6\end{bmatrix}$ as a linear combination of the basis vectors in d.

6. Let $A = \begin{bmatrix}2 & 1 & 1 & 0\\1 & 1 & 0 & 1\end{bmatrix} = [A_1, A_2, A_3, A_4]$, and let $S$ be the space spanned by the columns $A_i$ of $A$. Find the coordinates of each of the $A_i$ if the following bases are chosen for $S$.
   a. $\{A_3, A_4\}$ b. $\{A_1, A_4\}$ c. $\{A_1, A_2\}$

7. Fill in the blanks.
   a. In $\mathscr{M}_{2,1}$, let $U_1 = \begin{bmatrix}1\\4\end{bmatrix}$ and $U_2 = \begin{bmatrix}4\\2\end{bmatrix}$ so that $U = \begin{bmatrix} & \end{bmatrix}$.
   b. If $X = \begin{bmatrix}-2\\6\end{bmatrix}$ is in $\mathscr{M}_{2,1}$, then $X = UX_U$ where $X_U = \begin{bmatrix}\\ \end{bmatrix}$.
   c. If $X = \begin{bmatrix}1\\0\end{bmatrix}$ is in $\mathscr{M}_{2,1}$, then $X_U = \begin{bmatrix}\\ \end{bmatrix}$.
   d. If $X = \begin{bmatrix}\\ \end{bmatrix}$ is in $\mathscr{M}_{2,1}$ then $X_U = \begin{bmatrix}1\\1\end{bmatrix}$.

**New Terms**

coordinate set for a vector, 116
coordinate matrix, 117
basis matrix, 117

## 4.5 Vector Spaces and Homogeneous Linear Equations

We have seen that the problem of finding $X$ such that $AX = 0$ is equivalent to finding $x_i$ such that

$$x_1A_1 + x_2A_2 + \cdots + x_nA_n = 0,$$

where the $A_i$ are the columns of $A$. Thus we are asking for the set of all linear combinations of the $A_i$ which yield 0, if any. We recall from Section 4.1 that the solution set $H$ is a vector space. We shall call upon your knowledge of the Gauss-Jordan reduction process and the method of constructing $H$ to formulate some general results about the dimension of $H$. First we note that $H$ is a subspace of **n-space**, or $\mathcal{M}_{n,1}$.

We suppose, as before, that we are dealing with a situation in which we have $m$ equations in $n$ unknowns with $k < m$ of the equations redundant and $m - k \leqslant n$. By Gauss-Jordan reduction we obtain a row reduced echelon form as shown below in which the asterisks represent real numbers which may or may not be 0:

$$\begin{array}{c} \\ m-k \\ \\ \\ k \\ \\ \end{array}\overset{n}{\left[\begin{array}{cccc|c} * & * & \cdots & * & 0 \\ * & * & \cdots & * & 0 \\ \cdot & \cdot & \cdots & \cdot & \cdot \\ * & * & \cdots & * & 0 \\ \hline 0 & 0 & \cdots & 0 & 0 \\ \cdot & \cdot & \cdots & \cdot & \cdot \\ 0 & 0 & \cdots & 0 & 0 \end{array}\right]}.$$

There are two possibilities. Either $m - k = n$ or $m - k < n$. If $m - k = n$, then the matrix of asterisks is an $n \times n$ identity matrix implying that we have a unique solution 0. We say that $\{0\}$ is a vector space of dimension 0. On the other hand, if $m - k < n$, then there will be $d = n - (m - k)$ free variables. By assigning these as in Chapter 2, we obtain a solution set of of the form

$$H = \{X_H \mid X_H = a_1B_1 + a_2B_2 + \cdots + a_dB_d\},$$

in such a way that the set $\{B_1, B_2, \ldots, B_d\}$ is linearly independent. Thus $H$ is of dimension $d = n - (m - k)$. A typical example of such a row-reduced echelon matrix is

$$\left[\begin{array}{cccccc|c} 1 & 0 & -2 & 0 & -4 & -7 & 0 \\ 0 & 1 & -3 & 0 & -5 & -8 & 0 \\ 0 & 0 & 0 & 1 & -6 & -9 & 0 \\ 0 & 0 & 0 & 0 & 0 & 0 & 0 \end{array}\right].$$

In this particular case we know that we may freely assign values for $x_6$, $x_5$, and $x_3$. Thus, if we let $x_3 = a$, $x_5 = b$, and $x_6 = c$, we find that

$$H = \left\{X_H \;\middle|\; X_H = \begin{bmatrix} 2a + 4b + 7c \\ 3a + 5b + 8c \\ 1a + 0b + 0c \\ 0a + 6b + 9c \\ 0a + 1b + 0c \\ 0a + 0b + 1c \end{bmatrix}\right\}$$

is the general solution. We know that in this particular case $H$ is a vector space because we can rewrite it as

$$H = \left\{ X_H \mid X_H = a \begin{bmatrix} 2 \\ 3 \\ 1 \\ 0 \\ 0 \\ 0 \end{bmatrix} + b \begin{bmatrix} 4 \\ 5 \\ 0 \\ 6 \\ 1 \\ 0 \end{bmatrix} + c \begin{bmatrix} 7 \\ 8 \\ 0 \\ 9 \\ 0 \\ 1 \end{bmatrix} \right\}.$$

Since each $X_H$ is a linear combination of 3 linearly independent vectors from 6-space, we know that the solution space is a subspace of 6-space of dimension 3.

Now let us consider the original questions that we asked in Chapter 2 concerning linear equations as they apply in general to $AX = 0$.

1. Does a solution exist? Yes, it does, although it may only be the trivial one.
2. If so, is the solution unique? The solution will be unique only in case $n = m - k$, in which case there will remain no free variables. In this case the dimension of the solution space is 0.
3. If there is at least one solution, and it is not unique, then how many solutions are there? If there is one non-trivial solution, then there are certainly an infinite number, since every scalar multiple of the one solution is also a solution. Thus, the dimension of the solution space is at least 1.
4. What is the complete set of solutions? The vector space of dimension $d$ as constructed above constitutes the complete solution set. The vectors constructed serve as one possible basis for this space.

One of the words which you will see if you continue your study of matrices is the term "rank." We shall define the **rank of a matrix** $A$ to be the number $(m - k)$ of non-redundant rows of $A$ when $A$ is thought of as the coefficient matrix of a set of homogeneous linear equations. We use the symbol $r(A)$ to stand for the rank of $A$. Thus $r(A)$ can be calculated as the number of non-zero rows remaining after $A$ is transformed to row-reduced echelon form. The concept of rank has been introduced here so that some important theorems may be stated using that standard terminology. The subject will be covered in more detail in the following chapter.

For convenience, we now list some of the standard results from the study of homogeneous linear equations. The reader should have no trouble reconciling our phraseology here with that of our previous statements.

**THEOREM 4.5.1** *Consider the set $AX = 0$ of $m$ linear equations in $n$ unknowns in which $k$ of the equations are redundant. Then*

1. *There is at least one $X$, namely $0$, such that $AX = 0$.*
2. *The complete set of solutions is a subspace of $n$-space of dimension*

$$d = n - (m - k) = n - r(A).$$

3. *A necessary and sufficient condition that $X = 0$ be the unique solution is that $d = 0$; that is, $n = (m - k) = r(A)$.*
4. *If $n > m - k = r(A)$, then there are exactly $n - r(A)$ linearly independent solutions.*

## 4.5 Exercises

1. Suppose that $A$ is an $m \times n$ matrix with $m \leqslant n$. Decide whether the following statements are true or false.
   a. A system $AX = 0$ has a non-trivial solution if $m < n$.
   b. A system $AX = 0$ has a non-trivial solution only if $m < n$.
   c. If $A$ is non-singular, then 0 is the only solution of $AX = 0$.
   d. The solution set of $AX = 0$ forms a vector space.
   e. The dimension of $\{X \mid AX = 0\}$ is greater than or equal to $n - m$.
   f. 0 is in the set of vectors spanned by the columns of $A$.
   g. The set of $n - r(A)$ solution vectors created as a result of Gauss-Jordan reduction of $[A \mid 0]$ may be linearly dependent.
   h. A necessary condition that 0 be a solution of $AX = 0$ is that $m < n$.
   i. A homogeneous linear equation $AX = 0$ is always consistent.
   j. $r(A) \leqslant m$.
   k. $r(A) \leqslant n$.

2. Verify, from the definition of vector space, that $\{X \mid AX = 0\}$ is a vector space.

3. Consider the set of equations $AX = 0$ with $A$ an $m \times n$ matrix and $k < m$ redundant equations. Find the dimension of the solution space if
   a. $n = 5, m = 3, k = 1$.
   b. $n = m = 5, k = 0$.
   c. $n = 6, m = 4, k = 1$.
   d. $n = 10, m = 5 = r(A)$.
   e. $n = 4, m = 4, r(A) = 3$.

4. Use our results from the theory of homogeneous linear equations to show that the set $\left\{\begin{bmatrix}1\\2\end{bmatrix}, \begin{bmatrix}2\\1\end{bmatrix}, \begin{bmatrix}4\\3\end{bmatrix}\right\}$ is a linearly dependent set. (*Hint*: Find the dimension of the solution space of $AC = 0$, where $A = \begin{bmatrix}1 & 2 & 4\\2 & 1 & 3\end{bmatrix}$ and $C$ is the matrix of the coefficients in the definition of linear dependence.)

### New Terms

$n$-space, 121     rank of a matrix, 122

## 4.6 Vector Spaces and Non-homogeneous Linear Equations

We now use our results for $AX = 0$ to study the case where $AX = B$ and $B \neq 0$. Suppose first that we have a particular solution $X_P$, such that $AX_P = B$, and some other vector $X_H$, such that $AX_H = 0$. Consider the vector $Y = X_P + X_H$. Is $Y$ also a solution of $AX = B$? Since

$$AY = A(X_P + X_H) = AX_P + AX_H = B + 0 = B,$$

we see that $Y$ is such a solution. Therefore, if we know any particular solution to $AX = B$ and any homogeneous solution $X_H$, we can find another particular solution easily. Is every solution to $AX = B$ of the form $X_P + X_H$? If any solution $X_P$ exists, then certainly it is of this form since 0 is a solution to $AX = 0$ and $X_P = X_P + 0$. If $Y$ is any other solution, such that $AY = B$, then $Y$ can be expressed as

$$Y = X_P + (Y - X_P),$$

and

$$A(Y - X_P) = AY - AX_P = B - B = 0,$$

so that $Y - X_P$ is a solution of $AX = 0$. Thus $Y$ is of the form $X_P + X_H$, with $X_H = Y - X_P$. We have therefore proved that

1. If $Y = X_P + X_H$, then $Y$ is a solution of $AX = B$.
2. If $Y$ is a solution of $AX = B$, then $Y$ is of the form $Y = X_P + X_H$.

We may rephrase the preceding results as

**THEOREM 4.6.1** *A necessary and sufficient condition that $Y$ be a solution of $AX = B$ is that $Y = X_p + X_H$ where $AX_p = B$ and $AX_H = 0$.*

Reconsideration of the Gauss-Jordan process shows that it is a method which provides solutions to $AX = B$ and $AX = 0$ simultaneously. For, since the matrix $A$ is the same in both cases, and $AX = 0$ has only a constant column of zeros, we have, in reality, been solving a related homogeneous set of equations each time we solved a non-homogeneous set. At the same time, we have been solving for one particular solution of $AX = B$, namely $X_P$. Any vector in the complete set of solutions for $AX = B$ can then be formed as a sum of $X_P$ and some vector in the vector space of solutions $\{X_H\} = H$.

With these ideas in mind, we see that if there is at least one solution to $AX = B$, then there will be a unique solution only in case $\{X_H\}$ is $\{0\}$. Otherwise, there will be an infinite number of solutions. It will be instructional for you to consider whether or not the complete set of solutions to $AX = B$, $B \neq 0$ is a vector space itself.

The following theorem relates the existence of particular solutions to the concept of rank.

**THEOREM 4.6.2** *There will exist a solution of $AX = B$ if and only if the rank of the augmented matrix $[A \mid B]$ is equal to the rank of $A$.*

**Proof** When we attempt to solve $AX = B$ by Gauss-Jordan reduction we will reach an equivalent augmented matrix of the form

$$\begin{array}{c} \\ m - k \\ \\ \\ k \\ \\ \end{array} \overset{n}{\left[\begin{array}{cccc|c} * & * & \cdots & * & * \\ \cdot & \cdot & \cdots & \cdot & \cdot \\ * & * & \cdots & * & * \\ \hline 0 & 0 & \cdots & 0 & a_1 \\ \cdot & \cdot & \cdots & \cdot & \cdot \\ 0 & 0 & \cdots & 0 & a_k \end{array}\right]}$$

where the asterisks and $a_i$ indicate any real number. If there is a solution then $a_1 = a_2 = \cdots = a_k = 0$, and $r(A)=r([A \mid B])$. On the other hand, if $r(A) = r([A \mid B])$, then the $a_i$ are 0; and there will be a solution. ∎

## 4.6 Exercises

1. Show that if $X = X_P + X_H$, then $X$ is a solution of $AX = B$.

2. Show that if $AX = B$, then $X$ is of the form $X_P + X_H$.

3. Determine whether or not a particular solution exists in the following cases with augmented matrices $[A \mid B]$ as shown. If one exists, find the complete solution in the form $\{X_P + X_H\}$. Verify your solution by substitution in the original equations.

a. $\left[\begin{array}{cc|c} 1 & 2 & 1 \\ 0 & 0 & 1 \end{array}\right]$ b. $\left[\begin{array}{cc|c} 1 & 2 & 0 \\ 0 & 0 & 1 \end{array}\right]$

c. $\left[\begin{array}{cc|c} 1 & 3 & 1 \\ 2 & 1 & 1 \end{array}\right]$ d. $\left[\begin{array}{cc|c} 1 & 3 & 1 \\ 3 & 9 & 1 \end{array}\right]$

e. $\left[\begin{array}{cc|c} 0 & 0 & 0 \\ 1 & 2 & 1 \end{array}\right]$ f. $\left[\begin{array}{ccc|c} 1 & 0 & 1 & 1 \\ 0 & 1 & 0 & 2 \\ 0 & 0 & 0 & 3 \end{array}\right]$

g. $\left[\begin{array}{ccc|c} 1 & 1 & 2 & 0 \\ 0 & 1 & 0 & -1 \\ 1 & 0 & 2 & 1 \end{array}\right]$ h. $\left[\begin{array}{ccc|c} 1 & 1 & 1 & 1 \\ 0 & 1 & 0 & 1 \\ 1 & 0 & 0 & 1 \end{array}\right]$

## 4.7 Arrows and Vectors

In Section 4.1, we defined a vector of order $n$ to be an ordered $n$-tuple of real numbers—that is, a $1 \times n$ row matrix or an $n \times 1$ column matrix. The set of all $1 \times n$ row matrices forms a vector space under the usual addition and scalar multiplication of matrices described in Chapter 1; similarly, the set of all $n \times 1$ column vectors forms a vector space. We now describe a geometrical interpretation of vectors which we will find convenient to use throughout the remainder of the text.

In the three-dimensional space in which we live, it is common to adopt some sort of **rectangular coordinate system** to describe points in space. This is done by selecting an origin (a point denoted in the figures by 0) from which all distances are measured, coordinate directions in which distances are measured, and a convenient unit of length. Thus the top of the telephone pole across the street may be described (with one's self as origin) as 45 feet north, 20 feet east, and 25 feet up. We call the coordinate system in this example rectangular because the coordinate directions are at right angles to each other. One conventional coordinate system for 3-space is shown in Figure 4.7.1 with origin 0, and directions $x$, $y$, and $z$. A point $P$ in this space may be described in this system as being $a$ units in

Figure 4.7.1

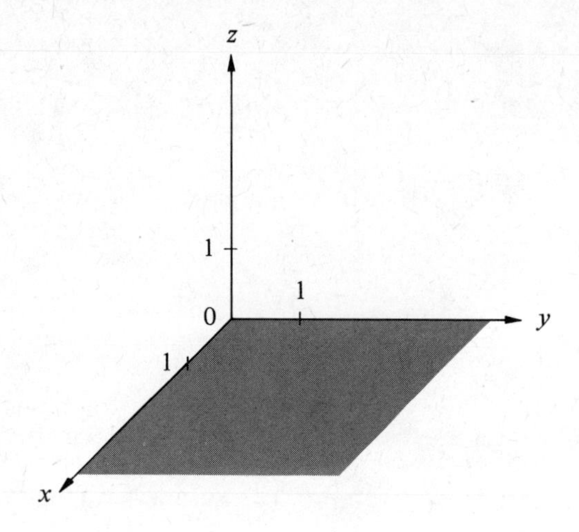


the $x$-direction, $b$ units in the $y$-direction, and $c$ units in the $z$-direction (see Figure 4.7.2). In this way we associate the point $P$ in space with the triple of numbers $(a, b, c)$. Thus the triple $(a, b, c)$ serves as a name for the point $P$ relative to the given coordinate system. Were we to change the coordinate system, the point $P$ would correspond to another triple, and the triple $(a, b, c)$ would correspond to a different point.

Figure 4.7.2

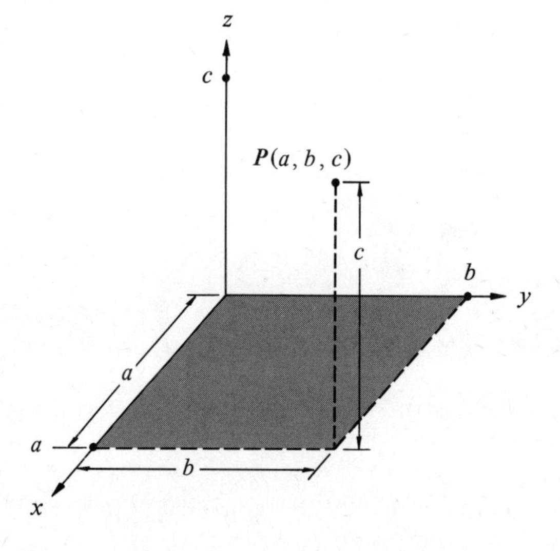


Now, with a fixed coordinate system, we interpret the vector $[a, b, c]$ as the **arrow** (directed line segment) whose tail is at the origin and whose head is at the point associated with the triple $(a, b, c)$. (see Figure 4.7.3.) In the same way we interpret the column vector

$$\begin{bmatrix} a \\ b \\ c \end{bmatrix}$$

as the arrow from the origin to the point $P$, associated with $(a, b, c)$.

Figure 4.7.3

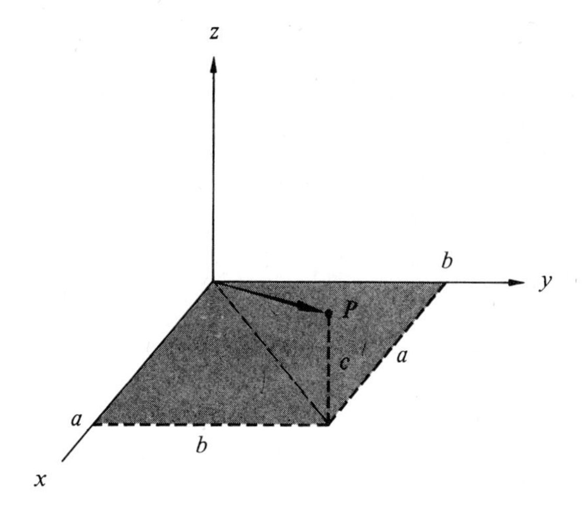


For our geometric interpretation of $1 \times 2$ row matrices, we take a fixed coordinate system in the plane; and, in terms of that system, we make correspond to every point an ordered pair $(a, b)$ such as can be seen in Figure 4.7.4. We then interpret the row vector $[a, b]$ as the arrow from the origin to the point associated with the pair $(a, b)$ (Figure 4.7.5). In a similar manner the column vector $\begin{bmatrix} a \\ b \end{bmatrix}$ is associated with the same arrow in the plane as is $[a, b]$.

We have already established in Chapter 1 an algebra for $1 \times 2$ matrices. We defined the sum $[a, b] + [c, d]$ to be $[a + c, b + d]$ and the scalar product $\alpha[a, b]$ to be $[\alpha a, \alpha b]$. In pictorial terms of arrows in a plane, or in

Figure 4.7.4

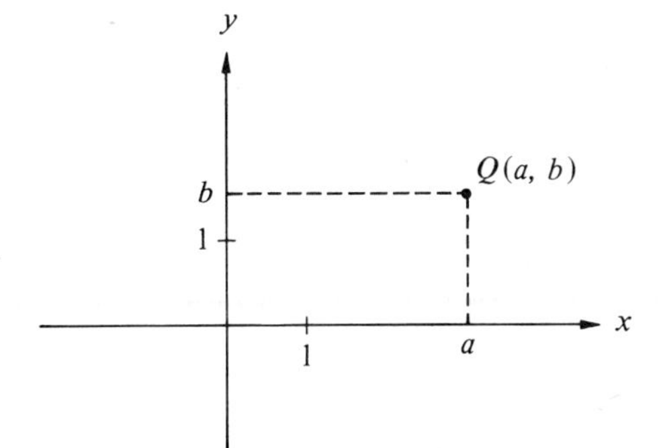


Figure 4.7.5

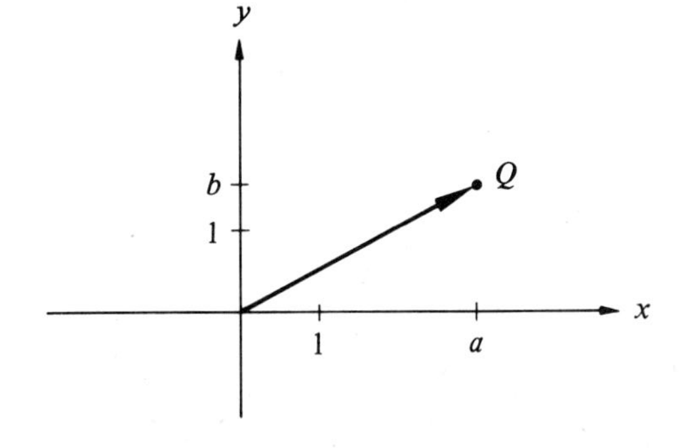

Figure 4.7.6

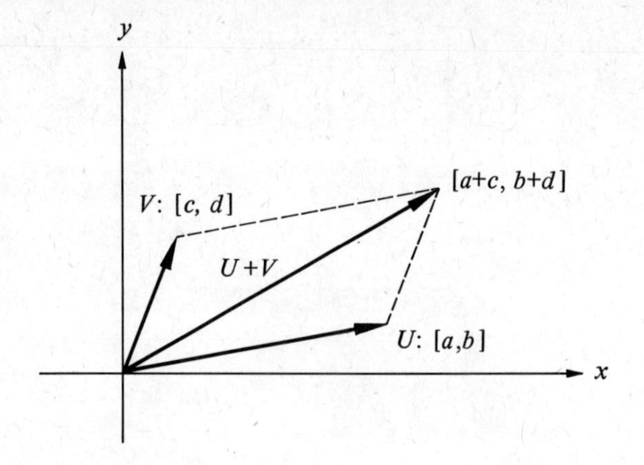


3-space, addition may be realized as shown in Figure 4.7.6. The sum, $U + V$, of vectors $U$, $V$ corresponds to the arrow whose tail is at the origin and whose head lies at the end of the diagonal (from the origin) of the parallelogram, two sides of which are the arrows corresponding to $U$ and $V$. This rule of addition of arrows is often referred to as the **Parallelogram Law**. One may verify that this addition is associative and commutative, that the degenerate arrow 0 (the arrow from the origin to the origin—corresponding to the vector $[0, 0]$)—acts as additive identity, and that, as shown in Figure 4.7.7 the additive inverse $(-U)$ of the arrow $U$ is an arrow of the same length as $U$ but in the opposite direction. Further, a scalar multiple $aV$ of an arrow $V$ is an arrow whose length is $|a|$ times the length of $V$ and is in the same direction as $V$ if $a > 0$ but in the opposite direction if $a < 0$ (Figure 4.7.8). This scalar multiplication of arrows may be easily shown to satisfy the same properties as our scalar multiplication of matrices. Thus the set of arrows from the origin is a vector space under these operations. Since the space of arrows is essentially the same as that of $1 \times 2$ vectors

Figure 4.7.7

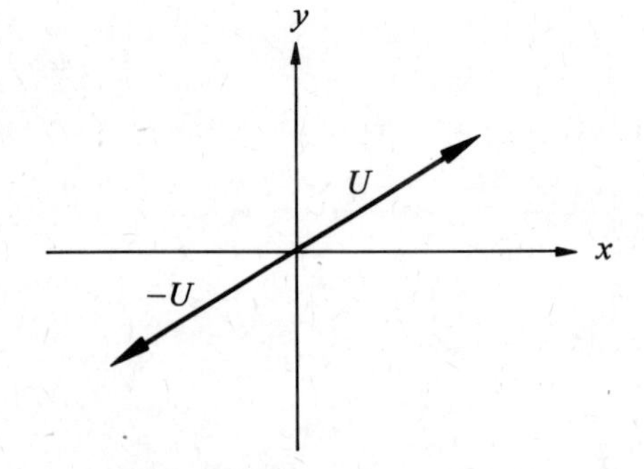


Figure 4.7.8

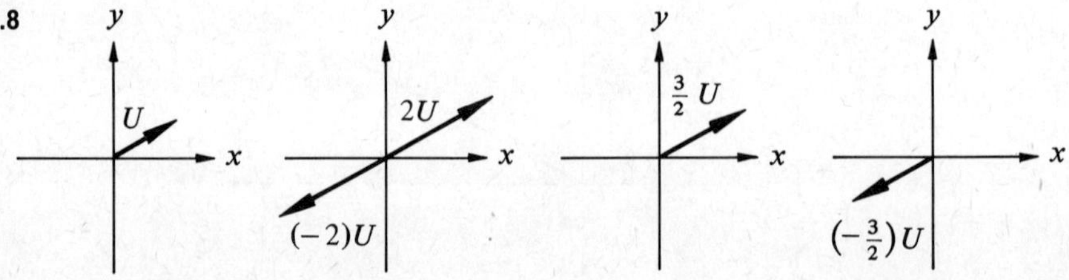

(matrices), we will henceforth drop the distinction and refer to the arrows as vectors also. Whether we interpret an arrow as a $1 \times 2$ row vector or as a $2 \times 1$ column vector depends entirely upon the context of a given discussion, and we shall have occasion to use both interpretations.

All that we have said about arrows in the plane may be repeated for arrows in 3-space. With each such arrow we associate a $1 \times 3$ row vector or a $3 \times 1$ column vector. The Parallelogram Law is the rule by which we add arrows in 3-space. Scalar multiplication is defined as before. As a result, the arrows in 3-space also form a vector space and we will henceforth drop the distinction between an arrow in 3-space and a $1 \times 3$ row vector (or column vector). We do not extend this discussion to "arrows in 4-space" or to "arrows in 17-space" because pictures of such things are not within our grasp.

Let us consider the space of all $1 \times 2$ row vectors and a particular vector $V \neq 0$ in that space, say $V = [-1, 2]$. The set $S$ of all linear combinations of $V$ (that is, of all scalar multiples of $V$) is a subspace of the space of $1 \times 2$ row vectors, and consists of all vectors which lie in the line through the origin which contains $V$ as shown in Figure 4.7.9. In this way we often think of the subspace spanned by $V$ as being the line containing $V$.

Figure 4.7.9

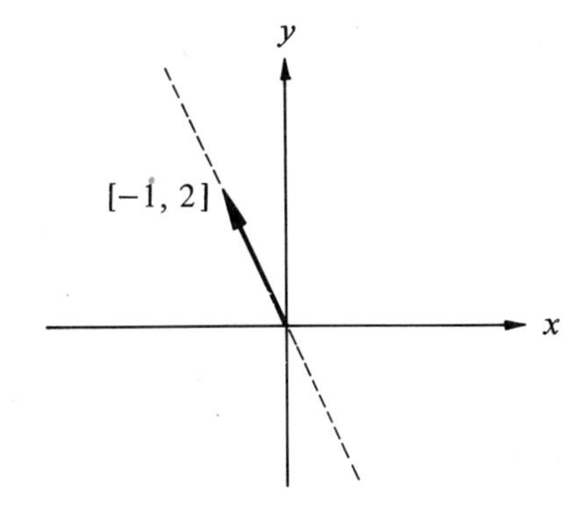


If $U$ and $V$ are $1 \times 2$ row vectors, then either they are linearly independent, or one of them is a scalar multiple of the other. In the latter case, the space spanned by them may be thought of as the line through the origin which contains them (Figure 4.7.10). In the former case, since the two are linearly independent, the vectors $U$ and $V$ form a basis for the space, so

Figure 4.7.10

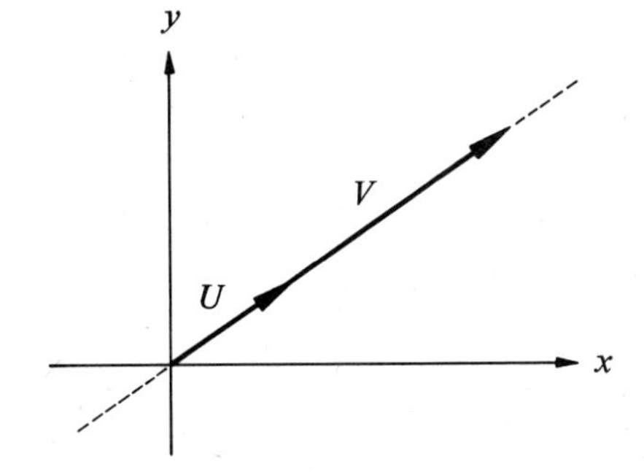

Figure 4.7.11

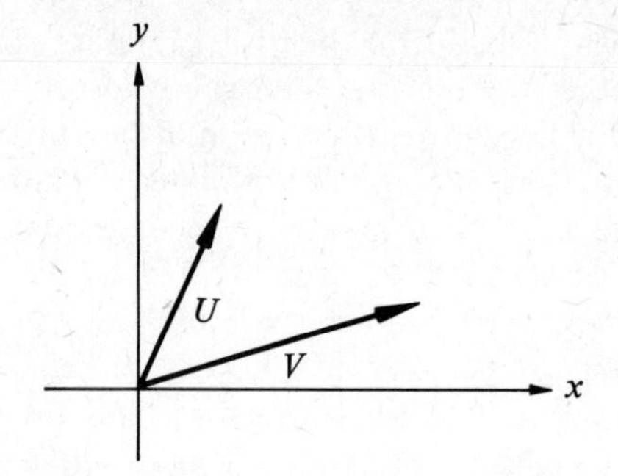


that the subspace spanned by them may be thought of as the entire plane (Figure 4.7.11).

If we consider the space of all $1 \times 3$ row vectors, and $U$ is a non-zero member of that space, then the subspace spanned by $U$ again may be thought of as the line through the origin containing $U$ (Figure 4.7.12). If $V$ and $W$ are linearly independent vectors, then the subspace spanned by them is the plane through the origin containing $V$ and $W$. If $X$ is a third vector, so that $\{V, W, X\}$ is linearly dependent, then $X$ is a linear combination of $V$ and $W$ and lies in the plane spanned by them (Figure 4.7.13). Finally, if $U$, $V$, $W$ are three linearly independent vectors, then the space spanned by them may be thought of as the whole of 3-space.

Figure 4.7.12

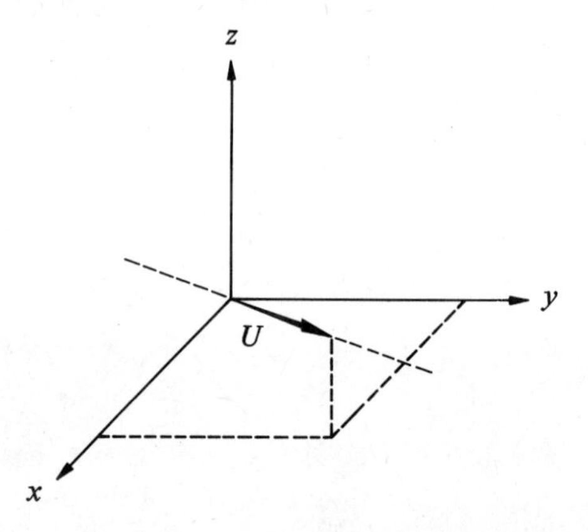


Figure 4.7.13

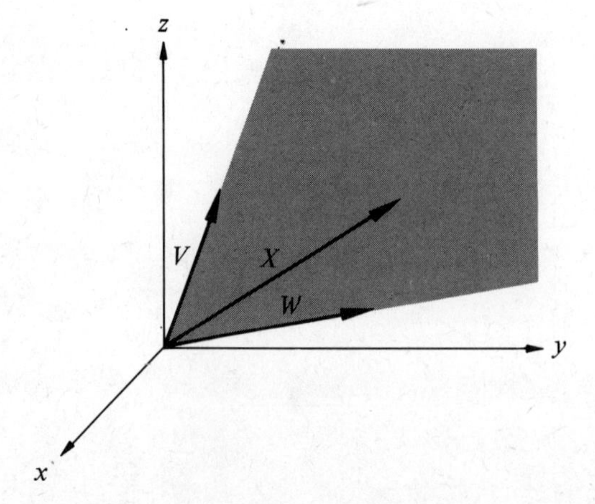

## 4.7 Exercises

1. For the following pairs of vectors $U$, $V$ in 2-space, sketch a coordinate system and draw the arrows corresponding to $U$, $V$, $U + V$, $U - V$, and $V - U$, and draw the parallelograms related to the sums (or differences).
   a. $U = [1, 1], V = [2, 1]$
   b. $U = [-1, -1], V = [1, 2]$
   c. $U = [1, -2], V = [-1, 3]$
   d. $U = [-2, -3], V = [2, -1]$
   e. $U = [1, 3], V = [0, 1]$
   f. $U = [2, 0], V = [0, -3]$

2. a. For the pair $U$, $V$ in Exercise 1*a*. sketch a coordinate system and draw the arrows corresponding to $2U$, $3V$, $2U + 3V$, $2U - 3V$, and the parallelograms corresponding to the sum and the difference.

   b. For the pair in Exercise lc., draw the arrows corresponding to $\frac{1}{3}U$, $\frac{1}{2}V$, $\frac{1}{3}U + \frac{1}{2}V$, $\frac{1}{3}U - \frac{1}{2}V$, and the parallelograms corresponding to the sum and the difference.

3. With coordinate system as in Figure 4.7.1, draw the arrows corresponding to $U = [1, 2, 3]$ and $V = [3, 1, 2]$. Sketch the parallelogram corresponding to the sum $U + V$.

4. In 2-space, sketch the vector $U$ and the subspace spanned by $U$.
   a. $U = [2, 2]$
   b. $U = [-1, 2]$
   c. $U = [-1, -2]$
   d. $U = [1, -3]$
   e. $U = [0, -3]$
   f. $U = [2, 0]$

5. In 3-space, sketch the vector $U$ and the space spanned by $U$.
   a. $U = [2, 1, 3]$
   b. $U = [1, 2, 1]$
   c. $U = [0, 2, 1]$
   d. $U = [1, 2, 0]$
   e. $[1, 1, -1]$

6. In a 3-dimensional coordinate system $x$, $y$, $z$:

   a. Draw the vectors

$$U = \begin{bmatrix} 1 \\ 0 \\ 1 \end{bmatrix}, \quad V = \begin{bmatrix} 2 \\ 3 \\ 3 \end{bmatrix}$$

   and sketch a portion of the plane determined by $U$ and $V$.

   b. Show algebraically that

$$\begin{bmatrix} 1 \\ 1 \\ 4/3 \end{bmatrix}$$

   is in the plane spanned by $U$ and $V$ and that

$$\begin{bmatrix} 1 \\ 1 \\ 1 \end{bmatrix}$$

   is not; that is, show that the former is a linear combination of $U$ and $V$ and that the latter is not. Use your sketch from 6a. to illustrate the situation.

7. Describe in your own words the subspace of 4-space spanned by the following.
   a. $U = [1, 1, 0, 0]$
   b. $U = [1, 0, 0, 0]$ and $V = [0, 1, 0, 0]$
   c. $U = [1, 1, 0, 0]$ and $V = [1, 2, 0, 0]$
   d. $U = [1, 1, 0, 0]$, $V = [1, 2, 0, 0]$ and $W = [1, 1, 1, 0]$

8. In 3-space, sketch the vectors $U$ and $V$ and the subspace spanned by them.
   a. $U = [1, 1, 1]$, $V = [0, 1, 2]$    b. $U = [1, 2, 0]$, $V = [0, 0, 1]$
   c. $U = [3, 1, 1]$, $V = [1, 3, 2]$

### New Terms

rectangular coordinate system, 125
arrow, 126
Parallelogram Law, 128

## 4.8 Dot Product, Length, Distance, and Angle

In this section we wish to give a definition for length of a vector and for the distance and angle between two vectors. The setting of our discussion is the 2-dimensional space $\mathcal{M}_{1,2}$ (or $\mathcal{M}_{2,1}$) which we use for the sake of simplicity. The ideas and results presented here may be generalized to 3-space or 30-space or to any $n$-dimensional space. The interested reader is encouraged to consult the references at the end of this section for a more general outlook.

Let $U = [a, b]$ and $V = [c, d]$ be two vectors in $\mathcal{M}_{1,2}$. We define the **dot product** $U \cdot V$ to be the matrix $UV^T$. Thus $U \cdot V = [a, b]\begin{bmatrix} c \\ d \end{bmatrix} = [ac + bd]$. (For vectors $U$, $V$ in $\mathcal{M}_{2,1}$ we define $U \cdot V = U^T V$.) You should note that the dot product of two vectors is a $1 \times 1$ matrix which has the same properties as a scalar. For this reason we shall not use the matrix symbolism and shall treat $U \cdot V$ as a scalar. Thus the dot product of $[2, -3]$ and $[-3, 1]$ is $2(-3) + (-3)(1) = -9$.

For any vector $U = [a, b]$ we have $U \cdot U \geqslant 0$, and $U \cdot U = 0$ only in case $U = [0, 0]$. To prove this, we simply note that $U \cdot U = UU^T = a^2 + b^2$ and that $a^2 + b^2 > 0$ except when both $a$ and $b$ are 0. You should now use the definition of dot product and the familiar matrix properties to prove (see Exercise 3) the first three parts of Theorem 4.8.1.

**THEOREM 4.8.1** *For any vectors $U$, $V$, $W$ in $\mathcal{M}_{1,2}$ and any scalar $c$, the dot product satisfies the following properties:*

1. $U \cdot V = V \cdot U$;
2. $U \cdot (V + W) = U \cdot V + U \cdot W$;
3. $U \cdot (cV) = c(U \cdot V)$;
4. $U \cdot U \geqslant 0$ *and* $U \cdot U = 0$ *only in case* $U = 0$.

We are now prepared to define the **length of a vector** $U = [a, b]^T$ as $(U \cdot U)^{1/2}$, or equivalently, as $(U^T U)^{1/2}$. We will use the symbol $\|U\|$ to stand for the length of $U$. Note that, by Property 4. of Theorem 4.8.1 the length of a vector $U \neq 0$ is always defined and is the (positive) square root of $U \cdot U$. For $U = 0$, we have $\|U\| = 0$. Further note that this definition of length is consistent with our representation of vectors as arrows, described in Section 4.7. Thus the arrow corresponding to the vector $[a, b]$ is the arrow from the origin to the point whose horizontal and vertical coordinates are $a$ and $b$, respectively, as in Figure 4.8.1. The length of the arrow repre-

Figure 4.8.1

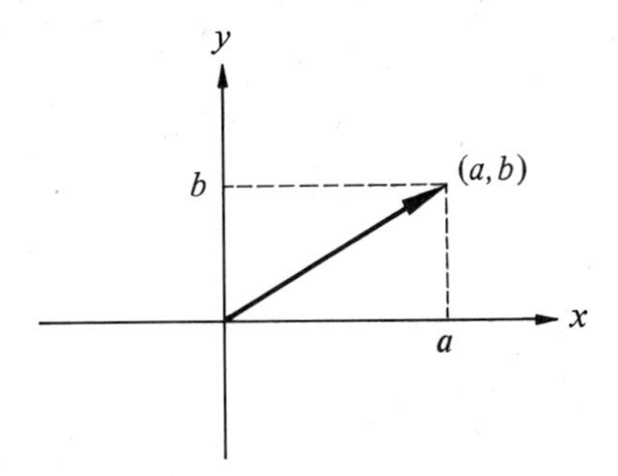


senting $[a, b]$ is $(a^2 + b^2)^{1/2}$. We say that a vector of length 1 is a **unit vector**. Among many examples, $[1, 0]$, $[0, 1]$, $[1/\sqrt{2}, 1/\sqrt{2}]$ are unit vectors.

Now let us consider two non-zero vectors $U = [a, b]^T$ and $V = [c, d]^T$. We define the **distance** between $U$ and $V$ to be the non-negative quantity

$$d(U, V) = \|U - V\| = [(U - V)^T(U - V)]^{1/2}.$$

Thus

$$d^2(U, V) = [a - c, b - d]\begin{bmatrix} a - c \\ b - d \end{bmatrix} = (a - c)^2 + (b - d)^2.$$

Similarly, we define the **angle** between $U$ and $V$ to be the angle $\theta$ such that $0° \leqslant \theta \leqslant 180°$ and

$$\cos\theta = \frac{U \cdot V}{\|U\|\,\|V\|}.$$

That this definition of angle between two vectors is consistent with the arrow representation of vectors may be seen as follows. We represent $U$, $V$, and $\theta$ as in Figure 4.8.2. Note that the length of one dotted line segment is the length of the vector $U - V$. Considering the triangle two sides of which are $U$ and $V$, we can find from the Law of Cosines that

$$\|U - V\|^2 = \|U\|^2 + \|V\|^2 - 2\|U\|\,\|V\|\cos\theta.$$

Solving this for $\cos\theta$ we have

$$\cos\theta = \frac{\|U\|^2 + \|V\|^2 - \|U - V\|^2}{2\|U\|\,\|V\|}.$$

Figure 4.8.2

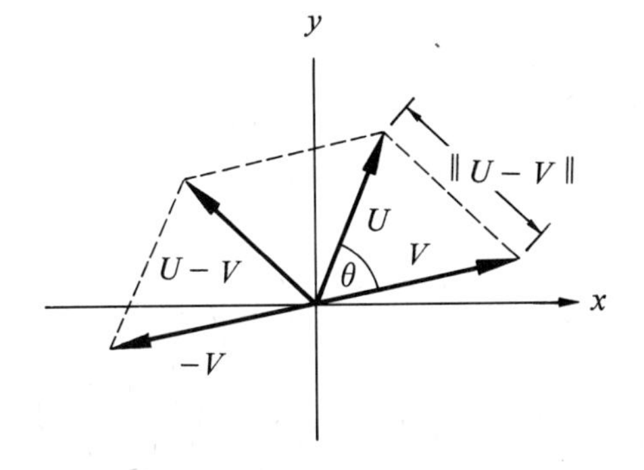

Now, a little algebra applied to the numerator of this fraction shows that

$$\begin{aligned}\|U\|^2 + \|V\|^2 - \|U - V\|^2 &= (a^2 + b^2) + (c^2 + d^2) \\ &\quad - ((a - c)^2 + (b - d)^2) \\ &= a^2 + b^2 + c^2 + d^2 \\ &\quad - (a^2 - 2ac + c^2 + b^2 - 2bd + d^2) \\ &= 2(ac + bd) \\ &= 2(U \cdot V).\end{aligned}$$

Thus we may write

$$\cos\theta = \frac{2U \cdot V}{2\|U\|\,\|V\|} = \frac{U \cdot V}{\|U\|\,\|V\|}.$$

**Example 4.8.1** Let $U = [-1, 2]^T$ and $V = [1, -1]^T$. Find the distance between $U$ and $V$ and the cosine of the angle between the two vectors.

First we see that

$$\begin{aligned}d^2(U, V) &= \|U - V\| \\ &= (U - V)^T(U - V) \\ &= [-2, 3]\begin{bmatrix}-2 \\ 3\end{bmatrix} \\ &= 13,\end{aligned}$$

so that

$$d(U, V) = \sqrt{13}.$$

To find $\cos\theta$ we note that $U \cdot V = [-1, 2] \cdot [1, -1] = -1 - 2 = -3$; and $\|U\| = (U \cdot U)^{1/2} = ((-1)^2 + (2)^2)^{1/2} = \sqrt{5}$; and $\|V\| = (V \cdot V)^{1/2} = ((1)^2 + (-1)^2)^{1/2} = \sqrt{2}$. Thus we see that

$$\cos\theta = \frac{U \cdot V}{\|U\|\,\|V\|} = \frac{-3}{\sqrt{5}\sqrt{2}} = -0.9487.$$

The angle $\theta$ can now be found by consulting a table of the cosine function. (If you have a table available you may verify that $\theta$ is approximately 161.5°; if you do not have a table, sketch the vectors and measure the angle $\theta$.) ∎

We note that if two vectors $U$ and $V$ in 2-space are at right angles (Figure 4.8.3) then the cosine of the angle between them is 0. Thus we must

**Figure 4.8.3**

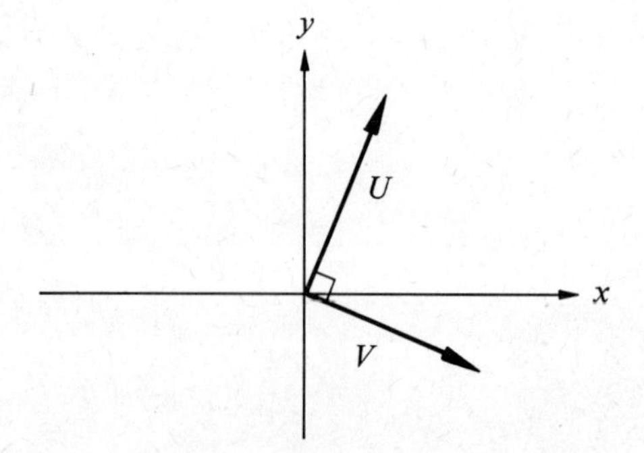

have $U \cdot V = 0$. Conversely, if $U \cdot V = 0$, and neither $U$ nor $V$ is 0, so that $\|U\| \neq 0$ and $\|V\| \neq 0$, then the cosine of the angle between them is 0 and the vectors are at right angles. In this case we use the terminology that $U$ and $V$ are **perpendicular** or that they are **orthogonal** to each other. The zero vector is considered to be orthogonal to every vector.

Example 4.8.2 Consider the vector $U = [1, 2]$. Find a unit vector $V$ which is orthogonal to $U$. We first find a vector $W$ which is orthogonal to $U$ as follows. If $W = [a, b]$, then $W$ is orthogonal to $U$ if and only if $U \cdot W = 0$. Thus we require that $1a + 2b = 0$, and thus we may choose any values of $a$ and $b$ (other than $a = b = 0$) which satisfy this equation; for example, we may take $a = 2$, and $b = -1$. Then $W = [2, -1]$ is orthogonal to $U$. Now to obtain a vector of unit length which is orthogonal to $U$, we simply divide $W$ by $\|W\|$ to obtain

$$V = \frac{1}{\|W\|} W = \frac{1}{\sqrt{5}} W = \left[\frac{2}{\sqrt{5}}, -\frac{1}{\sqrt{5}}\right].$$

Thus $V$ has unit length $\left(\text{since } \|V\| = \left\|\frac{1}{\|W\|} W\right\| = \frac{1}{\|W\|} \|W\| = 1\right)$, and $V$ is orthogonal to $U$ since by our properties for dot product,

$$V \cdot U = \left(\frac{1}{\|W\|} W\right) \cdot U = \frac{1}{\|W\|} \left(W \cdot U\right) = \frac{1}{\|W\|} (0) = 0. \quad ▮$$

Finally, we say that two vectors $U = [a, b]$ and $V = [c, d]$ have the **same direction** if the angle between them is $0°$, and have **opposite direction** if the angle between them is $180°$. Now the angle between any vector $U$ and itself is $0°$, so that $U$ has the same direction as $U$. Since $\|-U\| = ((-U) \cdot (-U))^{1/2} = ((-1)^2 U \cdot U)^{1/2} = (U \cdot U)^{1/2} = \|U\|$, we may compute the angle $\theta$ between $U$ and $-U$ as $180°$ since

$$\cos \theta = \frac{U \cdot (-U)}{\|U\| \, \|U\|} = \frac{-U \cdot U}{U \cdot U} = -1.$$

Thus $-U$ has the opposite direction to $U$.

Example 4.8.3 Find a unit vector $V_1$ with the same direction as $U = [1, 2]$ and a unit vector $V_2$ with the opposite direction. Clearly, $[1, 2]$ has the same direction as itself, so that we take $V_1 = U/\|U\| = [1, 2]/\|[1, 2]\| = [1/\sqrt{5}, 2/\sqrt{5}]$. Moreover $-[1, 2]$ has opposite direction to $[1, 2]$ so that $V_2 = [-1/\sqrt{5}, -2/\sqrt{5}]$ will suffice. (Are these unique?) ▮

## 4.8 Exercises

1. Sketch the vectors and find their dot product:

   a. $[1, 0], [0, 2]$ b. $[1, 1], [1, 3]$
   c. $[-1, 2], [1, 1/2]$ d. $[2, 1], [-1, 3]$
   e. $[1, 2], [2, 4]$ f. $[1, 2], [-3, -6]$

2. Find the length of
   a. [1, 0] b. [2, 1] c. [−2, 3]
   d. [3, 4] e. [−5, 12] f. 2[1, 2]

3. Sketch the vector $U = [3, 4]$ and find a unit vector in the direction of $U$.

4. Prove Theorem 4.8.1.

5. Find the cosine of the angle and the distance between the two vectors in each part of Exercise 1.

6. a. Show that if $[x, y]$ has the same direction as [1, 2], then $[x, y]$ is a scalar multiple of [1, 2].
   b. Show that if $[x, y]$ has the same direction as $[a, b]$, then $[x, y]$ is a scalar multiple of $[a, b]$

7. Find a unit vector which is orthogonal to [3, 4].

### Suggested Reading

G. Strang. *Linear Algebra and its Applications.* New York: Academic Press, 1976.
D. Finkbeiner. *Introduction to Matrices and Linear Transformations.* San Francisco: Freeman, 1966.
P. Halmos. *Finite-Dimensional Vector Spaces* 2nd ed. New York: Van Nostrand Reinhold, 1958.

### New Terms

dot product, 132
length of a vector, 132
unit vector, 133
distance, 133
angle, 133
perpendicular, 135
orthogonal, 135
same direction, 135
opposite direction, 135

# Determinants, Rank, and Change of Basis

# 5

## 5.1 Definition and Evaluation of Determinants

In this chapter we wish to describe an important function—a function which associates with each *square* matrix $A$ a real number, called the **determinant** of $A$. Theorems will be presented which facilitate the computation of determinants and which portray the relationship between the determinant of a matrix and the rank of the matrix. The relationship between the rank of a square matrix and the existence of an inverse will be used to establish a non-singular basis matrix that is helpful in studying the effects of a change of basis upon the coordinates of a given vector.

The value of the determinant function at a square matrix $A$ is commonly denoted as det $A$ or as $|A|$. In a case where the elements of $A$ are given explicitly as

$$A = \begin{bmatrix} a_{11} & \cdots & a_{1n} \\ a_{21} & \cdots & a_{2n} \\ \cdot & \cdots & \cdot \\ a_{n1} & \cdots & a_{nn} \end{bmatrix},$$

we may write

$$\det \begin{bmatrix} a_{11} & \cdots & a_{1n} \\ a_{21} & \cdots & a_{2n} \\ \cdot & \cdots & \cdot \\ a_{n1} & \cdots & a_{nn} \end{bmatrix} \quad \text{or} \quad \begin{vmatrix} a_{11} & \cdots & a_{1n} \\ a_{21} & \cdots & a_{2n} \\ \cdot & \cdots & \cdot \\ a_{n1} & \cdots & a_{nn} \end{vmatrix}$$

for the value of the determinant function at $A$. Our choice of the following defining properties is motivated by the Gauss-Jordan reduction operations of Chapter 2. By themselves, they provide a straightforward method for computation. [See Finkbeiner (Suggested Reading) for proof that there exists a unique determinant function with these properties.] The four defining properties of the determinant function are:

**Property 1** If any two rows of a square matrix are interchanged, then the determinant of the resulting matrix is the negative of the determinant of the original matrix.

Example 5.1.1

$$\begin{vmatrix} 1 & 2 & 3 \\ 4 & 5 & 6 \\ 7 & 8 & 9 \end{vmatrix} = -\begin{vmatrix} 4 & 5 & 6 \\ 1 & 2 & 3 \\ 7 & 8 & 9 \end{vmatrix}.$$ ∎

**Property 2** If any row of a square matrix is multiplied by a non-zero scalar (real number) $k$, then the determinant of the resulting matrix is $k$ times the determinant of the original matrix.

Example 5.1.2

$$\begin{vmatrix} 1 & 2 & 3 \\ 4k & 5k & 6k \\ 7 & 8 & 9 \end{vmatrix} = (k)\begin{vmatrix} 1 & 2 & 3 \\ 4 & 5 & 6 \\ 7 & 8 & 9 \end{vmatrix}.$$ ∎

**Property 3** If a scalar multiple of one row of a square matrix is added to another row, then the determinant of the resulting matrix is the *same* as the determinant of the original matrix.

Example 5.1.3

$$\begin{vmatrix} 1 & 2 & 3 \\ 4 & 5 & 6 \\ 7 & 8 & 9 \end{vmatrix} = \begin{vmatrix} 1 & 2 & 3 \\ 6 & 9 & 12 \\ 7 & 8 & 9 \end{vmatrix}.$$

In the first matrix we added twice the first row to the second row to obtain the second matrix. ∎

**Property 4** The determinant of an upper triangular matrix (all the elements below the main diagonal are zero) is the *product* of the diagonal elements.

Example 5.1.4

$$\begin{vmatrix} 2 & 5 \\ 0 & 3 \end{vmatrix} = 6; \qquad \begin{vmatrix} 0 & 0 & 0 \\ 0 & 0 & 0 \\ 0 & 0 & 0 \end{vmatrix} = 0;$$

$$\begin{vmatrix} 1 & 0 & 0 \\ 0 & 1 & 0 \\ 0 & 0 & 1 \end{vmatrix} = 1; \qquad \begin{vmatrix} 1 & 17 & 39 \\ 0 & -1 & 68 \\ 0 & 0 & -2 \end{vmatrix} = 2.$$ ∎

In the special case of a $1 \times 1$ matrix, its determinant is simply the single element of the matrix.

Example 5.1.5 Let us now, by way of illustration, find the value of the determinant function at the matrix

$$\begin{bmatrix} 1 & -1 & 2 \\ -2 & 4 & -3 \\ 3 & -3 & 9 \end{bmatrix}.$$

Applying Property 3, we add 2 times the first row to the second, and subtract 3 times the first row from the third, so that

$$\begin{vmatrix} 1 & -1 & 2 \\ -2 & 4 & -3 \\ 3 & -3 & 9 \end{vmatrix} = \begin{vmatrix} 1 & -1 & 2 \\ 0 & 2 & 1 \\ 0 & 0 & 3 \end{vmatrix}.$$

Now from Property 4, we find that since the last result is triangular, the determinant has the value 6. ∎

Example 5.1.6 As another example, let us find the determinant of

$$\begin{bmatrix} 1 & 2 & 3 \\ 4 & 5 & 6 \\ 7 & 8 & 9 \end{bmatrix}.$$

Using Property 3 twice we see that

$$\begin{vmatrix} 1 & 2 & 3 \\ 4 & 5 & 6 \\ 7 & 8 & 9 \end{vmatrix} = \begin{vmatrix} 1 & 2 & 3 \\ 0 & -3 & -6 \\ 0 & -6 & -12 \end{vmatrix}.$$

Now using Property 2, we see that

$$\begin{vmatrix} 1 & 2 & 3 \\ 0 & -3 & -6 \\ 0 & -6 & -12 \end{vmatrix} = (-3)\begin{vmatrix} 1 & 2 & 3 \\ 0 & 1 & 2 \\ 0 & -6 & -12 \end{vmatrix};$$

and now re-applying Property 3, we find that

$$(-3)\begin{vmatrix} 1 & 2 & 3 \\ 0 & 1 & 2 \\ 0 & -6 & -12 \end{vmatrix} = (-3)\begin{vmatrix} 1 & 2 & 3 \\ 0 & 1 & 2 \\ 0 & 0 & 0 \end{vmatrix} = (-3)(1 \cdot 1 \cdot 0) = 0. \quad ∎$$

Observe that in the foregoing examples it is only necessary to *triangularize* rather than reduce to echelon form; that is, we need not zero out a column above a pivotal element, but only below it.

We digress briefly to assure you that the determinant function described in Properties 1–4 above is the very same determinant often studied in high-school mathematics. There, the determinant of a $2 \times 2$ matrix $\begin{bmatrix} a & b \\ c & d \end{bmatrix}$ is usually defined as $\begin{vmatrix} a & b \\ c & d \end{vmatrix} = ad - bc$, while the determinant of a $3 \times 3$ matrix

$$\begin{bmatrix} a & b & c \\ d & e & f \\ g & h & i \end{bmatrix}$$

is given as $(aei - afh - bdi + bfg + cdh - ceg)$. Let us examine the $2 \times 2$ matrix $\begin{bmatrix} a & b \\ c & d \end{bmatrix}$. If $a \neq 0$, then by virtue of Property 2, $\begin{vmatrix} a & b \\ c & d \end{vmatrix} = a\begin{vmatrix} 1 & b/a \\ c & d \end{vmatrix}$; and by Property 3, $a\begin{vmatrix} 1 & b/a \\ c & d \end{vmatrix} = a\begin{vmatrix} 1 & b/a \\ 0 & d - cb/a \end{vmatrix}$. This, by Property 4, is equal to $a(d - cb/a) = ad - bc$, the familiar $2 \times 2$ determinant recalled above. If $a = 0$, then $\begin{vmatrix} a & b \\ c & d \end{vmatrix} = \begin{vmatrix} 0 & b \\ c & d \end{vmatrix} = -\begin{vmatrix} c & d \\ 0 & b \end{vmatrix}$. But the last result is upper triangular, so that

$$\begin{vmatrix} a & b \\ c & d \end{vmatrix} = -\begin{vmatrix} c & d \\ 0 & b \end{vmatrix} = -cb = ad - bc, \qquad \text{since } a = 0.$$

We leave it to you to verify from our definition that the $3 \times 3$ matrix

$$\begin{bmatrix} a & b & c \\ d & e & f \\ g & h & i \end{bmatrix}$$

does indeed have the determinant value shown above. If you are familiar with $3 \times 3$ determinants, you have probably learned certain patterns for obtaining their value. We regret to say that no such nice patterns exist for large matrices; or rather, patterns exist, but are so elaborately complicated that they are unsuited for practical calculations. An efficient way to find the determinant of a large matrix is to triangularize it using the analogues of the Gauss-Jordan reduction rules, although there are other methods which have theoretical and practical significance. We shall discuss some of these in the next section of this chapter.

We now wish to point out three related facts which are useful in evaluating determinants.

**Fact 1** *If in the Gauss-Jordan reduction of a determinant, a zero appears on the diagonal and all elements below it in that column are zero, then the determinant is zero.*

That the above statement is true may be seen from the following considerations. Suppose that we have reduced the determinant through $k$ steps, so that in the first $k$ diagonal positions there is a 1, and in the $(k + 1)$th diagonal position, and everywhere below it in the $(k + 1)$th column, there is a zero.

$$k\left\{\begin{matrix} \\ \\ \\ \\ \\ \end{matrix}\right. \overbrace{\begin{vmatrix} 1 & X & X & \cdots & X & X & \cdots & X \\ 0 & 1 & X & \cdots & X & X & \cdots & X \\ 0 & 0 & 1 & \cdots & X & X & \cdots & X \\ \cdot & \cdot & \cdot & \cdots & \cdot & \cdot & \cdots & \cdot \\ 0 & 0 & 0 & \cdots & 1 & X & \cdots & X \\ 0 & 0 & 0 & \cdots & 0 & 0 & \cdots & X \\ \cdot & \cdot & \cdot & \cdots & \cdot & \cdot & \cdots & \cdot \\ 0 & 0 & 0 & \cdots & 0 & 0 & \cdots & X \end{vmatrix}}^{k}$$

In the Gauss-Jordan process, as applied to systems of linear equations, we have learned that in this situation we merely skip over the $(k + 1)$th column and proceed to reduce the $(k + 2)$th column using the element in the $(k + 1, k + 2)$ position. But now observe that if we do that, then, when we have completed the reduction process and the determinant is reduced to triangular form, there will still be a zero in the $(k+1, k+1)$ position. No matter what the following reduction process has wrought, the triangular final matrix will have a 0 on the main diagonal and its determinant will have the value 0. Thus, in practice, we would have stopped our work at the $(k+1)$th step, since the determinant must be zero.

Example 5.1.7 Find the determinant of

$$\begin{bmatrix} 1 & -1 & 1 & -1 & 1 \\ 1 & 0 & 2 & 1 & 0 \\ -1 & 1 & -1 & 0 & 1 \\ 2 & -2 & 2 & -1 & 3 \\ 2 & -1 & 3 & 1 & 3 \end{bmatrix}.$$

The result of zeroing out the first column below the 1, 1 position is

$$\begin{vmatrix} 1 & -1 & 1 & -1 & 1 \\ 0 & 1 & 1 & 2 & -1 \\ 0 & 0 & 0 & -1 & 2 \\ 0 & 0 & 0 & 1 & 1 \\ 0 & 1 & 1 & 3 & 1 \end{vmatrix}.$$

Now using the 2, 2 position to zero out below it, we obtain

$$\begin{vmatrix} 1 & -1 & 1 & -1 & 1 \\ 0 & 1 & 1 & 2 & -1 \\ 0 & 0 & 0 & -1 & 2 \\ 0 & 0 & 0 & 1 & 1 \\ 0 & 0 & 0 & 1 & 2 \end{vmatrix}.$$

Now observe that the 3, 3 element *and all elements of the column below it are zeros.* Hence, by our preceding argument, the determinant must be zero. This is verified by continuing the reduction using the 3, 4 element to obtain

$$\begin{vmatrix} 1 & -1 & 1 & -1 & 1 \\ 0 & 1 & 1 & 2 & -1 \\ 0 & 0 & 0 & -1 & 2 \\ 0 & 0 & 0 & 0 & 3 \\ 0 & 0 & 0 & 0 & 4 \end{vmatrix},$$

which is in triangular form. The presence of the 0 in the 3, 3 position now assures us that the determinant (being the product of the diagonal elements) is zero. ∎

**Fact 2** *If a row of a matrix consists entirely of zeros, then the determinant of that matrix is 0.*

This also is easily seen to be true. For, if any row of the matrix $A$ is zero and we multiply that row by a non-zero scalar $k$, the result is still the matrix $A$. But by Property 2, the determinant is $k$ times the determinant of $A$. Thus $k\,|A| = |A|$ for any $k \neq 0$. It must then be true that $|A| = 0$.

Example 5.1.8
$$A = \begin{bmatrix} 1 & -1 & 1 & 2 \\ 0 & 0 & 0 & 0 \\ 2 & -1 & 1 & -3 \\ 1 & -1 & 3 & 7 \end{bmatrix} = \begin{bmatrix} 1 & -1 & 1 & 2 \\ k\cdot 0 & k\cdot 0 & k\cdot 0 & k\cdot 0 \\ 2 & -1 & 1 & -3 \\ 1 & -1 & 3 & 7 \end{bmatrix}.$$

Thus

$$\begin{vmatrix} 1 & -1 & 1 & 2 \\ 0 & 0 & 0 & 0 \\ 2 & -1 & 1 & -3 \\ 1 & -1 & 3 & 7 \end{vmatrix} = \begin{vmatrix} 1 & -1 & 1 & 2 \\ k\cdot 0 & k\cdot 0 & k\cdot 0 & k\cdot 0 \\ 2 & -1 & 1 & -3 \\ 1 & -1 & 3 & 7 \end{vmatrix}.$$

$$= k\begin{vmatrix} 1 & -1 & 1 & 2 \\ 0 & 0 & 0 & 0 \\ 2 & -1 & 1 & -3 \\ 1 & -1 & 3 & 7 \end{vmatrix}. \quad \blacksquare$$

The validity of our second fact may also be demonstrated in another way. Using the above example, and interchanging the zero row with the last row, we obtain

$$\begin{vmatrix} 1 & -1 & 1 & 2 \\ 0 & 0 & 0 & 0 \\ 2 & -1 & 1 & -3 \\ 1 & -1 & 3 & 7 \end{vmatrix} = -\begin{vmatrix} 1 & -1 & 1 & 2 \\ 1 & -1 & 3 & 7 \\ 2 & -1 & 1 & -3 \\ 0 & 0 & 0 & 0 \end{vmatrix}.$$

We now reduce the last result to triangular form, obtaining

$$-\begin{vmatrix} 1 & -1 & 1 & 2 \\ 0 & 0 & 2 & 5 \\ 0 & 1 & -1 & -7 \\ 0 & 0 & 0 & 0 \end{vmatrix} = \begin{vmatrix} 1 & -1 & 1 & 2 \\ 0 & 1 & -1 & -7 \\ 0 & 0 & 2 & 5 \\ 0 & 0 & 0 & 0 \end{vmatrix}.$$

But the presence of the zero in the 4, 4 position assures us that the determinant is zero.

**Fact 3** *If two rows of a matrix are proportional, the determinant of the matrix is zero.*

This is now very easy to see, for if some row is $k$ times another row we subtract $k$ times the one row from the other and produce a matrix with one row zero. By Fact 2, the determinant is 0.

These facts will be useful in evaluating the determinants in the following exercises.

## 5.1 Exercises

In Exercises 1–10, find the determinant of the given matrix.

1. $\begin{bmatrix} 1 & 2 \\ 3 & 4 \end{bmatrix}$

2. $\begin{bmatrix} -3 & 2 \\ 1 & 2 \end{bmatrix}$

3. $\begin{bmatrix} 1 & -1 & 0 \\ 2 & 1 & 3 \\ 1 & 0 & 1 \end{bmatrix}$

4. $\begin{bmatrix} 0 & 2 & 1 \\ -1 & 1 & -2 \\ 2 & 2 & 3 \end{bmatrix}$

5. $\begin{bmatrix} 2 & -4 & 2 & 0 \\ -1 & 0 & 1 & 2 \\ 1 & 0 & 1 & 3 \\ 3 & 2 & 1 & 0 \end{bmatrix}$

6. $\begin{bmatrix} 1 & 0 & 4 & -1 \\ -4 & 0 & 2 & 0 \\ 3 & 2 & 2 & 1 \\ 2 & 5 & 3 & 2 \end{bmatrix}$

7. $\begin{bmatrix} 1 & 0 & 1 & 1 & 0 \\ 0 & 1 & 0 & 0 & 1 \\ 1 & 0 & 1 & 1 & 0 \\ 0 & 1 & 0 & 1 & 0 \\ 0 & 0 & 1 & 0 & 1 \end{bmatrix}$

8. $\begin{bmatrix} 1 & 0 & 1 & 1 & 1 \\ -1 & 2 & -1 & 3 & 1 \\ 2 & 1 & 2 & 1 & 1 \\ 1 & -1 & 0 & 2 & 1 \\ 3 & 1 & 2 & 0 & 1 \end{bmatrix}$

9. $\begin{bmatrix} 0 & 0 & 2 \\ 0 & 3 & 0 \\ 4 & 0 & 0 \end{bmatrix}$

10. $\begin{bmatrix} 0 & 0 & 0 & 2 \\ 0 & 0 & 3 & 0 \\ 0 & 4 & 0 & 0 \\ 5 & 0 & 0 & 0 \end{bmatrix}$

11. Verify that $\begin{vmatrix} a & b & c \\ d & e & f \\ g & h & i \end{vmatrix} = aei - afh - bdi + bfg + cdh - ceg.$

12. In Chapter 1 we defined the result of multiplying a matrix $A$ by a scalar $s$ to be the matrix whose elements are the elements of $A$ each multiplied by $s$. With $s = 3$ and $A = \begin{bmatrix} 1 & 2 \\ 3 & 4 \end{bmatrix}$ what is $|sA|$ and how is it related to $|A|$? What is the result if $A$ is a $3 \times 3$ matrix?

In Exercises 13–20 use the facts developed in the latter part of the section to show that each determinant is zero.

13. $\begin{vmatrix} 1 & 2 & 1 \\ 1 & 2 & 4 \\ 2 & 4 & 4 \end{vmatrix}$

14. $\begin{vmatrix} 1 & 2 & 3 \\ 2 & 4 & 6 \\ 1 & 5 & 8 \end{vmatrix}$

15. $\begin{vmatrix} 1 & -1 & 2 & 1 & 3 \\ 1 & 0 & 4 & 2 & 2 \\ 1 & 0 & 4 & 3 & 3 \\ 2 & 0 & 8 & 4 & 7 \\ 3 & 0 & 12 & 8 & 10 \end{vmatrix}$

16. $\begin{vmatrix} 1 & 2 & 3 & 3 & 4 \\ -1 & -1 & -1 & 0 & -1 \\ 2 & 6 & 10 & 12 & 14 \\ 1 & 3 & 6 & 6 & 9 \\ 3 & 5 & 8 & 9 & 13 \end{vmatrix}$

17. $\begin{vmatrix} 1 & 2 & 1 & 3 \\ 2 & 5 & 1 & 8 \\ 3 & 7 & 2 & 11 \\ 3 & 5 & 1 & 1 \end{vmatrix}$

18. $\begin{vmatrix} 1 & -1 & 0 & 1 \\ 2 & 1 & -3 & 4 \\ 0 & 3 & -3 & 2 \\ 1 & 5 & 4 & 7 \end{vmatrix}$

19. $\begin{vmatrix} 2 & 4 & 6 & 8 \\ 1 & 2 & 3 & 4 \\ 5 & 1 & 3 & 7 \\ 6 & 1 & 5 & 4 \end{vmatrix}$

20. $\begin{vmatrix} a & b & c \\ a^2 - ab & ab - b^2 & ac - bc \\ a^2 - b^2 & a^2 + b^2 & a^3 + b^3 \end{vmatrix}$

### Suggested Reading

D. Finkbeiner. *Introduction to Matrices and Linear Transformations.* San Francisco: Freeman, 1966.

B. Jones. *Linear Algebra.* San Francisco: Holden-Day, 1973.

### New Terms

determinant, 137

## 5.2 Some Further Properties of Determinants

For the sake of completeness, and for some theoretical advantages, we present here some results of a more traditional approach to determinants.

To begin with, let us consider an $n \times n$ matrix $A = [a_{ij}]$, and a particular element in that matrix, say $a_{ks}$. We define the **cofactor** $A_{ks}$ of $a_{ks}$ to be

$$A_{ks} = (-1)^{k+s} D_{ks},$$

where $D_{ks}$ denotes the determinant of the $(n-1) \times (n-1)$ submatrix of $A$ found by deleting the $k$th row and the $s$th column from $A$.

Example 5.2.1 In the matrix

$$A = \begin{bmatrix} 1 & 2 & 3 \\ 4 & 5 & 6 \\ 7 & 8 & 9 \end{bmatrix}$$

the cofactor of the element 6, which is in the 2, 3 position is

$$A_{23} = (-1)^{2+3} \begin{vmatrix} 1 & 2 \\ 7 & 8 \end{vmatrix} = 6.$$

Here, $D_{23} = \begin{vmatrix} 1 & 2 \\ 7 & 8 \end{vmatrix}$ is seen to be the determinant of the submatrix obtained from $A$ by deleting the row and column of $A$ in which 6 lies. ∎

The cofactor of 5 in the above example is (since 5 is in the 2, 2 position)

$$A_{22} = (-1)^{2+2} \begin{vmatrix} 1 & 3 \\ 7 & 9 \end{vmatrix} = -12.$$

The cofactors of the other elements of $A$ are:

$A_{11} = -3$, $A_{12} = 6$, $A_{13} = -3$, $A_{21} = 6$, $A_{31} = -3$, $A_{32} = 6$, $A_{33} = -3$,

which you should now verify.

## 5.2 Exercises

Find the cofactors of each element of

1. $\begin{bmatrix} 3 & 1 & 5 \\ 2 & 0 & 3 \\ 1 & 1 & 2 \end{bmatrix}$.

2. $\begin{bmatrix} 3 & 0 & 1 \\ 2 & 1 & 1 \\ 1 & 2 & 3 \end{bmatrix}$.

Find the cofactors of the elements of the second row of

3. $\begin{bmatrix} 1 & 1 & 2 & 3 \\ -1 & 1 & 2 & 1 \\ 0 & 1 & 1 & 2 \\ 3 & 0 & -1 & 1 \end{bmatrix}$.

4. $\begin{bmatrix} 0 & 2 & 4 & 4 \\ 2 & 0 & 1 & 3 \\ 0 & 1 & 1 & 2 \\ 3 & 1 & 0 & 3 \end{bmatrix}$.

Find the cofactors of the elements of the second column of

5. $\begin{bmatrix} 2 & 1 & 0 & 3 \\ 1 & -1 & 1 & 2 \\ 1 & 1 & 2 & 1 \\ 2 & 2 & 3 & 4 \end{bmatrix}$.

6. $\begin{bmatrix} 3 & -1 & 1 & 1 \\ 3 & 1 & -4 & 1 \\ 2 & 2 & 6 & 4 \\ 1 & 2 & 7 & 1 \end{bmatrix}$.

7. Let
$$\begin{bmatrix} 1 & 2 & 3 \\ -1 & 0 & 4 \\ 2 & 1 & 2 \end{bmatrix} = A.$$
Find the cofactors of the elements of the first row of $A$.

8. Find the cofactors of the elements of the first column of $A^T$ with $A$ as in Exercise 7. Compare your results with those of Exercise 7.

Our first result is the following:

**THEOREM 5.2.1** *The determinant of a matrix $A = [a_{ij}]$ is the sum of the products of the elements of any row multiplied by the corresponding cofactors of those elements. That is, for any row, say the ith,*

$$|A| = a_{i1}A_{i1} + a_{i2}A_{i2} + \cdots + a_{in}A_{in}.$$

Example 5.2.2 Consider the matrix

$$A = \begin{bmatrix} 1 & 2 & 3 \\ 4 & 5 & 6 \\ 7 & 8 & 9 \end{bmatrix},$$

the cofactors of which we computed above. Let us calculate the determinant of $A$ (which we already know is zero from Example 5.1.6) by applying the above result to the first row of $A$:

$$|A| = a_{11}A_{11} + a_{12}A_{12} + a_{13}A_{13} = 1(-3) + 2(6) + 3(-3) = 0.$$

Let us now calculate the determinant of $A$ using the other rows, just to verify that it makes no difference which row we use to expand the determinant of $A$. Using the second row, for instance, we have

$$|A| = a_{21}A_{21} + a_{22}A_{22} + a_{23}A_{23} = 4(6) + 5(-12) + 6(6) = 0;$$

and if we use the third row, we have

$$|A| = a_{31}A_{31} + a_{32}A_{32} + a_{33}A_{33} = 7(-3) + 8(6) + 9(-3) = 0. \quad \blacksquare$$

Example 5.2.3 Consider the matrix of Example 5.1.5,

$$A = \begin{bmatrix} 1 & -1 & 2 \\ -2 & 4 & -3 \\ 3 & -3 & 9 \end{bmatrix}.$$

In that example we had already seen that $|A| = 6$. Let us now expand the determinant of $A$ using the first row. We find that $A_{11} = 27$, $A_{12} = 9$, and $A_{13} = -6$. Then

$$|A| = (1)A_{11} + (-1)A_{12} + (2)A_{13} = 27 - 9 + 2(-6) = 6.$$

You should now continue this example by expanding $|A|$ by its second row and by its third row. ∎

The next result, which is a companion theorem to the first, is interesting in itself, and, with the first, will be used in the next section to establish a connection between the determinant and the inverse of a square matrix.

**THEOREM 5.2.2** *The sum of the products of the elements of any row of a square matrix $A$ and the cofactors of the corresponding elements of a different row of $A$ is zero. In symbols, for $i \neq k$,*

$$a_{i1}A_{k1} + a_{i2}A_{k2} + \cdots + a_{in}A_{kn} = 0.$$

Example 5.2.4 Consider the matrix of the previous example. If we form the sum of the products of the elements of the second row ($i = 2$) with the cofactors of the elements of the first row ($k = 1$), we obtain

$$(-2)(27) + 4(9) - 3(-6) = -54 + 36 + 18 = 0.$$

Similarly, the third row elements along with the first row cofactors yield

$$3(27) - 3(9) + 9(-6) = 81 - 27 - 54 = 0. \quad \blacksquare$$

## 5.2 Exercises (continued)

In Exercises 9–12, find the determinant by applying Theorem 5.2.1. to the *first row*. Verify that the same value is obtained by applying Theorem 5.2.1 to the *third* row.

9. $\begin{vmatrix} 1 & -1 & 2 \\ 3 & 2 & -4 \\ 6 & -1 & 1 \end{vmatrix}$

10. $\begin{vmatrix} 1 & 4 & 7 \\ 2 & 5 & 8 \\ 3 & 6 & 9 \end{vmatrix}$

11. $\begin{vmatrix} 1 & -1 & 2 & 3 \\ 2 & 1 & 4 & -1 \\ 1 & 1 & 2 & -2 \\ -3 & 4 & 1 & 2 \end{vmatrix}$ 12. $\begin{vmatrix} 1 & 0 & 5 & 0 \\ 0 & 3 & 0 & 7 \\ 2 & 0 & 6 & 0 \\ 0 & 4 & 0 & 8 \end{vmatrix}$

13. Apply Theorem 5.2.2 to the first ($i = 1$) and third ($k = 3$) rows of each of the determinants in Exercises 9–12 above to obtain the value zero.

14. Find the determinants of the transposes of the matrices in Exercises 9 and 11 above and compare them with the determinants of the original matrices.

15. Prove Theorem 5.2.1 for arbitrary $2 \times 2$ matrices.

16. Prove Theorem 5.2.2 for arbitrary $2 \times 2$ matrices.

Before passing to our next two theorems, we ask you to consider the following: We evaluate the determinant of the $5 \times 5$ matrix below by the Gauss-Jordan triangularization process, listing the results of the reduction of the successive columns.

$$\begin{vmatrix} 1 & 1 & -1 & 1 & 1 \\ 2 & 3 & 0 & 1 & 3 \\ -1 & 1 & 6 & 0 & -1 \\ 3 & 5 & 3 & 8 & 3 \\ 1 & 0 & -3 & 7 & 3 \end{vmatrix} = \begin{vmatrix} 1 & 1 & -1 & 1 & 1 \\ 0 & 1 & 2 & -1 & 1 \\ 0 & 2 & 5 & 1 & 0 \\ 0 & 2 & 6 & 5 & 0 \\ 0 & -1 & -2 & 6 & 2 \end{vmatrix}$$

$$= \begin{vmatrix} 1 & 1 & -1 & 1 & 1 \\ 0 & 1 & 2 & -1 & 1 \\ 0 & 0 & 1 & 3 & -2 \\ 0 & 0 & 2 & 7 & -2 \\ 0 & 0 & 0 & 5 & 3 \end{vmatrix}$$

$$= \begin{vmatrix} 1 & 1 & -1 & 1 & 1 \\ 0 & 1 & 2 & -1 & 1 \\ 0 & 0 & 1 & 3 & -2 \\ 0 & 0 & 0 & 1 & 2 \\ 0 & 0 & 0 & 5 & 3 \end{vmatrix}$$

$$= \begin{vmatrix} 1 & 1 & -1 & 1 & 1 \\ 0 & 1 & 2 & -1 & 1 \\ 0 & 0 & 1 & 3 & -2 \\ 0 & 0 & 0 & 1 & 2 \\ 0 & 0 & 0 & 0 & -7 \end{vmatrix}$$

$$= -7.$$

In the above display, the first equality shows the reduction of the first column, the second equality shows the reduction of the second column, etc., so that the determinant is available as the product of the diagonal elements of the final triangular matrix.

Now, contrast this method of evaluating the above determinant with the method of cofactors. Using the latter method, we find that if we use the elements of the first row, for example, we obtain

$$\begin{vmatrix} 1 & 1 & -1 & 1 & 1 \\ 2 & 3 & 0 & 1 & 3 \\ -1 & 1 & 6 & 0 & -1 \\ 3 & 5 & 3 & 8 & 3 \\ 1 & 0 & -3 & 7 & 3 \end{vmatrix} = 1(-1)^{1+1}\begin{vmatrix} 3 & 0 & 1 & 3 \\ 1 & 6 & 0 & -1 \\ 5 & 3 & 8 & 3 \\ 0 & -3 & 7 & 3 \end{vmatrix}$$

$$+ 1(-1)^{1+2}\begin{vmatrix} 2 & 0 & 1 & 3 \\ -1 & 6 & 0 & -1 \\ 3 & 3 & 8 & 3 \\ 1 & -3 & 7 & 3 \end{vmatrix}$$

$$- 1(-1)^{1+3}\begin{vmatrix} 2 & 3 & 1 & 3 \\ -1 & 1 & 0 & -1 \\ 3 & 5 & 8 & 3 \\ 1 & 0 & 7 & 3 \end{vmatrix}$$

$$+ 1(-1)^{1+4}\begin{vmatrix} 2 & 3 & 0 & 3 \\ -1 & 1 & 6 & -1 \\ 3 & 5 & 3 & 3 \\ 1 & 0 & -3 & 3 \end{vmatrix}$$

$$+ 1(-1)^{1+5}\begin{vmatrix} 2 & 3 & 0 & 1 \\ -1 & 1 & 6 & 0 \\ 3 & 5 & 3 & 8 \\ 1 & 0 & -3 & 7 \end{vmatrix}.$$

In order to evaluate the original determinant we now have to evaluate each of the *five* $4 \times 4$ determinants. Each one, expanded by some row, is given in terms of *four* $3 \times 3$ determinants, each one of which is given in terms of *three* $2 \times 2$ determinants. Thus, to evaluate the original determinant by this method, we will need to calculate $5 \cdot 4 \cdot 3 = 60$ different $2 \times 2$ determinants. We recommend that you *not* complete the calculations!

Our next theorem is an interesting algebraic result.

**THEOREM 5.2.3** *If $A$ and $B$ are each $n \times n$ matrices, then $|AB| = |A|\,|B|$. That is, the determinant of the product of two square matrices is the product of their determinants.*

Example 5.2.5 Let $A = \begin{bmatrix} 1 & 2 \\ 3 & 4 \end{bmatrix}$, $B = \begin{bmatrix} 2 & -1 \\ 3 & 2 \end{bmatrix}$. Then $|A| = -2$, $|B| = 7$, so that $|A|\,|B| = -14$. And $|AB| = \begin{vmatrix} 8 & 3 \\ 18 & 5 \end{vmatrix} = 40 - 54 = -14$. ∎

Example 5.2.6

$$A = \begin{bmatrix} 1 & 2 & 3 \\ 1 & -1 & 2 \\ 2 & 1 & -3 \end{bmatrix}, \qquad B = \begin{bmatrix} 1 & 2 & 1 \\ 1 & 0 & -1 \\ 0 & -1 & 2 \end{bmatrix}.$$

Then
$$|A| = \begin{vmatrix} 1 & 2 & 3 \\ 0 & -3 & -1 \\ 0 & -3 & -9 \end{vmatrix} = \begin{vmatrix} 1 & 2 & 3 \\ 0 & -3 & -1 \\ 0 & 0 & -8 \end{vmatrix} = 24,$$

and

$$|B| = \begin{vmatrix} 1 & 2 & 1 \\ 0 & -2 & -2 \\ 0 & -1 & 2 \end{vmatrix} = -\begin{vmatrix} 1 & 2 & 1 \\ 0 & -1 & 2 \\ 0 & -2 & -2 \end{vmatrix} = -\begin{vmatrix} 1 & 2 & 1 \\ 0 & -1 & 2 \\ 0 & 0 & -6 \end{vmatrix} = -6,$$

so that $|A|\,|B| = -144$. In addition

$$|AB| = \begin{vmatrix} 3 & -1 & 5 \\ 0 & 0 & 6 \\ 3 & 7 & -5 \end{vmatrix} = -\begin{vmatrix} 3 & -1 & 5 \\ 3 & 7 & -5 \\ 0 & 0 & 6 \end{vmatrix}$$

$$= -\begin{vmatrix} 3 & -1 & 5 \\ 0 & 8 & -10 \\ 0 & 0 & 6 \end{vmatrix} = -144.$$ ∎

Unfortunately, no such nice result as Theorem 5.2.3 holds for addition or subtraction of matrices as the following example shows: Let

$$A = \begin{bmatrix} 1 & 0 & 0 \\ 0 & 0 & 0 \\ 0 & 0 & 1 \end{bmatrix}, \qquad B = \begin{bmatrix} 0 & 0 & 0 \\ 0 & 1 & 0 \\ 0 & 0 & 0 \end{bmatrix}.$$

Then $|A| = 0$, $|B| = 0$, but $|A + B| = 1$, while $|A - B| = -1$.

**Corollary 5.2.1** *For any $n \times n$ matrices $A$ and $B$,*

$$|AB| = |BA|.$$

**Proof** $|AB| = |A|\,|B| = |B|\,|A| = |BA|$. Then first and third equalities are true by Theorem 5.2.3. The second equality is also true, since $|A|$ and $|B|$ are scalars and hence commute. ∎

Our final theorem of this section is one which has both practical and theoretical significance.

**THEOREM 5.2.4** *The determinant of a matrix $A$ is equal to the determinant of the transpose $A^T$ of $A$.*

Example 5.2.7 Let $A = \begin{bmatrix} 1 & 2 \\ 3 & 4 \end{bmatrix}$, so that $A^T = \begin{bmatrix} 1 & 3 \\ 2 & 4 \end{bmatrix}$. Then $|A| = 4 - 6 = -2$ and $|A^T| = 4 - 6 = -2$. ∎

Example 5.2.8 With $A$ as in Example 5.1.5,

$$A^T = \begin{bmatrix} 1 & -2 & 3 \\ -1 & 4 & -3 \\ 2 & -3 & 9 \end{bmatrix},$$

so that
$$|A^T| = \begin{vmatrix} 1 & -2 & 3 \\ 0 & 2 & 0 \\ 0 & 1 & 3 \end{vmatrix} = 2\begin{vmatrix} 1 & -2 & 3 \\ 0 & 1 & 0 \\ 0 & 1 & 3 \end{vmatrix}$$
$$= 2\begin{vmatrix} 1 & -2 & 3 \\ 0 & 1 & 0 \\ 0 & 0 & 3 \end{vmatrix} = 6 = |A|.$$ ∎

The practical significance of this theorem to us at this stage is the following:

Properties 1–3 of determinants as listed in Section 5.1 are now valid if the word *row* is replaced by the word *column*; and Property 4 is valid if *upper triangular* is replaced by *lower triangular*.

Thus we may use column interchanges, divide or multiply a column by a non-zero scalar, and subtract a multiple of one column from another in reducing to triangular form. Moreover, Theorems 5.2.1 and 5.2.2 have valid column analogues, so that we may expand the determinant by the elements and cofactors of a column instead of by a row.

## 5.2 Exercises (continued)

In Exercises 17–22, verify Theorem 5.2.3 by separately calculating $|A|$, $|B|$ and $|AB|$.

17. $A = \begin{bmatrix} 1 & 2 \\ 3 & 4 \end{bmatrix}, B = \begin{bmatrix} 5 & 6 \\ 7 & 8 \end{bmatrix}$

18. $A = \begin{bmatrix} 1 & 2 \\ 1 & 3 \end{bmatrix}, B = \begin{bmatrix} -1 & 4 \\ -3 & 2 \end{bmatrix}$

19. $A = \begin{bmatrix} 1 & 0 & -1 \\ 2 & 1 & 3 \\ 1 & 4 & 1 \end{bmatrix}, B = \begin{bmatrix} 0 & 2 & -1 \\ 0 & 1 & 0 \\ 1 & 4 & 2 \end{bmatrix}$

20. $A = \begin{bmatrix} 4 & -1 & 2 \\ -1 & 0 & 2 \\ 2 & 1 & 1 \end{bmatrix}, B = \begin{bmatrix} 1 & 1 & 1 \\ 0 & 2 & 2 \\ 0 & 0 & 3 \end{bmatrix}$

21. $A = \begin{bmatrix} 1 & 1 & 2 & 1 \\ 1 & 0 & 1 & 1 \\ -1 & 1 & 3 & 0 \\ 2 & 4 & 1 & 1 \end{bmatrix}, B = \begin{bmatrix} 4 & 1 & -1 & 1 \\ 2 & 2 & 2 & 1 \\ 3 & 3 & 1 & 0 \\ 1 & 4 & 3 & 1 \end{bmatrix}$

22. $A = \begin{bmatrix} 1 & 0 & 1 & 1 \\ 0 & 1 & 1 & 0 \\ 2 & -1 & 1 & 2 \\ 2 & 2 & -1 & 1 \end{bmatrix}, B = \begin{bmatrix} 1 & 2 & 3 & 4 \\ -1 & 1 & -3 & 1 \\ 0 & 1 & 0 & 0 \\ 2 & 1 & 1 & 2 \end{bmatrix}$

In Exercises 23–28, verify Theorem 5.2.4 by finding $|A|$ and $|A^T|$ separately.

23. $A = \begin{bmatrix} 2 & 1 \\ 2 & 3 \end{bmatrix}$

24. $A = \begin{bmatrix} 1 & 4 \\ -1 & 2 \end{bmatrix}$

25. $A = \begin{bmatrix} 1 & 1 & 2 \\ -1 & 2 & 3 \\ 1 & 1 & 2 \end{bmatrix}$ 26. $A = \begin{bmatrix} 1 & 2 & 3 \\ 2 & 4 & 5 \\ 3 & 5 & 1 \end{bmatrix}$

27. $A = \begin{bmatrix} 1 & 1 & 4 & -1 \\ 2 & 1 & 5 & 3 \\ 1 & 1 & 2 & 1 \\ 1 & 0 & 1 & 2 \end{bmatrix}$ 28. $A = \begin{bmatrix} 1 & 2 & 3 & 4 \\ -1 & 1 & 2 & 1 \\ 0 & 1 & 1 & 0 \\ 1 & 0 & 0 & 1 \end{bmatrix}$

29. Prove Theorems 5.2.3 and 5.2.4 for arbitrary $2 \times 2$ matrices.

### New Terms

cofactor, 144

## 5.3 Some Applications of Determinants

We can now apply our foregoing theorems to find a "formula" for calculating the inverse of a square matrix, if it exists. Let us consider by way of example a $3 \times 3$ matrix

$$A = \begin{bmatrix} a_{11} & a_{12} & a_{13} \\ a_{21} & a_{22} & a_{23} \\ a_{31} & a_{32} & a_{33} \end{bmatrix};$$

and let us construct a new $3 \times 3$ matrix $B$ in the following way. To construct the first row of $B$, we write

$$B = \begin{bmatrix} A_{11} & A_{21} & A_{31} \\ \_\_\_ & \_\_\_ & \_\_\_ \\ \_\_\_ & \_\_\_ & \_\_\_ \end{bmatrix};$$

that is, the first *row* of $B$ consists of the cofactors of the elements of the first *column* of $A$. Similarly, the second *row* of $B$ consists of the cofactors of the second *column* of $A$, and the third row of $B$ consists of the cofactors of the third column of $A$, so that we have

$$B = \begin{bmatrix} A_{11} & A_{21} & A_{31} \\ A_{12} & A_{22} & A_{32} \\ A_{13} & A_{23} & A_{33} \end{bmatrix}.$$

Let us now form the matrix product $AB$ and apply Theorems 5.2.1 and 5.2.2 to obtain

$$\begin{bmatrix} a_{11} & a_{12} & a_{13} \\ a_{21} & a_{22} & a_{23} \\ a_{31} & a_{32} & a_{33} \end{bmatrix} \begin{bmatrix} A_{11} & A_{21} & A_{31} \\ A_{12} & A_{22} & A_{32} \\ A_{13} & A_{23} & A_{33} \end{bmatrix} = \begin{bmatrix} \det A & 0 & 0 \\ 0 & \det A & 0 \\ 0 & 0 & \det A \end{bmatrix}.$$

That this is actually the result may be seen from considering that the element in the 1, 1 position in the product matrix is

$$a_{11}A_{11} + a_{12}A_{12} + a_{13}A_{13},$$

which is, by Theorem 5.2.1, equal to det $A$. The element in the 1, 2 position of the product is $a_{11}A_{21} + a_{12}A_{22} + a_{13}A_{23}$, which is, by Theorem 5.2.2, equal to zero. The elements of the second and third rows of the product matrix are obtained in a similar manner. Thus

$$AB = \begin{bmatrix} \det A & 0 & 0 \\ 0 & \det A & 0 \\ 0 & 0 & \det A \end{bmatrix} = (\det A)\begin{bmatrix} 1 & 0 & 0 \\ 0 & 1 & 0 \\ 0 & 0 & 1 \end{bmatrix} = (\det A)I_3.$$

Moreover, from the column analogues of Theorems 5.2.1 and 5.2.2, we see that

$$BA = \begin{bmatrix} \det A & 0 & 0 \\ 0 & \det A & 0 \\ 0 & 0 & \det A \end{bmatrix} = (\det A)\begin{bmatrix} 1 & 0 & 0 \\ 0 & 1 & 0 \\ 0 & 0 & 1 \end{bmatrix} = (\det A)I_3.$$

Thus, if det $A \neq 0$, the preceding equations may be rewritten as

$$A\left(\frac{1}{\det A} B\right) = I_3 = \left(\frac{1}{\det A} B\right)A.$$

Thus we have the important result that

$$A^{-1} = \frac{1}{\det A} B.$$

Example 5.3.1 Let us calculate the inverse of the matrix $A = \begin{bmatrix} 1 & -1 & 2 \\ 2 & -1 & 3 \\ 1 & 1 & 2 \end{bmatrix}$. Here the determinant of $A$ is found to be 2. The cofactors of the elements are:

$$\begin{aligned} A_{11} &= -5 & A_{12} &= -1 & A_{13} &= 3 \\ A_{21} &= 4 & A_{22} &= 0 & A_{23} &= -2 \\ A_{31} &= -1 & A_{32} &= 1 & A_{33} &= 1. \end{aligned}$$

Thus our matrix $B$ is

$$\begin{bmatrix} -5 & 4 & -1 \\ -1 & 0 & 1 \\ 3 & -2 & 1 \end{bmatrix}.$$

and

$$A^{-1} = \frac{1}{\det A} B = 1/2\begin{bmatrix} -5 & 4 & -1 \\ -1 & 0 & 1 \\ 3 & -2 & 1 \end{bmatrix} = \begin{bmatrix} -5/2 & 2 & -1/2 \\ -1/2 & 0 & 1/2 \\ 3/2 & -1 & 1/2 \end{bmatrix}.$$

The reader may now verify by multiplication that this is, indeed, the inverse of $A$. ∎

The matrix $B$ in the foregoing discussion is often called the **adjoint** of the matrix $A$, and is denoted by adj $A$. Our result, exhibited for $3 \times 3$ matrices, is, of course, valid for any order square matrix $A$, and may be briefly stated as:

**THEOREM 5.3.1** *$A(adj\ A) = (adj\ A)A = (det\ A)I_n$.**

Thus we have seen that if det $A \neq 0$, then $A$ has an inverse—namely, $A^{-1} = (1/\det A)(\text{adj } A)$. The converse of this statement is true also. If $A$ has an inverse, then det $A \neq 0$. For if $A$ has an inverse, then using Theorem 5.2.3, we have $|A|\,|A^{-1}| = |AA^{-1}| = |I_n| = 1$, so that $|A| \neq 0$. As a side result, we also see that det $|A^{-1}| = 1/\det A$. Thus we have the following fundamental relationship between inverses and determinants:

**THEOREM 5.3.2** *A square matrix A has an inverse if and only if its determinant is non-zero.*

While the preceding theorem is useful, we hasten to point out that the method of calculating the inverse of a matrix, as illustrated in Example 5.3.1, is extremely time-consuming for large matrices, and is essentially useless as a method of calculation. Just consider that finding the inverse of a $10 \times 10$ matrix by that method would require computing the determinants of one $10 \times 10$ matrix and one hundred $9 \times 9$ matrices. Recall from Section 2.6 that the Gauss-Jordan reduction method only requires reduction of $[A \mid I_n]$ to $[I \mid A^{-1}]$. The computation is scarcely double that needed for finding $|A|$.

Exercises 11–15 below provide examples of the application of determinants to analytical geometry and plane geometry.

## 5.3 Exercises

In Exercises 1–4, verify Theorem 5.3.1 by calculating the products $A(\text{adj } A)$ and $(\text{adj } A)A$.

1. $\begin{bmatrix} 1 & 3 \\ 1 & 2 \end{bmatrix}$

2. $\begin{bmatrix} 3 & 1 \\ 6 & 2 \end{bmatrix}$

3. $\begin{bmatrix} 4 & 1 & 2 \\ 1 & 1 & -3 \\ 5 & 2 & -1 \end{bmatrix}$

4. $\begin{bmatrix} 1 & 2 & 1 \\ 1 & 1 & 0 \\ 0 & 2 & 3 \end{bmatrix}$

In Exercises 5–10, use the method of this section and the method of Section 2.6 to calculate the inverses of the given matrix.

5. $\begin{bmatrix} 1 & 2 \\ 3 & 4 \end{bmatrix}$

6. $\begin{bmatrix} -1 & 1 \\ 4 & 6 \end{bmatrix}$

7. $\begin{bmatrix} 1 & 2 & 1 \\ 1 & 3 & 1 \\ 2 & 1 & 4 \end{bmatrix}$

8. $\begin{bmatrix} 0 & 1 & 2 \\ 0 & 0 & 2 \\ 2 & 1 & 3 \end{bmatrix}$

9. $\begin{bmatrix} 4 & 2 & 0 & 1 \\ 1 & 1 & 1 & 1 \\ 3 & 1 & 0 & 1 \\ -1 & 2 & 3 & 2 \end{bmatrix}$

10. $\begin{bmatrix} 1 & 1 & 2 & 3 \\ 1 & 2 & 3 & 0 \\ 0 & 0 & 1 & 2 \\ 0 & 0 & 2 & 5 \end{bmatrix}$

* We remind the reader that adj $A$ is a square matrix, while det $A$ is a scalar.

11. Consider the vectors [2, 1] and [−3/2, 3]. Show that these are orthogonal. Compute the length of each vector, and show that the area of the rectangle they span is 15/2. Now compute the determinant $\begin{vmatrix} 2 & 1 \\ -3/2 & 3 \end{vmatrix}$ and see that it also has the value 15/2.

12. Can you show that the area of the triangle spanned by [1, 1] and [−2, 2] has area equal to $1/2\begin{vmatrix} 1 & -2 \\ 1 & 2 \end{vmatrix}$?

13. Show that the vectors [1, 1, 2], [3, −1, −1], and [−1, −7, 4] are mutually perpendicular. Find each of their lengths and compute the volume of the rectangular parallelopiped spanned by them. Compare this with the absolute value of the determinant

$$\begin{vmatrix} 1 & 1 & 2 \\ 3 & -1 & -1 \\ -1 & -7 & 4 \end{vmatrix}$$

14. Consider the line $L$ in $\mathbb{R}^2$ joining [2, −1] and [3, 5]. This line has slope $(5 - (-1))/(3 - 2) = 6$ and thus its equation is $y - 5 = 6(x - 3)$. Expand the determinant in the equation below and compare with the above result.

$$\begin{vmatrix} 1 & x & y \\ 1 & 2 & -1 \\ 1 & 3 & 5 \end{vmatrix} = 0.$$

We now consider **Cramer's Rule** which is a method for solving certain special types of systems of equations.

Consider the system of three equations in three unknowns

$$\begin{aligned} a_{11}x_1 + a_{12}x_2 + a_{13}x_3 &= b_1 \\ a_{21}x_1 + a_{22}x_2 + a_{23}x_3 &= b_2 \\ a_{31}x_1 + a_{32}x_2 + a_{33}x_3 &= b_3. \end{aligned}$$

Cramer's Rule states that if the determinant of the coefficient matrix is not zero—that is, if $|A| \neq 0$, then the (unique) solution of this system is given in terms of determinants by

$$x_1 = \frac{\begin{vmatrix} b_1 & a_{12} & a_{13} \\ b_2 & a_{22} & a_{23} \\ b_3 & a_{32} & a_{33} \end{vmatrix}}{|A|}, \quad x_2 = \frac{\begin{vmatrix} a_{11} & b_1 & a_{13} \\ a_{21} & b_2 & a_{23} \\ a_{31} & b_3 & a_{33} \end{vmatrix}}{|A|}, \quad x_3 = \frac{\begin{vmatrix} a_{11} & a_{12} & b_1 \\ a_{21} & a_{22} & b_2 \\ a_{31} & a_{32} & b_3 \end{vmatrix}}{|A|}.$$

In general, *for a system of n equations in n unknowns in which the coefficient matrix A has a non-zero determinant, the kth unknown is the quotient of two determinants. The denominator determinant is the determinant of A, and the numerator is the determinant of the matrix formed by replacing the kth column of A by B.*

You may have used Cramer's Rule to solve two equations in two unknowns, or to solve three equations in three unknowns. If not, there are several exercises at the end of this section on which you may try your hand. We illustrate Cramer's Rule with a system of four equations in four unknowns.

Example 5.3.2 Solve (by Cramer's Rule) the system

$$\begin{aligned} x_1 - x_2 + x_3 - x_4 &= 6 \\ 2x_1 + x_2 \qquad + x_4 &= -1 \\ x_1 + 2x_2 + x_3 - 2x_4 &= 5 \\ 3x_1 - x_2 - 2x_3 + x_4 &= -2. \end{aligned}$$

In matrix form, this system may be written as $AX = B$, where

$$A = \begin{bmatrix} 1 & -1 & 1 & -1 \\ 2 & 1 & 0 & 1 \\ 1 & 2 & 1 & -2 \\ 3 & -1 & -2 & 1 \end{bmatrix}, \quad X = \begin{bmatrix} x_1 \\ x_2 \\ x_3 \\ x_4 \end{bmatrix}, \quad B = \begin{bmatrix} 6 \\ -1 \\ 5 \\ -2 \end{bmatrix}.$$

The reader should now verify, by hook or by crook, that

$$|A| = \begin{vmatrix} 1 & -1 & 1 & -1 \\ 2 & 1 & 0 & 1 \\ 1 & 2 & 1 & -2 \\ 3 & -1 & -2 & 1 \end{vmatrix} = -32.$$

By Cramer's Rule, then,

$$x_1 = \frac{\begin{vmatrix} 6 & -1 & 1 & -1 \\ -1 & 1 & 0 & 1 \\ 5 & 2 & 1 & -2 \\ -2 & -1 & -2 & 1 \end{vmatrix}}{-32}, \qquad x_2 = \frac{\begin{vmatrix} 1 & 6 & 1 & -1 \\ 2 & -1 & 0 & 1 \\ 1 & 5 & 1 & -2 \\ 3 & -2 & -2 & 1 \end{vmatrix}}{-32},$$

$$x_3 = \frac{\begin{vmatrix} 1 & -1 & 6 & -1 \\ 2 & 1 & -1 & 1 \\ 1 & 2 & 5 & -2 \\ 3 & -1 & -2 & 1 \end{vmatrix}}{-32}, \quad \text{and} \quad x_4 = \frac{\begin{vmatrix} 1 & -1 & 1 & 6 \\ 2 & 1 & 0 & -1 \\ 1 & 2 & 1 & 5 \\ 3 & -1 & -2 & -2 \end{vmatrix}}{-32}.$$

You should verify that $x_1 = 1$, $x_2 = -1$, $x_3 = 2$, and $x_4 = -2$; moreover, you should now substitute these values in the original system to verify that this is, indeed, a solution. ∎

We note the obvious drawbacks to Cramer's Rule as a computational method.

1. It is only applicable directly to systems of equations with a square coefficient matrix.
2. It fails as a method if the coefficient matrix has a zero determinant.
3. For $n$ equations in $n$ unknowns, it requires the calculation of $n + 1$ determinants each of order $n$. For large matrices, a great deal more work is required than for Gauss-Jordan reduction.

Despite these serious objections, Cramer's Rule is a well-known method of solving small systems of equations, or of computing the values for one or two of the variables in larger systems. Moreover, it provides a sometimes convenient tool when an explicit representation of the solution is required.

## 5.3 Exercises (continued)

Solve the following systems by Cramer's Rule.

15. $$\begin{aligned} x_1 + 2x_2 &= 3 \\ 2x_1 - x_2 &= 4 \end{aligned}$$

16. $$\begin{aligned} 2x_1 + x_2 &= 1 \\ x_1 - 3x_2 &= 2 \end{aligned}$$

17. $$\begin{aligned} x_1 + 2x_2 + 3x_3 &= 5 \\ 2x_1 + 3x_2 - x_3 &= -3 \\ 3x_1 + 7x_2 + x_3 &= -2 \end{aligned}$$

18. $$\begin{aligned} x_1 + x_2 + x_3 &= 2 \\ x_1 - x_2 - x_3 &= -6 \\ x_1 + 2x_2 + 3x_3 &= 9 \end{aligned}$$

19. (for the masochist) Solve the following system of equations by Cramer's Rule, evaluating all determinants by the cofactor method.

$$\begin{aligned} x_1 - x_2 - x_3 - x_4 - x_5 - x_6 &= 2 \\ 2x_1 + x_2 + x_3 + x_4 + x_5 + x_6 &= 1 \\ x_1 - x_2 + x_3 - x_4 + x_5 - x_6 &= 0 \\ x_1 + 2x_2 + x_3 - x_4 - x_5 - x_6 &= 3 \\ 2x_1 - x_2 - x_3 + x_4 - x_5 + x_6 &= 1 \\ x_1 + x_2 + x_3 + x_4 + x_5 - x_6 &= 0 \end{aligned}$$

**New Terms**

adjoint, 152　　Cramer's Rule, 154

## 5.4 The Rank of a Matrix

In the previous chapter the notion of *rank* was discussed and some connections between rank and linear independence were explored. Here we use the same definition for rank but connect this notion with the primary subject of this chapter—determinants—with several interesting and informative results.

Recall that we considered $A$ to be the matrix of a homogeneous system of $m$ equations in $n$ unknowns, and let $k$ denote the number of rows of zeros in the final augmented matrix reduced by the Gauss-Jordan process. The presence of these last $k$ rows of zeros signalled to us that there were $k$ redundant equations in the original system; or, to put it another way, that there were $m - k$ non-redundant equations in the original system. We then defined the rank of $A$ to be the number $m - k$ of non-zero rows in the final augmented matrix. Thus, it is apparent that we may determine the rank of $A$ by first performing the Gauss-Jordan process and then simply counting the number of non-zero rows.

Example 5.4.1 The matrix

$$A = \begin{bmatrix} 1 & 1 & 1 & 3 \\ 2 & -1 & 2 & 3 \\ 4 & 1 & 4 & 9 \end{bmatrix} \text{ may be reduced to } \begin{bmatrix} 1 & 0 & 1 & 2 \\ 0 & 1 & 0 & 1 \\ 0 & 0 & 0 & 0 \end{bmatrix}.$$

Thus, in the matrix equation $AX = 0$, there are $k = 1$ redundant equations. Hence the rank of $A$ is 2. ▮

Example 5.4.2 Let $A$ be the matrix

$$A = \begin{bmatrix} 1 & 1 & 1 & 1 \\ 1 & 1 & 2 & 3 \\ 3 & 3 & 5 & 4 \end{bmatrix} \text{ which may be reduced to } \begin{bmatrix} 1 & 1 & 1 & 1 \\ 0 & 0 & 1 & 2 \\ 0 & 0 & 2 & 1 \end{bmatrix},$$

and then successively to

$$\begin{bmatrix} 1 & 1 & 0 & -1 \\ 0 & 0 & 1 & 2 \\ 0 & 0 & 0 & -3 \end{bmatrix}, \text{ and finally to } \begin{bmatrix} 1 & 1 & 0 & 0 \\ 0 & 0 & 1 & 0 \\ 0 & 0 & 0 & 1 \end{bmatrix},$$

which is in the row-echelon form. We thus see that $k = 0$ and that the rank of $A$ is $m - k = 3$. ▮

## 5.4 Exercises

By inspection determine the rank of each of the following matrices.

1. $\begin{bmatrix} 1 \\ 0 \end{bmatrix}$

2. $\begin{bmatrix} 1 & 1 \end{bmatrix}$

3. $\begin{bmatrix} 1 & 1 & 0 & 0 \\ 0 & 0 & 1 & 0 \\ 0 & 0 & 0 & 1 \\ 0 & 0 & 0 & 0 \end{bmatrix}$

4. $\begin{bmatrix} 1 & 0 & 0 & 0 & 0 \\ 0 & 0 & 0 & 0 & 0 \\ 0 & 1 & 0 & 0 & 0 \\ 0 & 0 & 0 & 0 & 0 \end{bmatrix}$

Find the rank of each of the following matrices.

5. $\begin{bmatrix} 1 & 2 & 0 & 1 \\ 3 & 1 & 5 & 2 \end{bmatrix}$

6. $\begin{bmatrix} 0 & 1 & 4 & 1 \\ 1 & 2 & 3 & 4 \end{bmatrix}$

7. $\begin{bmatrix} 1 & -1 & 2 & 4 \\ 0 & 2 & 4 & 6 \\ -1 & 3 & -4 & -4 \end{bmatrix}$

8. $\begin{bmatrix} 1 & -2 & -1 \\ 2 & 3 & 5 \\ -1 & 5 & 4 \end{bmatrix}$

9. $\begin{bmatrix} 1 & 2 & -3 & 4 \\ -2 & -3 & 0 & 4 \\ -1 & -1 & -3 & 8 \\ 1 & 3 & -9 & 16 \end{bmatrix}$

10. $\begin{bmatrix} 1 & 2 & -3 & 4 & 1 \\ -2 & -3 & 0 & 4 & 4 \\ -1 & -1 & -3 & 8 & 5 \\ 1 & 3 & -9 & 16 & 7 \end{bmatrix}$

We now proceed to justify the presence of our discussion of *rank* in this chapter on determinants. Let us first note that in order to determine the rank of a matrix, we do not have to completely reduce it to echelon form. That is, it is not necessary in a particular column to reduce the elements *above* the diagonal (or pivotal) element in the reduction process since this in no way affects the number of zero rows.

Example 5.4.3

$$A = \begin{bmatrix} 1 & 3 & 2 & 3 \\ 0 & 0 & 1 & 4 \\ 0 & 0 & 0 & 7 \end{bmatrix}$$

clearly has rank 3 even though it may be further reduced to the echelon form

$$\begin{bmatrix} 1 & 3 & 0 & 0 \\ 0 & 0 & 1 & 0 \\ 0 & 0 & 0 & 1 \end{bmatrix}.$$

∎

So let us now consider the $3 \times 4$ matrix $A$ in the partially reduced form, where

$$A = \begin{bmatrix} 1 & 2 & 1 & 2 \\ 0 & 1 & 2 & 3 \\ 0 & 0 & 0 & 0 \end{bmatrix}.$$

$A$ has rank 2, and we note that any $3 \times 3$ submatrix, found by selecting some three of the columns of $A$ must necessarily have determinant zero, since any such $3 \times 3$ submatrix will have its last row zero. Thus the submatrix

$$\begin{bmatrix} 1 & 1 & 2 \\ 0 & 2 & 3 \\ 0 & 0 & 0 \end{bmatrix}$$

formed from the first, third and fourth columns of $A$ has determinant 0. And in general, for the partially reduced $m \times n$ matrix

$$\begin{matrix} r \left\{ \begin{matrix} \\ \\ \\ \\ \end{matrix} \right. \\ k \text{ rows of zeros} \left\{ \begin{matrix} \\ \\ \\ \end{matrix} \right. \end{matrix} \begin{bmatrix} 1 & a_{12} & a_{13} & \cdots & a_{1r} & \cdots & a_{1n} \\ 0 & 1 & a_{23} & \cdots & a_{2r} & \cdots & a_{2n} \\ \cdot & \cdot & \cdot & \cdots & \cdot & \cdots & \cdot \\ 0 & 0 & 0 & \cdots & 1 & \cdots & a_{rn} \\ 0 & 0 & 0 & \cdots & 0 & \cdots & 0 \\ \cdot & \cdot & \cdot & \cdots & \cdot & \cdots & \cdot \\ 0 & 0 & 0 & \cdots & 0 & \cdots & 0 \end{bmatrix},$$

with rank $r = m - k$, it is certainly true that any square submatrix of order $r + 1$ or larger has a zero determinant, since any such submatrix must contain at least one of the last $k$ rows of zeros.

On the other hand, we see that in the matrix

$$A = \begin{bmatrix} 1 & 2 & 1 & 2 \\ 0 & 1 & 2 & 3 \\ 0 & 0 & 0 & 0 \end{bmatrix},$$

which has rank 2, it is possible to find *at least one* $2 \times 2$ submatrix which has non-zero determinant. For example, the matrix $\begin{bmatrix} 1 & 2 \\ 0 & 1 \end{bmatrix}$ lying in the first two rows and first two columns of $A$ has non-zero determinant. Similarly, in the $4 \times 5$ matrix $B$ of rank 3,

$$B = \begin{bmatrix} 1 & 2 & 1 & 2 & 3 \\ 0 & 1 & 2 & -1 & 2 \\ 0 & 0 & 0 & 1 & 4 \\ 0 & 0 & 0 & 0 & 0 \end{bmatrix},$$

each $4 \times 4$ submatrix has zero determinant, while we can find at least one $3 \times 3$ submatrix with non-zero determinant. Here such a matrix is found by choosing the submatrix lying in the first three rows and the first, second, and fourth columns, namely,

$$\begin{bmatrix} 1 & 2 & 2 \\ 0 & 1 & -1 \\ 0 & 0 & 1 \end{bmatrix}.$$

We choose the columns which contain the "first 1's" in each row, and thus construct, since there are exactly $m - k$ such columns, a triangular submatrix of order $m - k$ which has non-zero determinant; indeed, its diagonal elements are all 1's.

Thus we see that the rank of a *reduced matrix* $A$ can be realized as the order (size) of the highest order submatrix of $A$ with non-zero determinant.

So much for reduced matrices. But what about non-reduced ones? For them the answer is the same, as stated in the following:

**THEOREM 5.4.1** *The rank of a matrix is the order of the highest order square submatrix with non-zero determinant.*

The truth of this theorem may be made plausible by the following remarks. In the Gauss-Jordan reduction process used to reduce $A$ to echelon form, three types of operations are allowed:

1. *Row interchange.* This type of operation does not alter the zero-ness or non-zero-ness of the determinant of any submatrix containing the interchanged rows since it merely changes the algebraic sign of any such determinant.

2. *Multiplying a row by a non-zero scalar $c$.* Such an operation does not alter the zero-ness or non-zero-ness of the determinant of any submatrix containing the row multiplied since the operation merely multiplies such a determinant by the scalar $c$.

3. *Adding a multiple of one row to another row.* This operation does not alter the values of subdeterminants containing the rows involved.

While the foregoing is not a proof of Theorem 5.4.1, a proof can be constructed from this type of consideration. In the classical presentation of matrix algebra, Theorem 5.4.1 is often taken as the definition of rank, with our definition following as a theorem.

## 5.4 Exercises (continued)

Find the rank of each of the following matrices by finding an appropriate order submatrix with non-zero determinant.

11. $\begin{bmatrix} 1 & -1 & 2 \\ 2 & 1 & -1 \\ 3 & 0 & 1 \end{bmatrix}$

12. $\begin{bmatrix} 1 & 2 & 3 \\ -1 & 1 & 0 \\ 2 & -1 & 2 \end{bmatrix}$

13. $\begin{bmatrix} 2 & 2 & 3 \\ 0 & -1 & 2 \\ 2 & 2 & 3 \\ 2 & 0 & 6 \end{bmatrix}$

14. $\begin{bmatrix} 1 & 2 & 1 \\ 0 & 1 & -1 \\ 2 & 4 & 2 \\ 1 & 4 & -1 \end{bmatrix}$

We conclude this section with a few elementary observations concerning rank: first, a given submatrix of $A^T$ is the transpose of a corresponding submatrix of $A$. By our previous results, their determinants are equal. Thus the rank of $A^T$ is the same as the rank of $A$; that is, if $r$ is the rank of $A$, then each subdeterminant of $A$ of order $r + 1$ and larger is 0. Hence the same is true for $A^T$. In addition, there is at least one subdeterminant of order $r$ in $A$ which is not 0 and the obvious one corresponding to it in $A^T$ is non-zero also. Thus we have

**THEOREM 5.4.2** *The rank of a matrix $A$ is the same as the rank of its transpose $A^T$.*

Second, since rank is defined as the number of non-redundant rows, we know that if $A$ is an $m \times n$ matrix, then the rank of $A \leqslant m$. But by Theorem 5.4.2, we must also have the rank of $A \leqslant n$. Hence we have

**THEOREM 5.4.3** *If $A$ is an $m \times n$ matrix, then*

$$\text{rank of } A \leqslant \min\{m, n\}.$$

Third, two extremes occur for the rank of a matrix. If the matrix $A$ has all its entries zero, then we say that the rank of $A$ is *zero*. If $A$ is $m \times n$, and the rank of $A$ is equal to the minimum of $m$ and $n$, then we say that $A$ has **full rank**, that is, it has the maximum rank possible for its dimensions.

Example 5.4.4 The matrices

$$[0], \quad [0 \quad 0 \quad 0], \quad \begin{bmatrix} 0 & 0 & 0 & 0 & 0 \\ 0 & 0 & 0 & 0 & 0 \end{bmatrix} \quad \text{and} \quad \begin{bmatrix} 0 & 0 & 0 \\ 0 & 0 & 0 \\ 0 & 0 & 0 \end{bmatrix}$$

all have rank 0. The matrices

$$[7], \quad \begin{bmatrix} -4 & 0 \\ 2 & 0 \end{bmatrix}, \quad \begin{bmatrix} 2 \\ 1 \end{bmatrix}, \quad [1 \ \ 2 \ \ 3 \ \ 4 \ \ 5], \quad \text{and} \quad \begin{bmatrix} 1 & 2 & 3 \\ 0 & 0 & 0 \\ 2 & 4 & 6 \end{bmatrix}$$

all have rank 1. The matrices

$$[1], \quad \begin{bmatrix} 2 & 3 \\ 4 & 5 \end{bmatrix}, \quad \text{and} \quad \begin{bmatrix} 2 & 3 & 0 & 0 & 0 \\ 4 & 5 & 0 & 0 & 0 \end{bmatrix}$$

are all of *full rank*. ∎

Finally, suppose that $A$ is an $n \times n$ square matrix of rank $n$. Then the determinant of $A$ is non-zero so that by Theorem 5.3.2, $A$ has an inverse. Thus we have the following connection between rank and inverses:

**THEOREM 5.4.4** *The $n \times n$ matrix $A$ has an inverse if and only if the rank of $A$ is $n$.*

## 5.4 Exercises (continued)

In Exercises 15–20, row reduce separately the matrix and its transpose to find their ranks. Which matrices have full rank?

15. $\begin{bmatrix} 2 & 1 \\ 4 & 2 \\ 1 & 1 \end{bmatrix}$

16. $\begin{bmatrix} 2 & 2 & 1 \\ 3 & 3 & 2 \end{bmatrix}$

17. $\begin{bmatrix} 1 & -1 & 2 & 1 \\ 1 & 0 & 5 & 0 \\ 0 & -3 & 2 & 5 \end{bmatrix}$

18. $\begin{bmatrix} 1 & -2 & 1 \\ -2 & 6 & 2 \\ -1 & 7 & 9 \end{bmatrix}$

19. $\begin{bmatrix} 1 & 2 & 1 & -1 \\ 2 & 5 & 2 & -1 \\ 3 & 7 & 3 & -2 \\ 1 & 3 & 2 & 0 \end{bmatrix}$

20. $\begin{bmatrix} -1 & 2 & 1 & 3 & 4 \\ 1 & 1 & 2 & 1 & 3 \\ 2 & 1 & -1 & 1 & 2 \end{bmatrix}$

21. Let $A = \begin{bmatrix} 1 & -1 & 1 & 0 \\ 2 & -1 & 0 & 1 \end{bmatrix}$. Show that $AA^T$ has an inverse, and find $(AA^T)^{-1}$. Prove that $A^TA$ has no inverse.

22. a. In each of the indicated matrix products, find the rank of each factor and the rank of the product matrix.

$$\begin{bmatrix} 0 & 1 \\ 0 & 0 \end{bmatrix}\begin{bmatrix} 0 & 0 \\ 0 & 1 \end{bmatrix}; \quad \begin{bmatrix} 0 & 1 \\ 0 & 0 \end{bmatrix}\begin{bmatrix} 1 & 0 \\ 0 & 1 \end{bmatrix}; \quad \begin{bmatrix} 1 & 0 & 0 \\ 0 & 0 & 1 \\ 1 & 0 & 1 \end{bmatrix}\begin{bmatrix} 0 & 1 & 0 \\ 1 & 0 & 0 \\ 1 & 1 & 0 \end{bmatrix}.$$

What conclusions can you draw about the rank of $AB$ relative to the ranks of $A$ and of $B$?

b. In each of the products in a. above, calculate the product in reverse order and its rank. Can you draw a conclusion about the rank of $BA$ relative to the rank of $AB$?

c. Find some examples among $2 \times 2$ matrices which *disprove* each of the following non-theorems:
   i. rank of $(A + B)$ = rank of $A$ + rank of $B$.
   ii. rank of $(A - B)$ = rank of $A$ − rank of $B$.

**New Terms**

full rank, 160

## 5.5 Change of Basis

In this section we shall consider the effects of a change of basis upon the coordinates of a given vector $X$ in 2-space or 3-space. To illustrate the situation we suppose that $\{U_1, U_2\}$ and $\{V_1, V_2\}$ are each bases in $\mathscr{M}_{2,1}$. The problem has two aspects:

1. Given $X_U$, the coordinate matrix of $X$ with respect to the $U$-basis, find $X_V$, the coordinate matrix of $X$ with respect to the $V$-basis.
2. Given $X_V$, find $X_U$.

The key to the solution of both problems is the fact that we can express each element of one basis in terms of the elements of the other basis. For example, since $\{V_1, V_2\}$ is a basis for $\mathscr{M}_{2,1}$ we know that there exist scalars $a_{ij}$ such that

$$U_1 = a_{11}V_1 + a_{21}V_2$$
$$U_2 = a_{12}V_1 + a_{22}V_2.$$

Thus $U_1 = [V_1, V_2]\begin{bmatrix} a_{11} \\ a_{21} \end{bmatrix}$ and $U_2 = [V_1, V_2]\begin{bmatrix} a_{12} \\ a_{22} \end{bmatrix}$. More compactly, denoting the matrix $[V_1, V_2]$ by $V$, and letting $A_1 = \begin{bmatrix} a_{11} \\ a_{21} \end{bmatrix}$, $A_2 = \begin{bmatrix} a_{12} \\ a_{22} \end{bmatrix}$, we have

$$U_1 = VA_1 \quad \text{and} \quad U_2 = VA_2.$$

Putting $U_1$ and $U_2$ together in a matrix $U$, we have

$$U = [U_1, U_2] = [VA_1, VA_2] = V[A_1, A_2] = VA. \tag{1}$$

This is the result we need to solve our problem.

Consider the first aspect of our problem in which we know $X_U = \begin{bmatrix} u_1 \\ u_2 \end{bmatrix}$ such that

$$X = u_1U_1 + u_2U_2, \tag{2}$$

and wish to find $X_V = \begin{bmatrix} v_1 \\ v_2 \end{bmatrix}$ such that

$$X = v_1V_1 + v_2V_2. \tag{3}$$

It will be helpful to put (2) and (3) in an equivalent form, namely

(4) $$X = [U_1, U_2]\begin{bmatrix} u_1 \\ u_2 \end{bmatrix} = UX_U$$

and

(5) $$X = [V_1, V_2]\begin{bmatrix} v_1 \\ v_2 \end{bmatrix} = VX_V.$$

Thus we see that $$UX_U = VX_V$$

and using our earlier result (1) that $U = VA$, we have

$$(VA)X_U = VX_V$$

or $$V(AX_U) = VX_V.$$

The matrix $V$ is non-singular (see Theorems 4.5.1(3) and 5.4.4) so $V^{-1}$ must exist. Multiplying in our last result by $V^{-1}$ on the left we obtain

$$AX_U = X_V.$$

Example 5.5.1 Let

$$\{U_1, U_2\} = \left\{\begin{bmatrix} 2 \\ 0 \end{bmatrix}, \begin{bmatrix} 0 \\ 2 \end{bmatrix}\right\} \quad \text{and} \quad \{V_1, V_2\} = \left\{\begin{bmatrix} 1 \\ -1 \end{bmatrix}, \begin{bmatrix} 1 \\ 1 \end{bmatrix}\right\}.$$

Then

$$U_1 = \begin{bmatrix} 2 \\ 0 \end{bmatrix} = 1\begin{bmatrix} 1 \\ -1 \end{bmatrix} + 1\begin{bmatrix} 1 \\ 1 \end{bmatrix} = \begin{bmatrix} 1 & 1 \\ -1 & 1 \end{bmatrix}\begin{bmatrix} 1 \\ 1 \end{bmatrix} = VA_1$$

and

$$U_2 = \begin{bmatrix} 0 \\ 2 \end{bmatrix} = -1\begin{bmatrix} 1 \\ -1 \end{bmatrix} + 1\begin{bmatrix} 1 \\ 1 \end{bmatrix} = \begin{bmatrix} 1 & 1 \\ -1 & 1 \end{bmatrix}\begin{bmatrix} -1 \\ 1 \end{bmatrix} = VA_2,$$

so that $A = \begin{bmatrix} 1 & -1 \\ 1 & 1 \end{bmatrix}$. Suppose $X = \begin{bmatrix} 2 \\ 3 \end{bmatrix}$. Then

$$X = \begin{bmatrix} 2 \\ 3 \end{bmatrix} = 1\begin{bmatrix} 2 \\ 0 \end{bmatrix} + (3/2)\begin{bmatrix} 0 \\ 2 \end{bmatrix} = \begin{bmatrix} 2 & 0 \\ 0 & 2 \end{bmatrix}\begin{bmatrix} 1 \\ 3/2 \end{bmatrix} = UX_U,$$

and we see that $X_U = \begin{bmatrix} 1 \\ 3/2 \end{bmatrix}$. Then

$$X_V = AX_U = \begin{bmatrix} 1 & -1 \\ 1 & 1 \end{bmatrix}\begin{bmatrix} 1 \\ 3/2 \end{bmatrix} = \begin{bmatrix} -1/2 \\ 5/2 \end{bmatrix},$$

as we verify by computing

$$(-1/2)\begin{bmatrix} 1 \\ -1 \end{bmatrix} + 5/2\begin{bmatrix} 1 \\ 1 \end{bmatrix} = \begin{bmatrix} 2 \\ 3 \end{bmatrix} = X.$$

Since $U = VA$, and each of $V$ and $A$ is square, we have that $|U| = |V|\,|A|$. Moreover, $U$ is non-singular so that $|U| \neq 0$ and hence $|A| \neq 0$. Therefore $A^{-1}$ exists in general and we can solve for $X_U$ in the formula $AX_U = X_V$. Thus,

$$A^{-1}(AX_U) = A^{-1}X_V$$

so that

$$X_U = A^{-1}X_V.$$

Example 5.5.2 In the previous example, $A = \begin{bmatrix} 1 & -1 \\ 1 & 1 \end{bmatrix}$ and $A^{-1} = \begin{bmatrix} 1/2 & 1/2 \\ -1/2 & 1/2 \end{bmatrix}$.

For $X = \begin{bmatrix} 2 \\ 3 \end{bmatrix}$, the coordinate matrix $X_V = \begin{bmatrix} -1/2 \\ 5/2 \end{bmatrix}$, and

$$X_U = A^{-1}X_V = \begin{bmatrix} 1/2 & 1/2 \\ -1/2 & 1/2 \end{bmatrix}\begin{bmatrix} -1/2 \\ 5/2 \end{bmatrix} = \begin{bmatrix} 1 \\ 3/2 \end{bmatrix},$$

as we expected. Letting $Y = \begin{bmatrix} 1 \\ 3 \end{bmatrix} = -1\begin{bmatrix} 1 \\ -1 \end{bmatrix} + 2\begin{bmatrix} 1 \\ 1 \end{bmatrix}$, we have $Y_V = \begin{bmatrix} -1 \\ 2 \end{bmatrix}$. Thus

$$Y_U = A^{-1}Y_V = \begin{bmatrix} 1/2 & 1/2 \\ -1/2 & 1/2 \end{bmatrix}\begin{bmatrix} -1 \\ 2 \end{bmatrix} = \begin{bmatrix} 1/2 \\ 3/2 \end{bmatrix}.$$

We verify that $(1/2)\begin{bmatrix} 2 \\ 0 \end{bmatrix} + (3/2)\begin{bmatrix} 0 \\ 2 \end{bmatrix} = \begin{bmatrix} 1 \\ 3 \end{bmatrix} = Y$. ∎

Example 5.5.3 For the moment imagine yourself in command of a lunar-bound space vehicle. Consider the situation at a certain instant in time when the spacecraft is 1000 miles from the moon's surface. Much of the information that you need to properly control your craft to its proposed landing site may be available only in terms of vectors expressed in a local coordinate system (with the spacecraft as origin). It is important for ground control, at Houston, Texas, to know what these vectors are in terms of the directions that have been used to establish a Houston coordinate system. Let us say that the Houston system has basis $\{V_1, V_2, V_3\}$ where

$$V_1 = \begin{bmatrix} 1 \\ 0 \\ 0 \end{bmatrix}, \quad V_2 = \begin{bmatrix} 0 \\ 1 \\ 0 \end{bmatrix}, \quad V_3 = \begin{bmatrix} 0 \\ 0 \\ 1 \end{bmatrix},$$

and that the spacecraft system has basis $\{U_1, U_2, U_3\}$. We imagine that the origin of the spacecraft coordinate system is translated to the Houston origin so that the two systems appear as in Figure 5.5.1.

Figure 5.5.1

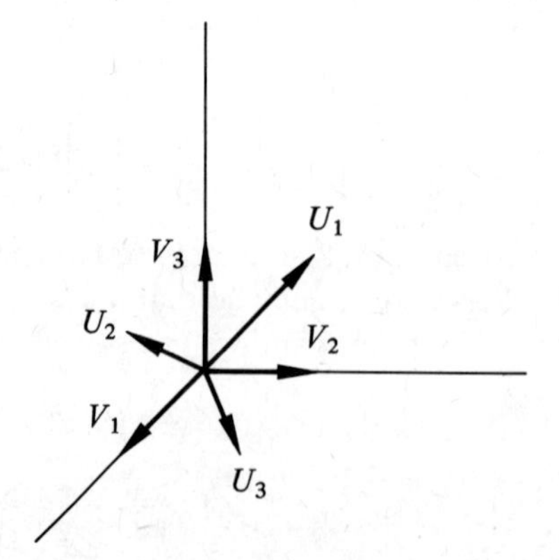

Further, at the instant in time of which we speak, the spacecraft basis vectors are known in terms of the Houston system to be

$$\begin{aligned} U_1 &= \phantom{V_1 +{}} V_2 + V_3 \\ U_2 &= V_1 \phantom{{}+ V_2} + V_3 \\ U_3 &= V_1 + V_2. \end{aligned}$$

Thus, $$U_1 = [V_1 \quad V_2 \quad V_3]\begin{bmatrix} 0 \\ 1 \\ 1 \end{bmatrix},$$

$$U_2 = [V_1 \quad V_2 \quad V_3]\begin{bmatrix} 1 \\ 0 \\ 1 \end{bmatrix},$$

and $$U_3 = [V_1 \quad V_2 \quad V_3]\begin{bmatrix} 1 \\ 1 \\ 0 \end{bmatrix}.$$

Then $$[U_1 \quad U_2 \quad U_3] = [V_1 \quad V_2 \quad V_3]\begin{bmatrix} 0 & 1 & 1 \\ 1 & 0 & 1 \\ 1 & 1 & 0 \end{bmatrix}$$

or $$U = VA.$$

If you observe the coordinate matrix $X_U = \begin{bmatrix} 1 \\ -1 \\ 2 \end{bmatrix}$ of a vector $X$, then the corresponding Houston representation of $X$ would be

$$X_V = AX_U = \begin{bmatrix} 0 & 1 & 1 \\ 1 & 0 & 1 \\ 1 & 1 & 0 \end{bmatrix}\begin{bmatrix} 1 \\ -1 \\ 2 \end{bmatrix} = \begin{bmatrix} 1 \\ 3 \\ 0 \end{bmatrix},$$

and the vector would be deciphered as $V_1 + 3V_2$. ∎

## 5.5 Exercises

1. a. Fill in the blanks in the following paragraph: Let $\{U_1, U_2\}$ and $\{V_1, V_2\}$ be bases for $\mathcal{M}_{2,1}$ with corresponding basis matrices $U$, $V$. If $X$ is in $\mathcal{M}_{2,1}$, then there exist coordinate matrices ______, ______ such that $X = U$______ and $X = V$______. If $A$ is the square matrix with the property that $U = VA$, then $X = VA$______ = ______$X_V$, from which we deduce that $AX_U =$ ______.

   b. Let $A = \begin{bmatrix} 1 & 2 \\ -1 & 3 \end{bmatrix}$ and $X_U = \begin{bmatrix} 1 \\ 2 \end{bmatrix}$. Find $X_V$.

   c. With $A$ as in b. above, let $X_V = \begin{bmatrix} 3 \\ -2 \end{bmatrix}$ and find $X_U$.

   d. By looking at $A$ we know that $U_1 =$ ______$V_1 +$ ______$V_2$, and $U_2 =$ ______$V_1 +$ ______$V_2$. Express $2U_1 + U_2$ in terms of the $V$ basis vectors.

2. How do we know that a basis matrix $U$ or $V$ in equation (1) of this section is non-singular?

3. Let $\{U_1, U_2\} = \left\{\begin{bmatrix}1\\1\end{bmatrix}, \begin{bmatrix}2\\1\end{bmatrix}\right\}$ and $\{V_1, V_2\} = \left\{\begin{bmatrix}1\\-2\end{bmatrix}, \begin{bmatrix}-1\\3\end{bmatrix}\right\}$.
   a. Find $A$ such that $U = VA$.
   b. Let $X = \begin{bmatrix}3\\2\end{bmatrix}$. Find $X_U$ by observation and use it and $A$ to compute $X_V$. Verify your result by checking to see that $VX_V = X$.
   c. Let $X = \begin{bmatrix}0\\1\end{bmatrix}$. Find $X_V$ by observation and use it and $A^{-1}$ to compute $X_U$. Verify your result.

4. Let $\{U_1, U_2\} = \left\{\begin{bmatrix}1\\2\end{bmatrix}, \begin{bmatrix}2\\1\end{bmatrix}\right\}$ and $\{V_1, V_2\} = \left\{\begin{bmatrix}1\\0\end{bmatrix}, \begin{bmatrix}0\\1\end{bmatrix}\right\}$.
   a. Find $A$ such that $U = VA$.
   b. Let $X = \begin{bmatrix}3\\3\end{bmatrix}$. Find $X_U$ by observation and use it to compute $X_V$. Verify your result by observation.
   c. Let $X = \begin{bmatrix}5\\4\end{bmatrix}$. Find $X_V$ by observation and use to compute $X_U$. Verify your result.

5. Let $\{U_1, U_2, U_3\} = \left\{\begin{bmatrix}1\\0\\1\end{bmatrix}, \begin{bmatrix}2\\1\\1\end{bmatrix}, \begin{bmatrix}1\\1\\2\end{bmatrix}\right\}$ and $\{V_1, V_2, V_3\} = \left\{\begin{bmatrix}1\\0\\0\end{bmatrix}, \begin{bmatrix}0\\1\\0\end{bmatrix}, \begin{bmatrix}0\\0\\1\end{bmatrix}\right\}$.
   a. Find $A$ such that $U = AV$.
   b. Find $X$ such that $X_U = \begin{bmatrix}1\\1\\1\end{bmatrix}$.
   c. Find $X_V$ by two methods.

## 5.6 Linear Transformations

In Chapter 1, we discussed the notion of a function and gave a variety of examples of functions from one set $S$ into another set $W$. The main topic of this section is a special type of function $T$ from a vector space $\mathscr{V}$ into itself; that is, $T$ associates with each vector $X$ in $\mathscr{V}$ another vector $Y$ in $\mathscr{V}$, and, in our usual functional notation, we write

$$Y = T(X).$$

As before, we shall often describe $Y$ as the image of $X$ under the function $T$.

The type of function $T$ which we wish to describe is called a **linear transformation** (*transformation* being a synonym for *function*). $T$ is linear in the following sense:

*A linear transformation $T$ from a vector space $\mathscr{V}$ into the space $\mathscr{V}$ is a function which associates with each vector $X$ in $\mathscr{V}$ an image vector $Y$*

*in $\mathscr{V}$ in such a way that for any vectors, $X_1$ and $X_2$, and any scalars, $a_1$ and $a_2$,*

$$T(a_1X_1 + a_2X_2) = a_1T(X_1) + a_2T(X_2).$$

There is an alternate definition of linear transformation as follows:

*$T$ is said to be a linear transformation provided that for any vectors $X_1$, $X_2$, and any scalar $a$,*

1. $T(X_1 + X_2) = T(X_1) + T(X_2)$
2. $T(aX_1) = aT(X_1)$.

In Exercise 6, the reader is asked to show that these two definitions are equivalent. We shall use the first formulation for most of our development.

Thus a linear transformation $T$ is linear in that the image of any linear combination $a_1X_1 + a_2X_2$ of vectors in $\mathscr{V}$ is that same linear combination $a_1T(X_1) + a_2T(X_2)$ of the images of $X_1$ and $X_2$.

Example 5.6.1 Let $T$ be the transformation on the space of all $2 \times 1$ vectors defined as

$$T\left(\begin{bmatrix} c \\ d \end{bmatrix}\right) = \begin{bmatrix} 2c \\ d \end{bmatrix}.$$

Then in particular, we have

$$T\left(\begin{bmatrix} 3 \\ 2 \end{bmatrix}\right) = \begin{bmatrix} 6 \\ 2 \end{bmatrix};$$

and

$$T\left(\begin{bmatrix} -4 \\ 7 \end{bmatrix}\right) = \begin{bmatrix} -8 \\ 7 \end{bmatrix};$$

and

$$T\left(3\begin{bmatrix} 3 \\ 2 \end{bmatrix} + 2\begin{bmatrix} -4 \\ 7 \end{bmatrix}\right) = T\left(\begin{bmatrix} 1 \\ 20 \end{bmatrix}\right) = \begin{bmatrix} 2 \\ 20 \end{bmatrix}.$$

On the other hand,

$$3T\left(\begin{bmatrix} 3 \\ 2 \end{bmatrix}\right) + 2T\left(\begin{bmatrix} -4 \\ 7 \end{bmatrix}\right) = 3\begin{bmatrix} 6 \\ 2 \end{bmatrix} + 2\begin{bmatrix} -8 \\ 7 \end{bmatrix} = \begin{bmatrix} 2 \\ 20 \end{bmatrix},$$

so that we have in this instance

$$T\left(3\begin{bmatrix} 3 \\ 2 \end{bmatrix} + 2\begin{bmatrix} -4 \\ 7 \end{bmatrix}\right) = 3T\left(\begin{bmatrix} 3 \\ 2 \end{bmatrix}\right) + 2T\left(\begin{bmatrix} -4 \\ 7 \end{bmatrix}\right).$$

In general for this transformation $T$, we have, for any scalars $a$ and $b$, and any vectors $\begin{bmatrix} c \\ d \end{bmatrix}$ and $\begin{bmatrix} e \\ f \end{bmatrix}$, that

$$T\left(a\begin{bmatrix} c \\ d \end{bmatrix} + b\begin{bmatrix} e \\ f \end{bmatrix}\right) = T\left(\begin{bmatrix} ac + be \\ ad + bf \end{bmatrix}\right) = \begin{bmatrix} 2(ac + be) \\ ad + bf \end{bmatrix} = \begin{bmatrix} 2ac \\ ad \end{bmatrix} + \begin{bmatrix} 2be \\ bf \end{bmatrix}$$

$$= a\begin{bmatrix} 2c \\ d \end{bmatrix} + b\begin{bmatrix} 2e \\ f \end{bmatrix} = aT\left(\begin{bmatrix} c \\ d \end{bmatrix}\right) + bT\left(\begin{bmatrix} e \\ f \end{bmatrix}\right).$$

Thus $T$ is linear. ▮

Example 5.6.2 Let $T$ be the transformation defined on the space of all $3 \times 1$ matrices as follows:

$$T\begin{bmatrix} a \\ b \\ c \end{bmatrix} = \begin{bmatrix} a \\ 0 \\ 0 \end{bmatrix}.$$

Then, for any scalars $x$ and $y$, and any vectors

$$\begin{bmatrix} a \\ b \\ c \end{bmatrix} \quad \text{and} \quad \begin{bmatrix} d \\ e \\ f \end{bmatrix},$$

we have

$$T\left(x\begin{bmatrix} a \\ b \\ c \end{bmatrix} + y\begin{bmatrix} d \\ e \\ f \end{bmatrix}\right) = T\left(\begin{bmatrix} xa + yd \\ xb + ye \\ xc + yf \end{bmatrix}\right) = \begin{bmatrix} xa + yd \\ 0 \\ 0 \end{bmatrix} = x\begin{bmatrix} a \\ 0 \\ 0 \end{bmatrix} + y\begin{bmatrix} d \\ 0 \\ 0 \end{bmatrix}$$

$$= xT\left(\begin{bmatrix} a \\ b \\ c \end{bmatrix}\right) + yT\left(\begin{bmatrix} d \\ e \\ f \end{bmatrix}\right).$$

Thus $T$ is a linear transformation. ∎

Example 5.6.3 The following is an example of a function from the space of all $2 \times 1$ matrices into itself which is *not* linear. Let $N$ be defined by:

$$N\left(\begin{bmatrix} a \\ b \end{bmatrix}\right) = \begin{bmatrix} 2a \\ b \end{bmatrix} + \begin{bmatrix} 1 \\ 1 \end{bmatrix}.$$

Then it is not true in general that $N(aX + bY) = aN(X) + bN(Y)$, for consider, with $a = b = 1$, and $X = \begin{bmatrix} 1 \\ 0 \end{bmatrix}$, $Y = \begin{bmatrix} 0 \\ 1 \end{bmatrix}$, that

$$N\left(\begin{bmatrix} 1 \\ 0 \end{bmatrix} + \begin{bmatrix} 0 \\ 1 \end{bmatrix}\right) = N\left(\begin{bmatrix} 1 \\ 1 \end{bmatrix}\right) = \begin{bmatrix} 2 \\ 1 \end{bmatrix} + \begin{bmatrix} 1 \\ 1 \end{bmatrix} = \begin{bmatrix} 3 \\ 2 \end{bmatrix},$$

while

$$N\left(\begin{bmatrix} 1 \\ 0 \end{bmatrix}\right) + N\left(\begin{bmatrix} 0 \\ 1 \end{bmatrix}\right) = \left(\begin{bmatrix} 2 \\ 0 \end{bmatrix} + \begin{bmatrix} 1 \\ 1 \end{bmatrix}\right) + \left(\begin{bmatrix} 0 \\ 1 \end{bmatrix} + \begin{bmatrix} 1 \\ 1 \end{bmatrix}\right)$$

$$= \begin{bmatrix} 3 \\ 1 \end{bmatrix} + \begin{bmatrix} 1 \\ 2 \end{bmatrix} = \begin{bmatrix} 4 \\ 3 \end{bmatrix}.$$

Hence $N$ is not linear, since $N\left(\begin{bmatrix} 1 \\ 0 \end{bmatrix} + \begin{bmatrix} 0 \\ 1 \end{bmatrix}\right) \neq N\left(\begin{bmatrix} 1 \\ 0 \end{bmatrix}\right) + N\left(\begin{bmatrix} 0 \\ 1 \end{bmatrix}\right)$. ∎

There are two very important linear transformations defined for every vector space $\mathscr{V}$ which require special mention. One is the transformation $E$ which maps every vector to itself:

$$E(X) = X \qquad \text{for each } X \text{ in } \mathscr{V}.$$

$E$ is called the **identity transformation**. The other special transformation we want to mention is usually denoted by 0, and is such that

$$0(X) = 0;$$

that is, 0 maps each vector in $\mathscr{V}$ to the zero vector in $\mathscr{V}$. 0 is called the **zero transformation**. We leave it as an exercise to verify that these transformations are indeed linear.

Finally, one may generalize from the definition of a linear transformation $T$ that, for any linear combination

$$V = a_1 V_1 + a_2 V_2 + \cdots + a_n V_n$$

of vectors $V_1, \ldots, V_n$, it is true that

$$T(V) = T(a_1 V_1 + \cdots + a_n V_n) = a_1 T(V_1) + \cdots + a_n T(V_n).$$

It is in this sense that one says, "a linear transformation preserves linear combinations"—meaning that the image of a linear combination of vectors is that same combination of the images of the separate vectors. We shall not present the general induction proof of the above assertion. It is, however, easily verified for three vectors. Thus,

$$\begin{aligned} T(a_1 V_1 + a_2 V_2 + a_3 V_3) &= T((a_1 V_1 + a_2 V_2) + a_3 V_3) \\ &= T(a_1 V_1 + a_2 V_2) + a_3 T(V_3) \\ &= a_1 T(V_1) + a_2 T(V_2) + a_3 T(V_3). \end{aligned}$$ ▮

We are now ready to develop a simple but fundamental property of linear transformations. Let $I = \begin{bmatrix} 1 \\ 0 \end{bmatrix}$ and $J = \begin{bmatrix} 0 \\ 1 \end{bmatrix}$ be a basis for the space of all $2 \times 1$ matrices. Let $T$ denote any linear transformation defined on that space. We note that any $2 \times 1$ matrix

$$\begin{bmatrix} a \\ b \end{bmatrix} = a \begin{bmatrix} 1 \\ 0 \end{bmatrix} + b \begin{bmatrix} 0 \\ 1 \end{bmatrix},$$

so that

$$T\left(\begin{bmatrix} a \\ b \end{bmatrix}\right) = T\left(a \begin{bmatrix} 1 \\ 0 \end{bmatrix} + b \begin{bmatrix} 0 \\ 1 \end{bmatrix}\right) = aT\left(\begin{bmatrix} 1 \\ 0 \end{bmatrix}\right) + bT\left(\begin{bmatrix} 0 \\ 1 \end{bmatrix}\right),$$

since $T$ is linear. Thus, if we are told what $T\left(\begin{bmatrix} 1 \\ 0 \end{bmatrix}\right)$ and $T\left(\begin{bmatrix} 0 \\ 1 \end{bmatrix}\right)$ are, then we know what $T\left(\begin{bmatrix} a \\ b \end{bmatrix}\right)$ is for any $\begin{bmatrix} a \\ b \end{bmatrix}$.

More generally, if $\{U, V\}$ is any basis for 2-space, then a linear transformation $T$ is completely determined when we know $T(U)$ and $T(V)$; since if $X$ is any vector, then for some scalars $a$ and $b$ we can write $X = aU + bV$, so that $T(X) = T(aU + bV) = aT(U) + bT(V)$.

In even more general terms, let $\mathscr{V}$ be a vector space with basis $\{B_1, B_2, \ldots, B_m\}$. If $T$ is a linear transformation on $\mathscr{V}$, then the value of $T$ at any $X$ in $\mathscr{V}$ is known when we know $T(B_1), \ldots, T(B_m)$. Thus if $X = a_1 B_1 + a_2 B_2 + \cdots + a_m B_m$, then

$$T(X) = T(a_1 B_1 + \cdots + a_m B_m) = a_1 T(B_1) + \cdots + a_m T(B_m).$$

Example 5.6.4 Let $T$ be the linear transformation which takes $\begin{bmatrix}1\\0\end{bmatrix}$ into $\begin{bmatrix}2\\0\end{bmatrix}$ and takes $\begin{bmatrix}0\\1\end{bmatrix}$ to $\begin{bmatrix}0\\1\end{bmatrix}$; that is,

$$T\left(\begin{bmatrix}1\\0\end{bmatrix}\right) = \begin{bmatrix}2\\0\end{bmatrix} \quad \text{and} \quad T\left(\begin{bmatrix}0\\1\end{bmatrix}\right) = \begin{bmatrix}0\\1\end{bmatrix}.$$

Then for any $\begin{bmatrix}a\\b\end{bmatrix}$, we first write $\begin{bmatrix}a\\b\end{bmatrix} = a\begin{bmatrix}1\\0\end{bmatrix} + b\begin{bmatrix}0\\1\end{bmatrix}$; then, since $T$ is linear,

$$T\left(\begin{bmatrix}a\\b\end{bmatrix}\right) = T\left(a\begin{bmatrix}1\\0\end{bmatrix} + b\begin{bmatrix}0\\1\end{bmatrix}\right) = aT\left(\begin{bmatrix}1\\0\end{bmatrix}\right) + bT\left(\begin{bmatrix}0\\1\end{bmatrix}\right)$$
$$= a\begin{bmatrix}2\\0\end{bmatrix} + b\begin{bmatrix}0\\1\end{bmatrix} = \begin{bmatrix}2a\\b\end{bmatrix}.$$

Thus $T$ is the same transformation as that in Example 5.6.1. ∎

Finally, we note that the set of all images of elements of $\mathscr{V}$ under a linear transformation $T$ is a *subspace* of $\mathscr{V}$. This is easily proved by showing that the set of all images, denoted by $T(\mathscr{V})$ and defined as

$$T(\mathscr{V}) = \{Y \mid Y = T(X) \text{ for some } X \text{ in } \mathscr{V}\},$$

is closed under vector addition and scalar multiplication. Thus, if $Y_1$ and $Y_2$ are in $T(\mathscr{V})$, then there are $X_1$ and $X_2$ in $\mathscr{V}$ such that $T(X_1) = Y_1$ and $T(X_2) = Y_2$. Then

$$Y_1 + Y_2 = T(X_1) + T(X_2) = T(X_1 + X_2),$$

so that the sum $Y_1 + Y_2$ is in $T(\mathscr{V})$ since it is the image of $X_1 + X_2$, which must be in $\mathscr{V}$ since $\mathscr{V}$ is closed under addition. Similarly, a scalar multiple $aY$ of any $Y$ in $T(\mathscr{V})$ is also in $T(\mathscr{V})$—since for some $X$ in $\mathscr{V}$, $Y = T(X)$ so that $aY = aT(X) = T(aX)$. (Why is $aX$ in $\mathscr{V}$?)

The set $T(\mathscr{V})$ is called the **range space** of the transformation $T$, or, more briefly, just the range of $T$. Moreover, if $\{B_1, \ldots, B_m\}$ is a basis for $\mathscr{V}$ then the set $S = \{T(B_1), T(B_2), \ldots, T(B_m)\}$ spans $T(\mathscr{V})$. Thus we claim that if $Y$ is in $T(\mathscr{V})$ then $Y$ is a linear combination of the elements of $\{T(B_1), T(B_2), \ldots, T(B_m)\}$. That this is so may be seen from the following argument. If $Y$ is in $T(\mathscr{V})$, then $Y$ is the $T$-image of some vector $X$; that is, $Y = T(X)$. Since $X$ is in $\mathscr{V}$, it is a linear combination of the basis vectors $B_1, \ldots, B_m$; let us say

$$X = a_1B_1 + \cdots + a_mB_m.$$

Then

$$Y = T(X) = T(a_1B_1 + a_2B_2 + \cdots + a_mB_m)$$
$$= a_1T(B_1) + a_2T(B_2) + \cdots + a_mT(B_m),$$

and $Y$ is indeed a linear combination of elements of $S$. Thus, $S$ spans $T(\mathscr{V})$.

Example 5.6.5 Let $T\left(\begin{bmatrix} a \\ b \\ c \end{bmatrix}\right) = \begin{bmatrix} a \\ 0 \\ 0 \end{bmatrix}$, as in Example 5.6.2. With basis

$$\left\{ I = \begin{bmatrix} 1 \\ 0 \\ 0 \end{bmatrix}, \quad J = \begin{bmatrix} 0 \\ 1 \\ 0 \end{bmatrix}, \quad \text{and} \quad K = \begin{bmatrix} 0 \\ 0 \\ 1 \end{bmatrix} \right\},$$

we have $T(I) = \begin{bmatrix} 1 \\ 0 \\ 0 \end{bmatrix}$, and $T(J) = \begin{bmatrix} 0 \\ 0 \\ 0 \end{bmatrix}$, and $T(K) = \begin{bmatrix} 0 \\ 0 \\ 0 \end{bmatrix}$. Thus the set of images $\left\{ \begin{bmatrix} 1 \\ 0 \\ 0 \end{bmatrix}, \begin{bmatrix} 0 \\ 0 \\ 0 \end{bmatrix}, \begin{bmatrix} 0 \\ 0 \\ 0 \end{bmatrix} \right\}$ of the basis $\{I, J, K\}$ is easily seen to span $T(\mathscr{V})$. But the set of images is certainly not a basis for $T(\mathscr{V})$, since it is not a linearly independent set. Of course, a basis for $T(\mathscr{V})$ could be extracted from this spanning set. Thus we observe that

$$\left\{ \begin{bmatrix} 1 \\ 0 \\ 0 \end{bmatrix} \right\}$$

is a basis for the range space of $T$. Thus the range space of $T$ is the space of all scalar multiples of $\begin{bmatrix} 1 \\ 0 \\ 0 \end{bmatrix}$ and may be thought of as the line containing the vector $\begin{bmatrix} 1 \\ 0 \\ 0 \end{bmatrix}$. ∎

Example 5.6.6 If $T\left(\begin{bmatrix} a \\ b \end{bmatrix}\right) = \begin{bmatrix} 2a \\ b \end{bmatrix}$, as in Example 5.6.1, then $T\left(\begin{bmatrix} 1 \\ 0 \end{bmatrix}\right) = \begin{bmatrix} 2 \\ 0 \end{bmatrix}$ and $T\left(\begin{bmatrix} 0 \\ 1 \end{bmatrix}\right) = \begin{bmatrix} 0 \\ 1 \end{bmatrix}$, so that $T(\mathscr{V})$ is spanned in this case by $\begin{bmatrix} 2 \\ 0 \end{bmatrix}$ and $\begin{bmatrix} 0 \\ 1 \end{bmatrix}$. But these vectors are linearly independent, so that they also form a basis for the range space of $T$. Thus, for this particular transformation, the range space $T(\mathscr{V})$ is $\mathscr{V}$ itself. ∎

## 5.6 Exercises

In Exercises 1–5, decide which of the defined transformations are linear and which are non-linear.

1. $T\left(\begin{bmatrix} a \\ b \end{bmatrix}\right) = \begin{bmatrix} a + b \\ 0 \end{bmatrix}$

2. $T\left(\begin{bmatrix} a \\ b \end{bmatrix}\right) = \begin{bmatrix} ab \\ 0 \end{bmatrix}$

3. $T\left(\begin{bmatrix} a \\ b \\ c \end{bmatrix}\right) = \begin{bmatrix} a + 1 \\ b \\ c \end{bmatrix}$

4. $T\left(\begin{bmatrix} a \\ b \\ c \end{bmatrix}\right) = \begin{bmatrix} a + b \\ a - b \\ 2a \end{bmatrix}$

5. $T\left(\begin{bmatrix} a \\ b \\ c \end{bmatrix}\right) = \begin{bmatrix} a \\ b \\ abc \end{bmatrix}$

6. Show that the two definitions of *linear transformation* given in this section are equivalent.

7. If $T$ is a linear transformation from the space $\mathscr{V}$ into itself, prove that $T(0) = 0$; that is, any linear transformation maps the zero vector to the zero vector.

8. Prove that if $T$ is a linear transformation from the vector space $\mathscr{V}$ into itself, then for each vector $X$ in $\mathscr{V}$, $T(-X) = -T(X)$; that is, $T$ maps the additive inverse of each $X$ into the additive inverse of $T(X)$.

9. Let $X = \begin{bmatrix} 1 \\ 1 \\ 1 \end{bmatrix}$ and define the transformation $T$ on $\mathscr{M}_{3,1}$ as

$$T\begin{bmatrix} a \\ b \\ c \end{bmatrix} = (a + b + c)X.$$

Is $T$ a linear transformation? Why?

10. Let $T\left(\begin{bmatrix} x_1 \\ x_2 \end{bmatrix}\right) = \begin{bmatrix} x_1 + x_2 \\ x_1 - x_2 \end{bmatrix} = T(X)$. By solving the appropriate set of homogeneous linear equations find $\{X \mid T(X) = 0\}$. This set is called the **null space** of the linear transformation $T$.

11. Let
$$T\left(\begin{bmatrix} x_1 \\ x_2 \\ x_3 \end{bmatrix}\right) = \begin{bmatrix} x_1 + x_2 - x_3 \\ 3x_1 + 2x_2 - x_3 \\ 2x_1 + x_2 \end{bmatrix}.$$

Find the null space of $T$ (see Exercise 10).

12. a. Prove that the null space of a linear transformation $T$ is a vector space.
    b. From the results of Exercise 11, can you formulate a conjecture about the relationship between the dimension of the null space of $T$ and the rank of the coefficient matrix of the set of homogeneous equations?

### New Terms

linear transformation, 166
identity transformation, 169
zero transformation, 149
range space, 170
null space, 172

## 5.7 Linear Transformations and Matrices

Let us consider the space $\mathscr{M}_{2,1}$ of all $2 \times 1$ matrices. We now show how one may use a matrix to define a linear transformation $T$. We first choose some $2 \times 2$ matrix, say $\begin{bmatrix} 1 & 2 \\ 3 & 4 \end{bmatrix}$. Then we *define* the transformation $T$ on

$\mathcal{M}_{2,1}$ by showing how to find the image of each vector $\begin{bmatrix} a \\ b \end{bmatrix}$. Thus

$$T\left(\begin{bmatrix} a \\ b \end{bmatrix}\right) = \begin{bmatrix} 1 & 2 \\ 3 & 4 \end{bmatrix}\begin{bmatrix} a \\ b \end{bmatrix} = \begin{bmatrix} a + 2b \\ 3a + 4b \end{bmatrix}.$$

$T$ associates with each vector in $\mathcal{M}_{2,1}$ another vector in $\mathcal{M}_{2,1}$. Moreover, this association is linear in that, because of the distributive property of matrix multiplication and the properties of scalar multiplication,

$$\begin{aligned} T\left(x\begin{bmatrix} a \\ b \end{bmatrix} + y\begin{bmatrix} c \\ d \end{bmatrix}\right) &= \begin{bmatrix} 1 & 2 \\ 3 & 4 \end{bmatrix}\left(x\begin{bmatrix} a \\ b \end{bmatrix} + y\begin{bmatrix} c \\ d \end{bmatrix}\right) \\ &= x\begin{bmatrix} 1 & 2 \\ 3 & 4 \end{bmatrix}\begin{bmatrix} a \\ b \end{bmatrix} + y\begin{bmatrix} 1 & 2 \\ 3 & 4 \end{bmatrix}\begin{bmatrix} c \\ d \end{bmatrix} \\ &= xT\left(\begin{bmatrix} a \\ b \end{bmatrix}\right) + yT\left(\begin{bmatrix} c \\ d \end{bmatrix}\right). \end{aligned}$$

In a similar manner, we may use a $3 \times 3$ matrix to define a linear transformation $T$ as

$$T\left(\begin{bmatrix} a \\ b \\ c \end{bmatrix}\right) = \begin{bmatrix} 1 & 0 & 2 \\ 2 & 0 & 4 \\ 1 & 1 & 2 \end{bmatrix}\begin{bmatrix} a \\ b \\ c \end{bmatrix} = \begin{bmatrix} a + 2c \\ 2a + 4c \\ a + b + 2c \end{bmatrix}.$$

Indeed, there is no reason to stop with 3. We can, in general, define a linear transformation on $\mathcal{M}_{n,1}$ by using an $n \times n$ matrix.

Now the foregoing comments are not so remarkable at this point, for they are really a restatement of the distributive property of matrix multiplication and the properties of scalar multiplication of a matrix. The remarkable connection between linear transformations and matrices is that every linear transformation $T$ on, say, $\mathcal{M}_{2,1}$ can be associated with a matrix $\begin{bmatrix} a_{11} & a_{12} \\ a_{21} & a_{22} \end{bmatrix}$ such that $T(X) = AX$ for each $X$ in $\mathcal{M}_{2,1}$.

We now make precise how we may associate a matrix with a given linear transformation. We consider this problem first in Example 5.7.1.

**Example 5.7.1** Let the transformation $T$ be defined as

$$T(X) = T\left(\begin{bmatrix} a \\ b \end{bmatrix}\right) = \begin{bmatrix} a + 2b \\ 3a + 4b \end{bmatrix}.$$

We then choose a basis for $\mathcal{M}_{2,1}$ and the basis which we select is

$$\left\{\begin{bmatrix} 1 \\ 0 \end{bmatrix}, \begin{bmatrix} 0 \\ 1 \end{bmatrix}\right\}.$$

We recall that $T$ is completely determined by its action on a basis, so we write down the $T$-images of our chosen basis elements, and we further express these images in the chosen basis. Thus

$$T\left(\begin{bmatrix} 1 \\ 0 \end{bmatrix}\right) = \begin{bmatrix} 1 \\ 3 \end{bmatrix} = 1\begin{bmatrix} 1 \\ 0 \end{bmatrix} + 3\begin{bmatrix} 0 \\ 1 \end{bmatrix}$$

$$T\left(\begin{bmatrix} 0 \\ 1 \end{bmatrix}\right) = \begin{bmatrix} 2 \\ 4 \end{bmatrix} = 2\begin{bmatrix} 1 \\ 0 \end{bmatrix} + 4\begin{bmatrix} 0 \\ 1 \end{bmatrix}.$$

We now write down the matrix $A = \begin{bmatrix} 1 & 2 \\ 3 & 4 \end{bmatrix}$, whose columns are respectively the coordinates of $T\left(\begin{bmatrix} 1 \\ 0 \end{bmatrix}\right)$ and the coordinates of $T\left(\begin{bmatrix} 0 \\ 1 \end{bmatrix}\right)$ in the chosen basis, as indicated. Now note that the transformation $S$ such that $S(X) = AX$ is the same as the transformation $T$ since each has the same effect upon a basis; indeed, for any $X = \begin{bmatrix} a \\ b \end{bmatrix}$ we have

$$S(X) = \begin{bmatrix} 1 & 2 \\ 3 & 4 \end{bmatrix} \begin{bmatrix} a \\ b \end{bmatrix} = \begin{bmatrix} a + 2b \\ 3a + 4b \end{bmatrix} = T(X).$$

This relationship, we note, is a property of the special basis chosen. ∎

Let us consider another example.

Example 5.7.2 Let $T$ be defined so that $T\left(\begin{bmatrix} a \\ b \end{bmatrix}\right) = \begin{bmatrix} 2a - b \\ a + b \end{bmatrix}$. Still using the basis of the previous example, we write

$$T\left(\begin{bmatrix} 1 \\ 0 \end{bmatrix}\right) = \begin{bmatrix} 2 \\ 1 \end{bmatrix} = 2\begin{bmatrix} 1 \\ 0 \end{bmatrix} + 1\begin{bmatrix} 0 \\ 1 \end{bmatrix}$$

$$T\left(\begin{bmatrix} 0 \\ 1 \end{bmatrix}\right) = \begin{bmatrix} -1 \\ 1 \end{bmatrix} = -1\begin{bmatrix} 1 \\ 0 \end{bmatrix} + 1\begin{bmatrix} 0 \\ 1 \end{bmatrix}.$$

Our associated matrix is then $\begin{bmatrix} 2 & -1 \\ 1 & 1 \end{bmatrix}$. Let us consider a particular vector $\begin{bmatrix} 3 \\ 4 \end{bmatrix}$ and see how we use matrix multiplication to find $T\left(\begin{bmatrix} 3 \\ 4 \end{bmatrix}\right)$. We note from our defining equation for $T$ that $T\left(\begin{bmatrix} 3 \\ 4 \end{bmatrix}\right) = \begin{bmatrix} 2 \\ 7 \end{bmatrix}$. We first express $\begin{bmatrix} 3 \\ 4 \end{bmatrix}$ in terms of the chosen basis, so that $\begin{bmatrix} 3 \\ 4 \end{bmatrix} = 3\begin{bmatrix} 1 \\ 0 \end{bmatrix} + 4\begin{bmatrix} 0 \\ 1 \end{bmatrix}$. If we multiply the matrix of coordinates $\begin{bmatrix} 3 \\ 4 \end{bmatrix}$, which, because of the basis we chose, is the vector $\begin{bmatrix} 3 \\ 4 \end{bmatrix}$, by the matrix $\begin{bmatrix} 2 & -1 \\ 1 & 1 \end{bmatrix}$ we obtain $\begin{bmatrix} 2 & -1 \\ 1 & 1 \end{bmatrix}\begin{bmatrix} 3 \\ 4 \end{bmatrix} = \begin{bmatrix} 2 \\ 7 \end{bmatrix}$ which is the matrix of coordinates of the image vector. Again, the matrix of coordinates of the image is the same as the image vector $\begin{bmatrix} 2 \\ 7 \end{bmatrix}$ because of the particular basis used. In general, in this example, $\begin{bmatrix} 2 & -1 \\ 1 & 1 \end{bmatrix}\begin{bmatrix} a \\ b \end{bmatrix} = \begin{bmatrix} 2a - b \\ a + b \end{bmatrix}$. ∎

The same procedure used in the preceding examples may be used to associate with each linear transformation $T$ on $\mathscr{M}_{3,1}$ a certain matrix

$A$ such that

$$T\left(\begin{bmatrix} a \\ b \\ c \end{bmatrix}\right) = A\begin{bmatrix} a \\ b \\ c \end{bmatrix},$$

if we use the usual $I$, $J$, $K$ basis to construct the matrix.

Example 5.7.3 Let $T$ be the transformation defined as

$$T\left(\begin{bmatrix} a \\ b \\ c \end{bmatrix}\right) = \begin{bmatrix} 2a - b + c \\ b + c \\ a - b \end{bmatrix}.$$

We use the basis

$$I = \begin{bmatrix} 1 \\ 0 \\ 0 \end{bmatrix}, \qquad J = \begin{bmatrix} 0 \\ 1 \\ 0 \end{bmatrix}, \qquad K = \begin{bmatrix} 0 \\ 0 \\ 1 \end{bmatrix},$$

and write

$$T\left(\begin{bmatrix} 1 \\ 0 \\ 0 \end{bmatrix}\right) = \begin{bmatrix} 2 \\ 0 \\ 1 \end{bmatrix} = 2\begin{bmatrix} 1 \\ 0 \\ 0 \end{bmatrix} + (0)\begin{bmatrix} 0 \\ 1 \\ 0 \end{bmatrix} + 1\begin{bmatrix} 0 \\ 0 \\ 1 \end{bmatrix}$$

$$T\left(\begin{bmatrix} 0 \\ 1 \\ 0 \end{bmatrix}\right) = \begin{bmatrix} -1 \\ 1 \\ -1 \end{bmatrix} = -1\begin{bmatrix} 1 \\ 0 \\ 0 \end{bmatrix} + 1\begin{bmatrix} 0 \\ 1 \\ 0 \end{bmatrix} - 1\begin{bmatrix} 0 \\ 0 \\ 1 \end{bmatrix}$$

$$T\left(\begin{bmatrix} 0 \\ 0 \\ 1 \end{bmatrix}\right) = \begin{bmatrix} 1 \\ 1 \\ 0 \end{bmatrix} = 1\begin{bmatrix} 1 \\ 0 \\ 0 \end{bmatrix} + 1\begin{bmatrix} 0 \\ 1 \\ 0 \end{bmatrix} + (0)\begin{bmatrix} 0 \\ 0 \\ 1 \end{bmatrix}.$$

Forming the matrix as before, we obtain

$$A = \begin{bmatrix} 2 & -1 & 1 \\ 0 & 1 & 1 \\ 1 & -1 & 0 \end{bmatrix}.$$

Using the results of the above example, suppose that we wish to find

$$T\left(\begin{bmatrix} 1 \\ 2 \\ -1 \end{bmatrix}\right).$$

We first express

$$\begin{bmatrix} 1 \\ 2 \\ -1 \end{bmatrix}$$

in the chosen basis $\{I, J, K\}$ and find that the coordinate matrix is

$$\begin{bmatrix} 1 \\ 2 \\ -1 \end{bmatrix}$$

also. We then find

$$A\begin{bmatrix}1\\2\\-1\end{bmatrix} = \begin{bmatrix}2 & -1 & 1\\0 & 1 & 1\\1 & -1 & 0\end{bmatrix}\begin{bmatrix}1\\2\\-1\end{bmatrix} = \begin{bmatrix}-1\\1\\-1\end{bmatrix}$$

which is the coordinate matrix for

$$T\left(\begin{bmatrix}1\\2\\-1\end{bmatrix}\right).$$

Thus
$$T\left(\begin{bmatrix}1\\2\\-1\end{bmatrix}\right) = -1\begin{bmatrix}1\\0\\0\end{bmatrix} + 1\begin{bmatrix}0\\1\\0\end{bmatrix} - 1\begin{bmatrix}0\\0\\1\end{bmatrix}.$$ ∎

The method may be extended, of course, to obtain a matrix multiplication representation for a linear transformation on $\mathcal{M}_{n,1}$. The reader interested in the situation that arises when a basis other than the simple ones we have chosen is used is encouraged to work through Exercises 13–15 below.

## 5.7 Exercises

In Exercises 1–6, use the basis $\left\{\begin{bmatrix}1\\0\end{bmatrix}, \begin{bmatrix}0\\1\end{bmatrix}\right\}$ and construct the matrix $A$ of the given linear transformation on $\mathcal{M}_{2,1}$ relative to this basis. Use the matrix to find $T\left(\begin{bmatrix}2\\3\end{bmatrix}\right)$ in each case.

1. $T\left(\begin{bmatrix}a\\b\end{bmatrix}\right) = \begin{bmatrix}a\\b\end{bmatrix}$

2. $T\left(\begin{bmatrix}a\\b\end{bmatrix}\right) = \begin{bmatrix}b\\a\end{bmatrix}$

3. $T\left(\begin{bmatrix}a\\b\end{bmatrix}\right) = \begin{bmatrix}2a + b\\a + 2b\end{bmatrix}$

4. $T\left(\begin{bmatrix}a\\b\end{bmatrix}\right) = \begin{bmatrix}0\\3b - a\end{bmatrix}$

5. $T\left(\begin{bmatrix}a\\b\end{bmatrix}\right) = \begin{bmatrix}a + 2b\\-2a\end{bmatrix}$

6. $T\left(\begin{bmatrix}a\\b\end{bmatrix}\right) = \begin{bmatrix}b - a\\b + a\end{bmatrix}$

In Exercises 7–12, use the basis

$$\left\{\begin{bmatrix}1\\0\\0\end{bmatrix}, \begin{bmatrix}0\\1\\0\end{bmatrix}, \begin{bmatrix}0\\0\\1\end{bmatrix}\right\}$$

and construct the matrix of the given transformation $T$ on $\mathcal{M}_{3,1}$ relative to this basis. Use the matrix to find

$$T\left(\begin{bmatrix}-1\\2\\3\end{bmatrix}\right)$$

in each case.

7. $T\left(\begin{bmatrix} a \\ b \\ c \end{bmatrix}\right) = \begin{bmatrix} a \\ b \\ c \end{bmatrix}$

8. $T\left(\begin{bmatrix} a \\ b \\ c \end{bmatrix}\right) = \begin{bmatrix} c \\ a \\ b \end{bmatrix}$

9. $T\left(\begin{bmatrix} a \\ b \\ c \end{bmatrix}\right) = \begin{bmatrix} a \\ b \\ a + b + c \end{bmatrix}$

10. $T\left(\begin{bmatrix} a \\ b \\ c \end{bmatrix}\right) = \begin{bmatrix} b + c \\ a + c \\ a + b \end{bmatrix}$

11. $T\left(\begin{bmatrix} a \\ b \\ c \end{bmatrix}\right) = \begin{bmatrix} a + b + c \\ a - b - c \\ a + b - c \end{bmatrix}$

12. $T\left(\begin{bmatrix} a \\ b \\ c \end{bmatrix}\right) = \begin{bmatrix} 3a \\ 2b \\ c \end{bmatrix}$

13. In the text of this section we have hinted that the matrix associated with a transformation depends upon the basis chosen. To illustrate what happens, use the basis $\left\{U = \begin{bmatrix} 1 \\ 2 \end{bmatrix}, V = \begin{bmatrix} 2 \\ 1 \end{bmatrix}\right\}$ and the transformation $T$ of Example 5.7.1 above and:

    a. Express the $T$-images of $\begin{bmatrix} 1 \\ 2 \end{bmatrix}$ and of $\begin{bmatrix} 2 \\ 1 \end{bmatrix}$ in the $U$, $V$ basis.

    b. Construct the matrix $B$ which represents $T$ in the basis $\{U, V\}$. $B$ is the matrix whose columns are the coordinates of $T(U)$ and $T(V)$ found in a.

    c. Find the coordinates of $\begin{bmatrix} 3 \\ 4 \end{bmatrix}$ in the $U$, $V$ basis.

    d. By matrix multiplication find the $U$, $V$ coordinates of $T\left(\begin{bmatrix} 3 \\ 4 \end{bmatrix}\right)$ by multiplying the matrix found in b. above by the coordinate matrix found in c. above.

    e. Using the coordinate matrix of $T\left(\begin{bmatrix} 3 \\ 4 \end{bmatrix}\right)$, construct $T\left(\begin{bmatrix} 3 \\ 4 \end{bmatrix}\right)$.

14. a. Find the matrices $A$ of the transformations of Exercises 7–12 relative to the basis

$$\left\{U = \begin{bmatrix} 0 \\ 1 \\ 1 \end{bmatrix}, V = \begin{bmatrix} 1 \\ 0 \\ 1 \end{bmatrix}, W = \begin{bmatrix} 1 \\ 1 \\ 0 \end{bmatrix}\right\}.$$

    b. Express $X = \begin{bmatrix} -1 \\ 2 \\ 3 \end{bmatrix}$ in terms of the $U$, $V$, $W$ basis. Call the coordinate matrix $C$.

    c. Find the $U$, $V$, $W$ coordinates of $T(X)$ by finding $AC$.

    d. Find $T(X)$ by using its coordinate matrix and the basis vectors.

15. a. Find the matrix $A$ which represents the linear transformation $T$ of Exercise 5 with respect to the basis $\left\{U = \begin{bmatrix} 1 \\ 2 \end{bmatrix}, V = \begin{bmatrix} 2 \\ 1 \end{bmatrix}\right\}$.

    b. Express $\begin{bmatrix} 3 \\ 3 \end{bmatrix} = X$ in terms of the $U$, $V$ basis. The coordinate matrix $C$ of $X$ is __________.

    c. Find the $U$, $V$ coordinates of $T(X)$ by computing $AC$.

    d. Find $T(X)$ by the formula in Exercise 5 and by using the coordinates in c.

## 5.8 Some Special Linear Transformations in 2-Space

There are several types of linear transformations which have a special geometric interest. We investigate these in the familiar setting of 2-space.

1. *Stretching or Shrinking Transformations.* If each vector in 2-space is multiplied by a fixed scalar $c$, then we call the related transformation $S$ a **stretching transformation** if $|c| > 1$ and a **shrinking transformation** if $|c| < 1$. Here $S(U) = cU$, and it is immediate that $S$ is linear, since

$$\begin{aligned} S(a_1U_1 + a_2U_2) &= c(a_1U_1 + a_2U_2) \\ &= a_1(cU_1) + a_2(cU_2) \\ &= a_1S(U_1) + a_2S(U_2). \end{aligned}$$

You may easily verify that the matrix of such a transformation with respect to the usual basis is $\begin{bmatrix} c & 0 \\ 0 & c \end{bmatrix} = c\begin{bmatrix} 1 & 0 \\ 0 & 1 \end{bmatrix}$.

2. *Projection on a Line.* Consider a line $L$ through the origin in 2-space as in Figure 5.8.1, and a vector $U$. By the **projection** of $U$ on the line $L$ we mean (Figure 5.8.2) a vector $V$ such that the head of $V$ is the foot of the perpendicular to $L$ from the head of $U$. If we let $W$ be a unit vector in the line $L$ (Figure 5.8.3) then we require that $V$ be a scalar multiple of $W$, say $V = kW$. Thus the length $\|V\|$ of $V$ must be $|k|$. But from the figure, assuming $\theta \leqslant 90°$, we see that $\|V\| = \|U\| \cos\theta$. Thus we have $V =$

Figure 5.8.1

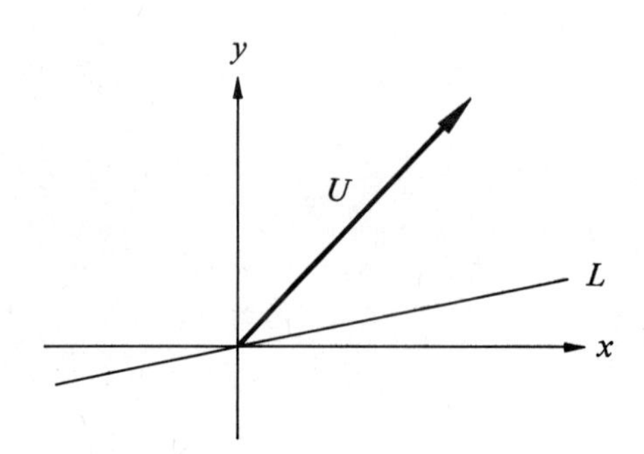


Figure 5.8.2

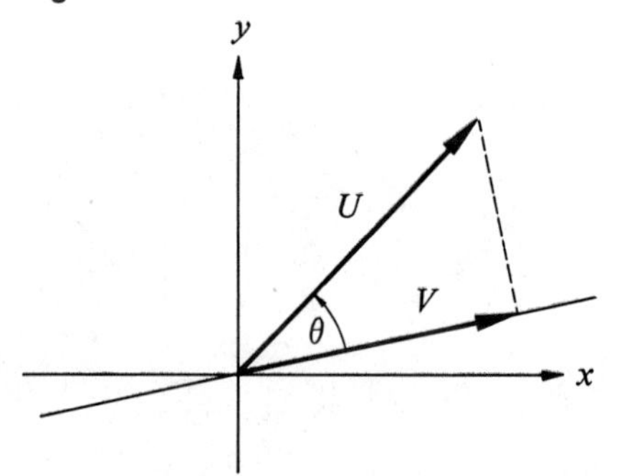


Figure 5.8.3

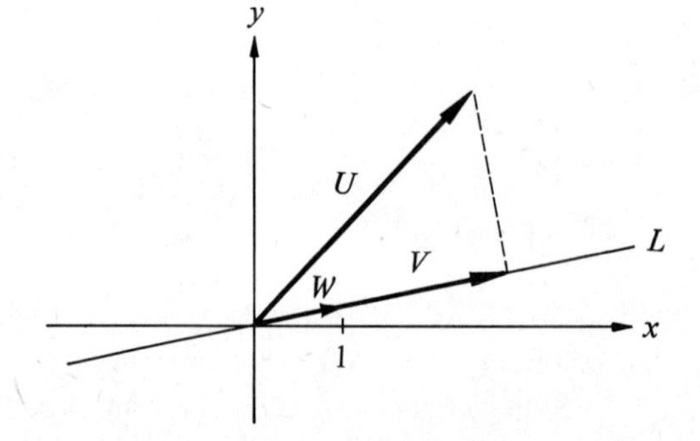

Figure 5.8.4

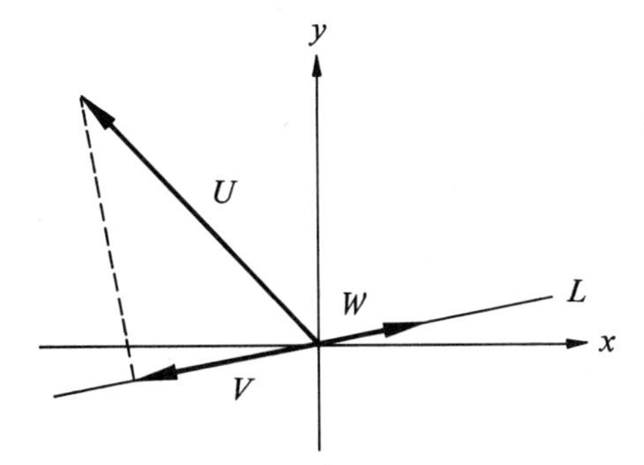


$(\|U\| \cos \theta)W$. But, since $\|W\| = 1$,

$$\cos \theta = \frac{U \cdot W}{\|U\| \, \|W\|} = \frac{U \cdot W}{\|U\|}$$

so that $\|U\| \cos \theta = U \cdot W$. Hence

$$V = (U \cdot W)W.$$

You should note that the final formulation is valid also in the case that $90^\circ \leqslant \theta \leqslant 180^\circ$; in that case $\cos \theta$ is negative and the projection $V$ is thus a negative multiple of $W$, which is as it should be (Figure 5.8.4).

The transformation $P$ which projects each vector in 2-space on the line $L$ is a linear transformation since, if $W$ is a unit vector in the line $L$, $P$ is defined as $P(U) = (U \cdot W)W$, a scalar multiple of $W$. Now let $a_1$, $a_2$ be scalars and $U_1$, $U_2$ any two vectors. From the definition of $P$ and the properties of the dot product,

$$\begin{aligned} P(a_1 U_1 + a_2 U_2) &= ((a_1 U_1 + a_2 U_2) \cdot W)W \\ &= (a_1 U_1 \cdot W + a_2 U_2 \cdot W)W \\ &= (a_1 U_1 \cdot W)W + (a_2 U_2 \cdot W)W \\ &= a_1((U_1 \cdot W)W) + a_2((U_2 \cdot W)W) \\ &= a_1 P(U_1) + a_2 P(U_2). \end{aligned}$$

Thus $P$ preserves linear combinations and is a linear transformation on 2-space.

Example 5.8.1 Let $P$ be the projection (Figure 5.8.5) on the line $L$ containing the vector $\begin{bmatrix} 1 \\ 1 \end{bmatrix}$. We should like to find the matrix of $P$ in the usual basis.

Figure 5.8.5

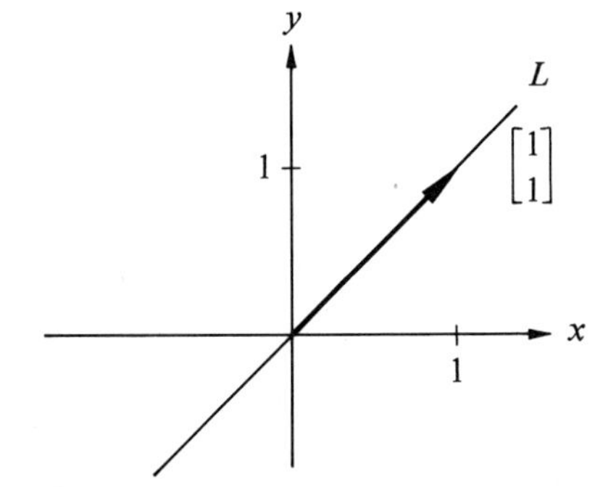

We take $W = \begin{bmatrix} 1/\sqrt{2} \\ 1/\sqrt{2} \end{bmatrix}$, which is a unit vector in the line $L$, and compute the effect of $P$ on the basis vectors.

$$P\left(\begin{bmatrix} 1 \\ 0 \end{bmatrix}\right) = \left(\begin{bmatrix} 1 \\ 0 \end{bmatrix} \cdot W\right)W = (1/\sqrt{2})W = \begin{bmatrix} 1/2 \\ 1/2 \end{bmatrix} = 1/2\begin{bmatrix} 1 \\ 0 \end{bmatrix} + 1/2\begin{bmatrix} 0 \\ 1 \end{bmatrix}$$

$$P\left(\begin{bmatrix} 0 \\ 1 \end{bmatrix}\right) = \left(\begin{bmatrix} 0 \\ 1 \end{bmatrix} \cdot W\right)W = (1/\sqrt{2})W = \begin{bmatrix} 1/2 \\ 1/2 \end{bmatrix} = 1/2\begin{bmatrix} 1 \\ 0 \end{bmatrix} + 1/2\begin{bmatrix} 0 \\ 1 \end{bmatrix}.$$

Thus the required matrix is $\begin{bmatrix} 1/2 & 1/2 \\ 1/2 & 1/2 \end{bmatrix}$. ∎

3. *Rotation Through a Fixed Angle.* Suppose that $T$ denotes the transformation that rotates each vector counterclockwise through a fixed angle $\theta$ (Figure 5.8.6). From Figure 5.8.7, it is clear that $T$ preserves scalar multiples so that $T(cU) = cT(U)$.

**Figure 5.8.6**

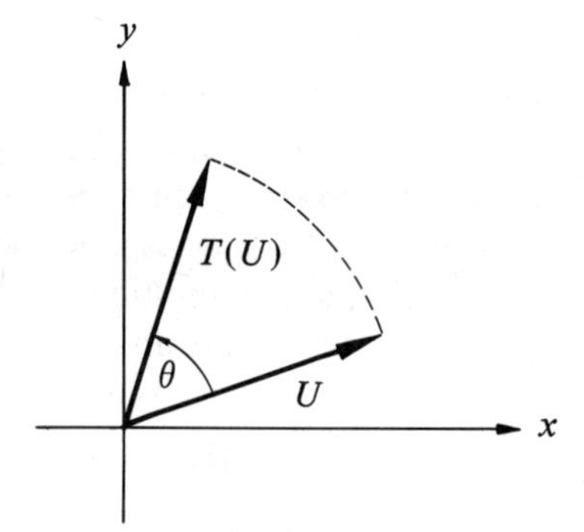


**Figure 5.8.7**

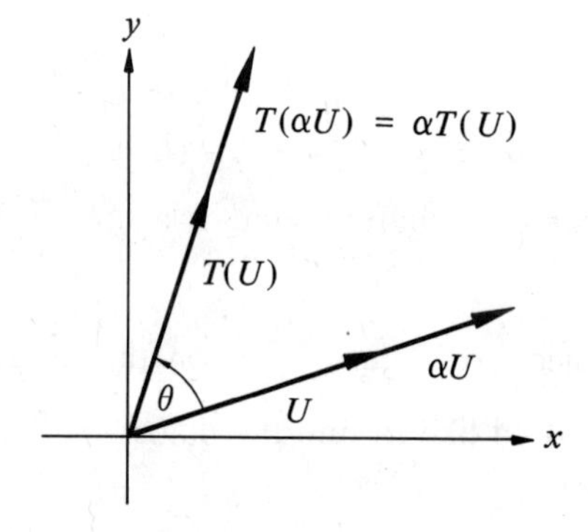


Similarly, from Figure 5.8.8, we see that $T$ preserves addition. Now since $T$ preserves sums and scalar products, $T(a_1U_1 + a_2U_2) = T(a_1U_1) + T(a_2U_2) = a_1T(U_1) + a_2T(U_2)$ so that $T$ is linear. Moreover, if we consider the effect of $T$ on the usual basis, we have (Figure 5.8.9)

$$T\left(\begin{bmatrix} 1 \\ 0 \end{bmatrix}\right) = \begin{bmatrix} \cos\theta \\ \sin\theta \end{bmatrix} = \cos\theta\begin{bmatrix} 1 \\ 0 \end{bmatrix} + \sin\theta\begin{bmatrix} 0 \\ 1 \end{bmatrix}$$

$$T\left(\begin{bmatrix} 0 \\ 1 \end{bmatrix}\right) = \begin{bmatrix} \cos(\theta + 90°) \\ \sin(\theta + 90°) \end{bmatrix} = -\sin\theta\begin{bmatrix} 1 \\ 0 \end{bmatrix} + \cos\theta\begin{bmatrix} 0 \\ 1 \end{bmatrix}.$$

Figure 5.8.8

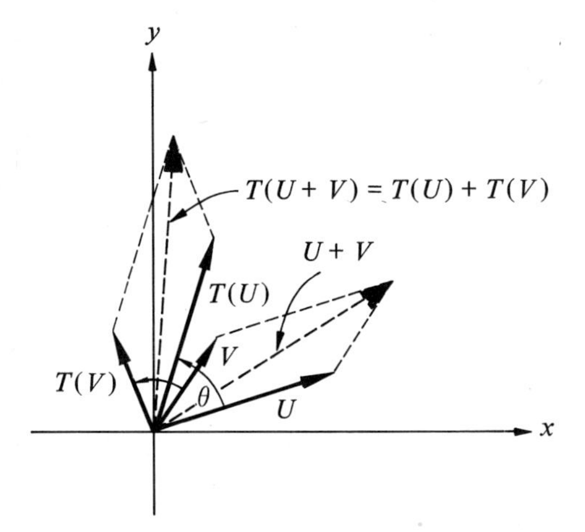


Figure 5.8.9

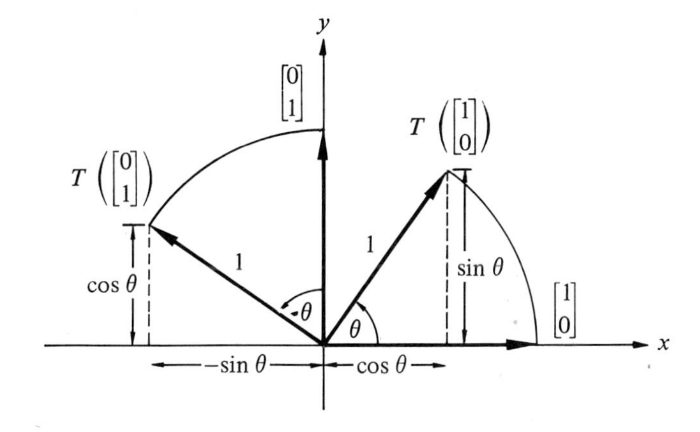


Thus the matrix of $T$ in the usual basis is

$$\begin{bmatrix} \cos\theta & -\sin\theta \\ \sin\theta & \cos\theta \end{bmatrix}.$$

Such a transformation is called a **rotation transformation**, or simply a **rotation**.

Example 5.8.2 Find, in the usual basis, the matrix of the rotation transformation through the angle $\theta = 60°$. From the foregoing discussion, the matrix is

$$\begin{bmatrix} \cos 60° & -\sin 60° \\ \sin 60° & \cos 60° \end{bmatrix} = \begin{bmatrix} 1/2 & -\sqrt{3}/2 \\ \sqrt{3}/2 & 1/2 \end{bmatrix}.$$

Thus for a typical vector $\begin{bmatrix} a \\ b \end{bmatrix}$ the coordinates of the image of $\begin{bmatrix} a \\ b \end{bmatrix}$ under this rotation are

$$\begin{bmatrix} 1/2 & -\sqrt{3}/2 \\ \sqrt{3}/2 & 1/2 \end{bmatrix}\begin{bmatrix} a \\ b \end{bmatrix} = \begin{bmatrix} a/2 - \sqrt{3}b/2 \\ \sqrt{3}a/2 + b/2 \end{bmatrix}.$$

Since we use the basis $\left\{\begin{bmatrix}1\\0\end{bmatrix}, \begin{bmatrix}0\\1\end{bmatrix}\right\}$, the $T$-image of $\begin{bmatrix}a\\b\end{bmatrix}$ is

$$(a/2 - \sqrt{3}b/2)\begin{bmatrix}1\\0\end{bmatrix} + (\sqrt{3}a/2 + b/2)\begin{bmatrix}0\\1\end{bmatrix} = \begin{bmatrix}a/2 - \sqrt{3}b/2\\ \sqrt{3}a/2 + b/2\end{bmatrix}. \blacksquare$$

## 5.8 Exercises

1. Let $T$ be the stretching transformation such that $T(V) = 2V$ for each $V$. Find the matrix $A$ of $T$ in the usual basis $\left\{\begin{bmatrix}1\\0\end{bmatrix}, \begin{bmatrix}0\\1\end{bmatrix}\right\}$.

2. Let $T$ be the stretching (or shrinking) transformation such that $T(V) = cV$ for each $V$. Find the matrix $A$ of $T$ in the usual basis.

3. Find the matrix $A$ (with respect to the usual basis) of the projection $P$ on the line $L$ containing the given vector. In each case find the image of a typical vector $\begin{bmatrix}a\\b\end{bmatrix}$.

   a. $\begin{bmatrix}1\\0\end{bmatrix}$ b. $\begin{bmatrix}0\\1\end{bmatrix}$ c. $\begin{bmatrix}-1\\1\end{bmatrix}$

   d. $\begin{bmatrix}-1\\-1\end{bmatrix}$ e. $\begin{bmatrix}1\\2\end{bmatrix}$ f. $\begin{bmatrix}2\\1\end{bmatrix}$

4. The projection $P$ on the line containing $\begin{bmatrix}1\\2\end{bmatrix}$ is a function from 2-space into 2-space. Is $P$ onto 2-space? Is $P$ one-to-one?

5. Find the matrix $A$ of the rotation transformation through the given angle $\theta$ and calculate the image of the vector $\begin{bmatrix}1\\1\end{bmatrix}$.

   a. $\theta = 30°$ b. $\theta = 45°$ c. $\theta = 240°$

   d. $\theta = 60°$ e. $\theta = 225°$ f. $\theta = 720°$

6. Calculate the matrix $A$ of the rotation through $\theta = 180°$. Is this particular rotational transformation also another kind of transformation?

**New Terms**

stretching transformation, 178
shrinking transformation, 178
projection, 178
rotation, 181

## 5.9 Eigenvectors and Eigenvalues

Suppose that we are given a linear transformation $T$ on, say, $\mathscr{M}_{3,1}$. We pose the following question: "Which vectors are only stretched or shrunk (in the same or opposite direction) by $T$?" That is, are there vectors $U$ and scalars $c$ such that $T(U) = cU$? To answer this question let us suppose that the given transformation $T$ is represented in the usual $I$, $J$, $K$ basis by the matrix

$$\begin{bmatrix} a_{11} & a_{12} & a_{13} \\ a_{21} & a_{22} & a_{23} \\ a_{31} & a_{32} & a_{33} \end{bmatrix}.$$

Our question may be restated, "Does there exist a scalar $c$ and a vector $U$, with coordinate matrix

$$\begin{bmatrix} x_1 \\ x_2 \\ x_3 \end{bmatrix},$$

relative to the $I$, $J$, $K$ basis, such that

$$\begin{bmatrix} a_{11} & a_{12} & a_{13} \\ a_{21} & a_{22} & a_{23} \\ a_{31} & a_{32} & a_{33} \end{bmatrix} \begin{bmatrix} x_1 \\ x_2 \\ x_3 \end{bmatrix} = c \begin{bmatrix} x_1 \\ x_2 \\ x_3 \end{bmatrix}?$$

We may rewrite the foregoing matrix equation using our matrix algebra, as

$$\begin{bmatrix} a_{11} - c & a_{12} & a_{13} \\ a_{21} & a_{22} - c & a_{23} \\ a_{31} & a_{32} & a_{33} - c \end{bmatrix} \begin{bmatrix} x_1 \\ x_2 \\ x_3 \end{bmatrix} = \begin{bmatrix} 0 \\ 0 \\ 0 \end{bmatrix}$$

or more simply as

$$(A - cI_3)X = 0.$$

In this form, the problem may be restated as, "Are there any values of $c$ and any matrices $X$ which satisfy the homogeneous system $(A - cI_3)X = 0$?" Recalling our study of homogeneous linear equations, the answer is that the zero matrix 0 is always such a matrix $X$, and further, for any scalar $c$ such that the rank of

$$(A - cI_3) = \begin{bmatrix} a_{11} - c & a_{12} & a_{13} \\ a_{21} & a_{22} - c & a_{23} \\ a_{31} & a_{32} & a_{33} - c \end{bmatrix}$$

is less than $n = 3$, there will always be such a non-zero $X$. On the other hand, if there is a non-zero $X$ satisfying the homogeneous equations, then the corresponding $c$ must be such that the rank of $(A - cI_3)$ is less than 3.

Now, one way of saying that the rank of $(A - cI_3)$ is less than 3, is to say that $\det(A - cI_3) = 0$. If we expand $\det(A - cI_3)$ we obtain a polynomial of degree 3 in the unknown $c$, and the real roots of this polynomial (remember that we are only working with real numbers here) are the values for which $\det(A - cI_3) = 0$. The polynomial $|A - cI_3|$ is called the **characteristic polynomial** of the matrix $A$ and the equation $|A - cI_3| = 0$ is called the **characteristic equation** of $A$. The real roots, if any, are called the **eigenvalues** (or characteristic values) of $A$, or of the original transformation $T$, represented by $A$ in the $I$, $J$, $K$ basis. For any such eigenvalue $c_1$ the homogeneous system $(A - c_1I_3)X = 0$ has non-zero solution vectors $X$, which are called

**eigenvectors** (or characteristic vectors) of $A$ corresponding to the eigenvalue $c_1$. Such eigenvectors may be found, of course, by our standard technique of solving systems of linear equations, once $c_1$ is known. Note that while the zero vector satisfies

$$(A - cI)0 = 0 \qquad \text{(indeed satisfies this for any value of } c\text{)}$$

we by definition exclude the vector 0 from the role of eigenvector. On the other hand, we do not exclude the number 0 from the role of eigenvalue. Thus, to take an extreme example, the zero matrix clearly has 0 as its only eigenvalue and has each non-zero vector in the space as an eigenvector.

Example 5.9.1 Let $T$ be the transformation whose matrix in the usual $I$, $J$, $K$ basis is

$$A = \begin{bmatrix} 1 & 1 & 0 \\ 0 & 2 & 1 \\ 0 & 0 & 3 \end{bmatrix}.$$

Then the characteristic equation of $A$ is

$$\begin{vmatrix} 1-c & 1 & 0 \\ 0 & 2-c & 1 \\ 0 & 0 & 3-c \end{vmatrix} = 0.$$

We expand this determinant by the elements and cofactors of the first row to obtain

$$(1-c)\begin{vmatrix} 2-c & 1 \\ 0 & 3-c \end{vmatrix} - 1\begin{vmatrix} 0 & 1 \\ 0 & 3-c \end{vmatrix} + 0\begin{vmatrix} 0 & 2-c \\ 0 & 0 \end{vmatrix}$$
$$= (1-c)(2-c)(3-c) = 0.$$

Thus the eigenvalues of $A$ are 1, 2, and 3. Corresponding to the eigenvalue 1, we find the eigenvectors by solving the homogeneous system $(A - 1I_3) = 0$. Thus we solve

$$\begin{bmatrix} 0 & 1 & 0 \\ 0 & 1 & 1 \\ 0 & 0 & 2 \end{bmatrix}\begin{bmatrix} x_1 \\ x_2 \\ x_3 \end{bmatrix} = \begin{bmatrix} 0 \\ 0 \\ 0 \end{bmatrix}$$

and find that $x_3 = 0$, $x_2 = 0$, and $x_1 = a$, for any choice of $a$. Thus the eigenvectors corresponding to the eigenvalue 1 are all those vectors of the form

$$a\begin{bmatrix} 1 \\ 0 \\ 0 \end{bmatrix}, \quad \text{with } a \neq 0 \text{ (remember, the zero vector is not an eigenvector!)}$$

You may now easily verify (by matrix multiplication) that

$$A\left(a\begin{bmatrix} 1 \\ 0 \\ 0 \end{bmatrix}\right) = a\begin{bmatrix} 1 \\ 0 \\ 0 \end{bmatrix}.$$

To find the eigenvectors corresponding to the eigenvalue 2, we solve the system $(A - 2I_3)X = 0$ which is

$$\begin{bmatrix} -1 & 1 & 0 \\ 0 & 0 & 1 \\ 0 & 0 & 1 \end{bmatrix} \begin{bmatrix} x_1 \\ x_2 \\ x_3 \end{bmatrix} = \begin{bmatrix} 0 \\ 0 \\ 0 \end{bmatrix}.$$

We find that in this case $x_3 = 0$, $x_2 = a$, $x_1 = a$ for any $a$. Thus the eigenvectors corresponding to the eigenvalue 2 are of the form

$$a\begin{bmatrix} 1 \\ 1 \\ 0 \end{bmatrix}, \quad \text{with } a \neq 0.$$

Again you may now verify that for any $a$,

$$A\left(a\begin{bmatrix} 1 \\ 1 \\ 0 \end{bmatrix}\right) = 2a\begin{bmatrix} 1 \\ 1 \\ 0 \end{bmatrix}.$$

Finally, we leave it to you to verify that the eigenvectors corresponding to the eigenvalue 3 are of the form

$$a\begin{bmatrix} 1/2 \\ 1 \\ 1 \end{bmatrix}, \quad \text{with } a \neq 0.$$

∎

Example 5.9.2 Let $A$ be the transformation whose matrix in the usual basis is $\begin{bmatrix} 0 & 2 \\ -3 & 5 \end{bmatrix}$.

Then the characteristic polynomial of $A$ is $\begin{vmatrix} -c & 2 \\ -3 & 5 - c \end{vmatrix} = c^2 - 5c + 6$.

Hence the eigenvalues of $A$ are 2, 3. To find the eigenvectors corresponding to the eigenvalue 2, we solve

$$\begin{bmatrix} -2 & 2 \\ -3 & 3 \end{bmatrix} \begin{bmatrix} x_1 \\ x_2 \end{bmatrix} = \begin{bmatrix} 0 \\ 0 \end{bmatrix}$$

and obtain the eigenvectors of the form $a\begin{bmatrix} 1 \\ 1 \end{bmatrix}$. To find the eigenvectors corresponding to the eigenvalue 3, we solve

$$\begin{bmatrix} -3 & 2 \\ -3 & 2 \end{bmatrix} \begin{bmatrix} x_1 \\ x_2 \end{bmatrix} = \begin{bmatrix} 0 \\ 0 \end{bmatrix}$$

to obtain eigenvectors of the form $a\begin{bmatrix} 2/3 \\ 1 \end{bmatrix}$. You should verify that the vectors $U$ found above are really eigenvectors by verifying that $AU = cU$. ∎

Example 5.9.3 Let $A$ be the transformation whose matrix in the usual $I$, $J$, $K$ basis is

$$\begin{bmatrix} 1 & 0 & 0 \\ 0 & 1 & 0 \\ 0 & 0 & 2 \end{bmatrix}.$$

Then the characteristic polynomial of $A$ is $(1 - c)(1 - c)(2 - c)$ and its eigenvalues are 1, 1, 2 with 1 as a root of multiplicity 2, or a double root. The eigenvectors corresponding to the eigenvalue 2 are of the form

$$a\begin{bmatrix}0\\0\\1\end{bmatrix}, \quad \text{with } a \neq 0. \qquad \blacksquare$$

To find the eigenvectors corresponding to the eigenvalue 1, we solve $(A - I_3)X = 0$

$$\begin{bmatrix}0 & 0 & 0\\0 & 0 & 0\\0 & 0 & 1\end{bmatrix}\begin{bmatrix}x_1\\x_2\\x_3\end{bmatrix} = \begin{bmatrix}0\\0\\0\end{bmatrix}$$

to find eigenvectors which are of the form

$$\begin{bmatrix}a\\b\\0\end{bmatrix}$$

for any choices of $a$ and $b$. Thus the set of all eigenvectors corresponding to 1 is the set of all non-zero linear combinations of

$$\begin{bmatrix}1\\0\\0\end{bmatrix} \quad \text{and} \quad \begin{bmatrix}0\\1\\0\end{bmatrix},$$

and, together with the zero vector forms a subspace of $\mathscr{M}_{3,1}$ of dimension 2.

Example 5.9.4 Let $A$ be the transformation whose matrix in the usual basis is $\begin{bmatrix}0 & -1\\1 & 0\end{bmatrix}$. Then the characteristic polynomial of $A$ is $\begin{vmatrix}-c & -1\\1 & -c\end{vmatrix} = c^2 + 1$. But, fortunately or otherwise, $c^2 + 1$ has no roots (we are only working with real numbers). Thus, no vectors are mapped by $A$ into scalar multiples of themselves (other than the zero vector, which is always mapped to itself). Had our underlying number system (field of scalars) been the complex numbers, then the transformation represented by $A$ would have eigenvalues $i$ and $-i$ (where $i^2 = -1$) and the corresponding eigenvectors are found to be of the forms

$$a\begin{bmatrix}i\\1\end{bmatrix} \quad \text{and} \quad a\begin{bmatrix}-i\\1\end{bmatrix}.$$

The point of our example is, however, that in our discussion, where the scalars are real numbers, some linear transformations simply do not have eigenvectors; thus some transformations alter the direction of every non-zero vector. $\blacksquare$

We close this section with the statement (but *not* the proof) of a curious theorem, and the application of that theorem to the calculation of the inverse of a matrix.

**THEOREM (HAMILTON-CAYLEY)**

*If $A$ is the $n \times n$ matrix of some transformation $T$ in the usual basis, then $A$ satisfies the characteristic polynomial of $A$ in the following sense: If $a_0 + a_1c + a_2c^2 + \cdots + a_nc^n$ is the characteristic polynomial of $A$, then $a_0I_n + a_1A + a_2A^2 + \cdots + a_nA^n = 0$.* (The 0 on the right hand side of this matrix equation is, of course, the symbol for the $n \times n$ matrix, all of whose elements are zero.)

Example 5.9.5 Let $A$ be the matrix of Example 5.9.2, $\begin{bmatrix} 0 & 2 \\ -3 & 5 \end{bmatrix}$, whose characteristic polynomial is $6 - 5c + c^2$. Then

$$\begin{aligned} 6I_2 - 5A + A^2 &= 6\begin{bmatrix} 1 & 0 \\ 0 & 1 \end{bmatrix} - 5\begin{bmatrix} 0 & 2 \\ -3 & 5 \end{bmatrix} + \begin{bmatrix} 0 & 2 \\ -3 & 5 \end{bmatrix}\begin{bmatrix} 0 & 2 \\ -3 & 5 \end{bmatrix} \\ &= \begin{bmatrix} 6 & 0 \\ 0 & 6 \end{bmatrix} + \begin{bmatrix} 0 & -10 \\ 15 & -25 \end{bmatrix} + \begin{bmatrix} -6 & 10 \\ -15 & 19 \end{bmatrix} \\ &= \begin{bmatrix} 0 & 0 \\ 0 & 0 \end{bmatrix}. \end{aligned}$$

∎

Example 5.9.6 Let $T$ be the transformation whose matrix in the usual basis is $\begin{bmatrix} 1 & 2 \\ 3 & 4 \end{bmatrix}$. The characteristic polynomial is $\begin{vmatrix} 1-c & 2 \\ 3 & 4-c \end{vmatrix} = c^2 - 5c - 2$. Then $\begin{bmatrix} 1 & 2 \\ 3 & 4 \end{bmatrix}$ satisfies this polynomial in that

$$\begin{aligned} A^2 - 5A - 2I_2 &= \begin{bmatrix} 1 & 2 \\ 3 & 4 \end{bmatrix}\begin{bmatrix} 1 & 2 \\ 3 & 4 \end{bmatrix} - 5\begin{bmatrix} 1 & 2 \\ 3 & 4 \end{bmatrix} - 2\begin{bmatrix} 1 & 0 \\ 0 & 1 \end{bmatrix} \\ &= \begin{bmatrix} 7 & 10 \\ 15 & 22 \end{bmatrix} + \begin{bmatrix} -5 & -10 \\ -15 & -20 \end{bmatrix} + \begin{bmatrix} -2 & 0 \\ 0 & -2 \end{bmatrix} \\ &= \begin{bmatrix} 0 & 0 \\ 0 & 0 \end{bmatrix}. \end{aligned}$$

∎

In order to apply this theorem to the calculation of the inverse of a matrix, we make one observation which you may verify in the examples presented so far; namely, the constant term $a_0$ in the characteristic polynomial of $A$ is the determinant of $A$. That this is true in general is seen from considering that the constant term of any polynomial in $c$ is the value of the polynomial when $c = 0$. The characteristic polynomial of $A$ is $|A - cI_n|$ and its value for $c = 0$ is simply $|A|$.

With this in mind, suppose that $A$ is a matrix which has an inverse (so that $|A| \neq 0$). By the Hamilton-Cayley Theorem, $A$ satisfies its characteristic polynomial, which is, let us say, $a_0 + a_1c + \cdots + a_nc^n$. Thus

$$a_0I_n + a_1A + a_2A^2 + \cdots + a_nA^n = 0.$$

Multiplying each side of the equation on the left by $A^{-1}$ we obtain

$$a_0A^{-1} + a_1I_n + a_2A + \cdots + a_nA^{n-1} = 0.$$

Since $a_0 = |A| \neq 0$, we solve for $A^{-1}$ and obtain

$$A^{-1} = \frac{-1}{a_0}(a_1 I_n + a_2 A + a_3 A^2 + \cdots + a_n A^{n-1}).$$

Example 5.9.7 For the matrix $A = \begin{bmatrix} 1 & 2 \\ 3 & 4 \end{bmatrix}$, the characteristic polynomial is $c^2 - 5c - 2$. Thus

$$\begin{bmatrix} 1 & 2 \\ 3 & 4 \end{bmatrix}^{-1} = \frac{-1}{-2}\left(-5\begin{bmatrix} 1 & 0 \\ 0 & 1 \end{bmatrix} + \begin{bmatrix} 1 & 2 \\ 3 & 4 \end{bmatrix}\right) = \begin{bmatrix} -2 & 1 \\ 3/2 & -1/2 \end{bmatrix}.$$ ∎

## 5.9 Exercises

1. Each of the following matrices is considered as the matrix of a certain linear transformation in the usual $\begin{bmatrix} 1 \\ 0 \end{bmatrix}, \begin{bmatrix} 0 \\ 1 \end{bmatrix}$ basis. Calculate the eigenvalues (if any) and the form of the corresponding eigenvectors.

   a. $\begin{bmatrix} 1 & 0 \\ 0 & 0 \end{bmatrix}$ b. $\begin{bmatrix} 1 & 1 \\ 2 & 2 \end{bmatrix}$ c. $\begin{bmatrix} 2 & 0 \\ 0 & 2 \end{bmatrix}$ d. $\begin{bmatrix} 2 & -4 \\ 1 & -3 \end{bmatrix}$

2. Each of the following matrices is considered as the matrix of a linear transformation in the usual

$$\left\{\begin{bmatrix} 1 \\ 0 \\ 0 \end{bmatrix}, \begin{bmatrix} 0 \\ 1 \\ 0 \end{bmatrix}, \begin{bmatrix} 0 \\ 0 \\ 1 \end{bmatrix}\right\}$$

   basis. Calculate the eigenvalues (if any) and the general form for the eigenvectors corresponding to each eigenvalue.

   a. $\begin{bmatrix} 1 & 2 & 3 \\ 0 & 2 & 3 \\ 0 & 0 & 3 \end{bmatrix}$ b. $\begin{bmatrix} 4 & 2 & -1 \\ -5 & -3 & 1 \\ 3 & 2 & 0 \end{bmatrix}$

   c. $\begin{bmatrix} 1 & 1 & 1 \\ 1 & 1 & 1 \\ 1 & 1 & 1 \end{bmatrix}$ d. $\begin{bmatrix} 5 & -3 & -2 \\ 1 & 1 & -2 \\ 1 & -3 & 2 \end{bmatrix}$

3. Verify the Hamilton-Cayley Theorem for each matrix in Exercise 1.

4. Verify the Hamilton-Cayley Theorem for each matrix in Exercise 2.

5. Find the characteristic polynomial for each of the following matrices and calculate the inverse of each matrix using the Hamilton-Cayley Theorem.

   a. $\begin{bmatrix} 1 & 2 \\ 0 & 3 \end{bmatrix}$ b. $\begin{bmatrix} 1 & 2 & 1 \\ 1 & 3 & 1 \\ 2 & 1 & 4 \end{bmatrix}$ c. $\begin{bmatrix} 0 & 1 & 2 \\ 0 & 0 & 2 \\ 2 & 1 & 3 \end{bmatrix}$

**New Terms**

characteristic polynomial, 183
characteristic equation, 183
eigenvalues, 183
eigenvectors, 184

## 5.10 Quadratic Forms

Consider the function $f$ from $\mathbb{R} \times \mathbb{R}$ into $\mathbb{R}$ such that

$$f(x_1, x_2) = a_{11}x_1^2 + a_{12}x_1x_2 + a_{21}x_2x_1 + a_{22}x_2^2.$$

Such a function is said to be a **quadratic function** and the sum on the right-hand side of the equation is called a **quadratic form** in the variables $x_1, x_2$. Such functions occur frequently in the study of statistics, economics and in several other mathematical applications.

Example 5.10.1 Let

$$f_1(x_1, x_2) = x_1^2 + x_2^2,$$
$$f_2(x_1, x_2) = 2x_1^2 - x_1x_2 + 3x_2x_1 + 5x_2^2,$$
$$f_3(x_1, x_2) = -3x_1^2 + x_2x_1 + 4x_2^2$$

and

$$f_4(x_1, x_2) = x_1^2 + 2x_1x_2 + 4x_2x_1 + 9x_2^2.$$

Each of the $f_i$ in Example 5.10.1 is a quadratic function. The significance for us of these functions and their associated forms lies in the fact that these functions can be conveniently represented by matrices. Consider $f_1$, above, for example. If we let $A_1 = \begin{bmatrix} 1 & 0 \\ 0 & 1 \end{bmatrix}$ and $X = \begin{bmatrix} x_1 \\ x_2 \end{bmatrix}$ and form the matrix product $X^TA_1X$ we have

$$f_1(X) = X^TA_1X = [x_1, x_2]\begin{bmatrix} 1 & 0 \\ 0 & 1 \end{bmatrix}\begin{bmatrix} x_1 \\ x_2 \end{bmatrix} = x_1^2 + x_2^2$$

—if we regard the $1 \times 1$ matrix $[x_1^2 + x_2^2]$ as the scalar $x_1^2 + x_2^2$.

Similarly for $f_2$,

$$f_2(X) = [x_1, x_2]\begin{bmatrix} 2 & -1 \\ 3 & 5 \end{bmatrix}\begin{bmatrix} x_1 \\ x_2 \end{bmatrix} = X^TA_2X$$

$$= [2x_1 + 3x_2, -x_1 + 5x_2]\begin{bmatrix} x_1 \\ x_2 \end{bmatrix}$$

$$= (2x_1^2 + 3x_2x_1) + (-x_1x_2 + 5x_2^2)$$
$$= 2x_1^2 - x_1x_2 + 3x_2x_1 + 5x_2^2,$$

since addition is commutative for the real numbers. You should verify that the matrices $A_3 = \begin{bmatrix} -3 & 0 \\ 1 & 4 \end{bmatrix}$ and $A_4 = \begin{bmatrix} 1 & 2 \\ 4 & 9 \end{bmatrix}$ will serve to represent $f_3$ and $f_4$. In general, then, the matrix

$$A = \begin{bmatrix} a_{11} & a_{12} \\ a_{21} & a_{22} \end{bmatrix}$$

will serve to represent the quadratic form

$$f(x_1, x_2) = a_{11}x_1^2 + a_{12}x_1x_2 + a_{21}x_2x_1 + a_{22}x_2^2$$

in that

$$f(x_1, x_2) = X^TAX.$$

Many terms which we have used for matrices may also be applied to quadratic forms. For example, we speak of the **rank of the quadratic form** $X^TAX$ as being the rank of any *symmetric matrix* $A$ such that $f(X) = X^TAX$.

It turns out, as you may have suspected, that the matrix which represents a given quadratic form is *not* unique. Consider $f_2(x_1, x_2)$ defined above. We found that $A_2 = \begin{bmatrix} 2 & -1 \\ 3 & 5 \end{bmatrix}$ represented $f_2$. However, $B_2 = \begin{bmatrix} 2 & 1 \\ 1 & 5 \end{bmatrix}$ also represents $f_2$ since

$$\begin{aligned} X^TB_2X = [x_1, x_2]\begin{bmatrix} 2 & 1 \\ 1 & 5 \end{bmatrix}\begin{bmatrix} x_1 \\ x_2 \end{bmatrix} &= 2x_1^2 + x_2x_1 + x_1x_2 + 5x_2^2 \\ &= 2x_1^2 + 2x_1x_2 + 5x_2^2 \\ &= 2x_1^2 - x_1x_2 + 3x_2x_1 + 5x_2^2, \end{aligned}$$

a phenomenon which occurs only because of commutativity of multiplication for the real numbers. $B_2$ is a symmetric matrix. For a given quadratic function $f$ such that

$$f(x_1, x_2) = a_{11}x_1^2 + a_{12}x_1x_2 + a_{21}x_2x_1 + a_{22}x_2^2,$$

it is always possible to find a unique symmetric matrix $B$ to represent $f$. We merely let $b_{11} = a_{11}$, $b_{22} = a_{22}$ and then average the coefficients of the cross-product terms $x_ix_j$ to let $b_{12} = b_{21} = (a_{12} + a_{21})/2$.

Example 5.10.2 For $f_3(x_1, x_2) = -3x_1^2 + x_2x_1 + 4x_2^2$ we have $B_3 = \begin{bmatrix} -3 & 1/2 \\ 1/2 & 4 \end{bmatrix}$. For

$f_4(x_1, x_2) = x_1^2 + 2x_1x_2 + 4x_2x_1 + 9x_2^2$, take $B_4 = \begin{bmatrix} 1 & 3 \\ 3 & 9 \end{bmatrix}$. ∎

The rank of $B_4$ is the rank of the form.

The quadratic function $f$, the quadratic form $X^TAX$, and the associated symmetric matrix $A$ which represents $f$ are said to be **positive definite** if $f(x_1, x_2) \geqslant 0$ for each pair $(x_1, x_2)$ and $f(x_1, x_2) = 0$ only if $x_1 = x_2 = 0$.

Example 5.10.3 $f_1(x_1, x_2) = x_1^2 + x_2^2$ has the associated symmetric matrix $\begin{bmatrix} 1 & 0 \\ 0 & 1 \end{bmatrix} = A$. Since $f_1 \geqslant 0$ for each pair $(x_1, x_2)$ and $f_1 = 0$ only if $x_1 = x_2 = 0$, then $f_1$, $X^TAX$, and $A$ are each said to be positive definite. ∎

Example 5.10.4 $f_4(x_1, x_2) = x_1^2 + 2x_1x_2 + 4x_2x_1 + 9x_2^2$ has the associated symmetric matrix $A = \begin{bmatrix} 1 & 3 \\ 3 & 9 \end{bmatrix}$. Since we may write $f_4(x_1, x_2) = (x_1 + 3x_2)^2$ it is clear that $f_4 \geqslant 0$. However, it is not true that $f_4 = 0$ implies that $x_1 = x_2 = 0$. For example, $f_4(3, -1) = 0$. Thus $f_4$ is not positive definite. ∎

The quadratic function $f$, the quadratic form $X^TAX$, and the associated symmetric matrix $A$ are said to be **positive semi-definite** if $f(x_1, x_2) \geqslant 0$ for each pair $(x_1, x_2)$, and there is at least one pair $(x_1, x_2) \neq (0, 0)$ such that $f(x_1, x_2) = 0$. Thus, $f_4$ above is seen to be positive semi-definite.

**Example 5.10.5** $f_3(x_1, x_2) = -3x_1^2 + x_2x_1 + 4x_2^2$ is neither positive definite nor positive semi-definite since $f_3(1, 0) = -3 < 0$. ∎

We could just as well have defined the terms **negative definite** and **negative semi-definite** above. To obtain these definitions we merely reverse the inequalities.

## 5.10 Exercises

1. Find a non-symmetric matrix representation for the following quadratic forms.
   a. $f(x_1, x_2) = x_1^2 + 2x_1x_2 + 3x_2x_1 + 5x_2^2$
   b. $f(x_1, x_2) = 2x_1^2 + x_1x_2 - x_2^2$
   c. $f(x_1, x_2) = 3x_1^2 + x_1x_2 - x_2x_1 + x_2^2$

2. Find a symmetric matrix representation of the functions in Exercise 1.

3. Show that $(X^TAX)^T = X^TA^TX$.

4. Find the quadratic form represented by the matrices listed below.

   a. $\begin{bmatrix} 1 & 2 \\ 3 & 4 \end{bmatrix}$ b. $\begin{bmatrix} 1 & 0 \\ 0 & 2 \end{bmatrix}$ c. $\begin{bmatrix} -1 & 0 \\ -2 & -3 \end{bmatrix}$

   d. $\begin{bmatrix} 1 & 3 \\ 3 & 1 \end{bmatrix}$ e. $\begin{bmatrix} 1 & 0 \\ 0 & -2 \end{bmatrix}$

5. State whether the following quadratic forms are positive definite, positive semi-definite, or neither. Justify your answer if $f$ is not positive definite by finding points required by the definition.
   a. $x_1^2 + 3x_2^2$ b. $x_1^2 + 2x_1x_2 + x_2^2$
   c. $x_1^2 - 2x_2^2$ d. $-x_1^2 + 2x_2^2$
   e. $x_1^2 - 3x_1x_2 - x_2x_1 + 4x_2^2$

6. On a graph in the plane, draw several of the level curves of the quadratic function $f(x_1, x_2) = x_1^2 + x_2^2$. That is, let $f(x_1, x_2) = 1, 4, 9$, say, and connect the points which satisfy the various equations with a smooth curve.

7. Refer to Exercise 6 and draw level curves for the quadratic function $f(x_1, x_2) = x_1^2 + 2x_2^2$.

### New Terms

# 6 Linear Programming

## 6.1 Introduction

One of the most useful applications of the concepts and techniques which we have developed is called **linear programming** (henceforth abbreviated *LP*). Contrary to what you may be thinking, the term "programming," as used here, does not refer to the programming of a computer. The term is used in the economic sense of *allocation* or *programming* of scarce resources as the examples in this section will show. The adjective "linear" refers to the linear equations and functions which arise in the study of many such problems. We should not, however, play down the importance of the computer as it pertains to LP. It was the advent of high-speed computing systems that kindled the great interest that has arisen in this subject over the past thirty years or so. Although much of the underlying theory has been known for some time, it was not until 1947 that George B. Dantzig, working on a United States Air Force project, introduced an efficient computational algorithm for the solution of a large class of problems. This procedure, known as the simplex algorithm, does require the use of a computer for the solution of meaningful business, industrial, and military problems. It has been recently asserted that the computer time spent in the solution of such problems amounts to one-fourth of that devoted to the solution of all scientific problems.*

In this section we shall formulate several problems which are classified as LP problems. In succeeding sections we shall investigate the geometry of LP problems and develop the simplex algorithm in some detail. We shall show why it is so important that the problems be "linear" in nature.

Example 6.1.1 Suppose that a university wishes to charter a sufficient number of aircraft to transport 500 students to an athletic event. Two types of aircraft

---

* *Encyclopedia Britannica*, 15th ed., Macropaedia, Vol. 15, p. 628. Chicago: Encyclopedia Britannica, Inc., 1975.

are available. $A_1$ carries 50 passengers, has a crew of 6 including 3 hostesses, and will cost $2000 for the trip. $A_2$ carries 100 passengers, has a crew of 8 with 4 hostesses, and will cost $5000 for the trip. On the day in question, only 24 hostesses are available. How many of each type of aircraft should be chartered to perform the airlift at the minimum cost?

The data may be compactly portrayed as in Table 6.1.1.

**Table 6.1.1**

| | *Totals* | $A_1$ | $A_2$ |
|---|---|---|---|
| *Cost* | *Minimize* | $2000 | $5000 |
| Passengers | 500 | 50 | 100 |
| Hostesses | 24 | 3 | 4 |

If we let $x_i$ denote the number of aircraft of type $i$ to be chartered, then we may describe the problem algebraically as follows:

Find $x_1, x_2$ such that

1. $x_i \geqslant 0, i = 1, 2$
2. $50x_1 + 100x_2 \geqslant 500$ (students)
3. $3x_1 + 4x_2 \leqslant 24$ (hostesses) and
4. $2000x_1 + 5000x_2 = f(x_1, x_2) = z$ is minimized.

As is the case in most allocation problems, negative values of the variables do not make sense, so **non-negativity restrictions** of the form $x_i \geqslant 0$ are included. Constraint 2 is included to insure that the number of aircraft chartered is sufficient to carry the required number of students. Constraint 3 prevents the number of hostesses required from being greater than the number available. $z = f(x_1, x_2)$ is called the **objective function** or **criterion function** or **cost function** which is to be minimized. ▮

Example 6.1.2

Consider the following problem—also one of interest to charter airlines. It provides a preview of a special type of LP problem, known as a "transportation problem," which, because of its special structure, will be studied in detail in the next chapter.

*XYZ* Airline has 6 aircraft, 2 of which are at New Orleans and the other 4 at New York. Six charter loads are to be picked up—3 at St Louis and 3 at Detroit. Each charter load is to be picked up by a different aircraft. Find the number of aircraft which should be sent to each pick-up site in order to minimize the total flight time involved in the pick-up deployment. Table 6.1.2 shows the flight times (in hours) from the cities in the first column to the cities in the top row, as well as the available aircraft and required aircraft. Available aircraft ($A$) and required aircraft ($R$) are depicted in Figure 6.1.1.

Table 6.1.2

| *Flight Time* | Detroit | St. Louis | *Aircraft Available* |
|---|---|---|---|
| New York | 2 | 3 | 4 |
| New Orleans | 3 | 2 | 2 |
| *Aircraft Required* | 3 | 3 | |

Figure 6.1.1

If we let $x_1$ denote the number of aircraft to be sent from New York to Detroit, $x_2$ the number to be sent from New York to St. Louis, $x_3$ the number from New Orleans to Detroit, and $x_4$ the number to be sent from New Orleans to St. Louis, then we may phrase the problem as follows:

---

Find $x_1, x_2, x_3, x_4$, each greater than or equal to zero, such that

1. $x_1 + x_2 = 4$
2. $x_3 + x_4 = 2$
3. $x_1 + x_3 = 3$
4. $x_2 + x_4 = 3$ and

$z = f(x_1, x_2, x_3, x_4) = 2x_1 + 3x_2 + 3x_3 + 2x_4$ is a minimum.

---

In Example 6.1.2, $f(x_1, x_2, x_3, x_4) = 2x_1 + 3x_2 + 3x_3 + 2x_4$ is the objective or criterion function. It represents the total flight time involved in the deployment. Equation 1 represents a constraint on the choices available for the $x_i$ and simply restricts us to choices of values for these variables that require the total number of aircraft sent from New York to be 4, since all four are needed to complete the deployment. Similarly, constraint 2 reflects a requirement on the number of aircraft departing New Orleans. Constraints 3 and 4 are equations which insure that the number of aircraft arriving at Detroit and St. Louis, respectively, are sufficient to handle the requirements there. Negative values of the variables do not make sense, so the non-negativity restrictions are included.

If we let $C = [2, 3, 3, 2]$,

$$A = \begin{bmatrix} 1 & 1 & 0 & 0 \\ 0 & 0 & 1 & 1 \\ 1 & 0 & 1 & 0 \\ 0 & 1 & 0 & 1 \end{bmatrix}, \quad X = \begin{bmatrix} x_1 \\ x_2 \\ x_3 \\ x_4 \end{bmatrix}, \quad \text{and} \quad R = \begin{bmatrix} 4 \\ 2 \\ 3 \\ 3 \end{bmatrix},$$

then we may phrase the problem of Example 6.1.2 in matrix notation:

---

$$\text{Minimize} \quad z = CX \quad \text{such that}$$
$$AX = R \quad \text{and}$$
$$X \geqslant 0.$$

---

Here we have used the notational device $X \geqslant 0$ to mean that each component of $X$ is non-negative. For any two matrices $A$ and $B$, we shall say that $A \leqslant B$ or $B \geqslant A$ if and only if $a_{ij} \leqslant b_{ij}$ for all $i$ and $j$.

We now consider an industrial problem which falls into the LP category.

Example 6.1.3 Company $Z$ blends two grades of gasoline, grade $A$ and grade $B$, to make two intermediate types of gasoline, regular and premium. Each gallon of regular gasoline requires 0.2 gallons of grade $A$ and 0.8 gallons of grade $B$. Each gallon of premium gasoline requires 0.8 gallons of grade $A$ and 0.2 gallons of grade $B$. The company makes 2 cents profit on each gallon of regular gasoline and 3 cents on each gallon of premium gasoline that it sells. If it has only 1000 gallons of each basic grade of gasoline, how much of each type of gasoline should be blended in order to maximize the company's profit, assuming that all can be sold?

In this case we let $x_1$ denote the amount of regular gasoline and $x_2$ denote the amount of premium gasoline to be blended from the basic grades on hand. Then we formulate the problem as follows:

---

Find $x_1, x_2$ such that $x_i \geqslant 0$ and

$$0.2x_1 + 0.8x_2 \leqslant 1000 \qquad \text{(grade } A\text{)}$$
$$0.8x_1 + 0.2x_2 \leqslant 1000 \qquad \text{(grade } B\text{)}$$

and $z = f(x_1, x_2) = 2x_1 + 3x_2$ is maximized.

---

In matrix format, letting

$$C = [2, 3], \qquad X = \begin{bmatrix} x_1 \\ x_2 \end{bmatrix}, \qquad A = \begin{bmatrix} 0.2 & 0.8 \\ 0.8 & 0.2 \end{bmatrix}, \qquad R = \begin{bmatrix} 1000 \\ 1000 \end{bmatrix}$$

we wish to

---

Find $X \geqslant 0$ such that $AX \leqslant R$ and $z = f(x_1, x_2) = CX$ is maximized.

---

In this problem the first constraint insures that the total amount of grade $A$ base to be utilized is not greater than the amount available. The second constraint performs the same function for the grade $B$ base. $f(x_1, x_2) = CX$ represents the company's profit for a given allocation of the available basic gasolines. ∎

The following small example of a problem historically called the "diet problem" provides another illustration of how an allocation problem may easily be put into LP form.

Example 6.1.4 The dining hall dietician wishes to provide at least 3 units of vitamin A and 4 units of vitamin B per serving. She has two choices of foods, $F_1$ and $F_2$, which contain vitamins A and B in the amounts shown in the table below.

**Table 6.1.3**

| *Food* | *Vitamin A* | *Vitamin B* |
|---|---|---|
| $F_1$ | 1 unit/oz. | 1 unit/oz. |
| $F_2$ | 1 unit/oz. | 2 unit/oz. |

If we let $x_1$ denote the number of ounces of $F_1$ per serving to be used and $x_2$ denote the number of ounces of $F_2$ per serving, then

$$1x_1 + 1x_2 \geqslant 3$$

is a constraint guaranteeing a sufficient amount of vitamin A per serving, and

$$1x_1 + 2x_2 \geqslant 4$$

is a constraint insuring a sufficient amount of vitamin B.

As in most dining halls, the manager wishes to provide this portion of the meal for as low a cost as possible. Suppose that $F_1$ costs 4 cents an ounce and that $F_2$ costs 5 cents an ounce. What is the price of the minimum cost serving satisfying the minimum vitamin requirements?

The cost of serving $x_1$ ounces of $F_1$ and $x_2$ ounces of $F_2$ is

$$z = f(x_1, x_2) = 4x_1 + 5x_2.$$

We thus have another type of linear programming problem:

---

Find $x_1, x_2$, each non-negative, such that

$$1x_1 + 1x_2 \geqslant 3$$

$$1x_1 + 2x_2 \geqslant 4$$

and $z = f(x_1, x_2) = 4x_1 + 5x_2$ is minimized.

---

In matrix format we wish to

Find $X \geqslant 0$ such that $AX \geqslant R$ and $z = CX$ is minimized.

Here

$$A = \begin{bmatrix} 1 & 1 \\ 1 & 2 \end{bmatrix}, \quad X = \begin{bmatrix} x_1 \\ x_2 \end{bmatrix}, \quad R = \begin{bmatrix} 3 \\ 4 \end{bmatrix} \quad \text{and} \quad C = [4, 5].$$ ∎

These examples should be sufficient to acquaint the reader with some of the types of problems which lend themselves to LP formulation. Examples 6.1.1 and 6.1.2 are of special interest. For practical purposes, the only values for the variables which would make sense would have to be non-negative *integers*. A solution in Example 6.1.2 of the form $X^T = [x_1, x_2, x_3, x_4] = [3/2, 5/2, 3/2, 1/2]$, calling for 3/2 aircraft to be sent from New York to Detroit would not be of much help to the operations officer in charge. Problems of this nature are called **integer programming** (IP) problems. Some IP problems can be solved by special algorithms requiring even less time than the **continuous** or non-integer problems. The general IP problem, although one of outstanding practical significance, still presents computational difficulties that researchers are trying to overcome. Two very interesting and intellectually satisfying algorithms for the solution of such problems were presented by Dr. Ralph Gomory in 1958 and 1960 (see Suggested Reading). Computer programs utilizing these algorithms have met with varying degrees of success. While some problems have been very readily solved, others take inordinate amounts of expensive computer time. For this reason, researchers have been looking at the problem from other points of view in an attempt to find methods of greater computational reliability.

The problems formulated in Examples 6.1.3 and 6.1.4 are continuous LP problems, that is, they are of a type for which any non-negative values of the variables will suffice. Except when we note otherwise, this is the type of problem in which we will be interested. Extremely efficient computer programs have been developed which are based upon the method to be developed in the following sections. Almost any computer center is certain to have a "canned" LP code available. The interested student may wish to try some of the problems on a local computer.

## 6.1 Exercises

Formulate Exercises 1–6 as LP problems.

1. A company makes two products—$P_1$ and $P_2$. The manufacturing process for $P_1$ requires 5 minutes on machine $M_1$, 10 minutes on machine $M_2$, and 3 minutes on machine $M_3$. Product $P_2$, however, requires 4 minutes on $M_1$, 8 minutes on $M_2$, and 6 minutes on $M_3$. The net profit on each unit of $P_1$ is \$4.00 while each unit of $P_2$ nets \$3.00. If there are 480 minutes of machine time available on $M_1$, 960 minutes on $M_2$, and 480 on $M_3$, how many units of each of $P_1$, $P_2$ should be made in order to maximize the firm's profit over the time period involved?

2. A furniture manufacturer makes chairs of types $F_1$, $F_2$ and tables of types $F_3$, $F_4$, $F_5$. Each item must pass through the wood-working shop and then go to the finishing department. The time unit requirements for each piece and the time available in each department are shown below, along with the unit profit on each piece. Assuming that orders which must be filled are on hand for $10F_1$ and $6F_4$. how many of each should be produced in order to maximize the firm's profit?

| *Dept.* \ *Furniture* | $F_1$ | $F_2$ | $F_3$ | $F_4$ | $F_5$ | *Time Available* |
|---|---|---|---|---|---|---|
| Woodworking | 2 | 3 | 2 | 4 | 5 | 200 |
| Finishing | 1 | 2 | 3 | 4 | 3 | 160 |
| Profit/unit | 10 | 20 | 25 | 30 | 30 | |

3. A supervisor wishes to assign each of his 3 men to one of 3 jobs in order to make the time required to accomplish all 3 jobs as small as possible. The time required for each man to complete each job is known and is recorded below.

| *Man* \ *Job* | *1* | 2 | 3 |
|---|---|---|---|
| 1 | 3 | 5 | 4 |
| 2 | 4 | 4 | 4 |
| 3 | 4 | 6 | 5 |

Let $x_{ij} = 1$ if man $i$ is assigned to job $j$, and 0 otherwise. Express the problem as an LP problem with 6 constraints (not including non-negativity restrictions). You should have 1 constraint for each man (to insure that he has a job) and 1 constraint for each job.

4. A truck is to be loaded with cases of products $P_1$, $P_2$ which it carries for $c_1$, $c_2$ dollars per case. A case of $P_i$ has volume $v_i$ and weight $w_i$. The truck limitations on freight weight and volume are $W$ and $V$, respectively. Find the number of cases of each type which should be carried in order to maximize net revenue.

5. A pharmaceutical firm makes 4 types of capsules, $C_1$, $C_2$, $C_3$, $C_4$. Each of the capsules uses ingredients $D_1$, $D_2$. The amounts of each of the ingredients required to make 1 unit of each type of capsule are shown in the table, along with the unit profit for each type of capsule, and the ingredient availability.

| *Ingredient* \ *Capsule* | $C_1$ | $C_2$ | $C_3$ | $C_4$ | *Units Available* |
|---|---|---|---|---|---|
| $D_1$ | 3 | 6 | 7 | 8 | 10,000 |
| $D_2$ | 2 | 5 | 8 | 10 | 14,000 |
| Unit Profit | 3 | 4 | 6 | 8 | |

Assuming that at least 250 units of $c_2$ and 500 units of $c_3$ must be produced, find the production plan which will maximize the firm's profit.

6. A chemical manufacturer has 3 warehouses ($W_i$) and 4 retail outlets ($S_j$) in a given district. A certain chemical is shipped by rail at a cost of $C$ dollars per pound-mile. The amount of the chemical on hand at the various warehouses and that required at the various outlets is shown in the table, along with the distances between the various points. Find the amounts which should be shipped from $W_i$ to $S_j$ in order to minimize the total shipping cost.

| $W_i$ \ $S_j$ | $S_1$ | $S_2$ | $S_3$ | $S_4$ | *Available* |
|---|---|---|---|---|---|
| $W_1$ | $d_{11}$ | $d_{12}$ | $d_{13}$ | $d_{14}$ | $a_1$ |
| $W_2$ | $d_{21}$ | $d_{22}$ | $d_{23}$ | $d_{24}$ | $a_2$ |
| $W_3$ | $d_{31}$ | $d_{32}$ | $d_{33}$ | $d_{34}$ | $a_3$ |
| Required | $b_1$ | $b_2$ | $b_3$ | $b_4$ | |

7. Express the LP problem of Exercise 1 in matrix format.

8. Express the LP problem of Exercise 3 in matrix format.

9. A men's shop wishes to purchase the following quantities of two types of men's suits.

| Type | 1 | 2 |
|---|---|---|
| Number | 200 | 300 |

Bids have been received from 3 different wholesalers, $W_1$, $W_2$, $W_3$ who have agreed to supply not more than the number of suits indicated below

| | |
|---|---|
| $W_1$ | 200 |
| $W_2$ | 250 |
| $W_3$ | 150 |

The owner has estimated that his profit/suit sold from each of the $W_i$ is given by the following table.

| *Wholesaler* | *Suit 1* | *Suit 2* |
|---|---|---|
| $W_1$ | 30 | 40 |
| $W_2$ | 38 | 35 |
| $W_3$ | 40 | 36 |

Let $x_{ij}$ denote the number of suits of type $i$ to be bought from wholesaler $j$ and set up an LP problem to maximize the owner's profit.

10. A charter outfit has two types of aircraft, $A_1$ and $A_2$ available. A firm wishes to rent a sufficient number of aircraft to transport 200 tons of equipment. $A_1$ can carry

20 tons while $A_2$ can carry only 10. $A_1$ rents for $400 and needs 4000 gallons of fuel for the round trip. $A_2$ rents for $300 and needs 1000 gallons of fuel for the trip. Assuming that each aircraft makes only 1 round-trip and that only 30,000 gallons of fuel are available, set up an LP problem to perform the move at minimum rental cost.

### Suggested Reading

M. M. Balinski. "Integer Programming: Methods, Uses, Computation." *Management Science* 12 (November, 1965): 253–312.

C. R. Carr and C. W. Howe. *Quantitative Decision Procedures in Management and Economics.* New York: McGraw-Hill, 1964.

A. Charnes and W. W. Cooper. *Management Models and Industrial Applications of Linear Programming.* Vols. I–II. New York: John Wiley and Sons, 1961.

G. B. Dantzig. *Linear Programming and Extensions.* Princeton, N.J.: Princeton University Press, 1963.

S. I. Gass. *Linear Programming.* 3d. ed. New York: McGraw-Hill, 1969.

S. I. Gass. *Illustrated Guide to Linear Programming.* New York: McGraw-Hill, 1970.

A. J. Hughes and D. E. Grawiog. *Linear Programming: An Emphasis on Decision Making.* Reading, Mass.: Addison-Wesley, 1973.

R. E. Gomory. "An Algorithm for Integer Solutions to Linear Programs." *Technical Report No. 1.* Princeton-IBM Mathematics Research Project. (November 17, 1958).

R. E. Gomory. "All-Integer Integer Programming Algorithm." *Report RC*-189. IBM Research Center, Yorktown Heights, N.Y. (January 29, 1960).

G. Hadley. *Linear Programming.* Reading, Mass.: Addison-Wesley, 1962.

T. H. Naylor, E. T. Byrne, and J. M. Vernon. *Introduction to Linear Programming: Methods and Cases.* Belmont, Ca.: Wadsworth, 1971.

D. R. Plane and G. A. Kochenberger. *Operations Research for Managerial Decisions.* Homewood, Ill.: Irwin, 1972.

D. R. Plane and C. McMillan. *Discrete Optimization.* Englewood Cliffs, N.J.: Prentice-Hall, 1971.

H. M. Wagner. *Principles of Operations Research.* 2d. ed. Englewood Cliffs, N.J.: Prentice-Hall, 1975.

### New Terms



## 6.2 Geometrical Solution of an LP Problem

We now examine a problem that is algebraically similar to Example 6.1.3. We shall call this problem "problem $P$" and shall refer to it many times throughout the remainder of this chapter. Here, we solve it geometrically and use it as a vehicle to illustrate some of the special properties of LP problems. We depend upon your intuitive understanding of the terms which we introduce in this section. In later sections we shall obtain a general formulation of the LP problem and define the terms more precisely.

---

Problem $P$: Find $x_1, x_2$ such that

1. $x_i \geqslant 0, i = 1, 2$
2. $\begin{cases} 2x_1 + x_2 \leqslant 6 \\ x_1 + x_2 \leqslant 4 \end{cases}$
3. $z = f(x_1, x_2) = 5x_1 + 3x_2$ is maximized.

---

We solve problem $P$ in 3 steps.

1. Out of all points $(x_1, x_2)$ in the $x_1, x_2$-plane, we find those which have the property that $x_1 \geqslant 0$ and $x_2 \geqslant 0$, and call this set $S_1$.
2. We then determine $S_2$, the set of all points $(x_1, x_2)$ in the $x_1, x_2$-plane that satisfy the remaining inequality constraints.
3. Out of $S_3 = S_1 \cap S_2$ we pick that point $(\bar{x}_1, \bar{x}_2)$, or points, at which $f(x_1, x_2)$ takes on its largest value, if there is one.

Let us first consider the $x_1, x_2$-plane as shown in Figure 6.2.1. Proceeding with Step 1, we see that $x_1 \geqslant 0$ for all points lying on or to the right of the $x_2$ axis. Similarly, $x_2 \geqslant 0$ for all points lying above or on the $x_1$-axis. Therefore, the set of points $S_1 = \{(x_1, x_2) \mid x_1 \geqslant 0, x_2 \geqslant 0\}$ is the set of points in the shaded first quadrant as shown, including the points on the non-negative $x_1$- and $x_2$-axes.

**Figure 6.2.1**

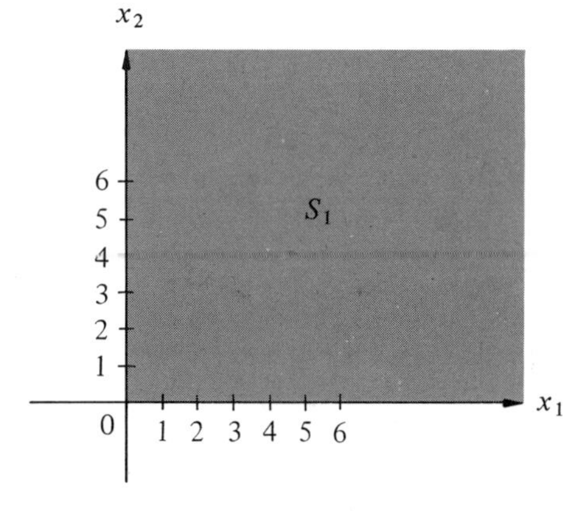


In Step 2 we need to determine $S_2$, the set of points which satisfy both $2x_1 + x_2 \leqslant 6$ and $x_1 + x_2 \leqslant 4$. Let us first determine $\{(x_1, x_2) \mid 2x_1 + x_2 \leqslant 6\}$. We know (Figure 6.2.2) that $\{(x_1, x_2) \mid 2x_1 + x_2 = 6\}$ forms a straight line in the $x_1, x_2$-plane. This straight line divides the $x_1, x_2$-plane into two separate or disjoint subsets, namely $P_1 = \{(x_1, x_2) \mid 2x_1 + x_2 < 6\}$ and $P_2 = \{(x_1, x_2) \mid 2x_1 + x_2 > 6\}$. Our only problem is to decide which side of the line is $P_1$ and which is $P_2$. The answer is quickly provided, however, by picking any point not on the line and testing it to see in which set

Figure 6.2.2

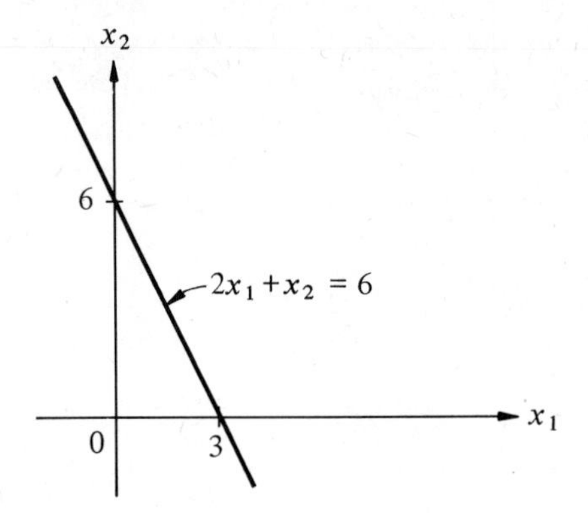


it belongs. The origin (0, 0) is usually a good point to test. Since $2(0) + 1(0) = 0 < 6$, (0, 0) lies in $P_1$, so we know that $P_1$ is the shaded half-plane in Figure 6.2.3.

Figure 6.2.3

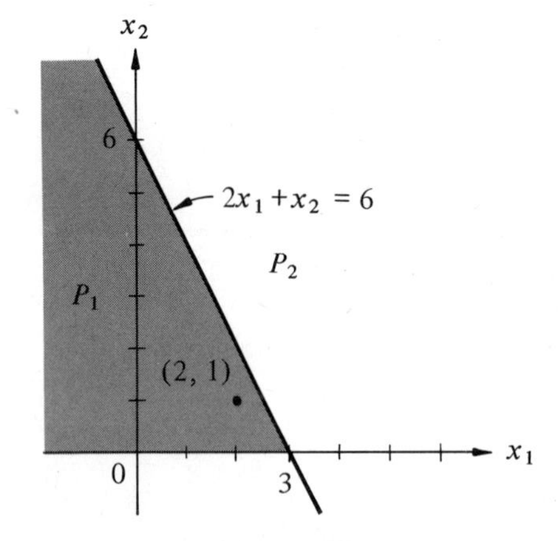


Figure 6.2.4

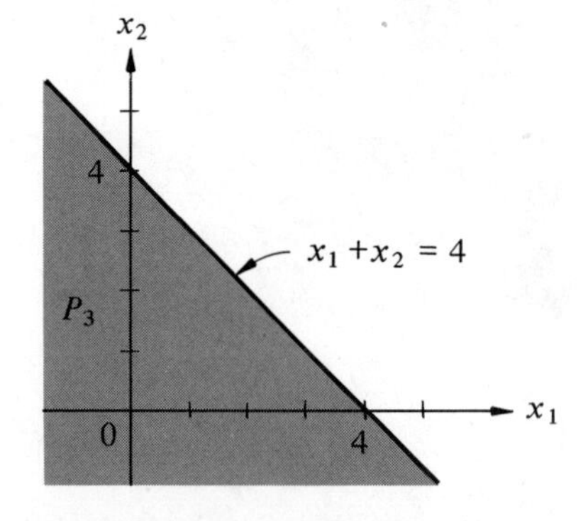


Similarly we use the line $x_1 + x_2 = 4$ to determine $P_3 = \{(x_1, x_2) \mid x_1 + x_2 \leqslant 4\}$, and find that $P_3$ is the shaded portion of Figure 6.2.4. $S_2 = \{(x_1, x_2) \mid 2x_1 + x_2 \leqslant 6 \text{ and } x_1 + x_2 \leqslant 4\}$ is thus the intersection of $P_1$ and $P_3$. In Figure 6.2.5, we show $S_2 = P_1 \cap P_3$ geometrically by exhibiting both lines on the same plane. The lightly shaded region represents $S_2 = P_1 \cap P_3$ while the darkly shaded regions represent portions of $P_1$ and $P_3$

Figure 6.2.5

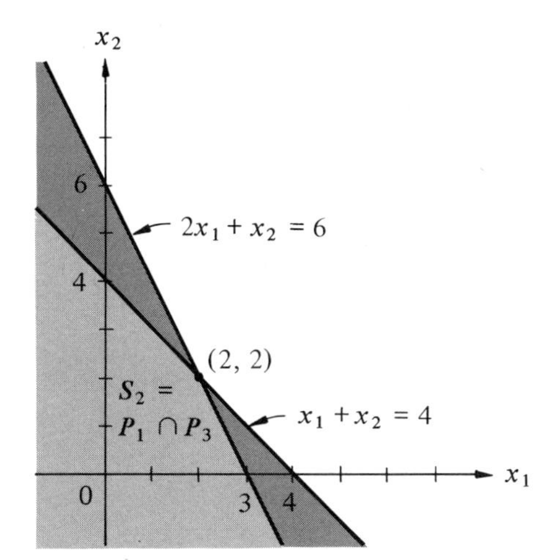


which do not lie in both. We obtain the point of intersection of the two lines by solving the two equations by Gauss-Jordan reduction.

In Step 3, we first find $S_1 \cap S_2$ by eliminating all points of $S_2$ not in $S_1$, the first quadrant, and obtain $S_3$ in Figure 6.2.6. $S_3$ is the set of all allowable or feasible points for an answer to our problem. Any point $(x_1, x_2)$ in $S_3$ has the property that it satisfies both the non-negativity restrictions and the other inequality constraints. The remaining problem is to pick that point $(\bar{x}_1, \bar{x}_2)$ or points in $S_3$ which provides the *best* answer, that is, that point which maximizes the function $f(x_1, x_2) = 5x_1 + 3x_2$.

Figure 6.2.6

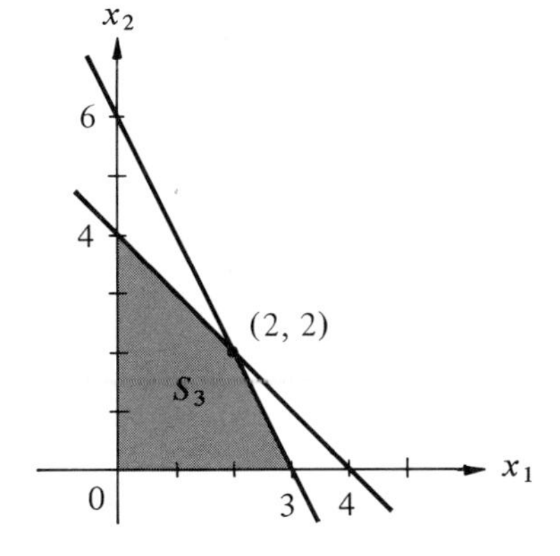


Let us arbitrarily pick some points in $S_3$ and compute their functional values. The results of such a computation are shown in Table 6.2.1 below where, for example, we computed $f(1, 0) = 5(1) + 3(0) = 5$ and $f(1, 1) = 5(1) + 3(1) = 8$.

Table 6.2.1

| $x_1$ | 0 | 1 | 2 | 3 | 0 | 0 | 0 | 0 | 2 | 1 | 2 | 1 |
|---|---|---|---|---|---|---|---|---|---|---|---|---|
| $x_2$ | 0 | 0 | 0 | 0 | 1 | 2 | 3 | 4 | 2 | 1 | 1 | 3 |
| $f(x_1, x_2)$ | 0 | 5 | 10 | 15 | 3 | 6 | 9 | 12 | 16 | 8 | 13 | 14 |

It would clearly be a never-ending process to try to compute the maximum value of $f$ by computing $f(x_1, x_2)$ for every point in $S_3$. Another method equivalent to such an enumeration must be found.

We begin to develop an equivalent procedure by examining the objective function $f$. For each value of the constant $a$, the function $f(x_1, x_2) = 5x_1 + 3x_2 = a$ represents a straight line in the $x_1, x_2$-plane. At each point on this fixed line, $f(x_1, x_2) = a$. Such lines for $a = 0, 5, 10, 16, 18$ are shown superimposed on $S_3$ in Figure 6.2.7.

**Figure 6.2.7**

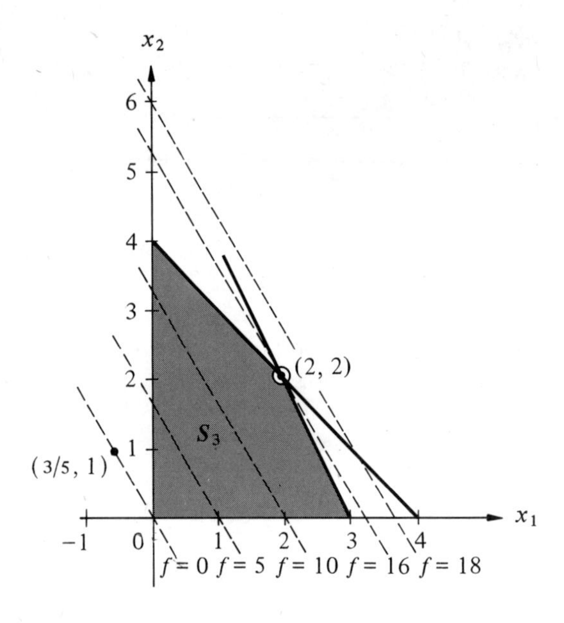


To draw the line $z = f(x_1, x_2) = 5x_1 + 3x_2 = 0$, we find two points on the line, say (0, 0) and (−3/5, 1), for example, by choosing a simple value for one of the variables and solving for the other. We know from Table 6.2.1 that (1, 0) is on the line $z = 5x_1 + 3x_2 = 5$. Letting $x_1 = 0$ we find that (0, 5/3) is also on this line. Similarly, (2, 0) and (0, 10/3) are on the line $z = 5x_1 + 3x_2 = 10$; (2, 2) and (0, 16/3) are on the line $z = 5x_1 + 3x_2 = 16$; and (18/5, 0) and (0, 6) are on the line $z = 5x_1 + 3x_2 = 18$. We use these points to plot the lines in Figure 6.2.7. The figure shows that each of these lines has the same slope which we compute as −5/3 by setting up the equation in the so-called slope-intercept form:

$$x_2 = \frac{-5x_1 + a}{3} = \frac{-5}{3}x_1 + \frac{a}{3}.$$

In this form we see that −5/3 is the slope of each of the lines and that for any particular value of $a$, the $x_2$ intercept is $a/3$, which we obtain by letting $x_1 = 0$.

Examination of the same graph shows that $z = f(x_1, x_2)$ takes on its maximum value in $S_3$ at a unique point, namely (2, 2), and that the maximum value of $z$ within $S_3$ is 16. (We shall refer to this result many times in this chapter.) To establish the maximum value of $z$, it may be helpful to think of taking a movable line through the origin with slope −5/3 and moving it, parallel to itself, in the direction of increasing $z$ as far as one can go while still encountering points within $S_3$. When you are forced to stop, you have found the optimum point (or points). In a minimization problem you would merely move as far as possible in the direction of decreasing $z$.

Having obtained the solution point for our problem, let us consider some ideas which will prove to be of interest later on.

1. The optimum value of $z$ occurred at a point on the boundary of $S_3$.
2. Even more specifically, the optimum value of $z$ occurred at a corner point of the boundary of $S_3$.
3. $S_3$ has the interesting geometrical property that if you choose any two points in $S_3$, the entire line segment connecting the two points lies in $S_3$.

To show that not all functions and plane point sets have these properties, consider the following examples.

Example 6.2.1 Suppose that we have $S_3$ as shown in Figure 6.2.8 and that we wish to find the point $(\bar{x}_1, \bar{x}_2)$ in $S_3$ which minimizes the so-called quadratic function

$$z = f(x_1, x_2) = (x_1 - 1)^2 + (x_2 - 1)^2.$$

Figure 6.2.8

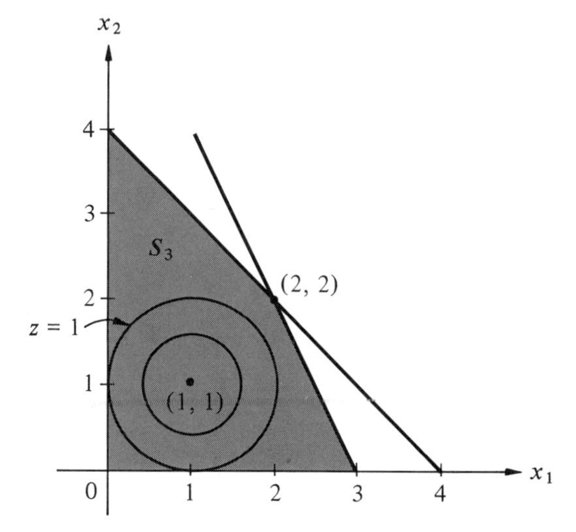


As a sum of squares, this function can never be negative. $f(1, 1) = 0$, and $f(x_1, x_2) > 0$ at any other point in the plane. Two of the equal-value circles of $z$ are shown. An equal-value circle has the property that $z$ is the same value at each point on it. For example, $z = 1$ at each point on the outer circle. The function takes on its minimum value 0 at (1, 1), an interior point of $S_3$—not a boundary point. Thus, for this non-linear function, we could not claim that any optimal solution, or even that at least one optimal solution, had to lie on the boundary of the region determined by the constraints. ∎

Figure 6.2.9

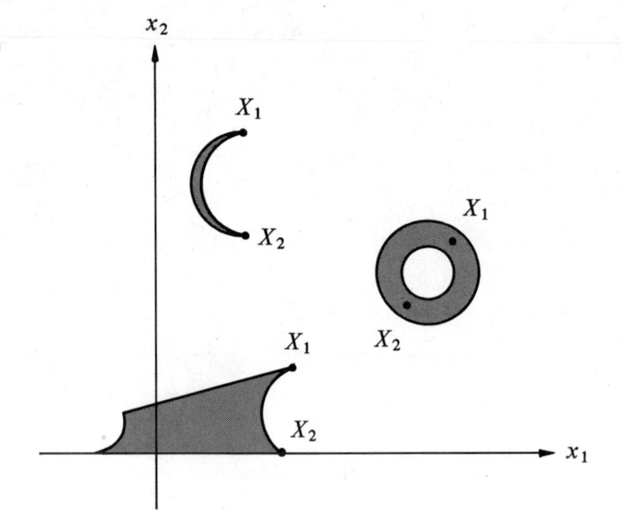


Example 6.2.2 Figure 6.2.9 shows three regions in the plane, each of which contains 2 points $X_1$ and $X_2$ such that the line segments joining the points do not lie entirely in the regions themselves. ∎

## 6.2 Exercises

1. a. Let $P_1 = \{(x_1, x_2) \mid x_1 + x_2 \leqslant 6\}$. Draw a graph to show $P_1$.
   b. On the same graph show $P_2 = \{(x_1, x_2) \mid 2x_1 + x_2 \leqslant 8\}$.
   c. Show $S_2 = P_1 \cap P_2$.
   d. Let $S_1 = \{(x_1, x_2) \mid x_1 \geqslant 0 \text{ and } x_2 \geqslant 0\}$. Show $S_1 \cap S_2 = S_3$.
   e. By solving the appropriate equations simultaneously determine the corner points of $S_3$.
   f. Evaluate $f(x_1, x_2) = x_1 + 2x_2$ at each corner point, going clockwise from the origin.
   g. Evaluate $f$ at interior points (1, 1), (1, 3), (3, 1), and at the boundary point (1, 5).

2. For each value of $a$, the equation $f(x_1, x_2) = 3x_1 + 2x_2 = a$ represents a ________ ________ in the plane. As $a$ varies we generate a family of ________ straight lines in the plane. If we desire to move to one of these lines with a higher value of $f$, we move in a direction ________ to that in which we would move if we desired to decrease $f$.

3. Consider the following LP problem:

Find $x_1, x_2$ such that $x_i \geqslant 0$ and

$$2x_1 + x_2 \leqslant 6$$
$$x_1 + 2x_2 \leqslant 6$$

and $f(x_1, x_2) = x_1 + x_2$ is maximized.

a. Find $S_1 = \{(x_1, x_2) \mid x_1 \geqslant 0 \text{ and } x_2 \geqslant 0\}$ graphically.
b. Find $S_2 = \{(x_1, x_2) \mid 2x_1 + x_2 \leqslant 6 \text{ and } x_1 + 2x_2 \leqslant 6\}$.

c. Find $S_3 = S_1 \cap S_2$.
d. On your graph show $\{(x_1, x_2) \mid f(x_1, x_2) = x_1 + x_2 = 2\}$.
e. Find the point in $S_3$ at which $f$ is a maximum.
f. Find the point in $S_3$ at which $f$ is a minimum.
g. Find the point or points in $S_3$ at which $g(x_1, x_2) = x_1 + 2x_2$ is maximized.

4. Solve Exercise 1 of Section 6.1 graphically.

5. Consider the LP problem:

---

Find $x_1, x_2$ such that $x_i \geqslant 0$ and

$$x_1 + 2x_2 \geqslant 3$$
$$2x_1 + x_2 \geqslant 3$$
$$-x_1 + x_2 \leqslant 1$$

and $f(x_1, x_2) = x_1 + x_2$ is minimized.

---

a. Show on a graph the set of points $S$ which satisfy the non-negativity restrictions and all of the constraints.
b. $S$ is an infinite or unbounded region, the corner points of which are ________, ________, and ________.
c. Evaluate $f$ at the corner points of $S$.
d. Show $\{(x_1, x_2) \mid f(x_1, x_2) = 3\}$ graphically.
e. The "best" point in $S$ is ________ at which $f =$ ________.
f. At which point in $S$ does $f$ take on its maximum?
g. Let $g(x_1, x_2) = -f(x_1, x_2)$ and find the maximum of $g$ over $S$. At which point of $S$ does it occur? What about the minimum value of $g$ over $S$?

6. a. Solve the following LP problem graphically:

---

Find $x_1, x_2$ such that $x_i \geqslant 0$ and

$$-x_1 + 2x_2 \leqslant 2$$

while $f(x_1, x_2) = -2x_1 + 4x_2$ is maximized.

---

b. How many points are there for which $f(x_1, x_2)$ is maximized?
c. Is there at least one corner point at which the maximum value is achieved?

7. a. Solve the following LP problem graphically:

---

Find $x_1, x_2$ such that $x_i \geqslant 0$ and

$$x_1 + x_2 \leqslant 2$$
$$2x_1 + x_2 \leqslant 4$$

and $f(x_1, x_2) = 2x_1 + 3x_2$ is maximized.

---

b. What part did the second constraint play in the solution?

8. a. Find the plane region which satisfies the following constraints graphically:

$$\begin{aligned} x_i &\geqslant 0, \quad i = 1, 2 \\ x_1 + x_2 &\leqslant 6 \\ x_1 - x_2 &\leqslant 1 \\ 2x_1 + x_2 &\geqslant 6 \\ -x_1 + 2x_2 &\leqslant 8. \end{aligned}$$

   b. Let $f(x_1, x_2) = (x_1 - 2)^2 + (x_2 - 3)^2$. Show that the minimum value of $f$ over the region in a. does not occur at a corner point of the region.

9. Company $Z$ makes products $A$ and $B$. Each product requires an operation to be performed by machine $M_1$ and machine $M_2$. Product $A$ requires 1 minute on $M_1$ and 3 minutes on $M_2$. Product $B$ requires 2 minutes on $M_1$ and 1 minute on $M_2$. There are 480 minutes available on each of $M_1$ and $M_2$. The profit on each unit of $A$ is 5 cents while the profit on each unit of $B$ is 4 cents.
   a. Set up an LP problem to maximize profit. Let $x_1$ denote the number of units of $A$ and $x_2$ the number of units of $B$ to be produced.
   b. Evaluate $f$ at each corner point, found graphically.
   c. Solve the LP problem graphically.

10. Solve the diet problem of Example 6.1.4 graphically.

11. Solve the aircraft charter problem of Example 6.1.1 graphically.

12. Solve the aircraft charter problem of Exercise 6.1.10 graphically.

## 6.3 The General Linear Programming Problem

Any problem of the form

---

Find $x_1, x_2, \ldots, x_r$ such that

1. $x_i \geqslant 0, i = 1, 2, \ldots, r$
2. $$\begin{aligned} a_{11}x_1 + a_{12}x_2 + \cdots + a_{1r}x_r &\lesseqqgtr r_1 \\ a_{21}x_1 + a_{22}x_2 + \cdots + a_{2r}x_r &\lesseqqgtr r_2 \\ &\cdots \\ a_{p1}x_1 + a_{p2}x_2 + \cdots + a_{pr}x_r &\lesseqqgtr r_p \end{aligned}$$
3. $z = c_1x_1 + c_2x_2 + \cdots + c_rx_r = f(x_1, x_2, \ldots, x_r)$ is optimized.

---

is called a **linear programming problem**. By **optimized** we mean either maximized or minimized. Our earlier LP examples fit into this general format in which we now use the symbol $\lesseqqgtr$ to indicate that any one of the symbols $\leqslant$, $=$, or $\geqslant$ may be present. As we have previously mentioned, the inequalities of the form $x_i \geqslant 0$ are called **non-negativity restrictions** after G. Hadley (see Suggested Reading). The remaining equations and/or inequalities are called **constraints**. The function $f$ is the **criterion** or **objective** function to be optimized. The coefficients $c_i$ of the $x_i$ in $f$ are called **criterion**, **objective function** or **cost coefficients**.

Recalling that we have built up a great deal of machinery to deal with equations rather than inequalities, we shall now demonstrate that, for any given LP problem, it is possible to formulate an equivalent version with only equality constraints. The non-negativity restrictions on the variables will remain, however, and will be dealt with separately in the final solution algorithm. The two versions of the LP problem are equivalent in the sense that a final solution for either one will provide a final solution for the other.

To make the method of constructing the equivalent problem clear, consider problem $P$:

---

Find $x_1, x_2$ such that

1. $x_i \geqslant 0, i = 1, 2$
2. $2x_1 + x_2 \leqslant 6$
   $x_1 + x_2 \leqslant 4$
3. $z = f(x_1, x_2) = 5x_1 + 3x_2$ is maximized.

---

We replace the first inequality constraint $2x_1 + x_2 \leqslant 6$ by an equality constraint

$$2x_1 + x_2 + x_3 = 6$$

obtained by adding a new variable, $x_3$, also required to be non-negative. As long as $x_3$ is non-negative, the original inequality constraint will be satisfied whenever the new equality is satisfied. For example, when $x_3 = 1$ and the equality is satisfied,

$$2x_1 + x_2 = 5,$$

and since $5 \leqslant 6$, $$2x_1 + x_2 \leqslant 6.$$

When $x_3 = 6$, and the equality is satisfied,

$$2x_1 + x_2 = 0 \leqslant 6,$$

and, similarly, when $x_3 = 0$,

$$2x_1 + x_2 = 6 \leqslant 6.$$

Thus, if we have non-negative values for $x_1, x_2, x_3$ such that

$$2x_1 + x_2 + x_3 = 6,$$

then $$x_3 = 6 - 2x_1 - x_2 \geqslant 0,$$

and, from the right-hand inequality we see that

$$2x_1 + x_2 \leqslant 6.$$

On the other hand, if we have non-negative values of $x_1, x_2$ such that

$$2x_1 + x_2 \leqslant 6$$

and let $x_3 = 6 - 2x_1 - x_2$, then $x_3 \geqslant 0$ and

$$2x_1 + x_2 + x_3 = 6.$$

A variable such as $x_3$ which is added merely to "take up the slack" between the left-hand side of a $\leqslant$ inequality and the constant term is called a **slack variable**. This slack variable may be thought of as serving as a "guard" for the original inequality constraint, in the sense that whenever it is nonnegative, the constraint is satisfied, and whenever it is negative the constraint is not satisfied. Thus, if values of $x_1, x_2, x_3$ are chosen that satisfy the *equation* but such that $x_3 < 0$, then the values for $x_1, x_2$ do not satisfy the original inequality constraint.

**Figure 6.3.1**

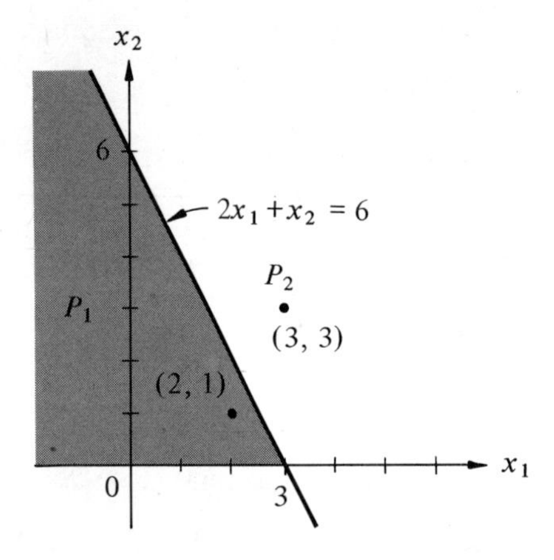


Here again, a look at the geometry of the situation will help to clarify matters. Consider Figure 6.3.1, which shows $P_1 = \{(x_1, x_2) \mid 2x_1 + x_2 \leqslant 6\}$. Suppose that we add the slack variable $x_3$ to form the equation

(i) $$2x_1 + x_2 + x_3 = 6.$$

Any point $(x_1, x_2, x_3)$ for which $x_3 = 0$ corresponds to a point $(x_1, x_2)$ in the plane which lies *on* the line, since $2x_1 + x_2 = 6$ if $x_3 = 0$. Any point $(x_1, x_2, x_3)$ which satisfies (i) and for which $x_3 = 1$ corresponds to a point $(x_1, x_2)$ which lies in the interior of $P_1$ since $2x_1 + x_2 = 5 < 6$. The point (2, 1, 1) in 3-space is such a point and it corresponds to (2, 1) in the plane as shown above. In general then, any point, $(x_1, x_2, x_3)$, for which $x_3 \geqslant 0$, has a corresponding point, $(x_1, x_2)$, which lies in $P_1$, since if $2x_1 + x_2 + x_3 = 6$ and $x_3 \geqslant 0$, then $2x_1 + x_2 \leqslant 6$ as desired. On the other hand, suppose, for example, that we have the point $(x_1, x_2, x_3)$ such that

$$2x_1 + x_2 + x_3 = 6$$

and $x_3 = -3$, say. Then

$$2x_1 + x_2 = 6 + 3 = 9 > 6,$$

and the original constraint is not satisfied. The point $(3, 3, -3)$ satisfies (i), and we note that the corresponding point (3, 3) lies outside of $P_1$. The sense in which we say that $x_3$ *serves as a guard* for the constraint should now be clear. Whenever we have a point $(x_1, x_2, x_3)$ for which the equation is satisfied and $x_3 \geqslant 0$, the original inequality is also satisfied. On the other hand, if we have a point $(x_1, x_2, x_3)$ for which $x_3 < 0$, then the point $(x_1, x_2)$ in the plane lies on the *wrong side* of the line in question and the original inequality constraint is not satisfied.

We have seen that when $x_3 = 0$, $2x_1 + x_2 = 6$, and the point $(x_1, x_2)$ lies on the line. Now, when $x_3 = 1$, we have $2x_1 + x_2 = 5$, also an equation of a straight line and one with the same slope as the original. In Figure 6.3.2 we show $\{(x_1, x_2) \mid 2x_1 + x_2 = 6 - x_3\}$ for various values of $x_3$. Those with $x_3 \geqslant 0$ lie in $P_1$ while the one with $x_3 < 0$ lies outside of $P_1$.

**Figure 6.3.2**

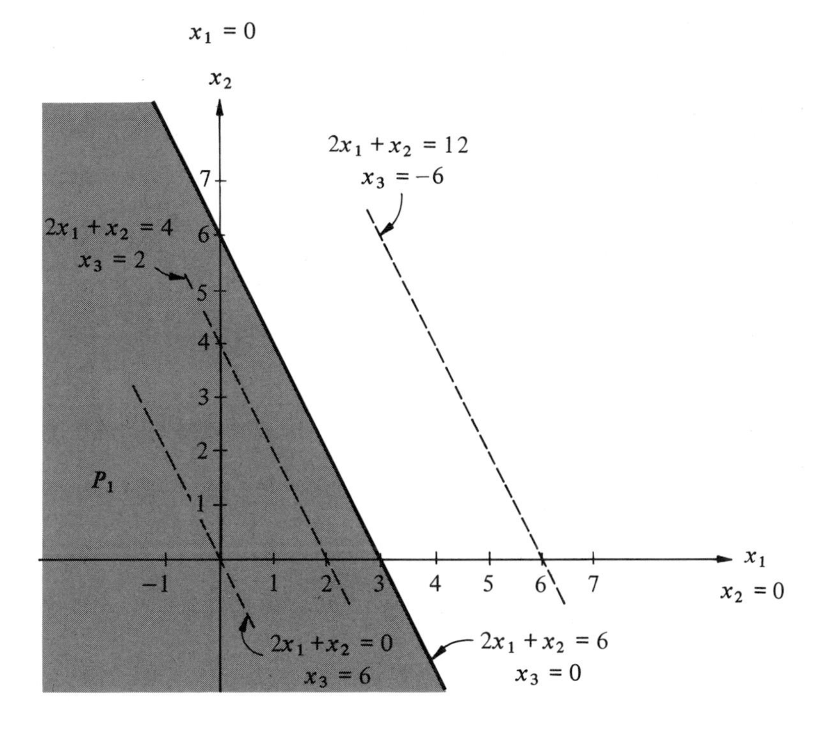


We now consider the second constraint of problem $P$, $x_1 + x_2 \leqslant 4$, and add another slack variable $x_4$ such that

$$x_1 + x_2 + x_4 = 4.$$

$x_4$ is a non-negative variable which will serve as a guard for the second constraint.

We use these new equations to rephrase our original problem $P$ in an equivalent form:

---

Find $x_1, x_2, x_3, x_4$ such that

1. $x_i \geqslant 0$ for $i = 1, 2, 3, 4$
2. $2x_1 + x_2 + x_3 \qquad = 6$

   $x_1 + x_2 \qquad + x_4 = 4$
3. $z = f(x_1, x_2, x_3, x_4) = 5x_1 + 3x_2$ is maximized.

---

Since we do not want the values of these new variables to affect the value of $z$, they do not enter into the objective function and are assigned objective function coefficients of 0.

Example 6.3.1 We can rephrase Example 6.1.3 in the new format as follows:

---

Find $x_1, x_2, x_3, x_4$ such that

1. $x_i \geqslant 0$ for $i = 1, 2, 3, 4$
2. $0.2x_1 + 0.8x_2 + x_3 \qquad = 1000$

   $0.8x_1 + 0.2x_2 \qquad + x_4 = 1000$
3. $z = f(x_1, x_2, x_3, x_4) = 2x_1 + 3x_2$ is maximized.

---

In matrix form, we wish to

---

Find $X$ such that

1. $X \geqslant 0$
2. $AX = R$
3. $z = CX = f(X)$ is maximized.

---

where

$$X^T = [x_1, x_2, x_3, x_4], \quad A = \begin{bmatrix} 0.2 & 0.8 & 1 & 0 \\ 0.8 & 0.2 & 0 & 1 \end{bmatrix}, \quad R = \begin{bmatrix} 1000 \\ 1000 \end{bmatrix},$$

$$C = [2 \quad 3 \quad 0 \quad 0].$$

∎

Constraints of the $\geqslant$ type may also be replaced by equality constraints. A particularly simple way of accomplishing this would be to multiply both sides of the original inequality by $-1$ and change the $\geqslant$ to $\leqslant$. In so doing we would have really only transferred the variables and the constant terms to opposite sides of the inequality. However, this would have the effect of replacing a positive constant on the right-hand side by a *negative* constant. For example, $x_1 + 2x_2 \geqslant 4$ would be replaced by $-x_1 - 2x_2 \leqslant -4$. Because of the particular way in which we will deal with the non-negativity restrictions on all of the variables, *we wish to have all of the right-hand constants non-negative to start with.* The situation can, however, be handled in a fashion similar to that used for the $\leqslant$ constant.

Consider the constraint

$$1x_1 + 2x_2 \geqslant 4$$

of the diet problem in Example 6.1.4, where the total amount of vitamin B to be served in $x_1$ ounces of $F_1$ and $x_2$ ounces of $F_2$ was to be at least 4 units. Suppose that a solution with $x_1 = 2$ and $x_2 = 2$ is considered. In this case, the number of units of vitamin B served is

$$1(2) + 2(2) = 6 > 4$$

and there is a surplus of vitamin B provided. If we let

$$x_3 = 1x_1 + 2x_2 - 4,$$

then $x_3 = 6 - 4 = 2$ in the case above and represents the surplus of vitamin B served. Such a variable is called a **surplus variable**. It is subtracted from the left side of the original $\geqslant$ constraint to form an equation. We

Figure 6.3.3

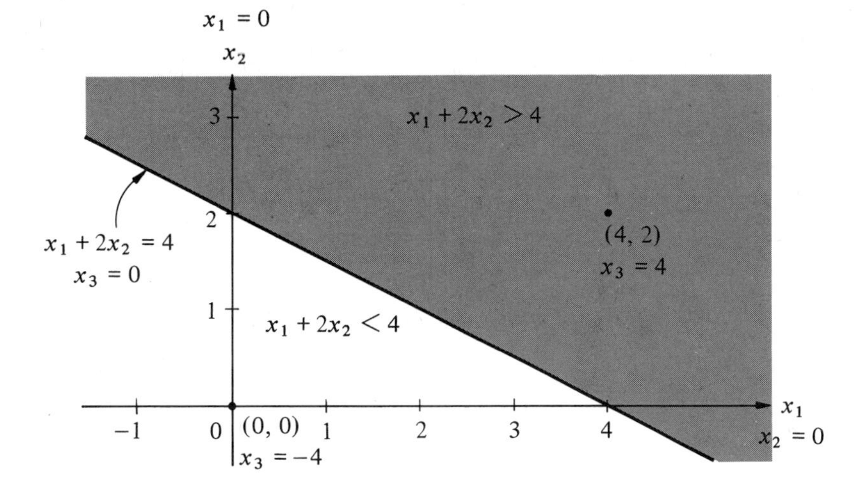


thus have the equation

$$1x_1 + 2x_2 - x_3 = 4.$$

The logical situation here is analogous to that of the $\leqslant$ constraint previously considered. If we obtain $(x_1, x_2, x_3)$ with $x_3 \geqslant 0$, then the original inequality is satisfied for $(x_1, x_2)$. If, however, we have a pair $(x_1, x_2)$ that does not satisfy the $\geqslant$ inequality, then any point $(x_1, x_2, x_3)$ such that

$$1x_1 + 2x_2 - x_3 = 4$$

must have $x_3 < 0$. Figure 6.3.3 depicts the situation in this case. Consider the point (0, 0). The point is not in the desired region, and from the equation

$$1(0) + 2(0) - x_3 = 4$$

we find that $x_3 = -4$. On the other hand, the point (4, 2) is in the desired region and we find from the equation

$$1(4) + 2(2) - x_3 = 4$$

that $x_3 = 4 \geqslant 0$.

Thus we see that the surplus variable which we subtract to create an equation acts as a *guard* for the $\geqslant$ constraint in the same sense that the slack variable acts as a guard for a $\leqslant$ constraint.

Example 6.3.2 In the problem in which we wish to

---

Find $x_1, x_2$ such that

1. $x_i \geqslant 0$, for $i = 1, 2$

2.a. $x_1 + 2x_2 \geqslant 4$

b. $-x_1 + 4x_2 \leqslant 3$

c. $3x_1 - 2x_2 \leqslant 6$

3. $f(x_1, x_2) = 2x_1 + 2x_2$ is maximized.

---

we shall add slack variables or subtract surplus variables to obtain a problem with equality constraints. In 2.a. we subtract the surplus variable $x_3$ to form $x_1 + 2x_2 - x_3 = 4$. In 2.b. and 2.c. we add slack variables $x_4$ and $x_5$ to form the equality constraints $-x_1 + 4x_2 + x_4 = 3$ and $3x_1 - 2x_2 + x_5 = 6$. We restrict the three new variables to be non-negative and obtain a new objective function $f'(x_1, x_2, x_3, x_4, x_5) = 2x_1 + 2x_2 = f(x_1, x_2)$ with coefficients of 0 assigned to each of the slack/surplus variables. We therefore have a new but equivalent problem:

---

Find $x_1, x_2, x_3, x_4, x_5$ such that

1. $x_i \geqslant 0$ for $i = 1, 2, 3, 4, 5$
2. $$\begin{aligned} x_1 + 2x_2 - x_3 \phantom{+ x_4 + x_5} &= 4 \\ -x_1 + 4x_2 \phantom{- x_3} + x_4 \phantom{+ x_5} &= 3 \\ 3x_1 - 2x_2 \phantom{- x_3 + x_4} + x_5 &= 6 \end{aligned}$$
3. $f(x_1, x_2) = 2x_1 + 2x_2$ is maximized.

---

∎

It should now be clear that we could in general formulate an equivalent LP problem as follows:

---

Find $x_1, \ldots, x_n$ such that

1. $x_i \geqslant 0$ for $i = 1, 2, \ldots, n$
2. $$\begin{aligned} a_{11}x_1 + a_{12}x_2 + \cdots + a_{1n}x_n &= r_1 \\ a_{21}x_1 + a_{22}x_2 + \cdots + a_{2n}x_n &= r_2 \\ \cdot \quad \cdot \quad \cdots \quad \cdot \quad & \cdot \\ a_{m1}x_1 + a_{m2}x_2 + \cdots + a_{mn}x_n &= r_m \end{aligned}$$
3. $z = c_1x_1 + c_2x_2 + \cdots + c_nx_n = f(x_1, x_2, \ldots, x_n)$ is optimized.

---

In matrix format we say that we wish to

---

Find $X \geqslant 0$

such that $AX = R$

and $z = f(X) = CX$ is optimized.

---

Henceforth, we shall assume that $m < n$ and that all redundant equations have been removed so that the rank of $A$ is $m$. If the rank of $A$ were $m$ and $m = n$, then there would only be one vector $X$ satisfying $AX = R$. Thus, no real optimization problem exists. We either use this $X$ or not, depending on whether each component of $X$ is non-negative.

Our reformulated problem $P$

---

Find $x_1, x_2, x_3, x_4$ such that

1. $x_i \geqslant 0$ for $i = 1, 2, 3, 4$
2. $$\begin{aligned} 2x_1 + x_2 + x_3 \phantom{+ x_4} &= 6 \\ x_1 + x_2 \phantom{+ x_3} + x_4 &= 4 \end{aligned}$$
3. $z = f(x_1, x_2, x_3, x_4) = 5x_1 + 3x_2$ is maximized.

---

may now be put into matrix format, with

$$C = [5 \ \ 3 \ \ 0 \ \ 0], \quad X^T = [x_1 \ \ x_2 \ \ x_3 \ \ x_4], \quad A = \begin{bmatrix} 2 & 1 & 1 & 0 \\ 1 & 1 & 0 & 1 \end{bmatrix}, \quad R = \begin{bmatrix} 6 \\ 4 \end{bmatrix}.$$

We now wish to

---

Find $X \geqslant 0$

such that $AX = R$

and $z = f(X) = CX$ is maximized.

---

In considering the relationship between the general problem with constraints in equality form and the original one, we note that the new problem has at least as many variables as the original, and that there will be an extra variable for each inequality constraint in the original formulation.

One could also start with a problem in equality form and obtain an equivalent problem with inequalities. Consider the equality constraint

$$2x_1 + 3x_2 = 4.$$

If we replace this one constraint with two inequalities, namely

$$2x_1 + 3x_2 \leqslant 4 \quad \text{and} \quad 2x_1 + 3x_2 \geqslant 4,$$

then any $X = [x_1, x_2]^T$ which can satisfy both of these constraints must satisfy the original and vice-versa.

## 6.3 Exercises

1. To create an equality constraint from $2x_1 + 3x_2 \leqslant 6$ we add the non-________ variable $x_3$ to form the equality constraint ________. We say that $x_3$ serves as a ________ for the constraint in the sense that the inequality constraint is satisfied when $x_3$ is ________, and the inequality constraint is ________ ________ when $x_3$ is negative. Because of the fact that $x_3$ $(=6 - 2x_1 - 3x_2)$ takes up the slack or makes up the difference between 6 and $2x_1 + 3x_2$, we call $x_3$ a ________ variable.

2. To create an equality constraint from $2x_1 + 3x_2 \geqslant 6$, we ________ the non-negative variable $x_3$ to form the equality constraint ________. Because of the fact that $x_3$ $(=2x_1 + 3x_2 - 6)$ ________, we call $x_3$ a surplus variable.

3. Express the problem of Exercise 6.2.3 in equality constraint form.

4. Describe the objective function, non-negativity restrictions, and constraints of the problem in Exercise 3 above both in inequality and equality constraint forms.

5. Express the problem of Exercise 6.2.5 in equality constraint form.

6. For later computational reasons, we wish to keep the constraints $r_i$ in our equality constraints ________.

7. Express the problem of Exercise 6.2.3 in matrix format with equality constraints (see Exercise 3 above).

8. Show graphically that

$$\{(x_1, x_2) \mid 2x_1 + x_2 = 3\} = \{(x_1, x_2) \mid 2x_1 + x_2 \leqslant 3\} \cap \{(x_1, x_2) \mid 2x_1 + x_2 \geqslant 3\}.$$

9. Express the problem of Exercise 6.2.5 in matrix format with equality constraints (see Exercise 5 above).

10. When we convert from constraints in inequality form to equation form, how many new variables are introduced?

11. Consider $S = \{(x_1, x_2) \mid 4x_1 + 3x_2 \leqslant 12\}$.
    a. Find the point $(x_1, x_2, x_3)$ corresponding to (1, 2) in the plane. Since $x_3$ is ________ we know that (1, 2) is in $S$.
    b. Find the point $(x_1, x_2, x_3)$ corresponding to (2, 2) in the plane. Since $x_3$ is ________ we know that (2, 2) ________ in $S$.
    c. Since $4(2) + 3(1) + 1(1) = 12$, we know that (________, ________, ________) is the point in 3-space corresponding to the point (________, ________) in $S$ in 2-space.

12. Express Exercise 8 of Section 6.1 in matrix inequality form.

### Suggested Reading

G. B. Dantzig. *Linear Programming and Extensions.* Princeton, N.J.: Princeton University Press, 1963.

S. I. Gass. *Linear Programming.* 3d. ed. New York: McGraw-Hill, 1969.

G. Hadley. *Linear Programming.* Reading, Mass.: Addison-Wesley, 1962.

H. M. Wagner. *Principles of Operations Research.* 2d. ed. Englewood Cliffs, N.J.: Prentice-Hall, 1975.

### New Terms

linear programming problem, 208
optimize, 208
non-negativity restriction, 208
constraints, 208
criterion function, 208
objective function, 208
criterion function coefficients, 209
objective function coefficients, 209
cost function coefficients, 209
slack variable, 210
surplus variable, 212

## 6.4 Basic Feasible Solutions

From our new formulation of the general LP problem

Find $X \geqslant 0$ such that

$AX = R$

and $z = f(X) = CX$ is optimized.

we proceed to define a few more of the terms which we have been using rather intuitively. We shall say that the vector $X$ is a **solution of the general LP**

**problem** in equality form provided that $AX = R$. Any solution $X$ with the additional property that $X \geqslant 0$ will be called a **feasible solution**. The set of all such feasible solutions will be called the **feasible region**. If $X$ is a feasible solution with the best possible value of the objective function, then $X$ is called an **optimal feasible solution**. For example, in a minimization problem if $X \geqslant 0$ and $f(X) \leqslant f(Y)$ for all feasible solutions $Y$, then $X$ is said to be an optimal feasible solution.

Example 6.4.1 Consider problem $P$ in matrix equality form as follows, with

$$C = [5, 3, 0, 0], \qquad A = \begin{bmatrix} 2 & 1 & 1 & 0 \\ 1 & 1 & 0 & 1 \end{bmatrix}, \qquad R = \begin{bmatrix} 6 \\ 4 \end{bmatrix},$$

and

$$X^T = [x_1, x_2, x_3, x_4].$$

For $X_1 = [0, 0, 6, 4]^T$ we have

$$AX_1 = \begin{bmatrix} 2 & 1 & 1 & 0 \\ 1 & 1 & 0 & 1 \end{bmatrix} \begin{bmatrix} 0 \\ 0 \\ 6 \\ 4 \end{bmatrix} = \begin{bmatrix} 6 \\ 4 \end{bmatrix} = R.$$

Thus, $X_1$ is a solution. Since $x_i \geqslant 0$ for each $i$, $X_1$ is a feasible solution. Now

$$f(X_1) = CX_1 = [5, 3, 0, 0] \begin{bmatrix} 0 \\ 0 \\ 6 \\ 4 \end{bmatrix} = 0 < 16.$$

Thus $X_1$ is not an optimal feasible solution. You should go back and verify from Figure 6.2.7 and the definitions above that $X_2 = [2, 2, 0, 0]^T$, $X_3 = [1, 1, 3, 2]^T$, and $X_4 = [0, 6, 0, -2]^T$ are each solutions; that $X_4$ is not feasible; and that $X_2$ is optimal and feasible. Recall that the optimum value of $z$ is 16. ∎

Let us continue to use problem $P$ as a vehicle to study the relationship between the geometry and the algebra of a typical LP problem.

In the equivalent new problem in equality form we wish to find a vector $X$ in 4-space. However, we may still portray the situation in two dimensions as before, as depicted in Figure 6.4.1. We point out the particular constraints that $x_3$ and $x_4$ are guarding by indicating that $x_3 = 0$ along the appropriate line defining its inequality and that $x_4 = 0$ along the other. Similarly, $x_2 = 0$ is another way of describing the $x_1$ axis, since $x_2 = 0$ for any point on it.

Each point $(x_1, x_2)$ in this 2-dimensional representation has a corresponding point $(x_1, x_2, x_3, x_4)$ in 4-space that may easily be computed from the equations defining the shaded region. For example, consider the 2-space origin (0, 0). If, in the constraint equations

$$\begin{aligned} 2x_1 + x_2 + x_3 \qquad &= 6 \\ x_1 + x_2 \qquad + x_4 &= 4 \end{aligned}$$

we let $x_1 = x_2 = 0$, we find that $x_3 = 6$ and $x_4 = 4$. Thus, in 4-space we have the point (0, 0, 6, 4) which we say corresponds to (0, 0) in 2-space.

Figure 6.4.1

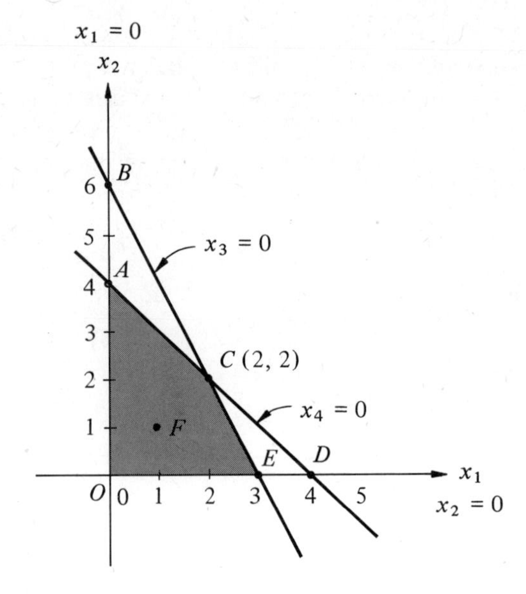


Now consider point $A$ in the figure. Point $A$ lies at the intersection of the lines $x_1 = 0$ and $x_4 = 0$. With $x_1 = x_4 = 0$ we have the resulting set of equations

$$x_2 + x_3 = 6$$
$$x_2 \qquad = 4$$

yielding $x_2 = 4$ and $x_3 = 2$, so that (0, 4, 2, 0) in 4-space corresponds to point $A$. At point $B$, $x_1 = x_3 = 0$. We thus have

$$x_2 \qquad = 6$$
$$x_2 + x_4 = 4$$

from which we conclude that $x_2 = 6$ and $x_4 = -2$. The corresponding 4-space point is (0, 6, 0, −2). *At any particular corner point of the feasible region, such as A, those variables, n − m in general, which are set equal to* 0 *are called the* **non-basic variables** *at that point. The remaining m variables, the values of which must be computed, are called the* **basic variables** *at the point in question.*

Proceeding clockwise in the figure, we obtain the information in Table 6.4.1. The reader should carefully verify each of the entries. An understanding of this method is vital in the sequel.

Table 6.4.1

| *Point* | *2-Space Solution* | *Non-basic Variables* | *Basic Variables* | *4-Space Solution* $X^T$ |
|---|---|---|---|---|
| *O* | [0, 0] | $x_1, x_2$ | $x_3, x_4$ | [0, 0, 6, 4] |
| *A* | [0, 4] | $x_1, x_4$ | $x_3, x_2$ | [0, 4, 2, 0] |
| *B* | [0, 6] | $x_1, x_3$ | $x_4, x_2$ | [0, 6, 0, −2] |
| *C* | [2, 2] | $x_4, x_3$ | $x_1, x_2$ | [2, 2, 0, 0] |
| *D* | [4, 0] | $x_4, x_2$ | $x_1, x_3$ | [4, 0, −2, 0] |
| *E* | [3, 0] | $x_3, x_2$ | $x_1, x_4$ | [3, 0, 0, 1] |
| *F* | [1, 1] | none | $x_1, x_2, x_3, x_4$ | [1, 1, 3, 2] |

Point $F$ is a special case. If $x_1$ and $x_2$ are each set equal to 1 then the original equations reduce to

$$x_3 = 3$$
$$x_4 = 2,$$

so that (1, 1, 3, 2) is the corresponding 4-space point.

We can learn several important ideas from the graph and the table in conjunction. First of all, each of the points, except $F$, has exactly $m = 2$ of the variables whose values are non-zero. In addition, each of these points which is in the shaded region has no negative components. Furthermore, each such point in the table is an extreme point or a corner point of the feasible region. Finally, the point $F$, an *interior* point of the feasible region, has more than $m = 2$ positive components; and it is not a corner point of the feasible region.

Referring to Figure 6.4.1, Table 6.4.1, our recent definitions, and our earlier results for problem $P$, we classify the solutions represented by the various points in Table 6.4.2.

**Table 6.4.2**

| *Point* | *Solution* | *Feasible Solution* | *Optimal Feasible Solution* |
|---|---|---|---|
| $O$ | yes | yes | no |
| $A$ | yes | yes | no |
| $B$ | yes | no | no |
| $C$ | yes | yes | yes |
| $D$ | yes | no | no |
| $E$ | yes | yes | no |
| $F$ | yes | yes | no |

It will be helpful to distinguish between feasible solutions like those at the corner points $O$, $A$, $C$, $E$ of the feasible region and other feasible solutions such as $F$. We recall that the difference arose because of the number of variables which were set equal to 0, thus causing the number of non-zero variables to be different also. To clarify the situation, we make the following additional definition, recalling that the rank of $A$ is $m$. We shall say that the solution $X$ is a **basic solution** of $AX = R$ provided that $n - m$ of the variables have been set equal to 0 and the remaining $m$ equations in $m$ unknowns solved simultaneously. The $n - m$ variables are the previously defined non-basic variables while the $m$ variables are the basic variables. If $n = 4$ and $m = 2$ as in problem $P$ then all possible sets of $4 - 2 = 2$ of the variables are set equal to 0 and the remaining sets of 2 equations in 2 unknowns are solved to obtain the various basic solutions.

If $X$ is a basic solution that is also feasible (that is, $X \geqslant 0$) then $X$ is called a **basic feasible solution** (*BFS*). If, in a basic solution $X$, some $x_i < 0$, then $X$ is called a **basic infeasible solution**. Points $B$ and $D$ in Figure 6.4.1 represent such solutions in problem $P$. Points $O$, $A$, $C$, $E$ correspond to the basic feasible solutions of problem $P$. It is no accident that points $O$, $A$, $C$, $E$ are the corner points of the feasible region. For each such corner point or extreme point of the feasible region of any LP problem it can be proved that there is *at least one* corresponding *BFS* which can be obtained by choosing the correct set of $n - m$ variables to set equal to 0.* Similarly, it can be proved that for each *BFS* there is a corresponding corner point of the feasible region.

We recall from our earlier study of linear equations of the form $AX = R$, where $A$ is an $m \times n$ matrix of rank $m$, that we would expect to find $n - m$ free variables, which we are at liberty to set equal to whatever values we like and to solve for the remaining $m$ variables in terms of these. Perhaps the easiest way to do this is to set $n - m$ of the variables equal to 0. Whenever we do this and obtain a solution, we obtain a basic solution.

**Example 6.4.2** Let us construct all six basic solutions of an LP problem with

$$A = \begin{bmatrix} 2 & 3 & 1 & 0 \\ 3 & 2 & 0 & 1 \end{bmatrix} \quad \text{and} \quad R = \begin{bmatrix} 6 \\ 6 \end{bmatrix}.$$

1. For $x_1 = x_2 = 0$, we have $x_3 = x_4 = 6$ and the *BFS* $X^T = [0, 0, 6, 6]$.
2. For $x_1 = x_3 = 0$, we have the remaining set of equations

$$\begin{aligned} 3x_2 \qquad &= 6 \\ 2x_2 + x_4 &= 6 \end{aligned}$$

from which we obtain $X^T = [0, 2, 0, 2]$ as a *BFS*.

3. For $x_1 = x_4 = 0$, we obtain the augmented matrix

$$\left[\begin{array}{cc|c} 3 & 1 & 6 \\ 2 & 0 & 6 \end{array}\right] \sim \left[\begin{array}{cc|c} 2 & 0 & 6 \\ 3 & 1 & 6 \end{array}\right] \sim \left[\begin{array}{cc|c} 1 & 0 & 3 \\ 0 & 1 & -3 \end{array}\right]$$

and $$X^T = [0, 3, -3, 0],$$

which is a basic *infeasible* solution.

4. From $x_2 = x_3 = 0$, we obtain $\left[\begin{array}{cc|c} 2 & 0 & 6 \\ 3 & 1 & 6 \end{array}\right]$ and find that $X^T = [3, 0, 0, -3]$ is another basic solution that is not feasible.
5. When $x_2 = x_4 = 0$, we have $\left[\begin{array}{cc|c} 2 & 1 & 6 \\ 3 & 0 & 6 \end{array}\right]$, from which we obtain $X^T = [2, 0, 2, 0]$, a *BFS*.
6. Finally, if $x_3 = x_4 = 0$, we obtain

$$\left[\begin{array}{cc|c} 2 & 3 & 6 \\ 3 & 2 & 6 \end{array}\right] \sim \left[\begin{array}{cc|c} 1 & 3/2 & 3 \\ 0 & -5/2 & -3 \end{array}\right] \sim \left[\begin{array}{cc|c} 1 & 0 & 6/5 \\ 0 & 1 & 6/5 \end{array}\right]$$

and see that $X^T = [6/5, 6/5, 0, 0]$ is a *BFS*.

You should sketch the graph of the 2-space feasible region represented by $AX = R$, treating $x_3$ and $x_4$ as slack variables, to see that these answers are reasonable. ∎

* G. Hadley, *Linear Programming* (Reading, Mass.: Addison-Wesley, 1962), p. 100ff.

## 6.4 Exercises

1. Consider the LP problem with constraints

$$x_1 + 2x_2 \leqslant 3$$
$$2x_1 + x_2 \leqslant 3.$$

   a. Add slack variables to put the constraints in equality form.
   b. Find $A$, $X$ and $R$ to express the equality constraints in the form $AX = R$.
   c. Sketch the feasible region for this problem (including non-negativity restrictions).
   d. Verify from the algebraic definition of a solution that

$$X_1 = [3/2, 0, 3/2, 0]^T,$$
$$X_2 = [0, 3, -3, 0]^T,$$
$$X_3 = [1/2, 1/2, 3/2, 3/2]^T$$

   are solutions.
   e. Which of the solutions in d. above are feasible?
   f. Which of the solutions in d. above are basic?
   g. Which solutions in d. are basic feasible solutions?

2. Consider the LP problem to maximize $f(X) = CX$, with $X \geqslant 0$ and $AX = R$.
   a. We say that $X$ is a solution provided that ____________.
   b. If $X$ is a solution and $X \geqslant 0$, we call $X$ a ____________.
   c. We call $\{X \mid AX = R \text{ and } X \geqslant 0\}$ the ____________.
   d. We say that the feasible solution $X$ is an optimal feasible solution provided that if $Y$ is another feasible solution, then ____________.

3. Consider the feasible region of Exercise 1 as shown in Figure 6.4.2.
   a. At point $O$, $x_1 = x_2 = 0$, while $x_3 = x_4 = 3$. We call $x_1$ and $x_2$ ________ variables at $O$, while $x_3$ and $x_4$ are called ________ variables at that point.
   b. The 4-space coordinates of point $F$ are ________. Is the solution at $F$ feasible? basic?

**Figure 6.4.2**

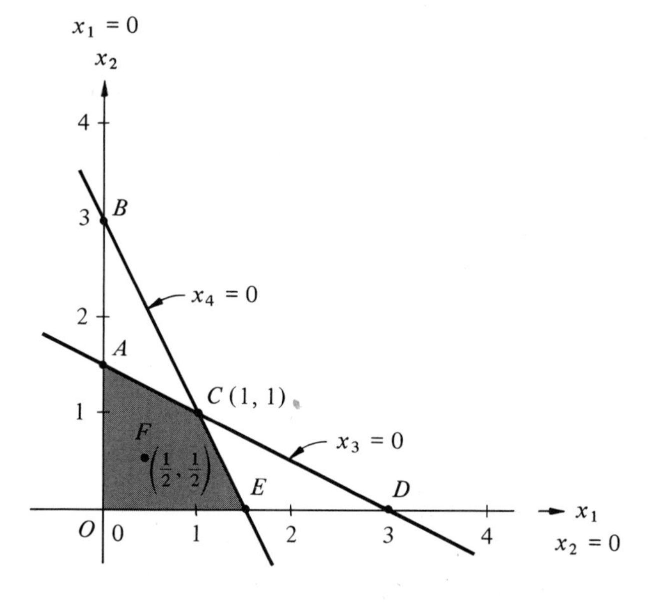

c. Fill in the following table in a fashion similar to that of Table 6.4.1.

**Table 6.4.3**

| *Point* | *2-Space Solution* | *Non-basic Variables* | *Basic Variables* | *4-Space Solution, $X^T$* | *Solution Feasible?* |
|---|---|---|---|---|---|
| *O* | | | | | |
| *A* | | | | | |
| *B* | | | | | |
| *C* | | | | | |
| *D* | | | | | |
| *E* | | | | | |

4. To obtain all of the basic solutions of the LP problem with constraint set $AX = B$ where $A$ is a $3 \times 7$ matrix, we would successively set ________ of the variables equal to 0 and solve the remaining ________ equations in ________ variables. We should obtain at most ________ basic solutions in so doing, not all of which need be ________.

5. Consider the constraint set

$$x_1 + x_2 \leqslant 2$$
$$2x_1 + x_2 \leqslant 4$$

of Exercise 6.2.7.

a. Add slack variables and put the constraints in the form $AX = R$.

$A_1 = \begin{bmatrix} \_\_\_ \\ \_\_\_ \end{bmatrix}$ and $A_4 = \begin{bmatrix} \_\_\_ \\ \_\_\_ \end{bmatrix}$, while $R = \begin{bmatrix} \_\_\_ \\ \_\_\_ \end{bmatrix}$.

b. Using the sketch of the feasible region provided in Figure 6.4.3, fill in Table

**Figure 6.4.3**

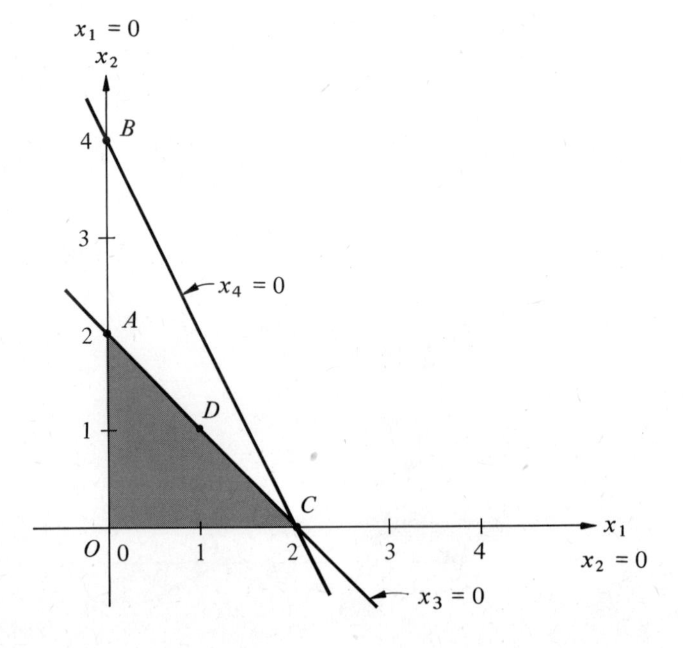

6.4.4. Note that the 2-space point $C$ has more than one corresponding 4-space point.

c. The 4-space point corresponding to $D$ is ________. The solution at $D$ is non-________ but ________. Note that $D$ is a boundary point of the feasible region but not a corner point.

**Table 6.4.4**

| *Point* | *2-Space Solution* | *Non-basic Variables* | *Basic Variables* | *4-Space Solution, $X^T$* | *Solution Feasible?* |
|---|---|---|---|---|---|
| $O$ | [0, 0] | | | [ ] | |
| $A$ | [0, 2] | | | [ ] | |
| $B$ | [ ] | $x_1, x_4$ | $x_3, x_2$ | [ ] | |
| $C$ | [2, 0] | $x_2, x_3$ | $x_1, x_4$ | [2, 0, 0, 0] | |
| $C$ | [2, 0] | $x_2, x_4$ | $x_3, x_1$ | [2, 0, 0, 0] | |
| $C$ | [2, 0] | | | [ ] | |

### Suggested Reading

G. B. Dantzig. *Linear Programming and Extensions.* Princeton, N.J.: Princeton University Press, 1963.

S. I. Gass. *Linear Programming.* 3d. ed. New York: McGraw-Hill, 1969.

G. Hadley. *Linear Programming.* Reading, Mass.: Addison-Wesley, 1962.

G. Hadley and M. C. Kemp. *Finite Mathematics in Business and Economics.* New York: American Elsevier, 1972.

H. M. Wagner. *Principles of Operations Research.* 2d. ed. Englewood Cliffs, N.J.: Prentice-Hall, 1975.

### New Terms

solution of the general LP problem, 216
feasible solution, 217
feasible region, 217
optimal feasible solution, 217
non-basic variables, 218
basic variables, 218
basic solution, 219
basic feasible solution, 219
basic infeasible solution, 219

## 6.5 A Matrix/Vector Approach to the LP Problem

In our equation version of the general LP problem where we wish to

Find $X$ such that

1. $X \geqslant 0$
2. $AX = R$ and
3. $z = f(X) = CX$ is optimized.

we note that the equality constraints could be written in the alternative form

$$x_1A_1 + x_2A_2 + \cdots + x_nA_n = R,$$

where the $A_i$ represent the columns of $A$ and are elements of $m$-space. We might then think of the LP problem as one asking us to find, in the set of all vectors $X$ in $n$-space, that subset of vectors with non-negative components which express $R$ as a linear combination of the $A_i$. Out of this subset we are to pick that vector or set of vectors which provide the best possible value of the objective function.

Let us consider problem $P$ in this light:

---

Find $X$ such that

1. $X \geqslant 0$
2. $2x_1 + x_2 + x_3 \phantom{+ x_4} = 6$

   $x_1 + x_2 \phantom{+ x_3} + x_4 = 4$ and
3. $z = f(X) = 5x_1 + 3x_2$ is maximized.

---

Out of all the vectors in 4-space we wish first to find

$$\left\{X \mid X = \begin{bmatrix} x_1 \\ x_2 \\ x_3 \\ x_4 \end{bmatrix} \text{ and } x_1\begin{bmatrix} 2 \\ 1 \end{bmatrix} + x_2\begin{bmatrix} 1 \\ 1 \end{bmatrix} + x_3\begin{bmatrix} 1 \\ 0 \end{bmatrix} + x_4\begin{bmatrix} 0 \\ 1 \end{bmatrix} = \begin{bmatrix} 6 \\ 4 \end{bmatrix} \text{ and } x_i \geqslant 0\right\}.$$

Here, $A_1 = \begin{bmatrix} 2 \\ 1 \end{bmatrix}$, $A_2 = \begin{bmatrix} 1 \\ 1 \end{bmatrix}$, $A_3 = \begin{bmatrix} 1 \\ 0 \end{bmatrix}$, and $A_4 = \begin{bmatrix} 0 \\ 1 \end{bmatrix}$, while $R = \begin{bmatrix} 6 \\ 4 \end{bmatrix}$. Geometrically, in 2-space, we now portray the situation as shown in Figure 6.5.1.

In the last section we demonstrated the manner in which one could construct all the possible *basic solutions* of this problem by successively choosing $n - m = 4 - 2 = 2$ of the variables to be set equal to 0 and

Figure 6.5.1

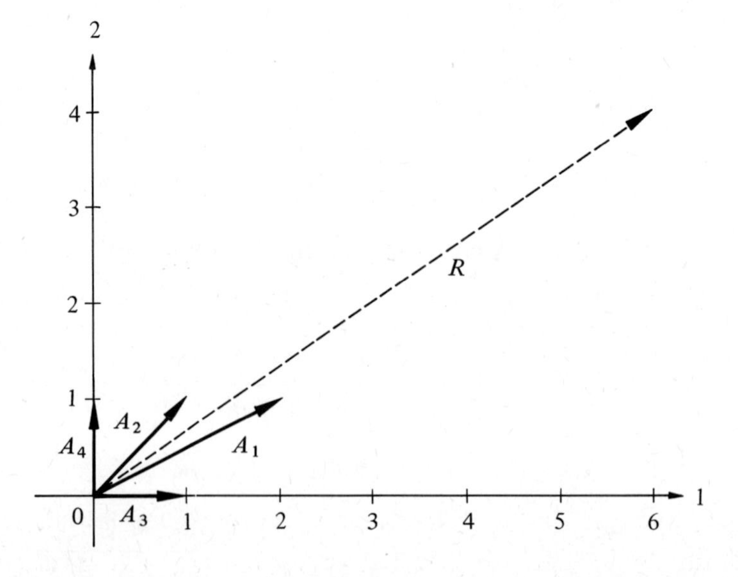

solving the resulting $m = 2$ equations for the remaining $m = 2$ variables. This idea is intimately connected with the idea of a basis for a vector space which we discussed in Chapter 4. Let us reconsider the graph (Figure 6.5.2) and part of the associated table (Table 6.5.1) for problem $P$.

**Figure 6.5.2**

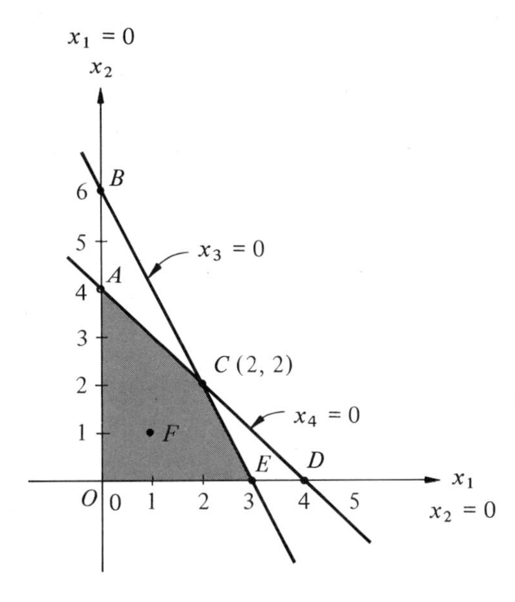


**Table 6.5.1**

| Point | Basic Variables | Non-basic Variables | Basic Solution, $X^T$ |
|---|---|---|---|
| $O$ | $x_3, x_4$ | $x_1, x_2$ | $[0, 0, 6, 4]$ |
| $A$ | $x_3, x_2$ | $x_1, x_4$ | $[0, 4, 2, 0]$ |
| $B$ | $x_4, x_2$ | $x_1, x_3$ | $[0, 6, 0, -2]$ |
| $C$ | $x_1, x_2$ | $x_4, x_3$ | $[2, 2, 0, 0]$ |
| $D$ | $x_1, x_3$ | $x_4, x_2$ | $[4, 0, -2, 0]$ |
| $E$ | $x_1, x_4$ | $x_3, x_2$ | $[3, 0, 0, 1]$ |

At point $O$ the basic variables are $x_3$ and $x_4$, and the set of equations which must be solved is

$$\begin{aligned} 2(0) + 1(0) + 1x_3 \quad\quad &= 6 \\ 1(0) + 1(0) \quad\quad + 1x_4 &= 4, \end{aligned}$$

or

$$x_3\begin{bmatrix}1\\0\end{bmatrix} + x_4\begin{bmatrix}0\\1\end{bmatrix} = \begin{bmatrix}6\\4\end{bmatrix}.$$

The vectors $A_3 = \begin{bmatrix}1\\0\end{bmatrix}$ and $A_4 = \begin{bmatrix}0\\1\end{bmatrix}$ form a basis in 2-space. Thus $R$ is some linear combination of the $A_i$ and in this case it is obvious that

$$R = \begin{bmatrix}6\\4\end{bmatrix} = 6\begin{bmatrix}1\\0\end{bmatrix} + 4\begin{bmatrix}0\\1\end{bmatrix},$$

so that $X = [x_1, x_2, x_3, x_4]^T = [0, 0, 6, 4]^T$ is our desired basic solution.

Since $x_i \geqslant 0$ for each $i$, the basic solution is also feasible. In this case, $R = 6A_3 + 4A_4 = [A_3, A_4]\begin{bmatrix} 6 \\ 4 \end{bmatrix} = \begin{bmatrix} 1 & 0 \\ 0 & 1 \end{bmatrix}\begin{bmatrix} 6 \\ 4 \end{bmatrix}$.

Consider the point $A$. At this point, $x_1 = x_4 = 0$, and we have the set of equations

$$x_2A_2 + x_3A_3 = x_2\begin{bmatrix} 1 \\ 1 \end{bmatrix} + x_3\begin{bmatrix} 1 \\ 0 \end{bmatrix} = \begin{bmatrix} 1 & 1 \\ 1 & 0 \end{bmatrix}\begin{bmatrix} x_2 \\ x_3 \end{bmatrix} = \begin{bmatrix} 6 \\ 4 \end{bmatrix}$$

to solve, and we already know that

$$R = \begin{bmatrix} 6 \\ 4 \end{bmatrix} = 4\begin{bmatrix} 1 \\ 1 \end{bmatrix} + 2\begin{bmatrix} 1 \\ 0 \end{bmatrix} = \begin{bmatrix} 1 & 1 \\ 1 & 0 \end{bmatrix}\begin{bmatrix} 4 \\ 2 \end{bmatrix}.$$

Since $\{A_2, A_3\}$ is a linearly independent subset of 2-space, these vectors also form a basis. For this reason the matrix $[A_2, A_3] = \begin{bmatrix} 1 & 1 \\ 1 & 0 \end{bmatrix}$ is a basis matrix.

In an LP problem with matrix $A$ we restrict our previous definition and call any $m \times m$ matrix whose columns are chosen from $A$ and form a linearly independent subset of $m$-space a **basis matrix*** for $A$. We will denote such a matrix by $B$. In the particular case above, $B = \begin{bmatrix} 1 & 1 \\ 1 & 0 \end{bmatrix} = [A_2, A_3]$. We shall find it convenient to use a special notation for the $m$-component matrix consisting of the values of the $m$ basic variables corresponding to the columns of $B$. We shall call this matrix $X_B$ since it does consist of a proper subset of the components of $X$, namely those corresponding to the $m$ linearly independent columns of $A$ selected to make up $B$. Thus for

$$B = [A_2, A_3] = \begin{bmatrix} 1 & 1 \\ 1 & 0 \end{bmatrix},$$

$$X_B = \begin{bmatrix} x_2 \\ x_3 \end{bmatrix} = \begin{bmatrix} 4 \\ 2 \end{bmatrix}.$$

The order of the columns in a basis matrix $B$ is important. To keep track of the order of these columns we shall number them as in $B = [B_1, B_2, \ldots, B_m]$. $B_1$ could be any one of the $n$ columns of $A$, as could any other $B_i$, as long as the entire set is linearly independent. (This rules out duplication or scalar multiples.) If $B = [A_2, A_3]$, then $B_1 = A_2$ and $B_2 = A_3$.

We shall further denote the component of $X_B$ that corresponds to column $i$ of $B$ by $x_{Bi}$, so that in general,

$$x_{B1}B_1 + x_{B2}B_2 + \cdots + x_{Bm}B_m = [B_1, B_2, \ldots, B_m]\begin{bmatrix} x_{B1} \\ x_{B2} \\ \vdots \\ x_{Bm} \end{bmatrix} = R.$$

Above, $4A_2 + 2A_3 = R$. Since $A_2 = B_1$ and $A_3 = B_2$, we have $4B_1 + 2B_2 = R$, so that

$$x_{B1} = 4 \quad \text{and} \quad x_{B2} = 2.$$

---

* We shall, for the most part, use the notation developed by G. Hadley (see Suggested Reading).

The entire matrix consisting of such components, namely

$$[x_{B1}, x_{B2}, \ldots, x_{Bm}]^T,$$

is just the matrix $X_B$. Since

$$[B_1, B_2, \ldots, B_m]\begin{bmatrix} x_{B1} \\ x_{B2} \\ \vdots \\ x_{Bm} \end{bmatrix} = R, \quad \text{and} \quad B = [B_1, B_2, \ldots, B_m],$$

we have the following formula:

$$BX_B = R.$$

Let us use points $B$ and $C$ of our graph to exercise our knowledge of the notation just introduced. First consider point $B$. At this point the corresponding basis consists of $A_4$ and $A_2$. Thus

$$B = [B_1, B_2] = [A_4, A_2] \quad \text{and} \quad B_1 = A_4 \quad \text{while} \quad B_2 = A_2.$$

Since $R = x_{B1}B_1 + x_{B2}B_2 = -2A_4 + 6A_2$, we find that $x_{B1} = -2$ and $x_{B_2} = 6$; and we verify that $BX_B = R$ since

$$\begin{bmatrix} 0 & 1 \\ 1 & 1 \end{bmatrix}\begin{bmatrix} -2 \\ 6 \end{bmatrix} = \begin{bmatrix} 6 \\ 4 \end{bmatrix}.$$

As a matter of fact, since $B$ is a basis matrix, $B^{-1}$ exists; and we see that we could have used $B^{-1}$ to solve for $X_B$ from the following formula:

$$X_B = B^{-1}R.$$

You should find $B^{-1}$ in the above example and verify that this is the case.

At point $C$, the basic variables are $x_1$, $x_2$, so that $B = [B_1, B_2] = [A_1, A_2]$. Since $B = \begin{bmatrix} 2 & 1 \\ 1 & 1 \end{bmatrix}$ and $BX_B = R$, we shall use $B^{-1}$ to find $X_B$.

Now $B^{-1} = \begin{bmatrix} 1 & -1 \\ -1 & 2 \end{bmatrix}$. Therefore

$$X_B = B^{-1}R = \begin{bmatrix} 1 & -1 \\ -1 & 2 \end{bmatrix}\begin{bmatrix} 6 \\ 4 \end{bmatrix} = \begin{bmatrix} 2 \\ 2 \end{bmatrix} = \begin{bmatrix} x_{B1} \\ x_{B2} \end{bmatrix} = \begin{bmatrix} x_1 \\ x_2 \end{bmatrix},$$

as we had found before.

You should now verify that Table 6.5.2 has been computed correctly.

**Table 6.5.2**

| *Point* | *Basic Variables* | *Basis Matrix B* | $X_B^T$ | $X^T$ |
|---|---|---|---|---|
| $O$ | $x_3, x_4$ | $[A_3, A_4]$ | $[6, 4]$ | $[0, 0, 6, 4]$ |
| $A$ | $x_3, x_2$ | $[A_3, A_2]$ | $[2, 4]$ | $[0, 4, 2, 0]$ |
| $B$ | $x_4, x_2$ | $[A_4, A_2]$ | $[-2, 6]$ | $[0, 6, 0, -2]$ |
| $C$ | $x_1, x_2$ | $[A_1, A_2]$ | $[2, 2]$ | $[2, 2, 0, 0]$ |
| $D$ | $x_1, x_3$ | $[A_1, A_3]$ | $[4, -2]$ | $[4, 0, -2, 0]$ |
| $E$ | $x_1, x_4$ | $[A_1, A_4]$ | $[3, 1]$ | $[3, 0, 0, 1]$ |

It should be emphasized that the ordering of the columns of $B$, although important, is quite arbitrary. At the point $A$, for example, we could have chosen to let $B = [A_2, A_3]$. This would merely have changed the order of the elements in $X_B$ so that $X_B$ would have been $\begin{bmatrix} 4 \\ 2 \end{bmatrix} = \begin{bmatrix} x_{B1} \\ x_{B2} \end{bmatrix} = \begin{bmatrix} x_2 \\ x_3 \end{bmatrix}$.

Example 6.5.1 Consider the problem

---

Find $X$ such that for $A = \begin{bmatrix} 2 & 1 & 1 & 0 \\ 1 & 1 & 0 & 1 \end{bmatrix}$, $R = \begin{bmatrix} 8 \\ 6 \end{bmatrix}$, $C = [3, 2, 0, 0]$

1. $X \geqslant 0$
2. $AX = R$ and
3. $z = f(X) = CX$ is maximized.

---

Fill in the blanks in Table 6.5.3 according to the notation established in Figure 6.5.3. The direction of increasing $f$ is indicated by the arrow

**Figure 6.5.3**

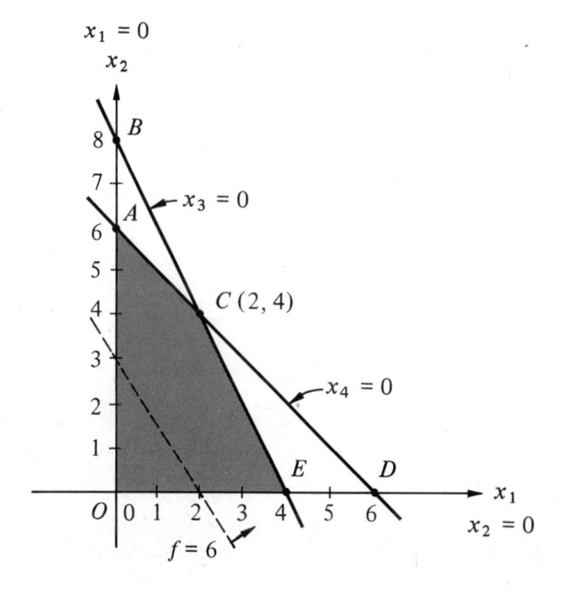


perpendicular to the line $3x_1 + 2x_2 = 6$. Use the fact that $X_B = B^{-1}R$ to find $X_B$. The answers, obtained by various methods, are shown in Table 6.5.4. Refer to the steps below only if you need help.

**Step 1** At point $O$ the basic variables are $x_3, x_4$ in that order. So $B = [A_3, A_4]$ and $BX_B = R$. We can either use Gauss-Jordan reduction or find $B^{-1}$. In the first method we have

$$[B \mid R] = [A_3, A_4 \mid R] = \left[\begin{array}{cc|c} 1 & 0 & 8 \\ 0 & 1 & 6 \end{array}\right],$$

so that $$X_B = \begin{bmatrix} 8 \\ 6 \end{bmatrix} \quad \text{and} \quad X^T = [0, 0, 8, 6].$$

Using the second method we would have

$$X_B = B^{-1}R = I_2^{-1}R = I_2R = R = \begin{bmatrix} 8 \\ 6 \end{bmatrix} \quad \text{and} \quad x_{B2} = x_4 = 6.$$

**Table 6.5.3**

| Point | Basic Variables | $B$ | $X_B^T$ | $X^T$ | $x_{B2}$ |
|---|---|---|---|---|---|
| $O$ | $x_3$, | [ ] | [ ] | [ ] | |
| $A$ | , $x_2$ | [ ] | [ ] | [ ] | |
| $B$ | $x_4$, | [ ] | [ ] | [ ] | |
| $C$ | $x_1$, | [ ] | [ ] | [ ] | |
| $D$ | , $x_3$ | [ ] | [ ] | [ ] | |
| $E$ | , $x_4$ | [ ] | [ ] | [ ] | |

**Step 2** At the point $A$ we have $x_3$ and $x_2$ as basic variables since $x_1 = x_4 = 0$. Thus $B = [A_3, A_2] = \begin{bmatrix} 1 & 1 \\ 0 & 1 \end{bmatrix}$.

$$[B \mid R] = [A_3, A_2 \mid R] = \left[\begin{array}{cc|c} 1 & 1 & 8 \\ 0 & 1 & 6 \end{array}\right] \sim \left[\begin{array}{cc|c} 1 & 0 & 2 \\ 0 & 1 & 6 \end{array}\right] \text{ so that}$$

$X_B = \begin{bmatrix} 2 \\ 6 \end{bmatrix}$ and $X^T = [0, 6, 2, 0]$. We verify that $2\begin{bmatrix} 1 \\ 0 \end{bmatrix} + 6\begin{bmatrix} 1 \\ 1 \end{bmatrix} = \begin{bmatrix} 8 \\ 6 \end{bmatrix}$ $= R$. Here $x_{B2} = x_2 = 6$.

**Step 3** At point $B$, the basis matrix $B = [A_4, A_2] = \begin{bmatrix} 0 & 1 \\ 1 & 1 \end{bmatrix}$.

$$B^{-1} = \frac{\begin{bmatrix} 1 & -1 \\ -1 & 0 \end{bmatrix}}{-1} = \begin{bmatrix} -1 & 1 \\ 1 & 0 \end{bmatrix},$$

and

$$X_B = B^{-1}R = \begin{bmatrix} -1 & 1 \\ 1 & 0 \end{bmatrix}\begin{bmatrix} 8 \\ 6 \end{bmatrix} = \begin{bmatrix} -2 \\ 8 \end{bmatrix}$$

so that $X^T = [0, 8, 0, -2]$ and $x_{B2} = 8$.

**Step 4** At point $C$, $B = [A_1, A_2] = \begin{bmatrix} 2 & 1 \\ 1 & 1 \end{bmatrix}$.

$$[B \mid R] = \left[\begin{array}{cc|c} 2 & 1 & 8 \\ 1 & 1 & 6 \end{array}\right] \sim \left[\begin{array}{cc|c} 1 & 1/2 & 4 \\ 0 & 1/2 & 2 \end{array}\right] \sim \left[\begin{array}{cc|c} 1 & 0 & 2 \\ 0 & 1 & 4 \end{array}\right],$$

so that $X_B = \begin{bmatrix} 2 \\ 4 \end{bmatrix}$ and $X^T = [2, 4, 0, 0]$, while $x_{B2} = 4$.

**Step 5** At point $D$, $B = [A_1, A_3] = \begin{bmatrix} 2 & 1 \\ 1 & 0 \end{bmatrix}$.

$$X_B = \begin{bmatrix} 6 \\ -4 \end{bmatrix}, \quad X^T = [6, 0, -4, 0], \quad \text{and} \quad x_{B2} = -4.$$

**Step 6** At point $E$, $B = [A_1, A_4]$.

$$X_B = \begin{bmatrix} 4 \\ 2 \end{bmatrix}, \quad X^T = [4, 0, 0, 2] \quad \text{and} \quad x_{B2} = 2.$$

This information is gathered together in Table 6.5.4.

**Table 6.5.4**

| *Point* | *Basic Variables* | $B$ | $X_B^T$ | $X^T$ | $x_{B2}$ |
|---|---|---|---|---|---|
| $O$ | $x_3, x_4$ | $[A_3, A_4]$ | $[8, 6]$ | $[0, 0, 8, 6]$ | 6 |
| $A$ | $x_3, x_2$ | $[A_3, A_2]$ | $[2, 6]$ | $[0, 6, 2, 0]$ | 6 |
| $B$ | $x_4, x_2$ | $[A_4, A_2]$ | $[-2, 8]$ | $[0, 8, 0, -2]$ | 8 |
| $C$ | $x_1, x_2$ | $[A_1, A_2]$ | $[2, 4]$ | $[2, 4, 0, 0]$ | 4 |
| $D$ | $x_1, x_3$ | $[A_1, A_3]$ | $[6, -4]$ | $[6, 0, -4, 0]$ | $-4$ |
| $E$ | $x_1, x_4$ | $[A_1, A_4]$ | $[4, 2]$ | $[4, 0, 0, 2]$ | 2 |

It should now be apparent that if we are able to represent the feasible region of an LP problem graphically, then it is possible to associate at least one basis with each extreme or corner point. We first determine a set of $n - m$ non-basic variables associated with the corner point. The columns of $A$ associated with the remaining $m$ variables make up $B$, in some order. When $B$ is known, it is possible to use either Gauss-Jordan elimination or $B^{-1}$ to find $X_B$, the values of the basic variables at the particular extreme or corner point under examination. Once $X_B$ is known, it is a simple matter to fill in the values on the non-basic variables (Why?) to find $X$.

The discerning reader may be wondering why we have said that it is possible to associate *at least one* basis with each corner point as opposed to *exactly one* basis. Although the latter statement would be true for all of the examples which we have studied so far, consider point $C$ in the so-called degenerate case depicted in Figure 6.5.4. One possible set of non-basic variables associated with $C$ is $\{x_2, x_3\}$. In this case $x_1$ and $x_4$ are basic. Thus $B_1 = [A_1, A_4]$ is one basis matrix associated with $C$. However, $x_4 = 0$ also, so that $\{x_2, x_4\}$ is another possible non-basic set for $C$. $B_2 = [A_1, A_3]$ is also a legitimate basis matrix for $C$. Is there another?

We should note especially from this example that it is possible for a basic variable to have a value 0. A basic feasible solution $X$ with the property that $x_i = 0$ for some *basic variable* $x_i$ is called a **degenerate basic feasible solution**. Unless otherwise stated, we shall suppose throughout the remainder of the text that we are discussing a non-degenerate case. Although degeneracy is fairly common in many types of LP problems, it seldom causes serious computational difficulties. The algorithm that will be developed will almost always handle the degenerate case without revision.

**Figure 6.5.4**

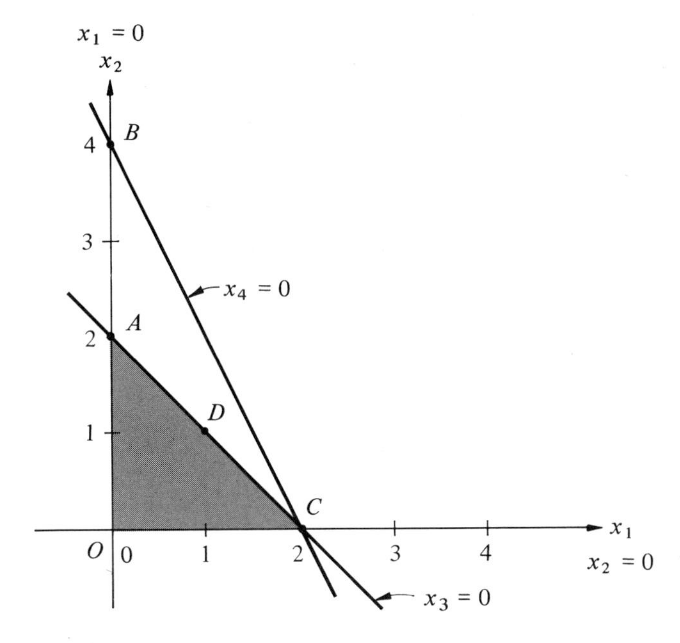


## 6.5 Exercises

1. Let $B^{-1} = \begin{bmatrix} 1 & 1 \\ 5 & 6 \end{bmatrix}$ and $R = \begin{bmatrix} 1 \\ 4 \end{bmatrix}$.
   a. Find $X_B$.
   b. What is $x_{B2}$?
   c. Find $B$.
   d. By solving the appropriate equations, find $a$, $b$ such that $R = aB_1 + bB_2$, where $B_i$ is column $i$ of $B$.
   e. Compare $[a, b]$ and $[x_{B1}\ x_{B2}]$.

2. With each corner point of the feasible region of an LP problem, we may associate at least one ________ matrix $B$. From each set of $m$ linearly independent columns of $A$ we can form a basis ________ and compute $X_B$ = ________. If $X_B \geqslant 0$, then the corresponding solution $X$ is a ________ ________. Otherwise, the corresponding solution is ________. How do we know that we have a solution at all when we pick these $m$ linearly independent vectors arbitrarily?

3. Consider the LP problem with $A = \begin{bmatrix} 1 & 3 & 1 & 0 \\ 3 & 1 & 0 & 1 \end{bmatrix}$, $R = \begin{bmatrix} 4 \\ 4 \end{bmatrix}$.
   a. $A_2 = [\quad]$.
   b. Let $B = [A_3, A_4] = B^{-1}$. Find $X_B$; $x_{B1} = x_{B2} =$ ________.
   c. Let $B = [A_1, A_4]$ and find $X_B$; $x_{B2} =$ ________.
   d. Let $B = [A_1, A_2]$ and find $X_B$; $x_{B1} =$ ________.
   e. Find $x_{B2}$ if $B = [A_4, A_1]$. Compare with c. above.

4. If $R = 2B_1 + 3B_2 + 4B_3$ where $B = [B_1, B_2, B_3]$, then $x_{B1} =$ ________; $x_{B2} =$ ________; and $x_{B3} =$ ________. If the columns of $B$ are rearranged to form a new matrix $B' = [B_3, B_1, B_2]$, then the new value of $x_{B2}$ is ________.

5. Consider the constraint set

$$\begin{aligned} x_1 + 2x_2 + x_3 \quad\quad &= 3 \\ 2x_1 + x_2 \quad\quad + x_4 &= 3 \end{aligned}$$

of Exercises 6.4.1 and 6.4.3.

a. Write the constraint equations in the new matrix format described in this section.
b. Plot the vectors $A_i$ and $R$ as in Figure 6.5.1.
c. Referring to Figure 6.4.2, fill in Table 6.5.5.

**Table 6.5.5**

| *Point* | *Basic Variables* | $B$ | $B^{-1}$ | $X_B$ | $X^T$ | $x_{B1}$ |
|---|---|---|---|---|---|---|
| $O$ | $x_3, x_4$ | $\begin{bmatrix} 1 & 0 \\ 0 & 1 \end{bmatrix}$ | $\begin{bmatrix} 1 & 0 \\ 0 & 1 \end{bmatrix}$ | $\begin{bmatrix} 3 \\ 3 \end{bmatrix}$ | [0, 0, 3, 3] | 3 |
| $A$ | | | | | | |
| $B$ | | | | | | |
| $C$ | | | | | | |
| $D$ | | | | | | |
| $E$ | | | | | | |

6. Consider the constraint set and figure associated with Exercise 6.4.5.

a. Write the constraints in the matrix format of this section.
b. Fill in Table 6.5.6 using appropriate data from Table 6.4.4 and Figure 6.4.3. $R = \begin{bmatrix} 2 \\ 4 \end{bmatrix}$.

**Table 6.5.6**

| *Point* | *Basic Variables* | $B$ | $B^{-1}$ | $X_B$ | $X^T$ | $x_{B2}$ | *Solution Degenerate?* |
|---|---|---|---|---|---|---|---|
| $O$ | | $[A_3, A_4] = \begin{bmatrix} 1 & 0 \\ 0 & 1 \end{bmatrix}$ | $\begin{bmatrix} 1 & 0 \\ 0 & 1 \end{bmatrix}$ | $\begin{bmatrix} 2 \\ 4 \end{bmatrix}$ | $[0, 0, 2, 4]$ | 4 | |
| $A$ | $x_2,$ | | | | | | |
| $B$ | $, x_2$ | $[A_3, A_2] = \begin{bmatrix} 1 & 1 \\ 0 & 1 \end{bmatrix}$ | | $\begin{bmatrix} -2 \\ 4 \end{bmatrix}$ | $[0, 4, -2, 0]$ | 4 | |
| $C$ | $x_1, x_4$ | | | | | | |
| $C$ | | | | | | | YES |
| $C$ | $x_1, x_2$ | $[A_1, A_2] = \begin{bmatrix} 1 & 1 \\ 2 & 1 \end{bmatrix}$ | $\begin{bmatrix} -1 & 1 \\ 2 & -1 \end{bmatrix}$ | $\begin{bmatrix} 2 \\ 0 \end{bmatrix}$ | $[2, 0, 0, 0]$ | 0 | YES |

### Suggested Reading

G. B. Dantzig. *Linear Programming and Extensions.* Princeton, N.J.: Princeton University Press, 1963.

S. I. Gass. *Linear Programming.* 3d. ed. New York: McGraw-Hill, 1969.

G. Hadley. *Linear Programming.* Reading, Mass.: Addison-Wesley, 1962.

### New Terms



## 6.6 A Possible Solution Technique

For our problem $P$ with

$$A = \begin{bmatrix} 2 & 1 & 1 & 0 \\ 1 & 1 & 0 & 1 \end{bmatrix}, \quad X = \begin{bmatrix} x_1 \\ x_2 \\ x_3 \\ x_4 \end{bmatrix}, \quad R = \begin{bmatrix} 6 \\ 4 \end{bmatrix}, \quad C = [5, 3, 0, 0],$$

we have the information shown in Figure 6.6.1 and Table 6.6.1. As before, the direction of increasing $f$ is indicated in Figure 6.6.1 by the arrow. We can

**Figure 6.6.1**

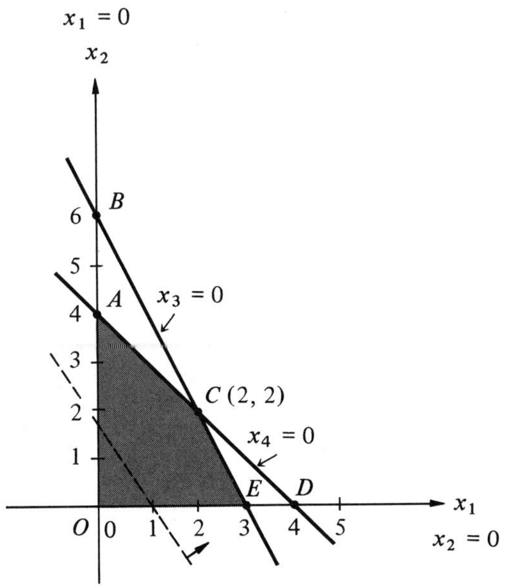

**Table 6.6.1**

| *Point* | $B$ | $X_B^T$ |
|---|---|---|
| $O$ | $[A_3, A_4]$ | $[6, 4]$ |
| $A$ | $[A_3, A_2]$ | $[2, 4]$ |
| $C$ | $[A_1, A_2]$ | $[2, 2]$ |
| $E$ | $[A_1, A_4]$ | $[3, 1]$ |

evaluate $z = f(X) = CX$ at each of the corner points listed in the table. We could use $CX$, but instead we choose to use the matrix $C_B$ which contains only the objective function coefficients of the basic variables in $X_B$ in the same order. For example, at point $O$, $X_B = \begin{bmatrix} x_{B1} \\ x_{B2} \end{bmatrix} = \begin{bmatrix} x_3 \\ x_4 \end{bmatrix} = \begin{bmatrix} 6 \\ 4 \end{bmatrix}$. Thus $C_B = [c_{B1}, c_{B2}] = [c_3, c_4] = [0, 0]$. Since, at $O$, $x_1 = x_2 = 0$, it really does not matter what $c_1$ and $c_2$ are. Thus $f(X) = CX = C_B X_B = [0, 0]\begin{bmatrix} 6 \\ 4 \end{bmatrix} = 0$. At $A$, $C_B = [c_3, c_2] = [0, 3]$ and $f(X) = C_B X_B = [0, 3]\begin{bmatrix} 2 \\ 4 \end{bmatrix} = 12$. Similarly, at $C$, $f(X) = [5, 3]\begin{bmatrix} 2 \\ 2 \end{bmatrix} = 16$; and at $E$, $f(X) = [5, 0]\begin{bmatrix} 3 \\ 1 \end{bmatrix} = 15$. So, as far as corner points are concerned, $f$ takes on its maximum value at $C$, as we saw previously, and the maximum value of $f$ is 16.

As we noted in Section 6.2, the geometry of the situation makes it clear that the optimal feasible solution occurs at the extreme point $C$. It may help, in considering larger LP problems, to consider the many "points" and faces of a many-faceted diamond. It is just such an enclosed solid (known as a polyhedron) which is created as a feasible region in a problem with planes rather than straight lines as boundaries.

We can state now, and we shall later attempt to make it seem more plausible, that for an LP problem which has an optimal feasible solution, *at least one optimal feasible solution will always occur at one of the corner points of the feasible region.* Thus, if we were to find all of the corner points, and then evaluate the objective function at each one, we could find the optimal feasible solution by exhaustively comparing these functional values.

Such a method is not computationally practical for problems of even moderate size. We have seen that (excluding the degenerate case) we get a different basis and thus a different extreme point for each possible *BFS*. We can choose the $m$ basic variables from the $n$ variables in $\binom{n}{m}$ ways.* To determine which do correspond to corner points (rather than points such as $B$ and $D$) the basic solutions must be computed and examined for feasibility. The objective function $f$ must then be evaluated for each remaining *BFS*. In even a comparatively small practical LP problem with $n = 20$ and $m = 10$, there are $\binom{20}{10} = 184{,}756$ basic solutions to compute, examine, evaluate, and compare.

The simplex technique, to which we now turn, is an ingenious method that examines only a subset of the entire set of basic feasible solutions (or associated extreme points), and usually only a small subset of them, until an optimum one is reached, at which time the algorithm terminates. The number of *BFS* examined is usually amazingly small, when compared with the total number which exist.

---

* The binomial coefficient $\binom{n}{m}$ is the standard notation for the number of different combinations of $n$ objects taken $m$ at a time. $\binom{n}{m} = \dfrac{n!}{(n-m)!m!}$.

## 6.6 Exercises

1. Consider the LP problem with $C = [1, 2, -1, 3]$. If $B = [A_1, A_3]$, then $C_B = [\ ,\ ]$. If $X_B = \begin{bmatrix} 1 \\ 2 \end{bmatrix}$, then $X^T = [\ ,\ ,\ ,\ ]$ and $f(X) = C_B X_B =$ ______.

2. Let $C = [3, -1, 4, 2]$ and $B = [A_3, A_4]$. Then, if ______ $= \begin{bmatrix} 5 \\ 6 \end{bmatrix}$, $f(X) = C_B X_B = [\ ,\ ]\begin{bmatrix} \ \\ \ \end{bmatrix} = 32$.

3. Let $A = \begin{bmatrix} 1 & 3 & 1 & 0 \\ 3 & 1 & 0 & 1 \end{bmatrix}$, $C = [1, 2, 0, 0]$, and $R = \begin{bmatrix} 4 \\ 4 \end{bmatrix}$.
   a. Find all possible $BFS$ by choosing 2 columns of $A$ as basic columns to form $B = [A_i, A_j]$ and then finding $X_B = B^{-1}R$. If $x_{Bi} < 0$ for some $i$, discard $X_B$.
   b. For each of the 4 $BFS$ in a., find $C_B$ and $f(X) = C_B X_B$.
   c. Assuming that we wish to maximize $f$ and that a corner point is optimal, find the optimal $BFS$.
   d. Find $C_B X_B$ for each $BFS$ if $C = [1, 0, 0, 0]$. What is the optimal $BFS$ now?
   e. Draw a sketch, and verify your results graphically.

4. In an LP problem with $m$ constraints and $n$ variables $C_B$ will have ______ components as will ______. To find $X$ from $X_B$ we must first know the columns of the ______ ______ $B$. If $A_3$ is in column 1 of $B$, then $x_{B1} = x_3$ and $c_{B1} =$ ______. We fill in the value of $x_{B1} = x_3$ in $X$ and continue to do so for each of the $m$ numbers $x_{Bi}$. We then merely fill in the remaining $n - m$ numbers $x_i$ with ______.

5. Suppose that $C = [-3, -1, 0, 0]$ in Exercise 6.5.5, and suppose that you wish to minimize $f$.
   a. Find $C_B X_B = f(X_B)$ for each feasible basis $B$ in Table 6.5.5.
   b. At what point $X$ does the optimum $BFS$ occur?
   c. What is it?

6. Let $C = [1, -1, 0, 0]$ in Exercise 5, and suppose that you wish to maximize $f$ Find $C_B X_B = f(X_B)$ for each feasible basis $B$ and find the optimum $BFS$.

7. Consider the maximization problem of Example 6.5.1 with

$$A = \begin{bmatrix} 2 & 1 & 1 & 0 \\ 1 & 1 & 0 & 1 \end{bmatrix}, \quad R = \begin{bmatrix} 8 \\ 6 \end{bmatrix}, \quad \text{and} \quad C = [3, 2, 0, 0].$$

   a. Using $X_B$ as shown in Table 6.5.4 compute $f(X_B)$ for each feasible basis.
   b. The optimal $X_B$ is ______, where $f(X_B) =$ ______.
   c. The solution $X$ associated with the optimum $X_B$ is [______, ______, ______, ______$]^T$.

8. Use the method of this section and the information in Table 6.5.6 to find three optimal basic feasible solutions if $C = [3, 1, 0, 0]$.

## 6.7 The Simplex Algorithm in Equation Form

We shall continue to assume that if an optimal feasible solution to an LP problem exists at all, then at least one optimal feasible solution will occur at an extreme or corner point of the feasible region. In this section we will show how it is possible to start at a given corner point and establish whether or not it would be beneficial, in terms of improvement of the value of the objective function, to move to an adjacent corner point. If there is an adjacent corner point (one connected to the original by an edge of the feasible region) to which we could move and improve the value of $z$, we will do so. If there is more than one such point, we will choose the one to move to by first establishing the rate of change of $z$ in each possible direction. We shall then move along the boundary of the feasible region to that adjacent corner point which offers the best rate of change of $z$. On reaching the new corner point we repeat the process, and keep moving from corner point to adjacent corner point until we reach an optimal corner point. The fact that we can tell when we have reached an optimal corner point is a happy phenomenon associated with the fact that we are dealing with a *linear* programming problem. If we ever reach a corner point from which it would not be profitable to move to any adjacent corner point, we can stop and know that no other point in the feasible region—corner point, boundary point, or interior point—has a better value of $z$ than the present one. This is proved in many of the references mentioned in this chapter. We shall attempt to make it seem plausible in Section 6.11.

Let us use problem $P$ to describe the procedure that we have in mind. The objective function is $f(x_1, x_2, x_3, x_4) = 5x_1 + 3x_2$, and our equality constraints are

$$\begin{aligned} 2x_1 + x_2 + x_3 \qquad &= 6 \\ x_1 + x_2 \qquad + x_4 &= 4. \end{aligned}$$

The other information which we need is contained in Figure 6.7.1 and Table 6.7.1.

Because of the ease in so doing, we shall start at $O$. At this point the basic variables are $x_3$ and $x_4$. Let us solve the constraint equations for

Figure 6.7.1

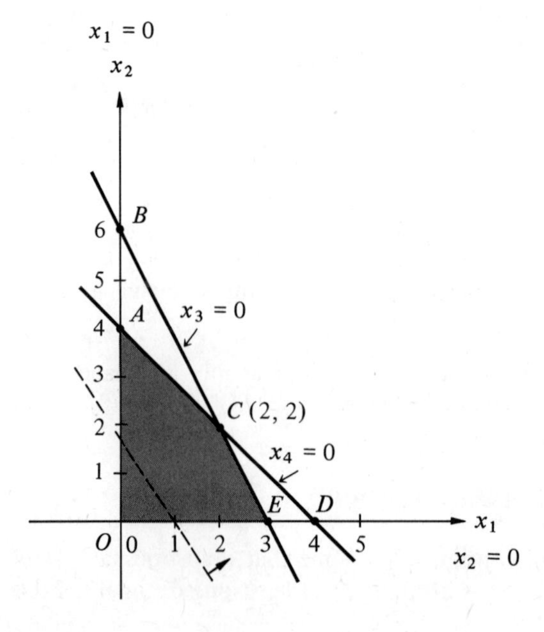

**Table 6.7.1**

| Point | Basic Variables | $X^T$ |
|---|---|---|
| $O$ | $x_3, x_4$ | [0, 0, 6, 4] |
| $A$ | $x_3, x_2$ | [0, 4, 2, 0] |
| $B$ | $x_4, x_2$ | [0, 6, 0, −2] |
| $C$ | $x_1, x_2$ | [2, 2, 0, 0] |
| $D$ | $x_1, x_3$ | [4, 0, −2, 0] |
| $E$ | $x_1, x_4$ | [3, 0, 0, 1] |

these variables. We obtain

$$\begin{aligned} x_3 &= 6 - 2x_1 - x_2 \\ x_4 &= 4 - \phantom{2}x_1 - x_2. \end{aligned}$$

Since, in reality, $x_1 = x_2 = 0$ at this point, it is clear that $x_3 = 6$ and $x_4 = 4$. We can proceed toward and eventually arrive at an adjacent extreme point by increasing *one* of the non-basic variables from 0 to an appropriate value and leaving the other non-basic variable at 0. Which non-basic variable should we choose to increase from 0? If we decide to increase $x_1$ and leave $x_2 = 0$, we will proceed from $O$ in the direction toward extreme points $E$ and $D$ along the line $x_2 = 0$. If we increase $x_2$ and leave $x_1 = 0$, we will move toward $A$ and $B$. We use the objective function to help us decide whether or not we are at an optimal corner point, and, if not, to choose which direction to go.

Now, $z = 5x_1 + 3x_2 + 0x_3 + 0x_4 = f(x_1, x_2, x_3, x_4)$. At $O$, $f(0, 0, 6, 4) = 0$. What will happen to $z$ if we increase $x_1$ and hold $x_2 = 0$? For each unit which we increase $x_1$, $z$ will increase by 5 units. Similarly, if we hold $x_1 = 0$ and increase $x_2$, $z$ will increase 3 units for each unit increase in $x_2$. Without looking at the graph or table, or making further computations, we do not know how much we will be able to increase either $x_1$ or $x_2$ *and still remain feasible.* We do know that $z$ will increase regardless of which of $x_1$ or $x_2$ we choose. Since the per unit rate of change (increase) of $z$ is higher for $x_1$, let us increase $x_1$ and move out along the $x_1$ axis.

We know from the geometry that we should stop at $E$ in order to remain in the feasible region. How does algebra solve this geometrical problem for us? Reconsider the basic equations at $O$, namely

$$\begin{aligned} x_3 &= 6 - 2x_1 - x_2 \\ x_4 &= 4 - \phantom{2}x_1 - x_2. \end{aligned}$$

If $x_2 = 0$ and is to remain at the value 0, we really have

$$\begin{aligned} x_3 &= 6 - 2x_1 \\ x_4 &= 4 - \phantom{2}x_1. \end{aligned}$$

What happens to the values of the previously basic variables $x_3$, $x_4$ as $x_1$ is increased *and these equations are required to hold*? The $(-2)$ coefficient of $x_1$ in the first equation tells us that $x_3$ will *decrease* 2 units from its present value of 6 for each unit that $x_1$ increases. That $x_3$ does decrease as $x_1$ increases from $O$ toward $E$ should be clear from the graph, since $x_3 = 6$ at

$O$ and $x_3 = 0$ at $E$. When $x_1$ is 3, $x_3$ will be down to 0 on the 2 for 1 basis indicated by the equation. We thus have one limitation on the increase of $x_1$. If there were no limitation on increasing $x_1$, of course, we would let $x_1$ increase without bound, which would allow $z$ to increase without bound also. Very few practical problems fall into this category, unfortunately, as we would then be able to obtain an infinite profit.

Consider the second equation,

$$x_4 = 4 - x_1.$$

As $x_1$ increases, $x_4$ *decreases* on a 1 for 1 basis from its present value of 4. Thus $x_4$ will be down to 0 when $x_1 = 4$. We have obtained a second limitation on the new value of $x_1$, that $x_1 \leqslant 4$. The prior limitation thus requires us to stop when $x_1 = 3$, at $E$. If we continued to let $x_1$ increase until $x_4 = 0$, we would have arrived at $D$ in the infeasible region. The algebra of the situation has indeed enabled us to stop at a corner of the boundary of the feasible region, thus fulfilling the non-negativity restrictions.

We now know that if we increase $x_1$, then $x_3$ and $x_4$ will both decrease, but that $x_3$ will reach 0 first, and thus become non-basic. We should note that the negative coefficients of $x_1$, along with the present values 6, 4 of the basic variables, as indicated in the original equations

$$\begin{aligned} x_3 &= 6 - 2x_1 - x_2 \\ x_4 &= 4 - \phantom{2}x_1 - x_2, \end{aligned}$$

were the clue to this fact. A coefficient of $(+2)$ instead of $(-2)$ would have indicated that $x_3$ would have increased on a 2 for 1 basis as $x_1$ increased. We would then have had no $x_3$ restriction on the increase of $x_1$.

Let us now re-solve the basic equations at $O$ to find the new basic equations that apply at point $E$. The new basic variable is $x_1$, and it replaces $x_3$. We therefore use the first equation to solve for $x_1$. Thus

$$2x_1 = 6 - x_2 - x_3$$

so that

$$x_1 = 3 - (1/2)x_2 - (1/2)x_3.$$

We use this value of $x_1$ to eliminate $x_1$ from the second equation. Thus

$$\begin{aligned} x_4 = 4 - x_1 - x_2 &= 4 - [3 - (1/2)x_2 - (1/2)x_3] - x_2 \\ x_4 &= 1 - (1/2)x_2 + (1/2)x_3. \end{aligned}$$

Our new basic equations at $E$ are thus

$$\begin{aligned} x_1 &= 3 - (1/2)x_2 - (1/2)x_3 \\ x_4 &= 1 - (1/2)x_2 + (1/2)x_3. \end{aligned}$$

Since $x_2 = x_3 = 0$, we see that $x_1 = 3$ and $x_4 = 1$ at $E$, values which we have determined earlier by other methods.

What is the value of $z$ at $E$? We had hoped to increase $z$ by 5 units for each unit increase in $x_1$. Since $x_1 = 3$ at $E$, $z$ should be 15. Since $f(3, 0, 0, 1) = 5(3) + 3(0) + 0(0) + 0(1) = 15$, we have reached our goal. We could have discovered this in another way, by expressing $z$ in terms of the present non-basic variables as follows:

$$z = 5x_1 + 3x_2$$

in terms of the non-basic variables at $O$. Since

$$x_1 = 3 - (1/2)x_2 - (1/2)x_3,$$

then

$$\begin{aligned} z &= 5 \cdot [3 - (1/2)x_2 - (1/2)x_3] + 3x_2 \\ &= 15 + (1/2)x_2 - (5/2)x_3 \end{aligned}$$

at $E$. Since $x_2 = x_3 = 0$, we have $z = 15$.

The plus and minus signs for the non-basic variables $x_2$ and $x_3$, now 0, are significant. We see that an increase in $x_2$ with $x_3 = 0$ will lead to an increase in $z$ of 1/2 unit for each units increase in $x_2$. On the other hand, an increase in $x_3$ will lead to a decrease in $z$. We might have expected the latter since we just decreased $x_3$ to increase $z$ in going from $O$ to $E$.

Our algebraic representation of $z$ at $E$ is thus sufficient to notify us that an increase in $x_2$ is appropriate. A look at the geometrical diagram would, of course, have told us the same thing. We still need to know where to stop, however, in order that we do not skip point $C$ and go all the way to the infeasible point $B$, along the line $x_3 = 0$. The constraint equations again tell us which of the basic variables to remove in order to remain feasible. Consider

$$\begin{aligned} x_1 &= 3 - (1/2)x_2 - (1/2)x_3 \\ x_4 &= 1 - (1/2)x_2 + (1/2)x_3, \end{aligned}$$

where we have decided to increase $x_2$ and hold $x_3 = 0$. In this case both $x_1$ and $x_4$ decrease as $x_2$ increases, and they decrease at the same rate, namely 1/2 unit for each unit increase in $x_2$. Since $x_4$ started out at the lesser value 1, it will arrive at 0 faster, that is, when $x_2$ has increased by 2 units, from 0 to 2. At that point, the new value of $x_1$, on the same basis, should be 2; and the new value of $z$ should be $15 + (1/2)(2) = 16$. (Why?)

We re-solve the old basic equations, replacing $x_4$ by $x_2$ in the basic set. We first solve for $x_2$ in terms of the new non-basic variables, $x_4$ and $x_3$, to obtain

$$(1/2)x_2 = 1 + (1/2)x_3 - x_4$$

or

$$x_2 = 2 + x_3 - 2x_4.$$

Using the latter equation to eliminate $x_2$ from the $x_1$-equation so as to express $x_1$ in terms of the non-basic variables alone, we have

$$x_1 = 3 - (1/2)[2 + x_3 - 2x_4] - (1/2)x_3$$

or

$$x_1 = 2 - x_3 + x_4$$

so that our new equations are

$$\begin{aligned} x_1 &= 2 - x_3 + x_4 \\ x_2 &= 2 + x_3 - 2x_4. \end{aligned}$$

Thus at $C$, $X = [2, 2, 0, 0]^T$, as we have seen before. Expressing $z$ in terms of the present non-basic variables, $x_3$ and $x_4$, we have

$$\begin{aligned} z &= 15 + (1/2)x_2 - (5/2)x_3 \qquad \text{(at } E) \\ &= 15 + (1/2)(2 + x_3 - 2x_4) - (5/2)x_3 \\ &= 16 - 2x_3 - x_4 \qquad \text{(at } C). \end{aligned}$$

The coefficients of each of the non-basic variables are now negative, indicating that an increase in either would lead to a decrease in $z$. Since $x_3 = x_4 = 0$, they cannot be decreased. Thus, it would do no good to move to either of the two adjacent extreme points that can be reached from $C$; so we stop. The optimal feasible solution is $X = [2, 2, 0, 0]^T$, and the maximum value of $z$ is 16.

You should recall that we started this exercise by increasing $x_1$ because it provided a greater *rate of increase* of $z$, while we realized that we could just as well have increased $x_2$ and proceeded toward $A$ with $x_1 = 0$. You should now go through the manipulations required to proceed to point $C$ through point $A$. You will note that exactly the same number of iterations (steps in our process) are required, namely 2.

## 6.7 Exercises

1. Consider the LP problem of Example 6.4.2 with

$$A = \begin{bmatrix} 2 & 3 & 1 & 0 \\ 3 & 2 & 0 & 1 \end{bmatrix}, \quad R = \begin{bmatrix} 6 \\ 6 \end{bmatrix}, \quad \text{and} \quad C = [4, 3, 0, 0]$$

where we wish to maximize $f(X) = CX$. Complete the blanks in the following solution. Draw a sketch of the feasible region.

a. The original constraint equations are

$$\begin{aligned} 2x_1 + 3x_2 + x_3 \quad\quad &= 6 \\ 3x_1 + 2x_2 \quad\quad + x_4 &= 6. \end{aligned}$$

The basic variables at the origin are $x_3$ and $x_4$. We solve the original equations and obtain

$$\begin{aligned} x_3 &= \underline{\hspace{4cm}} \\ x_4 &= \underline{\hspace{4cm}}, \end{aligned}$$

and $f(X) = 4x_1 + 3x_2 = 0$, in terms of the present basic variables. We see that $f$ will increase if either $x_1$ or $x_2$ is increased.

b. Since the rate of change of $f$ is 4 for $x_1$ and only 3 for $x_2$, we choose to let $x_1$ increase. Both ________ and ________ will decrease as $x_1$ increases, and ________ at a faster rate. We decide to make ________ non-basic since it reaches 0 first. We thus solve for $x_1$ in terms of $x_2, x_4$ to obtain and eliminate $x_1$ from the $x_3$ equation to obtain

$$\begin{aligned} x_3 &= 2 - (5/3)x_2 + (2/3)x_4 \\ x_1 &= 2 - (2/3)x_2 - (1/3)x_4. \end{aligned}$$

We now can express $f$ in terms of the new non-basic variables ________, ________ and obtain

$$\begin{aligned} f(X) &= 4x_1 + 3x_2 \\ &= 4(2 - (2/3)x_2 - (1/3)x_4) + 3x_2 \\ &= 8 + (1/3)x_2 - (4/3)x_4. \end{aligned}$$

c. $f$ will increase if ________ is increased. Both ________ and ________ will ________ as $x_2$ increases, but ________ will reach 0 first, so we make ________ non-basic. We re-solve our equations for the new basic set {________, ________} and obtain

$$x_2 = \_\_\_\_\_\_\_\_\_\_\_\_$$

$$x_1 = \_\_\_\_\_\_\_\_\_\_\_\_$$

and

$$f(X) = 8 + (1/3)(6/5 - (3/5)x_3 + (2/5)x_4) - (4/3)x_4$$
$$= 42/5 - (1/5)x_3 - (6/5)x_4.$$

d. Since $f(X)$ will decrease if either $x_3$ or $x_4$ is increased, we stop with optimal $BFS$ $X^T = [6/5, 6/5, 0, 0]$ and $f(X) = 42/5$.

2. Suppose we have solved for the basic variables $x_3$, $x_4$ in terms of the non-basic variables $x_1$, $x_2$ and have obtained

$$x_3 = 4 - 3x_1 + 2x_2$$
$$x_4 = 2 - 1x_1 - 5x_2$$

and $f(X) = 0 + 4x_1 + 3x_2$.

a. If we increase $x_1$, we will increase $f$ by ________ units for each unit $x_1$ increases, assuming that we hold $x_2$ at ________, and let $x_3$ and $x_4$ change as they must in order to keep the equations satisfied.

b. For each unit of increase of $x_1$, $x_3$ will ________ by ________ units and $x_4$ will ________ by ________ units. Since $4/3 < 2/1$ we must let $x_3$ become non-basic as ________ will have reached 0 when $x_1$ reaches 4/3. Since $x_1$ can only increase to 4/3 the new value of $f$ will be equal to the old value, 0, plus $4(4/3) = 16/3$.

c. The new set of basic equations is

$$x_1 = \_\_\_\_\_\_\_\_\_\_\_\_$$
$$x_4 = (2/3) - (17/3)x_2 + (1/3)x_3,$$

and $f(X) = 16/3 + (17/3)x_2 - (4/3)x_3$. We note that the new value of $x_1$ is 4/3 as expected and that $f$ has increased by 16/3 as predicted.

d. We now wish to increase ________.

e. If, instead of increasing $x_1$ at the beginning, we had chosen to increase $x_2$, we would have made ________ non-basic and achieved a change in $f$ of ________.

3. Use the method of this section to solve the maximization LP problem of Exercise 6.6.3, with $A = \begin{bmatrix} 1 & 3 & 1 & 0 \\ 3 & 1 & 0 & 1 \end{bmatrix}$, $R = \begin{bmatrix} 4 \\ 4 \end{bmatrix}$, and $C = [1, 2, 0, 0]$.

4. Use the method of this section to solve the minimization LP problem of Exercise 6.6.5, with $A = \begin{bmatrix} 1 & 2 & 1 & 0 \\ 2 & 1 & 0 & 1 \end{bmatrix}$, $R = \begin{bmatrix} 3 \\ 3 \end{bmatrix}$, and $C = [-3, -1, 0, 0]$.

5. In the following basic equations for a maximization problem

$$x_3 = 5 + ax_1 + bx_2$$
$$x_4 = 6 + cx_1 + dx_2$$

and $f(X) = 0 + ex_1 + gx_2$:

a. We wish to increase $x_1$ if $e$ is ________.

b. If $x_1$ is increased, $x_3$ will decrease if ________ is ________.

c. If $x_4$ is to be made non-basic when $x_1$ increases, the new value of $x_1$ will be ________, while the new value of $f$ will be ________. The new value of $x_3$, which remains basic, will be ________.

d. If we should remove $x_4$ in order to maintain a feasible solution and mistakenly remove $x_3$, how will the mistake show itself in the next set of equations?

e. Verify your conjecture in d. by letting $a = b = -1, c = -2, d = e = g = 3$ above and letting $x_1$ replace $x_3$ in the basic set.

### Suggested Reading

G. B. Dantzig. *Linear Programming and Extensions.* Princeton, N.J.: Princeton University Press, 1963.

D. R. Plane and G. A. Kochenberger. *Operations Research for Managerial Decisions.* Homewood, Ill.: Irwin, 1972.

## 6.8 The Simplex Algorithm in Tableau Form

The simplex algorithm of Dantzig is usually presented in a so-called tableau format that we shall introduce in this section. The tableaus developed in this version of the algorithm are very closely related to the sequence of sets of equations developed in the last section. For this reason, let us consider the graph for problem $P$ in Figure 6.8.1 and the equations associated with

Figure 6.8.1

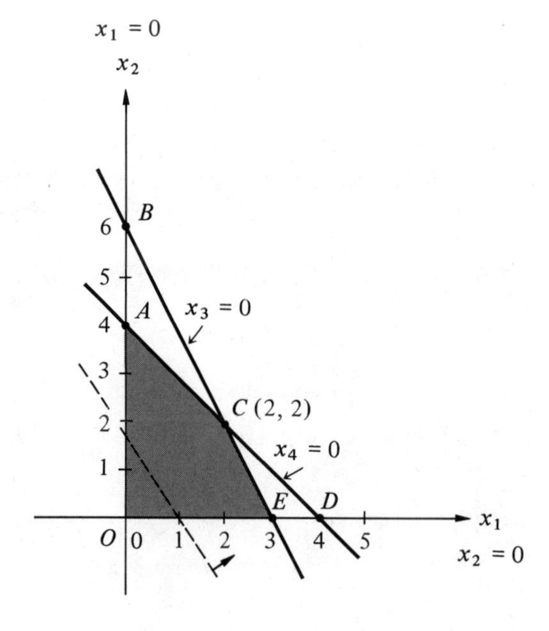


the solution. To emphasize the similarity between the ideas of the two sections, we shall rewrite the previous equations, putting all of the variables on the same side of the equation.

| Point | Equations | Basic Variables |
|---|---|---|
| $O$ | $6 = 2x_1 + x_2 + x_3$<br>$4 = x_1 + x_2 + x_4$<br>$z - 0 = 5x_1 + 3x_2$ | $x_3$<br>$x_4$ |
| $E$ | $3 = x_1 + (1/2)x_2 + (1/2)x_3$<br>$1 = (1/2)x_2 - (1/2)x_3 + x_4$<br>$z - 15 = (1/2)x_2 - (5/2)x_3$ | $x_1$<br>$x_4$ |
| $C$ | $2 = x_1 + x_3 - x_4$<br>$2 = x_2 - x_3 + 2x_4$<br>$z - 16 = -2x_3 - x_4$ | $x_1$<br>$x_2$ |

We note that the $z$ value for the appropriate point is obtained by setting the non-basic variables equal to 0. At point $C$, for example, $x_3 = x_4 = 0$, so that $z = 16$.

We now begin to develop the simplex tableau by writing down the augmented matrix for the equations at point $O$ in Tableau 6.8.1.

**Tableau 6.8.1**

| $R$ | $A_1$ | $A_2$ | $A_3$ | $A_4$ |
|---|---|---|---|---|
| 6 | 2 | 1 | 1 | 0 |
| 4 | 1 | 1 | 0 | 1 |

Conventionally in LP, the $R$ matrix is written on the left. We now add a column $B$ denoting the basis vectors at point $O$ and another column $C_B^T$ containing the cost coefficients of the variables associated with the basis vectors. Since $B = [A_3, A_4]$ at $O$, $C_B^T = \begin{bmatrix} c_3 \\ c_4 \end{bmatrix} = \begin{bmatrix} 0 \\ 0 \end{bmatrix}$. In order to keep track of the cost coefficients of all of the variables, we write $C$, the row matrix of such coefficients, along the top of the tableau. We then have Tableau 6.8.2.

**Tableau 6.8.2**

| | | $C$ | 5 | 3 | 0 | 0 |
|---|---|---|---|---|---|---|
| $B$ | $C_B^T$ | $R$ | $A_1$ | $A_2$ | $A_3$ | $A_4$ |
| $A_3$ | 0 | 6 | 2 | 1 | 1 | 0 |
| $A_4$ | 0 | 4 | 1 | 1 | 0 | 1 |

The tableau thus contains all of the information required to describe the $BFS$ corresponding to extreme point $O$, namely $X^T = [0, 0, 6, 4]$. In skeleton form, we have, in the tableau body, the quantities indicated in Tableau 6.8.3.

**Tableau 6.8.3**

| | | | |
|---|---|---|---|
| $B_1$<br>$B_2$ | $c_{B1}$<br>$c_{B2}$ | $R$ | $A$ |

As in the method described in the last section we wish to determine whether or not the present *BFS* at extreme point *O* is optimal. We create the last row of our tableau to provide us with this information. For reasons which we will make clear, we use the symbol $z_j$ and call this new row the $(c_j - z_j)$ row as shown in Tableau 6.8.4.

**Tableau 6.8.4**

| | | |
|---|---|---|
| $c_j - z_j$ | $z$ | $(c_1 - z_1)\,(c_2 - z_2)\,(c_3 - z_3)\,(c_4 - z_4)$ |

The second place in this new row is filled in with the value of $z$ at $O$, namely $C_B R = [0, 0]\begin{bmatrix} 6 \\ 4 \end{bmatrix} = 0$, which, as shown, can easily be computed from the second and third columns of Tableau 6.8.2. We thus have Tableau 6.8.5.

**Tableau 6.8.5**

| | | |
|---|---|---|
| $c_j - z_j$ | 0 | |

The elements in the remainder of the row are computed in the same manner as $z$. Under $A_1$ we wish to compute

$$c_1 - z_1 = c_1 - C_B A_1 = 5 - [0, 0]\begin{bmatrix} 2 \\ 1 \end{bmatrix} = 5;$$

again, this is a computation that is easily done with the matrices $C_B^T$ and $A_1$, conveniently placed as they are in Tableau 6.8.2. We now compute

$$c_2 - z_2 = c_2 - C_B A_2 = 3 - [0, 0]\begin{bmatrix} 1 \\ 1 \end{bmatrix} = 3,$$

$$c_3 - z_3 = c_3 - C_B A_3 = 0 - [0, 0]\begin{bmatrix} 1 \\ 0 \end{bmatrix} = 0,$$

and

$$c_4 - z_4 = c_4 - C_B A_4 = 0 - [0, 0]\begin{bmatrix} 0 \\ 1 \end{bmatrix} = 0.$$

The entire last row is shown in Tableau 6.8.6.

**Tableau 6.8.6**

| | | | | | |
|---|---|---|---|---|---|
| $c_j - z_j$ | 0 | 5 | 3 | 0 | 0 |

Tableau 6.8.7 is the completed starting tableau.

**Tableau 6.8.7**

| | | C | 5 | 3 | 0 | 0 |
|---|---|---|---|---|---|---|
| $B$ | $C_B^T$ | $R$ | $A_1$ | $A_2$ | $A_3$ | $A_4$ |
| $A_3$ | 0 | 6 | 2 | 1 | 1 | 0 |
| $A_4$ | 0 | 4 | 1 | 1 | 0 | 1 |
| $c_j - z_j$ | | 0 | 5 | 3 | 0 | 0 |

We recall that at $O$,

$$z - 0 = 5x_1 + 3x_2 + 0x_3 + 0x_4.$$

The newly-created row contains constants which are identical to these. We used these constants before to tell us whether or not the present extreme point or basic feasible solution was optimal. The (5) corresponding to $A_1$ and the (3) corresponding to $A_2$ tell us that we are not optimal at $O$, that is, that $z$ may be increased by moving to either adjacent extreme point that may be reached from $O$. Again, $z$ will be increased by 5 units for each unit that $x_1$ increases as we move to the appropriate adjacent extreme point, namely $E$. Similarly, $z$ will be increased by 3 units for each unit that $x_2$ increases as we move to $A$, if we so choose.

So as to conform to the procedure in the previous section, let us move to point $E$ by making $x_1$ basic or putting $A_1$ into the basis matrix $B$. We emphasize that this is a purely arbitrary decision and that we could just as well have chosen to increase $x_2$ and go to $A$. Most computer algorithms use this same criterion—the maximum rate of change of $z$—for choice of a variable to be made basic, or equivalently, a choice of vector to be brought into the basis matrix $B$. From computational experience it appears to be, *on the average*, the criterion which leads to an optimal *BFS* in the least number of steps.

From the old tableau, we construct a new tableau as follows:

1. Determine which vector (variable) should leave the basis (become non-basic) in order to maintain a basic *feasible* solution.
2. Perform Gauss-Jordan reduction on the augmented matrix portion of the tableau to obtain a new tableau with a new augmented matrix whose equations are those for point $E$.

Let us first determine which vector should leave the basis if $A_1$ enters. We use the elements of column $A_1$ to help us decide. We note from the ideas of the previous section* that the $(+2)$ in the $a_{11}$ spot means that $x_3$ will decrease 2 units for each unit that $x_1$ increases. Since $x_3$ is presently 6, $x_1$ could increase $6/2 = 3$ units before $x_3$ reaches 0. Similarly, the $(+1)$ in the $a_{21}$ position indicates that $x_4$ will *decrease* 1 unit for each unit that $x_1$ increases. Thus $x_1$ could increase $4/1 = 4$ units before $x_4$ reaches 0. Since the minimum of $\{3, 4\} = \min\{6/2, 4/1\} = 3$, we must replace $A_3$ with $A_1$

* It may be helpful to write down the entire equation with all of the non-basic variables transposed to the opposite side of the equation.

in order to remain feasible and prevent our going to point $D$ rather than $E$. Our general criterion will be derived later.

Having decided to let $A_1$ replace $A_3$ in the basis we merely use the Gauss-Jordan operations to (1) replace $x_3$ by $x_1$ as the basic variable in row 1 and (2) to eliminate $x_1$ from the set of non-basic variables in row 2, replacing it with the new non-basic $x_3$. We perform the first operation by designating the element in the $A_1$ column and the $A_3$ row as the **pivot element** and so indicate by circling it as shown in Tableau 6.8.8.

**Tableau 6.8.8**

| | | $C$ | 5 | 3 | 0 | 0 |
|---|---|---|---|---|---|---|
| $B$ | $C_B^T$ | $R$ | $A_1$ | $A_2$ | $A_3$ | $A_4$ |
| $A_3$ | 0 | 6 | (2) | 1 | 1 | 0 |
| $A_4$ | 0 | 4 | 1 | 1 | 0 | 1 |
| $c_j - z_j$ | | 0 | 5 | 3 | 0 | 0 |

We now obtain the first row of the new tableau by dividing each element of the old first row by the pivot element, 2, and obtain Tableau 6.8.9.

**Tableau 6.8.9**

| | | | | | | |
|---|---|---|---|---|---|---|
| | | | | | | |
| | | 3 | 1 | 1/2 | 1/2 | 0 |
| | | | | | | |

We then eliminate $x_1$ from row 2 by letting $R_2 \leftarrow -R_1 + R_2$ to obtain Tableau 6.8.10.

**Tableau 6.8.10**

| | | $C$ | 5 | 3 | 0 | 0 |
|---|---|---|---|---|---|---|
| | | | | | | |
| | | 3 | 1 | 1/2 | 1/2 | 0 |
| | | 1 | 0 | 1/2 | $-1/2$ | 1 |
| | | | | | | |

We now label Tableau 6.8.10 with the appropriate notation for the new extreme point. The new basis $B = [A_1, A_4]$. Thus $C_B$, the matrix of cost coefficients, is $[5, 0] = [c_1, c_4]$. In the old $R$ column we have $X_B$, the matrix or coordinates of $R$ in terms of the new basis $B = [A_1, A_4]$. We verify that

$$R = \begin{bmatrix} 6 \\ 4 \end{bmatrix} = 3\begin{bmatrix} 2 \\ 1 \end{bmatrix} + 1\begin{bmatrix} 0 \\ 1 \end{bmatrix} = 3A_1 + 1A_4 = \begin{bmatrix} 2 & 0 \\ 1 & 1 \end{bmatrix}\begin{bmatrix} 3 \\ 1 \end{bmatrix} = BX_B.$$

How should we label the columns which replaced the columns of $A$? Certainly the $A_i$ are no longer there. Suppose we let the matrix

$$\begin{bmatrix} 1 & 1/2 & 1/2 & 1 \\ 0 & 1/2 & -1/2 & 1 \end{bmatrix} = Y = [Y_1, Y_2, Y_3, Y_4],$$

and label the columns accordingly. Then, since

$$BY_1 = \begin{bmatrix} 2 & 0 \\ 1 & 1 \end{bmatrix}\begin{bmatrix} 1 \\ 0 \end{bmatrix} = \begin{bmatrix} 2 \\ 1 \end{bmatrix} = A_1,$$

$$BY_2 = \begin{bmatrix} 2 & 0 \\ 1 & 1 \end{bmatrix}\begin{bmatrix} 1/2 \\ 1/2 \end{bmatrix} = \begin{bmatrix} 1 \\ 1 \end{bmatrix} = A_2,$$

$$BY_3 = \begin{bmatrix} 2 & 0 \\ 1 & 1 \end{bmatrix}\begin{bmatrix} 1/2 \\ -1/2 \end{bmatrix} = \begin{bmatrix} 1 \\ 0 \end{bmatrix} = A_3,$$

and

$$BY_4 = \begin{bmatrix} 2 & 0 \\ 1 & 1 \end{bmatrix}\begin{bmatrix} 0 \\ 1 \end{bmatrix} = \begin{bmatrix} 0 \\ 1 \end{bmatrix} = A_4$$

we see that

$$\begin{aligned} BY = B[Y_1, Y_2, Y_3, Y_4] &= [BY_1, BY_2, BY_3, BY_4] \\ &= [A_1, A_2, A_3, A_4] = A. \end{aligned}$$

We then have the partial Tableau 6.8.11

**Tableau 6.8.11**

| | | $C$ | 5 | 3 | 0 | 0 |
|---|---|---|---|---|---|---|
| $B$ | $C_B^T$ | $X_B$ | $Y_1$ | $Y_2$ | $Y_3$ | $Y_4$ |
| $A_1$ | 5 | 3 | 1 | 1/2 | 1/2 | 0 |
| $A_4$ | 0 | 1 | 0 | 1/2 | −1/2 | 1 |
| | | | | | | |

which we see is related to the original tableau by the relation

$$B[X_B \mid Y] - [BX_B \mid BY] = [R \mid A].$$

We fill in the last row to find the present $z$ value at point $E$ and to see if we are optimal by the same method as before, except that we replace $A_j$ by $Y_j$ in our computation of $z_j$. Now,

$$z = C_B X_B = [5, 0]\begin{bmatrix} 3 \\ 1 \end{bmatrix} = 15,$$

$$c_1 - z_1 = c_1 - C_B Y_1 = 5 - [5, 0]\begin{bmatrix} 1 \\ 0 \end{bmatrix} = 0,$$

$$c_2 - z_2 = c_2 - C_B Y_2 = 3 - [5, 0]\begin{bmatrix} 1/2 \\ 1/2 \end{bmatrix} = 1/2,$$

$$c_3 - z_3 = c_3 - C_B Y_3 = 0 - [5, 0]\begin{bmatrix} 1/2 \\ -1/2 \end{bmatrix} = -5/2,$$

and $$c_4 - z_4 = c_4 - C_B Y_4 = 0 - [5, 0]\begin{bmatrix}0\\1\end{bmatrix} = 0.$$

We thus have the completed Tableau 6.8.12.

**Tableau 6.8.12**

| | | C | 5 | 3 | 0 | 0 |
|---|---|---|---|---|---|---|
| $B$ | $C_B^T$ | $X_B$ | $Y_1$ | $Y_2$ | $Y_3$ | $Y_4$ |
| $A_1$ | 5 | 3 | 1 | 1/2 | 1/2 | 0 |
| $A_4$ | 0 | 1 | 0 | 1/2 | −1/2 | 1 |
| $c_j - z_j$ | | 15 | 0 | 1/2 | −5/2 | 0 |

You should now compare the equations represented by the body of the tableau, that is, by $[X_B \mid Y]$, with those equations which we found to represent point $E$. They are exactly the same. We have merely used the Gauss-Jordan reduction process to move us from $O$ to $E$.

We now examine the last row to see if the present $BFS$ is optimal. Since there is a positive element, namely (1/2) in the $c_2 - z_2$ position, we know that we can increase $z$ by increasing $x_2$ or by bringing $A_2$ into the basis. Thus $Y_2$ is the new pivot column. The signs of both $y_{12} = 1/2$ and $y_{22} = 1/2$ are positive, indicating that $x_1$ and $x_4$ will decrease as $x_2$ increases and both at the same rate.

$$\frac{x_{B1}}{y_{12}} = \frac{x_1}{y_{12}} = \frac{3}{1/2} = 6$$

represents the maximum allowable increase in $x_2$ that can occur if $x_1$ is to remain feasible. On the other hand

$$\frac{x_{B2}}{y_{22}} = \frac{x_4}{y_{22}} = \frac{1}{1/2} = 2$$

represents the maximum allowable increase in $x_2$ that can occur if $x_4$ is to remain feasible. Since the minimum of

$$\left\{\frac{3}{1/2}, \frac{1}{1/2}\right\} = \min\left\{\frac{x_{B1}}{y_{12}}, \frac{x_{B2}}{y_{22}}\right\} = \min\{6, 2\} = 2,$$

we see that $x_4$ will reach 0 first when $x_2$ reaches 2. Thus $A_4$ must be removed from the basis to preserve feasibility and keep us from moving to point $B$, rather than $C$. The pivot element is thus $y_{22} = 1/2$ which we circle in Tableau 6.8.13.

**Tableau 6.8.13**

| | | C | 5 | 3 | 0 | 0 |
|---|---|---|---|---|---|---|
| $B$ | $C_B^T$ | $X_B$ | $Y_1$ | $Y_2$ | $Y_3$ | $Y_4$ |
| $A_1$ | 5 | 3 | 1 | 1/2 | 1/2 | 0 |
| $A_4$ | 0 | 1 | 0 | (1/2) | −1/2 | 1 |
| $c_j - z_j$ | | 15 | 0 | 1/2 | −5/2 | 0 |

The new basis will be $B = [A_1, A_2]$. In order to allay some confusion we will not label the various bases as $B^{(1)}$, $B^{(2)}$, and so forth. Similarly, we will not relabel the $X_B$ or $Y$ columns after the first iteration.

We now use the pivot element to make $x_2$ a basic variable in row 2 to create a 1 in the pivot position by multiplying the entire row by 2. We then use this element to eliminate $x_2$ in the first row and have the partial Tableau 6.8.14.

**Tableau 6.8.14**

| | | $C$ | 5 | 3 | 0 | 0 |
|---|---|---|---|---|---|---|
| $B$ | $C_B^T$ | $X_B$ | $Y_1$ | $Y_2$ | $Y_3$ | $Y_4$ |
| $A_1$ | 5 | 2 | 1 | 0 | 1 | $-1$ |
| $A_2$ | 3 | 2 | 0 | 1 | $-1$ | 2 |
| $c_j - z_j$ | | | | | | |

We now proceed as before to fill in the $c_j - z_j$ row, observing that

$$z = C_B X_B = 10 + 6 = 16,$$

$$c_1 - z_1 = c_1 - C_B Y_1 = 5 - [5, 3]\begin{bmatrix}1\\0\end{bmatrix} = 0,$$

$$c_2 - z_2 = c_2 - C_B Y_2 = 3 - [5, 3]\begin{bmatrix}0\\1\end{bmatrix} = 0,$$

$$c_3 - z_3 = c_3 - C_B Y_3 = 0 - [5, 3]\begin{bmatrix}1\\-1\end{bmatrix} = -2,$$

and

$$c_4 - z_4 = c_4 - C_B Y_4 = 0 - [5, 3]\begin{bmatrix}-1\\2\end{bmatrix} = -1.$$

The completed tableau representing the situation at point $C$ is shown in Tableau 6.8.15.

**Tableau 6.8.15**

| | | $C$ | 5 | 3 | 0 | 0 |
|---|---|---|---|---|---|---|
| $B$ | $C_B^T$ | $X_B$ | $Y_1$ | $Y_2$ | $Y_3$ | $Y_4$ |
| $A_1$ | 5 | 2 | 1 | 0 | 1 | $-1$ |
| $A_2$ | 3 | 2 | 0 | 1 | $-1$ | 2 |
| $c_j - z_j$ | | 16 | 0 | 0 | $-2$ | $-1$ |

The last row indicates that

$$z = 16 + 0x_1 + 0x_2 - 2x_3 - 1x_4.$$

An increase of either of the non-basic variables would lead to a decrease in $z$. Thus we know that we can stop the algorithm and that we have the optimal $BFS$. At this point $X_B = \begin{bmatrix} 2 \\ 2 \end{bmatrix} = \begin{bmatrix} x_1 \\ x_2 \end{bmatrix}$, so the optimal $BFS$ consists of

$$X = \begin{bmatrix} 2 \\ 2 \\ 0 \\ 0 \end{bmatrix},$$

with the optimum value of $z$ being 16. You should now compare equations in the last tableau, obtained by the Gauss-Jordan procedure, with the appropriate set of equations derived for point $C$ in the previous section.

The next section contains a recapitulation of the simplex algorithm as developed so far, and another example. Before proceeding, however, it would be instructive to recall from Chapter 2 that if a matrix $[B \mid I_m]$ is row-reduced to $[I_m \mid E]$, say, then $E = B^{-1}$. We note that we have used Gauss-Jordan reduction to replace, at each stage, the columns $A_i$ of $B$ by identity columns. Thus, the columns that were originally identity columns should now contain $B^{-1}$. In Tableau 6.8.7, $B = [A_3, A_4] = I_2$, and $B^{-1} = I_2$ is contained in columns $A_3$, $A_4$. In Tableau 6.8.12, $B = [A_1, A_4] = \begin{bmatrix} 2 & 0 \\ 1 & 1 \end{bmatrix}$, and $B^{-1} = \begin{bmatrix} 1/2 & 0 \\ -1/2 & 1 \end{bmatrix}$ appears in columns $Y_3$, $Y_4$, the original identity columns. In Tableau 6.8.15, $B = [A_1, A_2] = \begin{bmatrix} 2 & 1 \\ 1 & 1 \end{bmatrix}$, and $B^{-1} = \begin{bmatrix} 1 & -1 \\ -1 & 2 \end{bmatrix}$ again appears in columns $Y_3$, $Y_4$. In general, $B^{-1}$ will appear in the columns that contain the original identity matrix.

## 6.8 Exercises

1. Consider the LP maximization problem of Example 6.4.2 with

$$A = \begin{bmatrix} 2 & 3 & 1 & 0 \\ 3 & 2 & 0 & 1 \end{bmatrix}, \quad R = \begin{bmatrix} 6 \\ 6 \end{bmatrix}, \quad \text{and} \quad C = [4, 3, 0, 0].$$

a. Fill in the blanks in the first tableau.

| | | $C$ | | | | |
|---|---|---|---|---|---|---|
| $B$ | $C_B^T$ | $R$ | $A_1$ | $A_2$ | $A_3$ | $A_4$ |
| $A_3$<br>$A_4$ | | | | | | |
| $c_j - z_j$ | | | 4 | | | |

b. Is the present basis optimal? ________, $c_1 - z_1 = 4$ and $c_2 - z_2 =$ ________ are each ________.

c. Which vector should enter $B$ in accordance with our arbitrary criterion? ________.

d. Let $A_1$ enter the basis. In order to preserve _______, _______ must leave the basis since $\min\{6/2, 6/3\} = 2$. Thus _______ is the pivot element.

e. Construct the new tableau.

| | | $C$ | 4 | 3 | 0 | 0 |
|---|---|---|---|---|---|---|
| $B$ | $C_B^T$ | $X_B$ | $Y_1$ | $Y_2$ | $Y_3$ | $Y_4$ |
| $A_3$ | | | | | | |
| | | | | | | |
| $c_j - z_j$ | | | | | | |

Compare the constants obtained with those in Exercise 1 b. of Section 6.7. In this tableau, $x_{B1} =$ _______, $y_{12} =$ _______, and $y_{24} =$ _______.

f. Since $c_2 - z_2 =$ _______ $> 0$, we may improve on the present value of $f$ by bringing _______ into the basis. Since $\min\{$_______, _______$\} = 6/5$, $A_3$ must leave the basis in order to preserve feasibility. If we were to remove $A_1$ instead, the new value of _______ would be negative. If $A_2$ replaces $A_3$ the pivot element is _______.

g. Construct the new tableau.

| | | $C$ | 4 | 3 | 0 | 0 |
|---|---|---|---|---|---|---|
| $B$ | $C_B^T$ | | | | | |
| | | | | | | |
| $A_1$ | | | | | | |
| $c_j - z_j$ | | | | | | |

Compare the constants obtained with those in Exercise 1 c. of Section 6.7. In this tableau $c_{B2} =$ _______, $y_{11} =$ _______, and $y_{23} =$ _______.

h. Since $c_j - z_j \leqslant 0$ for all $j$, the optimal solution is $X^T = [6/5, 6/5, 0, 0]$ with $f = 42/5$.

2. Consider the last two columns of the tableaus obtained in Exercise 1.

a. In Tableau 1, $B = I_2$ and $B^{-1}$ is contained in the last two columns.

b. In Tableau 2, $B = [A_3, A_1] = \begin{bmatrix} 1 & 2 \\ 0 & 3 \end{bmatrix}$. Verify that $B^{-1}$ is contained in columns $[Y_3, Y_4]$.

c. In Tableau 3, $B = [$_______, _______$] = \begin{bmatrix} __ & __ \\ __ & __ \end{bmatrix}$. $B^{-1} = \begin{bmatrix} \_\_ & \_\_ \\ \_\_ & \_\_ \end{bmatrix}$ is contained in $[Y_3, Y_4]$.

d. For any basis matrix $B$, $BY_j = A_j$. Suppose that you only had the information in Tableau 3 and no longer knew the original tableau. How could you find that matrix $A$?

3. Solve the LP maximization problem of Exercise 6.3.5. with $A = \begin{bmatrix} 1 & 2 & 1 & 0 \\ 2 & 1 & 0 & 1 \end{bmatrix}$,
$R = \begin{bmatrix} 3 \\ 3 \end{bmatrix}$, and $C = [3, 1, 0, 0]$. (Also, see Exercise 6.6.5 and Exercise 6.7.4.)

### Suggested Reading

S. I. Gass. *Linear Programming.* 3d. ed. New York: McGraw-Hill, 1969.

G. Hadley. *Linear Programming.* Reading, Mass.: Addison-Wesley, 1962.

T. H. Naylor, E. T. Byrne, and J. M. Vernon. *Introduction to Linear Programming: Methods and Cases.* Belmont, Ca. Wadsworth, 1971.

### New Terms



## 6.9 The Simplex Algorithm and an Example

We have completely developed the computational aspects of the tableau form of the simplex algorithm in the case where each of the equations contains a slack variable. The remaining cases will be discussed in the following section. We recall that we start with a tableau in which we have an augmented matrix $[R \mid A] = [R, A_1, A_2, \ldots, A_n]$ and then proceed through a sequence of tableaus each of which has an augmented matrix of the form $[X_B \mid Y] = X_B, Y_1, Y_2, \ldots, Y_n]$, with associated basis matrix $B$, where, since $BY_j = A_j$, we have $Y_j = B^{-1}A_j$ for each $j$. In general, $BY = B[Y_1, Y_2, \ldots, Y_n] = [BY_1, \ldots, BY_n] = [A_1, \ldots, A_n] = A$, and $BX_B = R$. If we start at the origin with $B = I_m$, then

$$I_m Y = Y = A \quad \text{and} \quad I_m X_B = X_B = R,$$

so that in the first tableau, even, it is valid to refer to $a_{12}$ as $y_{12}$ and $r_1$ as $x_{B1}$.

The quantity $z_j$ which contributes to the $c_j - z_j$ row is defined in general by the equation

$$z_j = C_B Y_j = C_B B^{-1} A_j.$$

Thus we could form a matrix

$$Z = [z_1, z_2, \ldots, z_n]$$

such that the last row of the tableau is merely $C - Z = C - C_B Y$. For obvious reasons, this matrix is sometimes called the **revised-cost matrix**.

### The Simplex Algorithm

**Step 1** Find a *BFS*. We let the slack variables be basic and start at the origin with $B = I_m$.

**Step 2** Is the BFS optimal? We create the $C - Z$ row and examine it. For a max problem, a *BFS* will be optimal if $c_j - z_j \leqslant 0$ for all $j$. In the case of a min problem, a given *BFS* will be optimal if $c_j - z_j \geqslant 0$ for all $j$. If the *BFS* is optimal we stop. If the *BFS* is not optimal, then we proceed to Step 3.

**Step 3** Toward which adjacent extreme point should we proceed to obtain the greatest rate of increase in $z$? We choose a vector $A_k$ to enter the basis $B$ by examining those elements of the $c_j - z_j$ row which have the correct sign for the type of problem (max or min). For a max problem we choose $A_k$, such that $c_k - z_k = \text{Max}_j\{c_j - z_j \mid c_j - z_j > 0\}$. For a min problem we choose $A_k$, such that $c_k - z_k = \text{Min}_j\{c_j - z_j \mid c_j - z_j < 0\}$. In this case $c_j - z_j$ will be a set of negative numbers, and we pick the most negative. In the event of a tie, we arbitrarily pick the tied column $Y_k$ with lowest subscript, and let $A_k$ enter the basis.

**Step 4** Where should we stop in order to remain feasible? We choose a vector $B_s$ to leave the basis in such a way as to obtain a new *BFS*. By dividing the present values of the old basic variables by their rates of change, $y_{ik}$, that will occur if $A_k$ enters, we could obtain a set

$$S_1 = \left\{\frac{x_{B1}}{y_{1k}}, \frac{x_{B2}}{y_{2k}}, \ldots, \frac{x_{Bm}}{y_{mk}}\right\}.$$

Since only those basic variables $x_{Bi}$ for which $y_{ik}$ is positive will decrease in value if $A_k$ enters, we consider only

$$S = \left\{\frac{x_{Bi}}{y_{ik}} \mid y_{ik} > 0, i = 1, 2, \ldots, m\right\}.$$

We choose $s$ such that $x_{Bs}/y_{sk}$ is the smallest non-negative number in $S$ and let $B_s$ leave the basis, whichever one of the columns of $A$ it was. In case of a tie, we arbitrarily choose the lower-numbered basic column to leave. If it should happen that any other column were chosen to leave, then the next tableau would represent a basic, *infeasible* solution. (Which variables would have negative values?)

**Step 5** Let the element in row $s$ and column $k$ of $Y$, namely $y_{sk}$, be the pivot element. Divide each element of row $s$ by $y_{sk}$ to make $x_k$ basic. Use Gauss-Jordan elimination to eliminate $x_k$ from all rows except row $s$. Label the new tableau, replacing $B_s$, whatever $A$ column it was, by $A_k$. Replace the old value of $c_{Bs}$ with $c_k$. Return to Step 2.

Let us now use the algorithm to solve the problem of Example 6.5.1 which we repeat here for convenience.

---

Example 6.9.1

Find $X$ such that for $A = \begin{bmatrix} 2 & 1 & 1 & 0 \\ 1 & 1 & 0 & 1 \end{bmatrix}$, $R = \begin{bmatrix} 8 \\ 6 \end{bmatrix}$, $C = [3, 2, 0, 0]$,

1. $X \geqslant 0$
2. $AX = R$
3. $z = CX$ is maximized.

---

The extreme points are shown in Figure 6.9.1

**Figure 6.9.1**

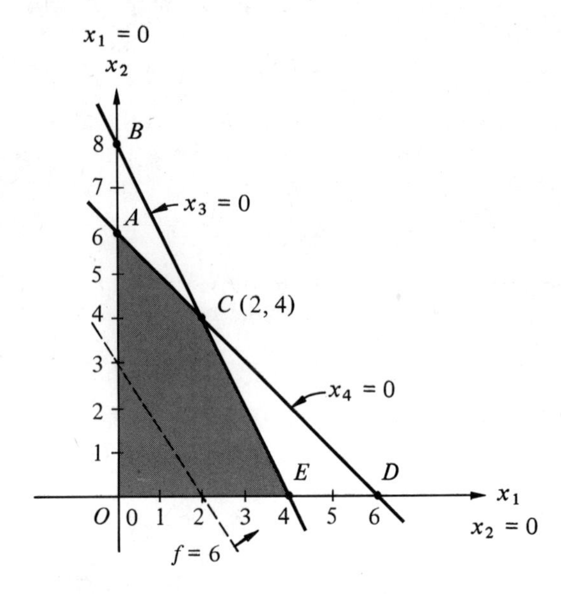


**Step 1** The partial first tableau is shown in Tableau 6.9.1. We have completed Step 1. The *BFS* is $X^T = [0, 0, 8, 6]$.

**Tableau 6.9.1**

| | | $C$ | 3 | 2 | 0 | 0 |
|---|---|---|---|---|---|---|
| $B$ | $C_B^T$ | $R$ | $A_1$ | $A_2$ | $A_3$ | $A_4$ |
| $A_3$ | 0 | 8 | 2 | 1 | 1 | 0 |
| $A_4$ | 0 | 6 | 1 | 1 | 0 | 1 |
| $C - Z$ | | | | | | |

**Step 2** Are we optimal? We create the $C - Z$ row by finding

$$z = C_B X_B = C_B R = 0(8) + 0(6) = 0,$$

$$c_1 - z_1 = c_1 - C_B Y_1 = c_1 - C_B A_1 = 3 - 0 = 3,$$

$$c_2 - z_2 = c_2 - C_B Y_2 = c_2 - C_B A_2 = 2 - 0 = 2,$$

$$c_3 - z_3 = c_3 - C_B Y_3 = c_3 - C_B A_3 = 0 - 0 = 0,$$

and $$c_4 - z_4 = c_4 - C_B Y_4 = c_4 - C_B A_4 = 0 - 0 = 0.$$

The columns of the tableau have been arranged to make these calculations as simple as possible. Since $c_1 - z_1 = 3$ and $c_2 - z_2 = 2$ are each greater than 0, we are not optimal. The completed first tableau, Tableau 6.9.2, is shown below. (Disregard the arrow and the circle for the moment.)

**Tableau 6.9.2**

| | | $C$ | 3 | 2 | 0 | 0 |
|---|---|---|---|---|---|---|
| $B$ | $C_B^T$ | $R$ | $A_1$ | $A_2$ | $A_3$ | $A_4$ |
| $A_3$ | 0 | 8 | (2) | 1 | 1 | 0 |
| $A_4$ | 0 | 6 | 1 | 1 | 0 | 1 |
| $C - Z$ | | 0 | 3 | 2 | 0 | 0 |
| | | | ↑ | | | |

**Step 3** Which vector should enter? We consider $\{c_1 - z_1, c_2 - z_2\} = \{3, 2\}$. Since $\max\{3, 2\} = 3$, we let $A_1$ enter $B$, and so indicate by putting an arrow under the column in Tableau 6.9.2 above.

**Step 4** Which vector must leave? We consider

$$\min\{8/2, 6/1\} = \min\left\{\frac{x_{B1}}{y_{11}}, \frac{x_{B2}}{y_{21}}\right\} = 4.$$

Thus $s = 1$ and $B_1 = A_3$ must leave the basis.

**Step 5** $y_{11} = 2$ is the pivot element which we circle in Tableau 6.9.2. We create a partial tableau as shown in Tableau 6.9.3.

**Tableau 6.9.3**

| | | $C$ | 3 | 2 | 0 | 0 |
|---|---|---|---|---|---|---|
| $B$ | $C_B^T$ | $X_B$ | $Y_1$ | $Y_2$ | $Y_3$ | $Y_4$ |
| $A_1$ | 3 | 4 | 1 | 1/2 | 1/2 | 0 |
| $A_4$ | 0 | 2 | 0 | 1/2 | −1/2 | 1 |
| $C - Z$ | | | | | | |

**Step 2** Are we optimal?

$$z = C_B X_B = 12,$$

$$c_1 - z_1 = c_1 - C_B Y_1 = 3 - 3 = 0,$$

$$c_2 - z_2 = c_2 - C_B Y_2 = 2 - 3/2 = 1/2,$$

$$c_3 - z_3 = c_3 - C_B Y_3 = 0 - 3/2 = -3/2,$$

and $$c_4 - z_4 = c_4 - C_B Y_4 = 0 - 0 = 0.$$

Since $c_2 - z_2 = 1/2 > 0$, we are not optimal. The completed tableau is Tableau 6.9.4.

**Tableau 6.9.4**

| | | $C$ | 3 | 2 | 0 | 0 |
|---|---|---|---|---|---|---|
| $B$ | $C_B^T$ | $X_B$ | $Y_1$ | $Y_1$ | $Y_3$ | $Y_4$ |
| $A_1$ | 3 | 4 | 1 | 1/2 | 1/2 | 0 |
| $A_4$ | 0 | 2 | 0 | (1/2) | −1/2 | 1 |
| $C - Z$ | | 12 | 0 | 1/2 | −3/2 | 0 |
| | | | | ↑ | | |

**Step 3** Which vector should enter? Since $c_2 - z_2$ is the only positive element in the $C - Z$ row, $A_2$ must enter.

**Step 4** Which vector must leave? We consider

$$\min\left\{\frac{x_{B1}}{y_{12}}, \frac{x_{B2}}{y_{22}}\right\} = \min\left\{\frac{4}{1/2}, \frac{2}{1/2}\right\} = \min\{8, 4\} = 4.$$

Thus $B_2 = A_4$ must leave.

**Step 5 and Step 2** $y_{22} = 1/2$ is the new pivot element. The complete new tableau is shown as Tableau 6.9.5.

**Tableau 6.9.5**

| | | $C$ | 3 | 2 | 0 | 0 |
|---|---|---|---|---|---|---|
| $B$ | $C_B^T$ | $X_B$ | $Y_1$ | $Y_2$ | $Y_3$ | $Y_4$ |
| $A_1$ | 3 | 2 | 1 | 0 | 1 | −1 |
| $A_2$ | 2 | 4 | 0 | 1 | −1 | 2 |
| $C - Z$ | | 14 | 0 | 0 | −1 | −1 |

Since $c_j - z_j \leqslant 0$ for each $j$, the tableau is optimal. $X^T = [2, 4, 0, 0]$, with $z = 14$, is the optimal $BFS$ shown at point $C$ in the graph in Figure 6.9.1. ∎

## 6.9 Exercises

1. We begin the simplex algorithm in Tableau 1 with $[R \mid A]$ and proceed through a sequence of tableaus of the form [________ | ________] where $BX_B =$ ________ and $BY_j =$ ________. Since $Y_j = B^{-1}A_j$ and $z_j = C_B Y_j$, we see that $(c_j - z_j)$ could be computed as $c_j - C_B Y_j = c_j - C_B B^{-1} A_j$. Verify this formula by using it to compute $c_3 - z_3$ in Tableau 6.9.4.

2. Verify that the information in the $C - Z$ row of Tableau 6.9.5 (with the exception of $z$) could have been obtained by performing Gauss-Jordan reduction on the $C - Z$ row of Tableau 6.9.4 just as you did on row 1; that is, you wish to create a 0 in the $c_2 - z_2$ position.

3. a. Solve the LP maximization problem with $A = \begin{bmatrix} 1 & 3 & 1 & 0 \\ 3 & 1 & 0 & 1 \end{bmatrix}$, $R = \begin{bmatrix} 4 \\ 4 \end{bmatrix}$, and $C = [1, 2, 0, 0]$. See Exercise 6.7.3.
   b. Verify in the final tableau that $A_3 = y_{13}B_1 + y_{23}B_2 = (3/8)A_2 - (1/8)A_1$.

4. Consider Tableaus 6.9.2 and 6.9.4 in this section.
   a. In Tableau 6.9.2 we see that $A_1 = \begin{bmatrix} 2 \\ 1 \end{bmatrix}$, $A_2 = \begin{bmatrix} 1 \\ 1 \end{bmatrix}$, and that $B = I_2$ is the basis matrix. In Tableau 6.9.4 the new basis matrix $B = [A_1, A_4] = \begin{bmatrix} 2 & 0 \\ 1 & 1 \end{bmatrix}$. Verify that the $Y$ matrix $\begin{bmatrix} 1 & 1/2 & 1/2 & 0 \\ 0 & 1/2 & -1/2 & 1 \end{bmatrix}$ is nothing more than the matrix of

coordinates of the original vectors $A_j$ with respect to the new basis. For example, $Y_2 = \begin{bmatrix} 1/2 \\ 1/2 \end{bmatrix}$ is the coordinate matrix of $A_2$ since

$$A_2 = \begin{bmatrix} 1 \\ 1 \end{bmatrix} = 1/2\begin{bmatrix} 2 \\ 1 \end{bmatrix} + 1/2\begin{bmatrix} 0 \\ 1 \end{bmatrix} = 1/2A_1 + 1/2A_4$$

$$= [A_1, A_4]\begin{bmatrix} 1/2 \\ 1/2 \end{bmatrix} = BY_2.$$

b. Verify that $Y_j = B^{-1}A_j$ for each $j$ in Tableau 6.9.4.

c. Verify that $C - Z = C - C_BY = C - C_BB^{-1}A$ in Tableau 6.9.4.

d. Show that $c_j - z_j = 0$ if $A_j$ is in $B$.

5. a. How would the criterion for the choice of the vector to enter the basis change in a minimization problem?

   b. How would the criterion for the choice of the vector to leave the basis change in a minimization problem?

6. In general, $C - Z = C - C_B$________.

7. a. Solve the LP maximization production problem of Exercise 6.2.9, with

$$A = \begin{bmatrix} 1 & 2 & 1 & 0 \\ 3 & 1 & 0 & 1 \end{bmatrix}, \quad R = \begin{bmatrix} 480 \\ 480 \end{bmatrix}, \quad \text{and} \quad C = [5, 4, 0, 0].$$

   b. Verify that the inverse of the final basis is contained in the last two columns of the final tableau. Note that $B = [A_2, A_1]$.

8. Show graphically for one of your solved problems that max $f = -(\min(-f))$. Since this is valid in general, any minimization problem may be solved by converting it to a maximization problem by changing the signs of the criterion coefficients, solving the problem as a maximization problem, and changing the sign of the optimum criterion value. The values of $X$ which solve the different problems are the same and need not be changed.

9. a. Use the simplex algorithm to solve the LP problem

---

Find $X$ such that $X \geqslant 0$,

$$x_1 + x_2 \leqslant 4$$
$$-x_1 + x_2 \leqslant 1$$
$$x_1 + 2x_2 \leqslant 5,$$

and $z = 2x_1 + 3x_2$ is maximized.

---

   b. Solve the problem graphically and note that fewer iterations would have been required if we had let $A_1$ rather than $A_2$ enter the basis at Step 1.

   c. The last basis matrix is $B = [B_1, B_2, B_3] = [$______, ______, ______$]$.

Verify that $B^{-1} = [Y_3, Y_4, Y_5] = \begin{bmatrix} 3 & 1 & -2 \\ -1 & 0 & 1 \\ 2 & 0 & -1 \end{bmatrix}$ from the last three columns of the last tableau.

d. Since $Y_3 = \begin{bmatrix} 3 \\ -1 \\ 2 \end{bmatrix}$, we know that

$$A_3 = \underline{\qquad\qquad} A_4 + \underline{\qquad\qquad} A_2 + \underline{\qquad\qquad} A_1.$$

**Suggested Reading**

S. I. Gass. *Linear Programming.* 3d. ed. New York: McGraw-Hill, 1969.

G. Hadley. *Linear Programming.* Reading, Mass.: Addison-Wesley, 1962.

H. M. Wagner. *Principles of Operations Research.* 2d. ed. Englewood Cliffs, N.J.: Prentice-Hall, 1975.

**New Terms**

revised-cost matrix, 252 simplex algorithm, 252

## 6.10 Artificial Variables and Alternate Optimum Solutions

The ease of finding a starting *BFS* in the case where each constraint has a slack variable is evident from the previous section. We merely let the slack variables be the basic variables. Since there are $m$ such variables and each one corresponds to a column of the identity matrix, we have $B = I_m$ as the starting basis. The question arises as to what should be done in the case where some of the constraints are of the $\geqslant$ or $=$ variety. For example, suppose that we have the problem

Find $x_1, x_2$ such that

1. $x_i \geqslant 0 \qquad i = 1, 2$
2. $x_1 + 2x_2 \geqslant 3$
   $-x_1 + 3x_2 \geqslant 4$
3. $z = x_1 + 2x_2$ is minimized.

and that we subtract surplus variables $x_3$, $x_4$ to form the set of equations

$$\begin{aligned} x_1 + 2x_2 - x_3 \qquad &= 3 \\ -x_1 + 3x_2 \qquad - x_4 &= 4. \end{aligned}$$

There is clearly no identity matrix to be chosen as an initial basis. In addition, if $x_3$ were made basic in the first equation, then $x_3 = -3$ would be infeasible as would $x_4 = -4$. In order to handle this situation we add an **artificial variable** to each equation. Thus the set of equations

$$\begin{aligned} x_1 + 2x_2 - x_3 \qquad + x_5 \qquad &= 3 \\ -x_1 + 3x_2 \qquad - x_4 \qquad + x_6 &= 4 \end{aligned}$$

provides a matrix

$$A = \begin{bmatrix} 1 & 2 & -1 & 0 & 1 & 0 \\ -1 & 3 & 0 & -1 & 0 & 1 \end{bmatrix}$$

with an $I_2$ sub-matrix, where $I_2 = [A_5, A_6]$, which may be used as an initial basis. If $I_2$ is chosen as the initial basis, then $x_5 = 3$ and $x_6 = 4$ are the initial basic variables. Then

$$\begin{aligned} x_1 + 2x_2 - x_3 \quad &= 0 \\ -x_1 + 3x_2 \quad - x_4 &= 0 \end{aligned}$$

and the original equations are clearly *not* satisfied. Thus, although we have a *BFS* for the artificial problem we have created, we do not have a *BFS* for the original problem. To achieve a *BFS* for the latter, we merely need to insure that the artificial variables $x_5$ and $x_6$ eventually have the value 0; that is, that they either become non-basic variables or have the value 0 even though basic. We can insure, in a minimization problem, that the variables will eventually leave the basis (if there exists a *BFS* to the original problem) by attaching an extraordinarily high cost coefficient to these variables. They will then naturally be chosen to leave the basis as soon as possible. After Charnes et al. (see Suggested Reading), we will use $M > 0$ to represent an arbitrarily large number to be used in such cases. For a maximization problem we would want a negative number whose absolute value is very large. We thus use $(-M)$ in that case. If it is impossible ultimately to remove one of the artificial vectors from the basis, then, if the corresponding basic variable is positive, it can be shown that there is no feasible solution to the original problem.

We shall now use this new technique to solve the equivalent problem

---

Minimize $z = x_1 + 2x_2 + Mx_5 + Mx_6$ such that

$$\begin{aligned} x_1 + 2x_2 - x_3 \quad + x_5 \quad &= 3 \\ -x_1 + 3x_2 \quad - x_4 \quad + x_6 &= 4 \end{aligned}$$

and $x_i \geq 0$.

---

with the provision that $x_5$ and $x_6$ must be 0 at the conclusion of the algorithm. The graph of the feasible region for the original problem is shown in Figure 6.10.1.

**Figure 6.10.1**

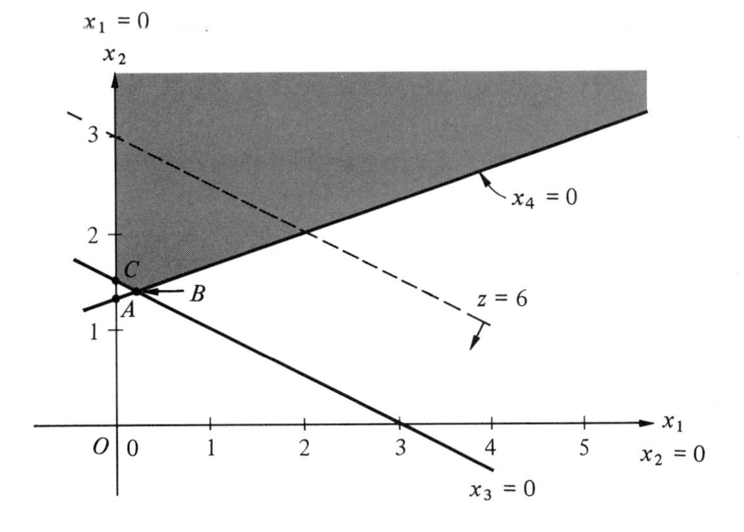

The completed first tableau is Tableau 6.10.1. For convenience, the $C - Z$ row has been split into two levels, the top row being the constants and the bottom row the $M$ part of the objective function.

**Tableau 6.10.1**

| | | $C$ | 1 | 2 | 0 | 0 | $M$ | $M$ |
|---|---|---|---|---|---|---|---|---|
| $B$ | $C_B^T$ | $R$ | $A_1$ | $A_2$ | $A_3$ | $A_4$ | $A_5$ | $A_6$ |
| $A_5$ | $M$ | 3 | 1 | 2 | $-1$ | 0 | 1 | 0 |
| $A_6$ | $M$ | 4 | $-1$ | (3) | 0 | $-1$ | 0 | 1 |
| $C - Z$ | | 0 | 1 | 2 | 0 | 0 | 0 | 0 |
| | | $7M$ | 0 | $-5M$ | $M$ | $M$ | 0 | 0 |
| | | | | ↑ | | | | |

**First Iteration**

**Step 1** $B = I_2$.

**2** $c_2 - z_2 = -5M + 2 < 0$; not optimal

**3** $A_2$ enters.

**4** $\text{Min}\{3/2, 4/3\} = 4/3$. $B_2 = A_6$ leaves.

**5** $y_{22} = 3$ is pivot. Use Gauss-Jordan reduction to obtain Tableau 6.10.2.

**Tableau 6.10.2**

| | | $C$ | 1 | 2 | 0 | 0 | $M$ | $M$ |
|---|---|---|---|---|---|---|---|---|
| $B$ | $C_B^T$ | $X_B$ | $Y_1$ | $Y_2$ | $Y_3$ | $Y_4$ | $Y_5$ | $Y_6$ |
| $A_5$ | $M$ | 1/3 | (5/3) | 0 | $-1$ | 2/3 | 1 | $-2/3$ |
| $A_2$ | 2 | 4/3 | $-1/3$ | 1 | 0 | $-1/3$ | 0 | 1/3 |
| $C - Z$ | | 8/3 | 5/3 | 0 | 0 | 2/3 | 0 | $-2/3$ |
| | | $M/3$ | $-5M/3$ | 0 | $M$ | $-2M/3$ | 0 | $5M/3$ |
| | | | ↑ | | | | | |

**Step 2** $c_1 - z_1 = -5M/3 + 5/3$ and $c_4 - z_4 = -2M/3 + 2/3$ are each negative; not optimal.

**3** $A_1$ enters.

**4** $B_1 = A_5$ leaves.

**5** $y_{11} = 5/3$ is pivot. Obtain Tableau 6.10.3.

**Tableau 6.10.3**

| | | $C$ | 1 | 2 | 0 | 0 | $M$ | $M$ |
|---|---|---|---|---|---|---|---|---|
| $B$ | $C_B^T$ | $X_B$ | $Y_1$ | $Y_2$ | $Y_3$ | $Y_4$ | $Y_5$ | $Y_6$ |
| $A_1$ | 1 | 1/5 | 1 | 0 | $-3/5$ | 2/5 | 3/5 | $-2/5$ |
| $A_2$ | 2 | 7/5 | 0 | 1 | $-1/5$ | $-1/5$ | 1/5 | 1/5 |
| $C - Z$ | | 3 | 0 | 0 | 1 | 0 | $-1$ | 0 |
| | | 0 | 0 | 0 | 0 | 0 | $M$ | $M$ |

**Step 2** $c_j - z_j \geqslant 0$ for all $j$. Optimal solution, since both $x_5$ and $x_6$ are 0.

For the original problem in equation form, the solution is $X = [1/5, 7/5, 0, 0]^T$ with $z = 3$. In the diagram (Figure 6.10.1) we have proceeded from point $O$ to point $A$ to point $B$. You may have noticed that $c_4 - z_4 = 0$ but that $A_4$ is not in the final basis. Thus the rate of change of $z$, if $A_4$ enters the basis, will be 0; and $z$ will not change.

From Figure 6.10.1, we see that if we are at $B$ and $A_4$ enters the basis, $A_1$ will leave the basis in order to remain feasible; we will move to point $C$ along a boundary which has the same slope as the objective function. The $BFS$ at point $C$ has an equal value of $z$ and is called an **alternate optimum** $BFS$. Such alternates will always be indicated by the presence of more than $m$ zeros in the $C - Z$ row of an optimal tableau.

In Tableau 6.10.4 we show the tableau that results if $A_4$ replaces $A_1$ in $B$. The pivot element was $y_{14} = 2/5$.

**Tableau 6.10.4**

| | | $C$ | 1 | 2 | 0 | 0 | $M$ | $M$ |
|---|---|---|---|---|---|---|---|---|
| $B$ | $C_B^T$ | $X_B$ | $Y_1$ | $Y_2$ | $Y_3$ | $Y_4$ | $Y_5$ | $Y_6$ |
| $A_4$ | 0 | 1/2 | 5/2 | 0 | $-3/2$ | 1 | 3/2 | $-1$ |
| $A_2$ | 2 | 3/2 | 1/2 | 1 | $-1/2$ | 0 | 1/2 | 0 |
| $C - Z$ | | 3 | 0 | 0 | 1 | 0 | $-1$ | 0 |
| | | 0 | 0 | 0 | 0 | 0 | $M$ | $M$ |

We thus have an alternate optimum solution $X = [0, 3/2, 0, 1/2]^T$ with $z = 3$. We repeat that in this exceptional case one of the constraint-defining equations represents a straight line with the same slope as that of the objective function. It is that phenomenon which enabled the alternate optimum solution to exist. Any point between $B$ and $C$ would also be an alternate optimum feasible solution, but it would not be a basic feasible solution since more than two variables would be positive.

## 6.10 Exercises

1. Use slack, surplus, and artificial variables where necessary to put the following constraints in proper form for the simplex algorithm. Number artificial variables last.

$$x_1 + x_2 \leqslant 3$$
$$-x_1 + 2x_2 \geqslant 4.$$

2. An ________ ________ is added to an equation arising from an ________ constraint or an inequality constraint of the ________ type in order to obtain an ________ ________ as a starting basis matrix.

3. Use the method of this section to solve the aircraft LP problem of Example 6.1.1. See Exercise 6.2.11 for a graphical solution. Number any artificial variables last.

For simplicity, divide each constant in the first constraint by 50 and each cost coefficient by 1000.

4. Illustrate the procedure used to find alternate optima in the solution of the LP problem with

$$A = \begin{bmatrix} 2 & 1 & 1 & 0 \\ 1 & 2 & 0 & 1 \end{bmatrix}, \quad R = \begin{bmatrix} 6 \\ 6 \end{bmatrix}, \quad \text{and} \quad C = [2, 1, 0, 0].$$

Draw a sketch to illustrate what occurs algebraically.

5. Use the method of this section to solve the LP problem you solved graphically in Exercise 6.2.5. Use your previously prepared sketch to follow the algorithm's progress toward the optimal *BFS*. For uniformity, number all slack/surplus variables before you number any artificial variables. Let $x_5$ denote the slack variable, and let $A_2$ enter the basis at the first step.

6. Solve the LP diet problem of Example 6.1.4. (See Exercise 6.2.10 for a graphical solution).

### Suggested Reading

A. Charnes, W. Cooper, and A. Henderson. *An Introduction to Linear Programming.* New York: John Wiley and Sons, 1953.

S. I. Gass. *Linear Programming.* 3d. ed. New York: McGraw-Hill, 1969.

G. Hadley. *Linear Programming.* Reading, Mass.: Addison-Wesley, 1962.

### New Terms

artificial variable, 258 alternate optimum BFS, 261

## 6.11 Why Does the Simplex Algorithm Work?

In this section we shall attempt to make plausible some of the underlying theory of the simplex algorithm. In equation form we shall discuss the problem:

Find $X$ such that

1. $X \geqslant 0$
2. $AX = R$ and
3. $f(X) = CX$ is maximized.

We assume that an optimal feasible solution exists and degeneracy does not occur. We stated in Section 6.2 that $z$ is a linear function and that each of the constraints is linear. We also showed an example of a programming problem (with linear constraints but a non-linear objective function) with the property that the unique optimum solution occurs at an interior point of the feasible region. Thus an algorithm which searched only extreme points of the feasible region would be doomed to failure. Such is not the case for an LP problem.

We shall first show that the feasible region $S$ of an LP problem is a so-called "convex set." We then illustrate by an example a method whereby any point $X$ of $S$ may be expressed as a special linear combination of the finite number of extreme points of $S$. Using this, we prove that $f$ must achieve its optimum value at one of the extreme points of $S$. Thus a method which examines only extreme points must succeed. The simplex algorithm provides a method of examining a subset of these extreme points, with each succeeding point having an improved value of $z$ (barring degeneracy). Since there are only a finite number of such points, the algorithm must terminate. We conclude by showing that if a $BFS$ is obtained with the property that for all $j$, $c_j - z_j \leqslant 0$, then the $BFS$ and its associated extreme point are optimum.

A subset $T$ of $n$-space is said to be a **convex subset of $n$-space** provided that if $X_1$ and $X_2$ are in $T$ and $c_1$ and $c_2$ are non-negative numbers such that $c_1 + c_2 = 1$, then the point $X = c_1X_1 + c_2X_2$ is also in $T$. $X$ is said to be a **convex combination** of $X_1$ and $X_2$. The set formed by taking all possible values for $c_1$ and $c_2$ is merely the straight line segment connecting $X_1$ and $X_2$. A particular example in 2-space for the case where $c_1 = c_2 = 1/2$ is shown in Figure 6.11.1.

**Figure 6.11.1**

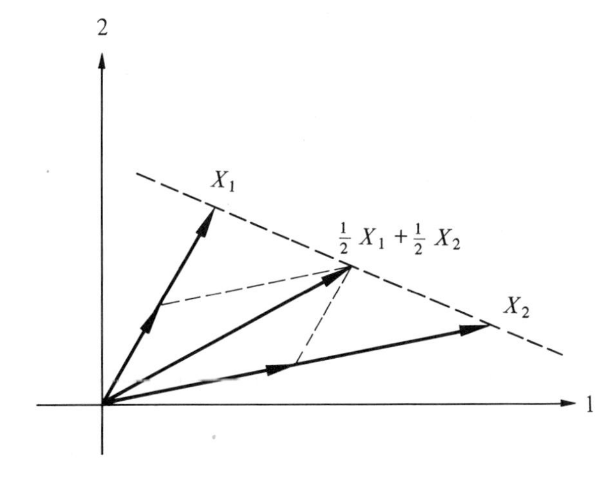


The term **extreme point of a convex set** $S$, which we have been using intuitively, can be precisely defined to be a point $X$ in $S$ that is not a convex combination of any *two* other points of $S$. Thus $X$ is not an interior point of any line segment in $S$. Examples of convex and non-convex subsets of 2-space are shown in Figure 6.11.2.

**THEOREM 6.11.1** *The feasible region $S$ of an LP problem is a convex subset of n-space.*

**Proof** $S = \{X \mid X \geqslant 0, AX = R\}$ is clearly a subset of $n$-space. Suppose $X_1$ and $X_2$ are in $S$. Then $X_1 \geqslant 0$, and $AX_1 = R$. Further, $X_2 \geqslant 0$; and

Figure 6.11.2

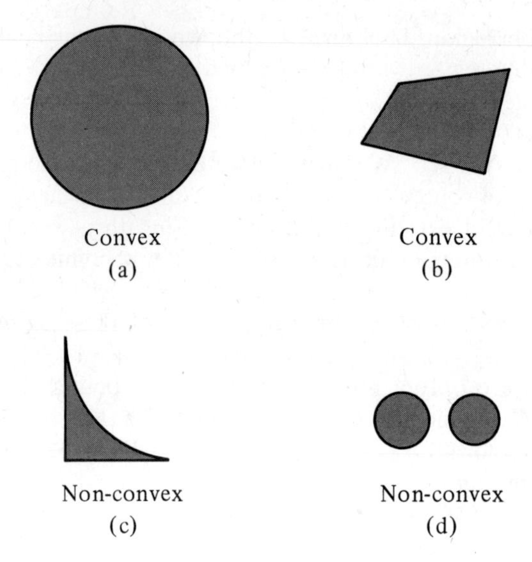


$AX_2 = R$. Consider the convex combination $Y = c_1X_1 + c_2X_2, c_1 + c_2 = 1$, $c_i \geqslant 0$. Then $Y \geqslant 0$ since $X_i \geqslant 0$ and $c_i \geqslant 0$. In addition,

$$AY = A(c_1X_1 + c_2X_2) = c_1(AX_1) + c_2(AX_2)$$
$$= c_1R + c_2R = (c_1 + c_2)R = R.$$

Thus $Y$ is in $S$ and $S$ is convex. ∎

Consider the convex set $S$ shown in Figure 6.11.3. We would like to illustrate the procedure by which it is possible to express any point of such a *bounded* feasible region $S$ as a convex combination of the four corner or extreme points 0, $X_1$, $X_2$, $X_3$. Consider first $X_2$, as an example of an extreme point of $S$. Since

$$X_2 = (0)0 + (0)X_1 + (1)X_2 + (0)X_3,$$

$X_2$ is trivially a convex combination of the entire set of extreme points of

Figure 6.11.3

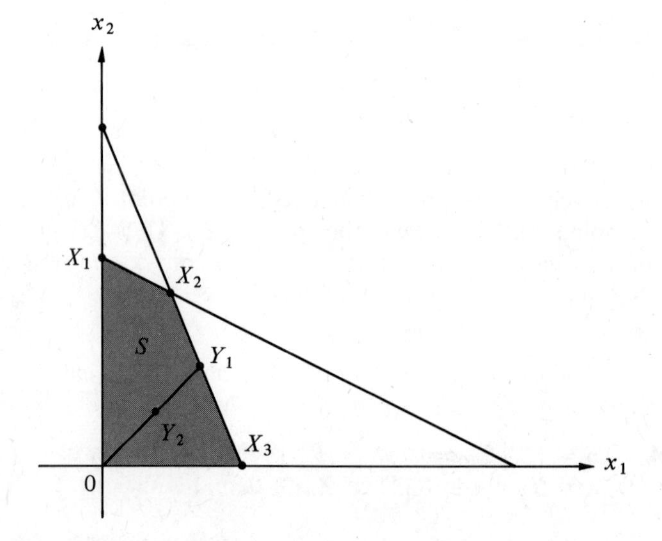

$S$. Suppose that $Y_1$ is a non-extreme boundary point of $S$. We wish to show how $Y_1$ may also be expressed as a convex combination of the extreme points of $S$. $Y_1$ lies on the line between $X_2$ and $X_3$. Thus $Y_1 = c_1X_2 + c_2X_3$ for some choice of $c_1$, $c_2$ such that $c_1 + c_2 = 1$ and $c_i \geqslant 0$. Finally, we wish to express the interior point $Y_2$, on the line between 0 and $Y_1$, as the convex combination of the four extreme points. As shown in Figure 6.11.3, for some choice of $d_1$, $d_2$,

$$Y_2 = d_1 0 + d_2 Y_1, \qquad d_1 + d_2 = 1, \qquad d_i \geqslant 0.$$

Thus $\quad Y_2 = d_1 0 + d_2(c_1X_2 + c_2X_3), \qquad c_1 + c_2 = 1, \qquad c_i \geqslant 0,$

so that

$$Y_2 = d_1 0 + (d_2c_1)X_2 + (d_2c_2)X_3.$$

Since $d_1$, $d_2c_1$, and $d_2c_2$ are each non-negative, and $d_1 + d_2c_1 + d_2c_2 = d_1 + d_2(c_1 + c_2) = d_1 + d_2 = 1$, then $Y_2$ is a convex combination of the extreme points of $S$.

We present the above only as an example of a method whereby we may express any point $X$ of $S$ as a convex combination of the extreme points of $S$. If there are $t$ such points for our general problem, then for any $X$ in $S$,

$$X = c_1X_1 + c_2X_2 + \cdots + c_tX_t$$

for some choice of scalars $c_i$ such that $c_1 + \cdots + c_t = 1$ and $c_i \geqslant 0$ for $i = 1, 2, \ldots, t$.

We now define a **linear function** $f$ to be a function with the properties that

$$f(X_1 + X_2) = f(X_1) + f(X_2)$$

and

$$f(cX_1) = cf(X_1)$$

for all $X_1$, $X_2$ in the domain of $f$, and for all scalars $c$. All of our objective functions $f(X) = CX$ are linear because

$$C(X_1 + X_2) = CX_1 + CX_2$$

and

$$C(cX_1) = cC(X_1).$$

**THEOREM 6.11.2** *If the linear function $f(X) = CX$ defined over the convex set $S$ with extreme points $X_1, X_2, \ldots, X_t$ has an optimal feasible solution $X_0$, then at least one of the extreme points is optimal.*

**Proof*** Assume that the $t$ extreme points have been numbered in an increasing order in terms of their respective functional values, that is,

$$f(X_1) \leqslant f(X_2) \leqslant \cdots \leqslant f(X_t).$$

$X_0$, as a point in $S$, may be expressed as a convex combination of the $t$ extreme points. Thus

$$X_0 = c_1X_1 + c_2X_2 + \cdots + c_tX_t,$$

where $c_1 + c_2 + \cdots + c_t = 1$ and $c_i \geqslant 0$. Since $f$ is a linear function,

$$\begin{aligned} f(X_0) &= f(c_1X_1 + c_2X_2 + \cdots + c_tX_t) \\ &= c_1f(X_1) + c_2f(X_2) + \cdots + c_tf(X_t). \end{aligned}$$

* The essential steps of this proof are due to Gass (see Suggested Reading).

Replacing $f(X_1), f(X_2), \ldots, f(X_{t-1})$ by $f(X_t)$, which in each case can be no smaller, we have,

$$\begin{aligned} f(X_0) &\leqslant c_1 f(X_t) + c_2 f(X_t) + \cdots + c_t f(X_t) \\ &= (c_1 + c_2 + \cdots + c_t) f(X_t) = f(X_t), \end{aligned}$$

since $S$ is convex. Since $f(X_0) \leqslant f(X_t)$ and $X_0$ is given to be optimal, we must have $f(X_0) = f(X_t)$. Thus $X_t$ is an extreme point which is optimal. ∎

We shall recall some notation before stating the final theorem. We have started with an augmented matrix $[R \mid A]$ and developed an augmented matrix $[X_B \mid Y]$ in which $BY = A$ and $BX_B = R$. Since $B$ is non-singular, $Y = B^{-1}A$ and $X_B = B^{-1}R$. We have defined the matrix $Z = [z_1, z_2, \ldots, z_n]$ so that $C - Z = [c_1 - z_1, c_2 - z_2, \ldots, c_n - z_n]$. Since $z_j = C_B Y_j$ for each $j$, we have $Z = C_B Y$.

**THEOREM 6.11.3** *Suppose that $X$ is a BFS such that for all $j$ we have $c_j - z_j \leqslant 0$. Then, for any feasible $X_0$, we have $f(X) \geqslant f(X_0)$.*

**Proof** In matrix notation $C - Z \leqslant 0$, or $C \leqslant Z$. Let $B$ be the basis matrix associated with $X$. Thus, since $X_0 \geqslant 0$,

$$\begin{aligned} f(X_0) &= CX_0 \leqslant ZX_0 = C_B Y X_0 = C_B(B^{-1}A)X_0 \\ &= C_B B^{-1}(AX_0) = C_B B^{-1} R = C_B X_B = f(X). \end{aligned}$$

So we see that $f(X) \geqslant f(X_0)$.* ∎

Because of this fact, we know that we have reached an optimum *BFS* when we reach a *BFS* with $C - Z \leqslant 0$ for a maximization problem.

## 6.11 Exercises

1. Find four different convex combinations of the vectors $\begin{bmatrix} 1 \\ 2 \end{bmatrix}$ and $\begin{bmatrix} 2 \\ 1 \end{bmatrix}$. Sketch your vectors in the plane and show that they lie on the line segment connecting the tip of $\begin{bmatrix} 1 \\ 2 \end{bmatrix}$ and $\begin{bmatrix} 2 \\ 1 \end{bmatrix}$.

2. a. Let $C = [1, 2, 3, 4]$, $Z = [2, 3, 4, 5]$ and suppose that $X_0 = [1, 1, 2, 3]^T$. Show that $CX_0 \leqslant ZX_0$.
   b. Let $C = [-4, 3, 0, 0]$, $Z = [-3, 4, 0, 0]$ and $X_0 = [-1, -2, 3, 0]^T$. Show that although $C \leqslant Z$, $CX_0 > ZX_0$. What hypothesis of Theorem 6.11.3 is not satisfied?

3. Let $C = [c_1, c_2]$, $Z = [z_1, z_2]$ and $X = \begin{bmatrix} x_1 \\ x_2 \end{bmatrix}$. Prove that if $C \leqslant Z$ and $X \geqslant 0$ then $CX \leqslant ZX$.

---

* The proof of this theorem is essentially a matrix version of the scalar proof given by Hadley (see Suggested Reading).

4. Why did we assume that our problem to be examined in this section was non-degenerate?

5. Show that if $f$ is a linear function then

$$f(c_1x_1 + c_2x_2 + \cdots + c_nx_n) = c_1f(x_1) + c_2f(x_2) + \cdots + c_nf(x_n).$$

6. Convince yourself that $z_j = C_BY_j$ represents that portion of the per unit rate of change of $z$ that will occur because of the changes in the values of the present basic variables that must occur if $A_j$ enters the basis.

### Suggested Reading

C. R. Carr and C. W. Howe. *Quantitative Decision Procedures in Management and Economics.* New York: McGraw-Hill, 1964.

G. B. Dantzig. *Linear Programming and Extensions.* Princeton, N.J.: Princeton University Press, 1963.

S. I. Gass. Linear Programming. 3d. ed. New York: McGraw-Hill, 1969.

G. Hadley. *Linear Programming.* Reading, Mass.: Addison-Wesley, 1962.

H. M. Wagner. *Principles of Operations Research.* 2d. ed. Englewood Cliffs, N.J.: Prentice-Hall, 1975.

### New Terms

convex subset of $n$-space, 263
convex combination, 263
extreme point of a convex set, 263
linear function, 265

# 7 The Transportation Problem

## 7.1 Introduction

One of the first LP problems ever formulated is known as "the transportation problem." Because of the special algebraic structure of such problems, it is possible to utilize an individually tailored solution algorithm that enables users to solve very large problems faster and with less computer space than would otherwise be required. In this chapter we shall not only see how the standard transportation problem (TP) is formulated and solved, but we shall also demonstrate how other important practical problems can be expressed in the TP format. This situation in which an algorithm developed for a restricted class of problems turns out to have a wider real-world application occurs frequently in the development of mathematical models.

## 7.2 Formulation of the General Transportation Problem

The first so-called "transportation problems" did involve transportation; normally the circumstances concerned the shipment of many units of a given uniform product from several possible shipping points (sources) to several possible destinations. The sources could be warehouses, for example, and the destinations retail outlets. If the sources and destinations are depicted in columns, as shown in Figure 7.2.1, the existence of a potential shipping route from source $i$ to destination $j$ is shown by an arrow from source $i$ to destination $j$. Following existing custom, we let the doubly subscripted variables $x_{ij}$ represent the number of units of the product to be sent from source $i$ to destination $j$. Where a potential route is shown, there is an $x_{ij}$ to be determined. The integers $S_k$ to the left of the source circles in Figure 7.2.1 indicate the number of items available at the sources. The integers $D_j$ to

**Figure 7.2.1**

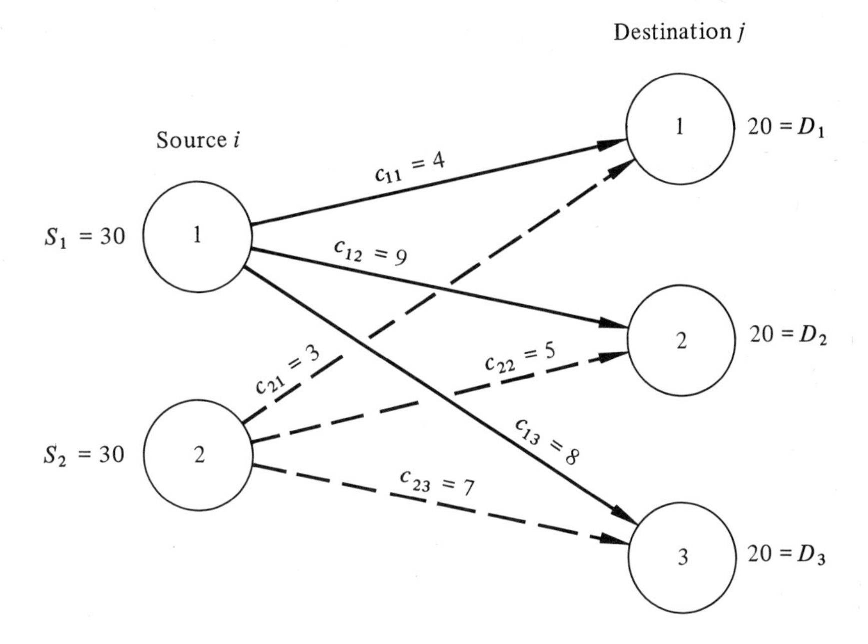


the right of the destination circles indicate the numbers of units demanded at the destinations. In this example, we have 30 units available at each source and 20 units required at each destination. Their sums will always be equal.* The integers portrayed over the arrows represent the known cost of shipping one unit of the product from source $i$ to destination $j$, that is, $c_{ij}$.

Given the $c_{ij}$, $S_i$, and $D_j$, the task is to find values for the $x_{ij}$ which will accomplish the transportation at the minimum total cost.

In order to formulate this practical problem in LP terms, we first regard the simple nature of the constraints involved. It is clear that all of the units available at each source must be shipped somewhere. The algebraic result is that, *for each source*, we have a constraint which insures that the entire available supply is shipped. Thus, we have the constraints $x_{11} + x_{12} + x_{13} = 30 = S_1$ and $x_{21} + x_{22} + x_{23} = 30 = S_2$. Likewise, the demands at each destination must be satisfied; thus, for each destination, we have a constraint to accomplish this task. To insure that destination 1 receives its required demand from some source, we formulate the constraint $x_{11} + x_{21} = 20 = D_1$. Similarly, for destinations 2 and 3 we have

$$x_{12} + x_{22} = 20 = D_2 \quad \text{and} \quad x_{13} + x_{23} = 20 = D_3.$$

Any non-negative set of values for the defined $x_{ij}$ which satisfies these five constraints will accomplish the required transportation task. The associated cost of any feasible solution is found by computing and totaling the cost for each possible route to obtain the value of the objective function $f = 4x_{11} + 9x_{12} + 8x_{13} + 3x_{21} + 5x_{22} + 7x_{23}$. With the constraints and the

* We note that the special algorithm to be developed requires that the total amount available be equal to the total amount required, as is the case in the example above. As we shall see in the exercises, this is not a significant restriction, since any problem for which this is not valid can be made into such by a simple subterfuge.

objective function formulated, we can express the entire LP problem as follows:

---

Find $x_{ij} \geqslant 0$ such that

$$\left.\begin{array}{l} x_{11} + x_{12} + x_{13} = 30 \\ x_{21} + x_{22} + x_{23} = 30 \end{array}\right\} \text{source constraints}$$

$$\left.\begin{array}{l} x_{11} + x_{21} = 20 \\ x_{12} + x_{22} = 20 \\ x_{13} + x_{23} = 20 \end{array}\right\} \text{destination constraints}$$

and $f(x_{ij}) = 4x_{11} + 9x_{12} + 8x_{13} + 3x_{21} + 5x_{22} + 7x_{23}$ is minimized.

---

This problem will be designated Problem $T_1$ and will be used for demonstration purposes throughout this chapter. The LP formulation of the **general transportation problem** with $m$ sources and $n$ destinations would appear as follows:

---

Find $x_{ij} \geqslant 0$ such that

$$\left.\begin{array}{l} x_{11}+x_{12}+\cdots+x_{1n} = S_1 \\ x_{21}+x_{22}+\cdots+x_{2n} = S_2 \\ \vdots \\ x_{m1}+x_{m2}+\cdots+x_{mn} = S_m \end{array}\right\} m \text{ sources}$$

$$\left.\begin{array}{l} x_{11} + x_{21} + \cdots + x_{m1} = D_1 \\ x_{12} + x_{22} + \cdots + x_{m2} = D_2 \\ \vdots \\ x_{1n} + x_{2n} + \cdots + x_{mn} = D_n \end{array}\right\} n \text{ destinations}$$

and $f(x_{ij}) = c_{11}x_{11} + \cdots + c_{1n}x_{1n} + c_{21}x_{21} + \cdots + c_{2n}x_{2n} + \cdots + c_{m1}x_{m1} + \cdots + c_{mn}x_{mn}$ is minimized

---

Example 7.2.1 Let us formulate the transportation problem illustrated in Figure 7.2.2. This problem will be known as Problem $T_2$.

**Figure 7.2.2**

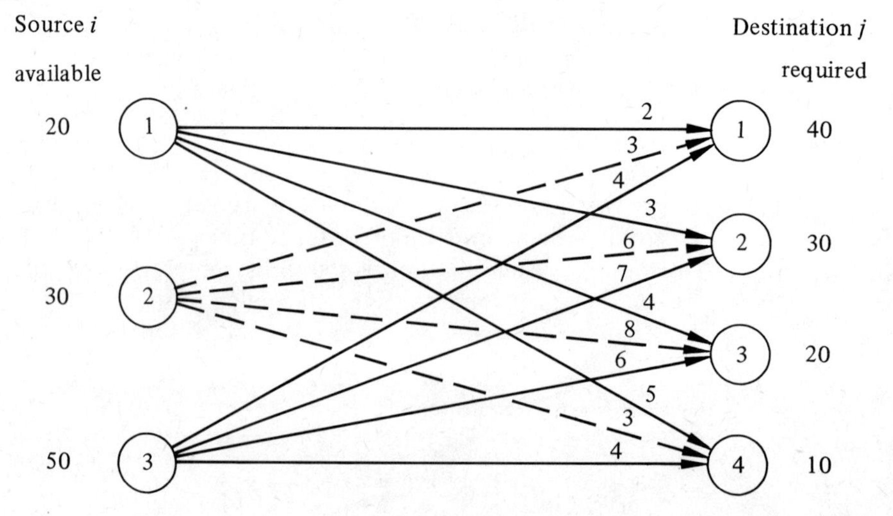

As before, we note that the total amount required equals the total amount available. We thus seek a structure with 3 source constraints and 4 destination constraints and obtain the following:

---

Find $x_{ij} \geqslant 0$ such that

$$\begin{array}{llllllllllllll}
x_{11} + x_{12} + x_{13} + x_{14} & & & & & = 20 \\
& x_{21} + x_{22} + x_{23} + x_{24} & & & & = 30 \\
& & x_{31} + x_{32} + x_{33} + x_{34} & & & = 50 \\
\hline
x_{11} & + x_{21} & + x_{31} & & & = 40 \\
x_{12} & + x_{22} & + x_{32} & & & = 30 \\
x_{13} & + x_{23} & + x_{33} & & & = 20 \\
x_{14} & + x_{24} & + x_{34} & & & = 10
\end{array}$$

and $f(x_{ij}) = 2x_{11} + 3x_{12} + 4x_{13} + 5x_{14} + 3x_{21} + 6x_{22} + 8x_{23} + 3x_{24} + 4x_{31} + 7x_{32} + 6x_{33} + 4x_{34}$ is minimized. ∎

---

The context of Problem $T_2$ need not have been transportation. Suppose that in Figure 7.2.2 the source circles represent 3 worker occupations that have been made partially irrelevant by advanced technology. Then, the 4 demand circles would represent viable present-day occupations into which it would be possible to retrain the various workers at a training cost which is shown. The problem would be to find how many workers in the various categories should be retrained into the various new categories. The algebraic format is clearly the same as before. We merely think of the various workers as being "shipped" from one job to another.

## 7.2 Exercises

1. Consider the transportation situation represented in Figure 7.2.3. The various shipping costs are shown above the connecting arrows.

**Figure 7.2.3**

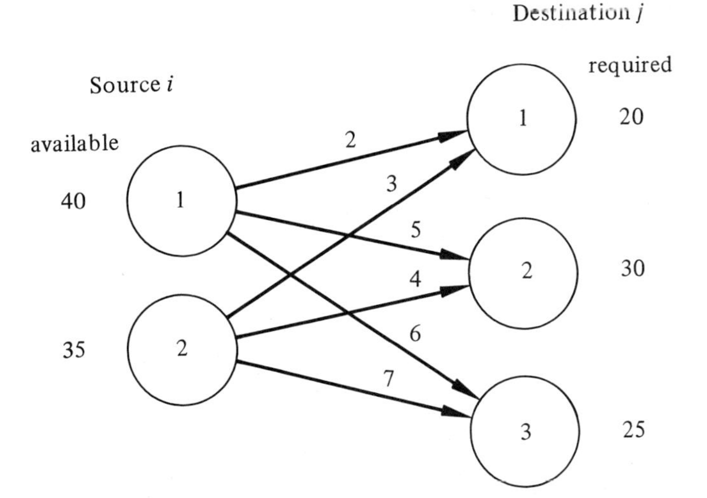

a. How many variables are involved? What are they?
b. Write down a constraint that will insure that all of the supply at Source 1 will be shipped.
c. Write down a constraint that will insure that all of the units required at Destination 2 will arrive.
d. Formulate a TP which will tell how many units should be shipped from each of the 2 sources to each of the 3 destinations in order to minimize the total shipping cost.

2. a. Set up the first tableau of a *linear programming* solution for the problem in Exercise 1. Six artificial vectors will be required in order to get an initial identity matrix.
   b. The matrix of the intial LP tableau in Exercise 2a contains ________ elements, not counting those in the constant column.

3. Company $A$ distributes a certain product from 2 warehouses ($W_1$, $W_2$) to 3 retail outlets ($R_1$, $R_2$, $R_3$). The shipping costs, supplies, and demands are contained in Table 7.2.1.

**Table 7.2.1**

| *From* \ *To* | $R_1$ | $R_2$ | $R_3$ | *Available* |
|---|---|---|---|---|
| $W_1$ | 5 | 8 | 6 | 60 |
| $W_2$ | 4 | 9 | 6 | 40 |
| *Required* | 35 | 30 | 35 | |

a. Formulate a TP which will tell the company management how many product units should be shipped from each of the 2 warehouses to each of the 3 retail stores.
b. Express the data from Table 7.2.1 in the pictorial format of Figure 7.2.1.

4. Because of skyrocketing maintenance costs, company B has decided to close its 3 oldest manufacturing plants. Each shutdown will result in the transfer of 50 assembly-line personnel to one of the 4 remaining plants ($N_1$, $N_2$, $N_3$, $N_4$), which have openings for 30, 50, 40, and 30 new employees, respectively. The total cost data for transferring displaced employees from the various old plants to the newer plants is contained in Table 7.2.2.

a. Set up a TP whose solution will tell company management how many

**Table 7.2.2**

| *From* \ *To* | $N_1$ | $N_2$ | $N_3$ | $N_4$ |
|---|---|---|---|---|
| $O_1$ | 5 | 4 | 6 | 7 |
| $O_2$ | 3 | 6 | 8 | 5 |
| $O_3$ | 7 | 6 | 7 | 9 |

employees should be sent from each of the older plants to each of the newer plants in such a way that the total transfer cost will be minimized.

b. Express the data for this exercise in the pictorial network format of Figure 7.2.1.

5. A rental car agency operates 7 rental offices in a given geographical region. After a certain busy weekend, 3 of the offices ($A_1$, $A_2$, $A_3$) have a surplus of cars for the expected week's needs, while the remaining 4 rental offices ($B_1$, $B_2$, $B_3$, $B_4$) have a deficit. The costs of shipping cars from the various $A_i$ to the $B_j$ are shown in Table 7.2.3, along with the number of cars needed and available.

**Table 7.2.3**

| From \ To | $B_1$ | $B_2$ | $B_3$ | $B_4$ | *Surplus* |
|---|---|---|---|---|---|
| $A_1$ | 20 | 30 | 40 | 35 | 10 |
| $A_2$ | 15 | 20 | 25 | 40 | 6 |
| $A_3$ | 25 | 35 | 20 | 35 | 8 |
| *Deficit* | 6 | 8 | 7 | 3 | |

a. Express the data for this exercise in the pictorial network format of Figure 7.2.1.
b. Formulate a TP which will provide the numbers of cars to be sent from the 3 sources to the 4 destinations.

6. Using the notation developed in this section, that is, $c_{ij}$, $S_i$, $D_j$, and $x_{ij}$, fill in the following blanks, as they apply to the data of Exercise 5.

a. $c_{14}$ = ________, $c_{23}$ = ________.
b. $S_2$ = ________, $D_3$ = ________.
c. $x_{23}$ represents the amount to be shipped from source ________ to destination ________.

7. With the passage of time in a given city men in 3 occupations ($O_1$, $O_2$, $O_3$) have found their jobs rendered obsolete by mechanized processes. During this period, however, there has arisen an increased need for welders ($W$), electricians ($E$), and electronic technicians ($ET$). The costs of retraining the men from the various $O_i$ into the new occupations are given in Table 7.2.4, along with the numbers of men in the original occupations (the supply, $S_i$) and the numbers of vacancies to be filled in the new jobs (the demands, $D_j$).

**Table 7.2.4**

| $i$ \ $j$ | ($W$) 1 | ($E$) 2 | ($ET$) 3 | $S_i$ |
|---|---|---|---|---|
| 1 | 8 | 6 | 4 | 20 |
| 2 | 5 | 4 | 5 | 15 |
| 3 | 6 | 7 | 9 | 25 |
| $D_j$ | 30 | 15 | 15 | |

Using the data in the table, formulate a TP which will provide an appropriate retraining program if the total retraining cost is to be minimized. (We must assume, of course, that each man is willing to retrain into any one of the new jobs.)

8. In this exercise we demonstrate how to handle the case when the total supply is *less* than the total demand, rather than equal to it. The question to be answered is "which demand should go unsatisfied?"

   Consider the TP of Exercise 4 above. For this case we change the number of openings at the new plants to 50 each, for a total of 200. The total number of transferees, or the total supply, remains 150. Thus there is an excess demand of $200 - 150 = 50$.

   To set up a TP, the solution to which will provide the desired answers, we create a fictitious source (known as a *dummy source*) called $O_4$ with a fictitious supply of $200 - 150 = 50$. We allow the 50 fictitious units to be sent to any of the 4 destinations. To solve the problem of excess demand using the standard TP algorithm to be developed we need cost coefficients $c_{41} - c_{44}$. We merely set each $c_{4j} = 0$.

   Formulate the new TP arising as a result of these changes. We will solve it later. For now we note that if we were to find that $x_{41} = 30$ in the optimal solution, we would know that only $20 = 50 - 30$ of the potential jobs at $N_1$ would be filled by the transferees.

9. Important meetings are to be held at each of the 2 high schools in a given school district (one meeting at each high school). Every student from the district's 4 junior high schools is to attend exactly 1 of these meetings. The distances between the various junior high schools and the 2 high schools is shown in Table 7.2.5, along with the populations of the junior highs and the auditorium seating capacities of the 2 high schools.

**Table 7.2.5**

| *From* \ *To* | $HS_1$ | $HS_2$ | *JH Population* |
|---|---|---|---|
| $JH_1$ | 10 | 5 | 500 |
| $JH_2$ | 4 | 9 | 400 |
| $JH_3$ | 15 | 7 | 800 |
| $JH_4$ | 6 | 2 | 650 |
| *HS Seating* | 1000 | 1350 | |

Formulate a TP which will tell how many students should be transported from $JH_i$ to $HS_j$ in order to minimize the total student-miles traveled.

## Suggested Reading

G. B. Dantzig. *Linear Programming and Extensions.* Princeton, N.J.: Princeton University Press, 1963.

S. I. Gass. *Linear Programming.* 3d. ed. New York: McGraw-Hill, 1969.

G. Hadley. *Linear Programming.* Reading, Mass.: Addison-Wesley, 1962.

D. R. Plane and G. A. Kochenberger. *Operations Research for Managerial Decisions.* Homewood, Ill.: Irwin, 1972.

H. M. Wagner. *Principles of Operations Research.* 2d. ed. Englewood Cliffs, N.J.: Prentice-Hall, 1975.

## 7.3 The Special Structure of Transportation Problems

If we think of phrasing Problem $T_2$ in the following form:

---

"Find $X \geqslant 0$ such that $AX = R$ and $f(X) = CX$ is minimized," then

$$A = [A_{11}, A_{12}, A_{13}, A_{14}, A_{21}, A_{22}, A_{23}, A_{24}, A_{31}, A_{32}, A_{33}, A_{34}]$$

$$= \begin{bmatrix} 1 & 1 & 1 & 1 & 0 & 0 & 0 & 0 & 0 & 0 & 0 & 0 \\ 0 & 0 & 0 & 0 & 1 & 1 & 1 & 1 & 0 & 0 & 0 & 0 \\ 0 & 0 & 0 & 0 & 0 & 0 & 0 & 0 & 1 & 1 & 1 & 1 \\ \hline 1 & 0 & 0 & 0 & 1 & 0 & 0 & 0 & 1 & 0 & 0 & 0 \\ 0 & 1 & 0 & 0 & 0 & 1 & 0 & 0 & 0 & 1 & 0 & 0 \\ 0 & 0 & 1 & 0 & 0 & 0 & 1 & 0 & 0 & 0 & 1 & 0 \\ 0 & 0 & 0 & 1 & 0 & 0 & 0 & 1 & 0 & 0 & 0 & 1 \end{bmatrix} \begin{matrix} \left.\begin{matrix} \\ \\ \\ \end{matrix}\right\} \text{source rows} \\ \left.\begin{matrix} \\ \\ \\ \\ \end{matrix}\right\} \text{destination rows} \end{matrix},$$

$$R^T = [20, 30, 50, 40, 30, 20, 10]$$

and $$C = [2, 3, 4, 5, 3, 6, 8, 3, 4, 7, 6, 4]$$

---

The matrix $A$ is obviously of a special nature; in particular, we note that

1. Each element is either a zero or a one.
2. Each column has precisely 2 ones.
3. The ones in column $A_{ij}$, corresponding to $x_{ij}$, are in rows $i$ and $3 + j$ ($m + j$, in general).
4. The $A_{ij}$ corresponding to any one source or destination are clearly linearly independent.

This special structure has certain implications for the solution algorithm, among which are the following:

1. Out of the 12 (i.e., $mn$) $A_{ij}$, exactly 6 (i.e., $m + n - 1$) are linearly independent. Thus the rank of $A$ is 6 (i.e., $m + n - 1$).
2. It is very easy to obtain an all-integer initial basic feasible solution (BFS) using only the $3 \times 4 = 12$ vectors of $T_2$. Otherwise it would be necessary to add 7 (i.e., $m + n$) artificial vectors and proceed with the standard simplex algorithm. For large $m$ and $n$ this could be a significant factor.
3. A very compact tableau may be used to solve the TP.
4. It is relatively simple to determine the revised cost coefficients ($c_{ij} - z_{ij}$) for each non-basic variable $x_{ij}$. Thus the question whether the present

BFS is optimal can be readily answered by checking to see if $c_{ij} - z_{ij} \geqslant 0$ for each non-basic $x_{ij}$.

5. If the present BFS is not optimal, and a vector $A_{ij}$ is chosen to enter the basis, it is easy to recognize its coordinate vector $Y_{ij}$ which will consist entirely of zeros and positive or negative ones. These coordinates will be used to choose the correct vector to leave the basis; to calculate the new value of $x_{ij}$, once it becomes basic; and also to calculate the new values of the remaining basic variables in the revised compact tableau.

We shall consider these statements throughout the remainder of this chapter and shall try to make them intuitively plausible and to relate associated computations to the special structure of the matrix $A$. For convenience, we consider the first assertion here, and delay the remainder until later sections.

**Assertion 1** The rank of $A$ is 6 (i.e., $m + n - 1$).

The vectors $A_{11}, A_{12}, A_{13}, A_{14}, A_{21}$, and $A_{31}$ can be seen to be linearly independent by virtue of the position of their non-zero elements. Thus the rank of $A$ is at least 6. Since there are only 7 rows, the rank of $A$ is at most 7. However, the row vector obtained by summing the 3 source rows is equal to the row vector obtained by summing the 4 destination rows. Thus the 7 rows are linearly dependent, and the rank of $A$ is no greater than 6. These two observations combine to prove that the rank of $A$ is exactly 6.

The extension to the general case is clear. There are always at least $n + (m - 1)$ linearly independent vectors in $A$. Choose the first $n$ vectors (that is, those associated with source 1) and then the $m - 1$ vectors obtained by using the first column in each of the remaining $m - 1$ source groups. The entire set would consist of the $n + (m - 1)$ or $m + n - 1$ vectors

$$A_{11}, A_{12}, \ldots, A_{1n}, A_{21}, A_{31}, \ldots, A_{m1}. \qquad \blacksquare$$

For the remainder of this section we will examine other aspects of the special structure of $A$ and illustrate how one may visually determine the coordinates of a given $A_{ij}$ in terms of an appropriate set of linearly independent vectors from $A$. For these purposes let us consider a subset of the column vectors of $T_2$, namely

$$A_{11}, A_{12}, A_{13}, A_{14}, A_{21}, \quad \text{and} \quad A_{22}.$$

Recalling that $T_2$ involves 3 sources and 4 destinations, the above vectors can be constructed as vectors in 7-space, partitioned in two parts—a $3 \times 1$ source vector, and a $4 \times 1$ destination vector. Vector $A_{ij}$ has a unit element (1) which indicates the source in row $i$ of the source vector portion and a unit element which indicates the destination in row $j$ of the destination portion. The 6 vectors mentioned above can thus be written as follows, with $S_i$ denoting source $i$, and $D_j$ denoting destination $j$.

| | $A_{11}$ | $A_{12}$ | $A_{13}$ | $A_{14}$ | $A_{21}$ | $A_{22}$ |
|---|---|---|---|---|---|---|
| $S_1$ | 1 | 1 | 1 | 1 | 0 | 0 |
| $S_2$ | 0 | 0 | 0 | 0 | 1 | 1 |
| $S_3$ | 0 | 0 | 0 | 0 | 0 | 0 |
| $D_1$ | 1 | 0 | 0 | 0 | 1 | 0 |
| $D_2$ | 0 | 1 | 0 | 0 | 0 | 1 |
| $D_3$ | 0 | 0 | 1 | 0 | 0 | 0 |
| $D_4$ | 0 | 0 | 0 | 1 | 0 | 0 |

It was shown earlier that the first 5 vectors are linearly independent. Let us suppose that we know that $A_{22}$ is some linear combination of these vectors, and we wish to find its coordinates. To see how this can be done visually, you should follow the arrows in Figure 7.3.1 as you read the description below.

**Figure 7.3.1**

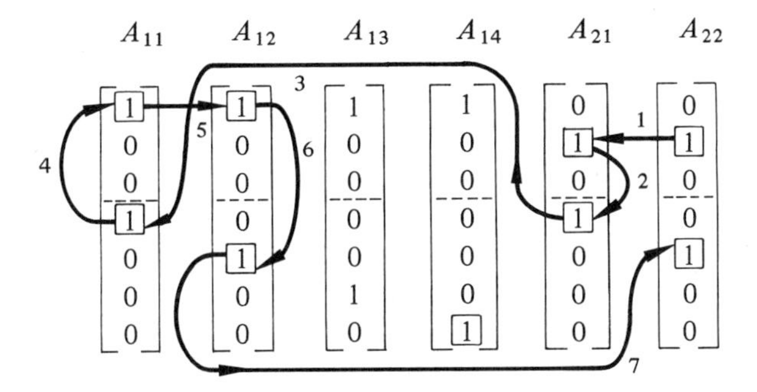


First we note that each of the 2 ones in $A_{22}$ must be *produced* by linear combinations of non-zero elements in the other vectors. Considering the first unit element of $A_{22}$ (in row 2), we note that $A_{21}$ is the only original vector that has a unit element in row 2. Thus $A_{21}$ must be included in the final linear combination. To illustrate this relationship we join the two units by a labelled arrow (Step 1). This is the first link in a chain we will create which starts with one of the unit elements in $A_{22}$ and terminates with the other one.*

Having discovered that $A_{21}$ must be included because of the 1 in row 2, we note that $A_{21}$ has another 1 in row 4 (Step 2), and that $A_{22}$ has no unit element in row 4. Thus we must search for a second vector which has a 1 in row 4 that can be subtracted from $A_{21}$ to eliminate the superfluous 1. There is only one such vector, namely $A_{11}$. Thus $A_{11}$ must also be included in the final set (Step 3). But, as before, $A_{11}$ brings with it an unwanted unit element in row 1 (Step 4). An additional vector must be found which contains a 1 in row 1, so that it may be added to $A_{11}$ to remove the extraneous unit. In this case, however, there are 3 candidates, $A_{12}$, $A_{13}$, and $A_{14}$. Although it might seem difficult to resolve this ambiguity, such is not the case.

Suppose, for example, that $A_{13}$ were chosen because of the row 1 unit. We would need to get rid of its row 6 unit, but there is no other vector with

* Although it may not be obvious at this point, the chain that we will create is unique, that is, there is only one such chain which starts with one of the two unit elements and ends with the other one.

a row 6 unit with which of eliminate this unit. A similar situation occurs if $A_{14}$ were chosen. The only choice which will work is $A_{12}$ (Step 5). Fortunately, and this will always eventually be the case, $A_{12}$ has a unit in row 5 (Step 6) which is exactly what we need to "close the loop" and complete the generation of $A_{22}$, since $A_{22}$ also has a unit, as yet unaccounted for, in row 5 (Step 7). By following the arrows in Figure 7.3.1, we see that $A_{22}$ is a linear combination of $A_{21}$, $A_{11}$, and $A_{12}$; and from the reasoning outlined above, we see that

$$A_{22} = A_{21} - A_{11} + A_{12},$$

that is,

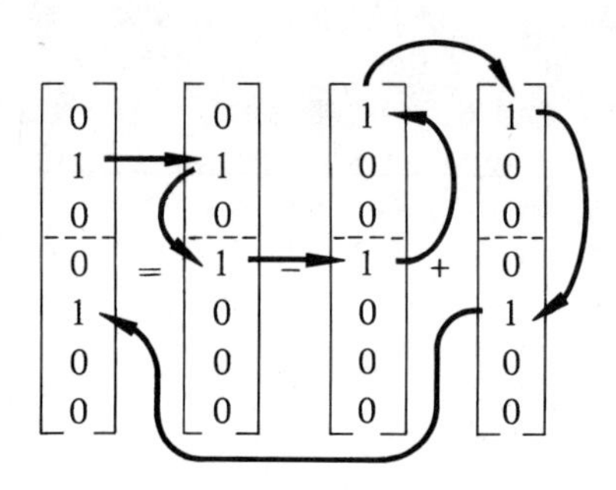

In terms of the linearly independent set $\{A_{11}, A_{12}, A_{13}, A_{14}, A_{21}, A_{31}\}$,

$$A_{22} = (-1)A_{11} + (1)A_{12} + (0)A_{13} + (0)A_{14} + (1)A_{21} + (0)A_{31}.$$

Thus the coordinate vector for $A_{22}$ with respect to this basis is $Y_{22} = [-1, 1, 0, 0, 1, 0]^T$, a vector which consists entirely of zeros and positive or negative units. Although we shall not prove it here, *this is always the case for any non-basic vector.* The implication is that if the simplex algorithm were used to solve such problems, the pivot elements would always be $+1$.* No division is required in any tableau, only addition or subtraction. *If we start with an R vector of integers, then the succession of $X_B$ vectors, including the optimal solution, will also consist entirely of integers.* We thus have an example of an LP problem which, because of its unique structure, always has an optimal *integer* solution; and we can always use the simplex algorithm to find it. Unfortunately, there are other LP problems that must have integer solutions but have no integer optimal solutions. Although you probably had not given the situation much thought, suppose that the items we were shipping in a TP were Rolls Royces, and the optimal solution demanded that you ship 0.5 cars from Norfolk to Detroit and 0.5 cars from Norfolk to Atlanta. The optimal real-world shipping plan is not at all clear.

In subsequent sections we shall return to the concept of the vector chains created above to find the coordinates of a given vector.

## 7.3 Exercises

1. Consider Exercise 7.2.1.
   a. Fill in the matrix $A$: $A_{12}^T = [\qquad\qquad]$.
   b. $R_T = [\qquad\qquad]$; $C = [\qquad\qquad]$.



* We recall from Section 6.9 that pivot elements are always positive.

c. The set $\{A_{11}, A_{12}, A_{13}\}$ is linearly ________.

d. Use Gauss-Jordan reduction to express $A_{22}$ as a linear combination of $A_{11}$, $A_{12}$, $A_{21}$.

e. Write down the vectors $A_{11}$, $A_{12}$, $A_{21}$, and $A_{22}$, in that order, and use the method of Figure 7.3.1 to write $A_{22}$ as a linear combination of $A_{11}$, $A_{12}$, and $A_{21}$.

f. The rank of $A$ is ________.

g. The set $\{A_{11}, A_{12}, A_{22}, A_{23}\}$ is linearly independent. Find $Y_{13}$, the vector of coordinates of $A_{13}$, with respect to this set.

2. The unit elements in $A_{23}$ above are in row ________ and row 2 + ________. If $A$ is an $m \times n$ matrix, then $A_{23}$ would have unit elements in row ________ and row ________.

3. Each column of the matrix $A$ of a TP contains precisely ________ unit elements, one associated with its source, and a second associated with its ________.

4. (*True* or *false*) Every TP with 2 sources and 3 destinations has the same matrix $A$.

5. Consider Exercise 7.2.5.

a. Write down the matrix $A$. $A_{23}^T = [\qquad\qquad]$.

b. $R^T = [\qquad\qquad]$; $C = [\qquad\qquad]$.

c. Write down the vectors $A_{11}$, $A_{12}$, $A_{22}$, and $A_{23}$. These vectors are linearly ________.

d. Use Gauss-Jordan reduction to express $A_{13}$ as a linear combination of the vectors in c.

e. Use the method of Figure 7.3.1 to express $A_{13}$ as a linear combination of the vectors in c.

f. Use the method of Figure 7.3.1 to express $A_{14}$ as a linear combination of $A_{11}$, $A_{12}$, $A_{22}$, $A_{23}$, $A_{33}$, and $A_{34}$. With respect to these vectors, $Y_{14} = [\qquad\qquad]^T$.

6. Show that the rank of $A$ in Exercise 5 is 6.

7. If the matrix $A$ of a TP has 6 rows and 7 columns, then it has rank ________. The matrix sum of the first ________ rows is equal to the matrix sum of the last ________ rows. If the sum of the supplies ($S_i$) equals the sum of the demands ($D_j$), then the augmented matrix $[A \mid R]$ has rank ________.

## 7.4 The Compact Transportation Tableau

The structure of the tableau customarily used for solution of transportation problems is shown in Tableau 7.4.1, as it would appear for Problem $T_2$. Since there are 3 sources and 4 destinations, the tableau contains 3 rows and 4 columns. Row $i$ contains information pertinent to source $i$, while column $j$ contains information for destination $j$. Listed to the right of row $i$ is the number of units available at source $i$ ($S_i$). $D_j$, the amount demanded at destination $j$, is shown at the bottom of column $j$. The $3 \times 4 = 12$ cells provide space to write in the values of the various basic variables ($x_{ij}$) as they are determined. Cost information for cell $(i, j)$, namely $c_{ij}$, is contained

Tableau 7.4.1

| $i \backslash j$ | 1 | 2 | 3 | 4 | $S_i$ |
|---|---|---|---|---|---|
| 1 | [2] $x_{11}$ | [3] $x_{12}$ | [4] $x_{13}$ | [5] $x_{14}$ | 20 |
| 2 | [3] $x_{21}$ | [6] $x_{22}$ | [8] $x_{23}$ | [3] $x_{24}$ | 30 |
| 3 | [4] $x_{31}$ | [7] $x_{32}$ | [6] $x_{33}$ | [4] $x_{34}$ | 50 |
| $D_j$ | 40 | 30 | 20 | 10 | |

in the small block in the upper left corner of the cell. For any set of $x_{ij}$, the corresponding $f$ value is found by summing the 12 $(c_{ij}x_{ij})$ terms.*

All of the special structure information that is contained in the original matrix $A$ is contained in Tableau 7.4.1. Each of the 3 source (destination) constraints can be re-created by taking the necessary information from the appropriate row (column). For example, row 2 information allows us to write the second source constraint

$$x_{21} + x_{22} + x_{23} + x_{24} = 30.$$

Similarly, column 3 helps us to write the third destination constraint

$$x_{13} + x_{23} + x_{33} = 20.$$

Not only are all 7 constraints reproducible, but the position of the all-important unit elements is also available. Those vectors $(A_{11}, A_{12}, A_{13}, A_{14})$ with a 1 in source row 1, for example, can be determined by the fact that their associated variables appear in row 1 cells. Likewise, the vectors with a 1 in destination row 2 $(A_{12}, A_{22}, A_{32})$ can be determined by the fact that the associated variables are contained in cells in column 2. Analogous statements hold for any $i$ or $j$.

With this information about the compact tableau freshly in mind, let us now reconsider the Problem $T_2$ situation described in Section 7.3 concerning the coordinates of $A_{22}$ with respect to the vectors $A_{11}, A_{12}, A_{13}, A_{14}$, and $A_{21}$. In that discussion we introduced the notion of a "closed path" connecting the various unit elements first to find the vectors with non-zero coefficients and then actually write down the $\pm 1$ coordinates. We will now show how to do the same thing with the compact tableau, so that we will not need to use the entire matrix $A$.

For this purpose, we introduce the partial Tableau 7.4.2, in which blank circles identify cells associated with the 5 known linearly independent vectors. A hexagon has been placed in the (2, 2) cell to indicate that $A_{22}$ is the vector under scrutiny. *Our goal now is to make it intuitively clear that if one is able to create a closed path made up of alternating horizontal and vertical arrows that starts with the hexagon, proceeds to various circles, and then returns to the hexagon, then $A_{22}$ is indeed a linear combination of the vectors; and the appropriate coordinates may be found by inspecting the path.* With a little

* For any BFS, of course, only 6 terms can be non-zero.

**Tableau 7.4.2**

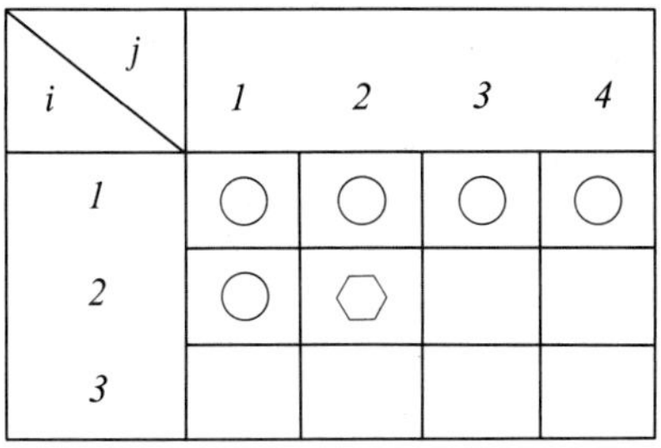

| i \ j | 1 | 2 | 3 | 4 |
|---|---|---|---|---|
| 1 | ○ | ○ | ○ | ○ |
| 2 | ○ | ⬡ | | |
| 3 | | | | |

practice it will be seen that the visual procedure is even easier to understand than the verbal one presented previously.

Although the notion of a *closed path between cells* may not yet be obvious to you, it will become conspicuous once your eye is trained to see it. The desired closed path connecting the (2, 2) cell and the circle cells of Tableau 7.4.2 consists of the square whose corners are the three circles and the hexagon in the upper left corner. Similarly, the (1, 3) cell, associated with $A_{13}$, and the (1, 4) cell, associated with $A_{14}$, are not included in the subset required to generate $A_{22}$; this fact is made visually evident by the isolation and exclusion of these cells from the path.

We cannot arrive at the above conclusions, however, until we proceed in a step-by-step manner to create the closed path of interest. The logic that goes into the steps is equivalent to that presented earlier, when the unit elements of the vectors were used to create a closed path connecting the "ones" in the various vectors.

To generate the path, we start with the hexagon and observe that it could either be linked horizontally with the (2, 1) cell, or vertically with the (1, 2) cell; there are no other choices. Although we can progress in either direction, let us first connect the (2, 2) cell horizontally with the (2, 1) cell as we have earlier in Figure 7.3.1.* The partial path thus achieved is shown in Tableau 7.4.3. From the (2, 1) position there can be no choice of direction since there is only one other circle in column 1. The (2, 1) cell must be connected vertically with the (1, 1) cell, as shown in Tableau 7.4.4. It is conceivable to link up the (1, 1) cell with either the (1, 2), (1, 3), or (1, 4) cells. If, however, either of the latter cells are chosen, the path would be stopped; there are no other circles in those columns. Thus, the next cell on the path is (1, 2) as shown in Tableau 7.4.5. At this point, again we cannot choose a

**Tableau 7.4.3**

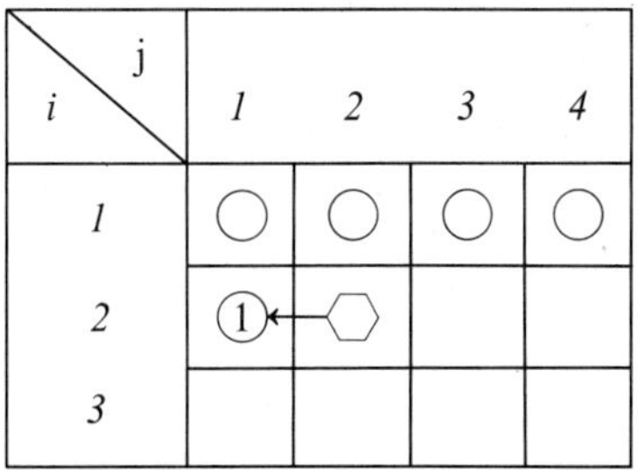

| i \ j | 1 | 2 | 3 | 4 |
|---|---|---|---|---|
| 1 | ○ | ○ | ○ | ○ |
| 2 | (1) ← | ⬡ | | |
| 3 | | | | |

* As shown in Figure 7.3.1, we first linked $A_{22}$ with $A_{21}$ because we chose to start with the first "one." If we had started with the unit in row 5, we would first have linked $A_{22}$ with $A_{12}$ and then proceeded to generate the same path in reverse order.

Tableau 7.4.4

| i \ j | 1 | 2 | 3 | 4 |
|---|---|---|---|---|
| 1 | ② ↑ | ○ | ○ | ○ |
| 2 | ① ← | ⬡ | | |
| 3 | | | | |

Tableau 7.4.5

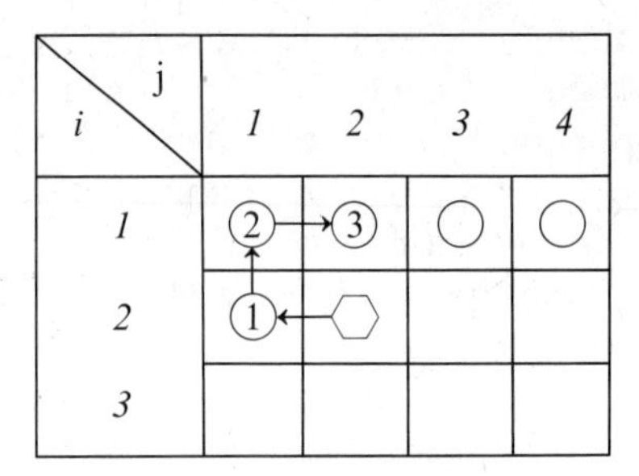

Tableau 7.4.6

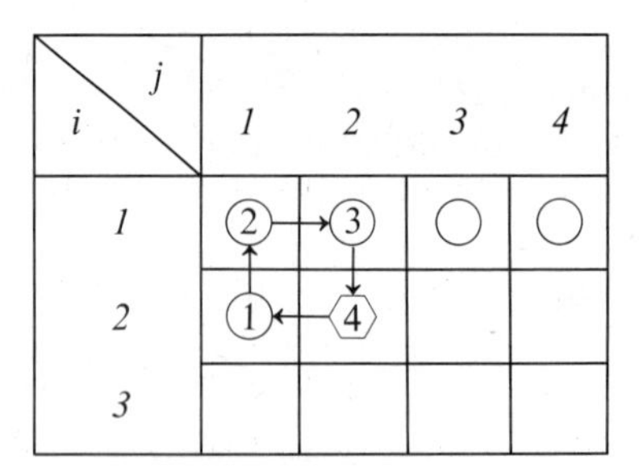

direction; we must proceed vertically (downward) to the (2, 2) cell—completing the cycle and forming the desired closed path. The closed path is visually evident in Tableau 7.4.6.

Thus we have established, with the closed path, which vectors $A_{22}$ "depends upon." It now remains to identify the precise relationship which exists. To do so, it is merely necessary to express $A_{22}$ as a linear combination of the vectors on the path in the order generated, starting with a positive unit coordinate and successively alternating signs. Thus

$$A_{22} = A_{21} - A_{11} + A_{12},$$

which is the same result we obtained earlier.

If we had not obtained the same result, it would have been an indication that at least one of the methods was faulty. We know from our previous work with bases that the coordinates of any given vector, with respect to a particular basis, are *unique*. This also means, of course, that *there is exactly one closed path joining cells representing any linearly independent set of vectors and any other particular cell representing a vector which happens to be a linear combination of those vectors.* Therefore a method which would seem to allow for an enormous number of choices does not lead to several alternate solutions at all. At any point where either a horizontal or a vertical choice (one or the other, not both) is to be made, there is exactly one cell to be chosen that will eventually lead to a closed path. An erroneous choice will identify itself sooner or later in terms of an incomplete path.

We now attempt to solidify these ideas by finding the appropriate paths in the following two examples.

**Example 7.4.1** In Tableau 7.4.7, circles have been placed in 6 (that is, $m + n - 1$) cells. Thus, we have a set of cells associated with the known maximum number of linearly independent vectors. Since it is visually obvious that there is no closed path containing any subset of these six vectors, they are linearly independent. We shall now determine the coordinates of the vector $A_{23}$ with respect to the prescribed basis, after first finding a closed path that contains the (2, 3) cell and the appropriate basic (circle) cells. The path generated is traced in Tableau 7.4.8. The confident reader should attempt to trace the path on Tableau 7.4.7 prior to reading the explanation below.

**Tableau 7.4.7**

| $i$ \ $j$ | 1 | 2 | 3 | 4 |
|---|---|---|---|---|
| 1 | ○ | ○ | ○ | ○ |
| 2 | ○ | | ⬡ | |
| 3 | ○ | | | |

**Tableau 7.4.8**

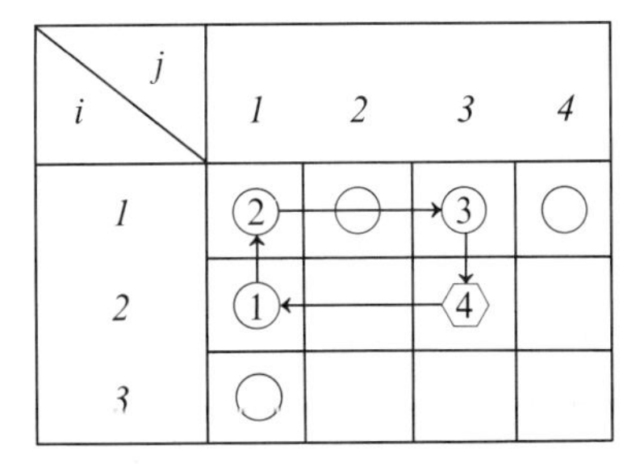


**Step 1** Starting with cell (2, 3) we proceed horizontally to (2, 1); there is no other cell in row 2.*

**Step 2** At this point we make a choice between (1, 1) and (3, 1). If we were to choose (3, 1), the path would stop prematurely. We proceed vertically to (1, 1).

**Step 3** Of the three potential horizontal choices, (1, 2) and (1, 4) do not lead anywhere. Therefore we proceed to (1, 3).

**Step 4** The path can now be completed by advancing vertically to the original (2, 3) cell.

---

* We could have proceeded vertically to (1, 3) with the same final result.

**Step 5** Observing the closed path created above, we see that $A_{21}$, $A_{11}$, and $A_{13}$ are involved in the closed path. Thus we can write

$$A_{23} = A_{21} - A_{11} + A_{13}$$

which we verify as

$$\begin{bmatrix} 0 \\ 1 \\ 0 \\ \hdashline 0 \\ 0 \\ 1 \\ 0 \end{bmatrix} = \begin{bmatrix} 0 \\ 1 \\ 0 \\ \hdashline 1 \\ 0 \\ 0 \\ 0 \end{bmatrix} - \begin{bmatrix} 1 \\ 0 \\ 0 \\ \hdashline 1 \\ 0 \\ 0 \\ 0 \end{bmatrix} + \begin{bmatrix} 1 \\ 0 \\ 0 \\ \hdashline 0 \\ 0 \\ 1 \\ 0 \end{bmatrix}$$ ∎

Example 7.4.2 In Tableau 7.4.9, another set of six linearly independent vectors is represented by circles in the appropriate cells. It should be visually evident here also that no closed path exists which connects any subset of the circle cells.

**Tableau 7.4.9**

| $i \setminus j$ | 1 | 2 | 3 | 4 |
|---|---|---|---|---|
| 1 | ○ | ○ | | |
| 2 | | ○ | ○ | |
| 3 | ⬡ | | ○ | ○ |

Let us proceed to find a closed path that contains cell (3, 1). The solution is shown in Tableau 7.4.10.

**Step 1** We proceed vertically and have no choice but to reach (1, 1).

**Step 2** We proceed horizontally and likewise reach (1, 2).

**Step 3** As in Step 1 we reach (2, 2).

**Step 4** As in Step 2 we reach (2, 3).

**Step 5** As in Step 1 we reach (3, 3).

**Step 6** The path can now be completed by returning to the (3, 1) cell.

**Tableau 7.4.10**

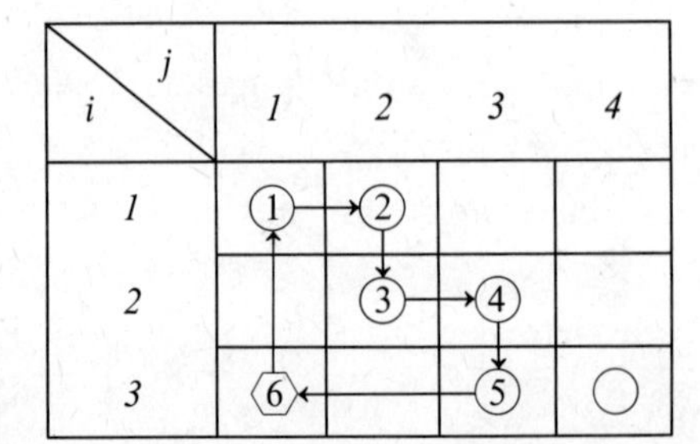

**Step 7** Observing the closed path obtained above we write

$$A_{31} = A_{11} - A_{12} + A_{22} - A_{23} + A_{33}$$

which we verify from

$$\begin{bmatrix} 0 \\ 0 \\ 1 \\ \hline 1 \\ 0 \\ 0 \\ 0 \end{bmatrix} = \begin{bmatrix} 1 \\ 0 \\ 0 \\ \hline 1 \\ 0 \\ 0 \\ 0 \end{bmatrix} - \begin{bmatrix} 1 \\ 0 \\ 0 \\ \hline 0 \\ 1 \\ 0 \\ 0 \end{bmatrix} + \begin{bmatrix} 0 \\ 1 \\ 0 \\ \hline 0 \\ 1 \\ 0 \\ 0 \end{bmatrix} - \begin{bmatrix} 0 \\ 1 \\ 0 \\ \hline 0 \\ 0 \\ 1 \\ 0 \end{bmatrix} + \begin{bmatrix} 0 \\ 0 \\ 1 \\ \hline 0 \\ 0 \\ 1 \\ 0 \end{bmatrix}$$

## 7.4 Exercises

1. Using the information in Tableau 7.4.1 re-create the algebraic representation of the associated TP.

2. Construct a tableau like Tableau 7.4.1 for the TP of Exercise 7.2.1.

3. (*True* or *false*)

   a. The $S_i$ and $D_j$ values must be known before one can determine whether $A_{11}$ (in Exercise 2) is a linear combination of any given subset of $A_{ij}$ vectors.

   b. The $c_{ij}$ values are irrelevant for determining linear independence of a set of $A_{ij}$ vectors.

4. Use Tableau 7.4.2 to find a closed path connecting $A_{11}, A_{13}, A_{21}$, and $A_{23}$. Express $A_{23}$ as a linear combination of the other 3 vectors. Verify your solution by appropriately adding and subtracting the involved $A_{ij}$ vectors.

5. Use Tableau 7.4.7 to find a closed path connecting $A_{11}, A_{13}, A_{31}$ and $A_{33}$. Express $A_{33}$ as a linear combination of the other 3 vectors.

6. Consider Exercise 7.2.5.

   a. Construct a tableau similar to Tableau 7.4.7 with circles in the (1, 1), (1, 2), (2, 2), and (2, 3) positions. Since there is no closed path connecting these circles, the associated vectors are linearly ________.

   b. Put a hexagon in the (1, 3) cell, and find a closed path connecting this cell with the circle cells. Express $A_{13}$ as a linear combination of the $A_{ij}$ represented by the circles. (Refer to Exercise 7.3.5 e after completing your solution.)

   c. Use the method of b. above to express $A_{14}$ as a linear combination of the circle cell vectors $A_{ij}$. (Compare your result with Exercise 7.3.5 f).

   d. Use the circle cell method of this section to express $A_{14}$ as a linear combination of $A_{11}, A_{12}, A_{22}, A_{23}, A_{33}$, and $A_{34}$.

7. Consider Exercise 7.2.7.

   a. Construct a circle cell tableau with circles in the (1, 1), (1, 4), (2, 1) and (2, 4) positions. The associated vectors are linearly ________.

b. Use a tableau similar to that in a. to represent $A_{13}$ as a linear combination of $A_{11}$, $A_{21}$, $A_{22}$, $A_{32}$, and $A_{33}$. Verify your solution by appropriately adding and subtracting the involved $A_{ij}$ vectors.

## 7.5 An Initial Basic Feasible Solution—the Northwest Corner Rule

Having described the compact solution tableau that may be used because of the special structure of TP, we now turn to a study of the solution procedure itself. Although the specific techniques are different, we will really just be performing the steps of the simplex algorithm, as follows:

**Step 1** Find an initial basic feasible solution (BFS).

**Step 2** Is the present BFS optimal? If yes, we stop. If no, we proceed to Step 3.

**Step 3** Which non-basic vector should enter the basis?

**Step 4** Which basic vector should leave the basis?

**Step 5** Complete the new tableau, and return to Step 2.

The purpose of this section is to explain Step 1. The special procedures for the remaining steps will be explained in subsequent sections.

One of the simplest techniques for finding an initial BFS is known as the "Northwest Corner Rule." We shall demonstrate its use for Problem $T_2$ in a partial tableau presented in Tableau 7.5.1. Note that no cost information is presented. Such data are not required at this stage.*

**Tableau 7.5.1**

| $i$ \ $j$ | 1 | 2 | 3 | 4 | $S_i$ |
|---|---|---|---|---|---|
| 1 | $x_{11}$ | $x_{12}$ | $x_{13}$ | $x_{14}$ | 20 |
| 2 | $x_{21}$ | $x_{22}$ | $x_{23}$ | $x_{24}$ | 30 |
| 3 | $x_{31}$ | $x_{32}$ | $x_{33}$ | $x_{34}$ | 50 |
| $D_j$ | 40 | 30 | 20 | 10 | |

As we have already established, a BFS for the TP should have $m + n - 1$ basic variables $x_{ij}$. In this case $m + n - 1 = 3 + 4 - 1 = 6$. Thus our problem is to choose 6 out of the 12 cells that correspond to linearly independent vectors, $A_{ij}$, such that the solution is feasible so that $x_{ij} \geqslant 0$. In order to establish feasibility, it is necessary to know the right-hand side vector $R$ which contains the constants in the various equations. Therefore the $S_i$, $D_j$ data are needed. In order to insure that we choose a linearly independent set of vectors, it is sufficient that the technique guarantee that no subset of

* There are other techniques for finding an initial BFS which consider cost information and will normally provide a BFS with a lower value of $f$. On the average, fewer iterations of the solution algorithm would then be required.

the $m + n - 1$ basic cells form a closed loop of the type discussed in Section 7.4.

The name "Northwest Corner Rule" was derived from the fact that the first basic vector chosen is always $A_{11}$, that is, we always choose to make $x_{11}$ a basic variable. Since the $x_{11}$ cell is in the *northwest* corner of the tableau, the name is appropriate.* The cells chosen will generally fall into a northwest-southeast pattern.

Since $x_{11}$ is to be non-zero, the only question remaining is, "How large can $x_{11}$ be?" Referring to Tableau 7.5.1, we see that $x_{11}$ is involved in two equations, namely

$$x_{11} + x_{12} + x_{13} + x_{14} = 20,$$

and

$$x_{11} + x_{21} + x_{31} = 40.$$

Even if each of the remaining variables is zero, $x_{11}$ can be no larger than 20 and still satisfy the first equation. Thus, we let $x_{11} = 20$, and indicate this in Tableau 7.5.2 by writing "20" in the (1, 1) cell. The circle indicates that $x_{11}$ is a basic variable. $S_1$ and $D_1$ are reduced by 20 before considering further allocations. By this first decision we have satisfied exactly one of the 7 equations by a judicious choice of one of the 6 basic variables.

**Tableau 7.5.2**

| $i \backslash j$ | 1 | 2 | 3 | 4 | $S_i$ |
|---|---|---|---|---|---|
| 1 | (20) | | | | ~~20~~ 0 |
| 2 | | | | | 30 |
| 3 | | | | | 50 |
| $D_j$ | ~~40~~ 20 | 30 | 20 | 10 | |

*This is exactly the procedure that will be followed at each step. We will choose a value for a particular basic variable that will satisfy exactly one of the remaining constraints.* Because of the fact that the sum of the supplies equals the sum of the demands, that is,

$$S_1 + S_2 + S_3 = D_1 + D_2 + D_3 + D_4,$$

our last choice of a linearly independent vector (the sixth) will satisfy not only the sixth equation, but also the seventh. Let us proceed.

When we set $x_{11} = 20$, we satisfy the first row equation; but the first column equation remains unsatisfied, that is, $D_1' = 20$. In TP terms, we shipped out all 20 of the items available at source 1, but destination 1 still has an unfulfilled demand of 20 units.

We proceed to satisfy equation 2 by considering the second variable in this equation, $x_{21}$. Since $x_{21}$ is also involved in two equations, namely

$$x_{21} + x_{22} + x_{23} + x_{24} = 30,$$

* Any other corner could have been chosen to generate the initial BFS.

and a revised first column equation (with $x_{11} = 20$)

$$x_{21} + x_{31} = 40 - 20 = 20,$$

we see that the maximum permissible value for $x_{21}$ is 20, an amount which satisfies the first column equation but leaves the second row equation unsatisfied. The latter equation is the one which must be satisfied next. The situation so far is shown in Tableau 7.5.3.

**Tableau 7.5.3**

| i \ j | 1 | 2 | 3 | 4 | $S_i$ |
|---|---|---|---|---|---|
| 1 | (20) ↓ | | | | ~~20~~ 0 |
| 2 | (20) | | | | ~~30~~ 10 |
| 3 | | | | | 50 |
| $D_j$ | ~~40~~ ~~20~~ 0 | 30 | 20 | 10 | |

The revised row 2 equation (with $x_{21} = 20$),

$$x_{22} + x_{23} + x_{24} = 10,$$

and the column 2 equation

$$x_{12} + x_{22} + x_{23} = 30$$

are examined next. Since $\min\{10, 30\} = 10$, we satisfy the row 2 equation by letting $x_{22} = 10$ as shown in Tableau 7.5.4.

The revised column 2 equation (with $x_{22} = 10$) is $x_{12} + x_{32} = 20$. Two facts concerning the possible use of $x_{12}$ can be illustrated. If $x_{12}$ were chosen to be a basic variable, the set $\{A_{11}, A_{12}, A_{21}, A_{22}\}$ would be linearly dependent, as is made clear by the existence of a closed loop including those cells. In addition, of course, $x_{12}$ is also involved in the row 1 equation. If we were to let $x_{12}$ be positive, we would, correspondingly, have to decrease $x_{11}$ from its present value of 20 to keep the row 1 equation satisfied. As far as

**Tableau 7.5.4**

| i \ j | 1 | 2 | 3 | 4 | $S_i$ |
|---|---|---|---|---|---|
| 1 | (20) ↓ | | | | ~~20~~ 0 |
| 2 | (20) → | (10) | | | ~~30~~ ~~10~~ 0 |
| 3 | | | | | 50 |
| $D_j$ | ~~40~~ ~~20~~ 0 | ~~30~~ 20 | 20 | 10 | |

the revised column 2 equation is concerned it is really only $x_{32}$ which may be utilized to satisfy it.* Since $\min\{20, 50\} = 20$, we let $x_{32} = 20$, as shown in Tableau 7.5.5. The row equation in which $x_{32}$ is also involved is not yet satisfied. At this point we have assigned positive values to 4 variables and have satisfied at the same time exactly 4 constraints. As you have probably noticed the fact that a constraint is satisfied is indicated in the tableau margin when the revised $S_i$ or $D_j$ is zero.

**Tableau 7.5.5**

| i \ j | 1 | 2 | 3 | 4 | $S_i$ |
|---|---|---|---|---|---|
| 1 | (20) | | | | ~~20~~ 0 |
| 2 | (20) → | (10) | | | ~~30~~ ~~10~~ 0 |
| 3 | | (20) | | | ~~50~~ 30 |
| $D_j$ | ~~40~~ ~~20~~ 0 | ~~30~~ ~~20~~ 0 | 20 | 10 | |

Since we left the row 3 equation incompletely satisfied by our last choice ($x_{32} = 20$), we consider the revised row 3 equation

$$x_{31} + x_{33} + x_{34} = 30.$$

It is clear that $x_{31}$ cannot be increased without disturbing the values of $x_{11}$ and $x_{21}$, in addition to the fact that $\{A_{21}, A_{22}, A_{31}, A_{32}\}$ is linearly dependent. (Why?) Thus we choose $x_{33}$ as our new basic variable and let $x_{33} = \min\{20, 30\} = 20$, satisfying the column 3 equation but leaving the row 3 equation still to be finished. The situation after 5 choices is shown in Tableau 7.5.6.

**Tableau 7.5.6**

| i \ j | 1 | 2 | 3 | 4 | $S_i$ |
|---|---|---|---|---|---|
| 1 | (20) | | | | ~~20~~ 0 |
| 2 | (20) → | (10) | | | ~~30~~ ~~10~~ 0 |
| 3 | | (20) → | (20) | | ~~50~~ ~~30~~ 10 |
| $D_j$ | ~~40~~ ~~20~~ 0 | ~~30~~ ~~20~~ 0 | ~~20~~ 0 | 10 | |

We recall that we were searching for 6 basic variables and realize that there is only one more to be chosen. The further revised row 3 equation is now

$$x_{31} + x_{34} = 10.$$

---

* In practice, what this really means is that once a row or column has been *left*, we need not consider further any of its variables for the initial BFS.

Really, only $x_{34}$ may be used (Why?), so we must set $x_{34} = 10$ to satisfy this equation. However, it turns out that this choice of $x_{34}$ also satisfies the column 4 equation at the same time. The final set of 6 basic variables is shown in Tableau 7.5.7. Non-basic variables are made evident by their blank cells.

**Tableau 7.5.7**

| $i$ \ $j$ | 1 | 2 | 3 | 4 | $S_i$ |
|---|---|---|---|---|---|
| 1 | 20 | | | | ~~20~~ 0 |
| 2 | 20 | 10 | | | ~~30~~ ~~10~~ 0 |
| 3 | | 20 | 20 | 10 | ~~50~~ ~~30~~ ~~10~~ 0 |
| $D_j$ | ~~40~~ ~~20~~ 0 | ~~30~~ ~~20~~ 0 | ~~20~~ 0 | ~~10~~ 0 | |

The procedure for the Northwest Corner Rule should now be intuitively clear. We begin in the upper left corner and assign the largest possible value to $x_{11}$, namely $\min\{S_1, D_1\}$. Depending upon whether it is the row 1 or column 1 constraint which is binding, we proceed to use either $x_{12}$ or $x_{21}$, as appropriate, to fulfill the unsatisfied constraint. The algorithm continues in a stair-step fashion, each time letting a particular variable be $\min\{S_i', D_j'\}$, satisfying one constraint and leaving one to be considered.* Here, $S_i'$, $D_j'$ denote the revised source and destination values. When we arrive at the assignment of basic variable $m + n - 1$, however, the choice of basic variable at that time will satisfy two constraints and complete the initial BFS. Since there are no closed loops generated by this pattern of choices, the associated vectors are linearly independent, as desired.

Although we used several tableaus to complete the explanation above, all of the calculations can be done on one tableau, as shown in Tableau 7.5.7.

Example 7.5.1 Let us use the above procedure to find an initial BFS for Problem $T_1$, as shown in Tableau 7.5.8. Use the tableau to write in the appropriate variable values.

**Tableau 7.5.8**

| $i$ \ $j$ | 1 | 2 | 3 | $S_i$ |
|---|---|---|---|---|
| 1 | | | | 30 |
| 2 | | | | 30 |
| $D_j$ | 20 | 20 | 20 | |

* The degenerate case, in which a choice of a single variable happens to satisfy two constraints, will not be considered here. The situation is easily handled and is discussed fully in the references listed in the Suggested Reading.

The completed Tableau 7.5.9 is shown below.

Since $\min\{S_1, D_1\} = \min\{20, 30\} = 20$, we let $x_{11} = 20$ and reduce $D_1$ to 0 and $S_1$ to 10, as shown. Column 1 is correct, so we move to column 2 to satisfy row 1.

Since $\min\{10, 20\} = 10$, we let $x_{12} = 10$ and reduce $S_1'$ to 0 and $D_2$ to 10. Row 1 is correct so we move to row 2 to satisfy column 2.

Since $\min\{10, 30\} = 10$, we let $x_{22} = 10$ and reduce $D_2'$ to 0 and $S_2$ to 20. Column 2 is correct and row 2 is still unsatisfied. We move to column 3.

Since $\min\{20, 20\} = 20$, we let $x_{33} = 20$, noting that row 3 and column 3 are both satisfied. The initial BFS is

$$[x_{11}, x_{12}, x_{13}, x_{21}, x_{22}, x_{23}] = [20, 10, 0, 0, 10, 20].$$

In this case $m = 2$, $n = 3$, so $m + n - 1 = 4$. We sought 4 linearly independent vectors. The compact tableau shows no closed loops, so the vectors are linearly independent. Those vectors are

$$A_{11} = \left[\begin{array}{c} 1 \\ 0 \\ \hline 1 \\ 0 \\ 0 \end{array}\right], \quad A_{12} = \left[\begin{array}{c} 1 \\ 0 \\ \hline 0 \\ 1 \\ 0 \end{array}\right], \quad A_{22} = \left[\begin{array}{c} 0 \\ 1 \\ \hline 0 \\ 1 \\ 0 \end{array}\right], \quad A_{23} = \left[\begin{array}{c} 0 \\ 1 \\ \hline 0 \\ 0 \\ 1 \end{array}\right],$$

and we note from the positions of 1's that these vectors are indeed linearly independent. ∎

**Tableau 7.5.9**

| $i$ \ $j$ | 1 | 2 | 3 | $S_i$ |
|---|---|---|---|---|
| 1 | (20) → | (10) ↓ | | ~~30~~ ~~10~~ 0 |
| 2 | | (10) → | (20) | ~~30~~ ~~20~~ 0 |
| $D_j$ | ~~20~~ 0 | ~~20~~ ~~10~~ 0 | ~~20~~ 0 | |

Example 7.5.2 Finally, let us find an initial BFS for the TP shown in Tableau 7.5.10. In this case $m + n - 1 = 7$, so we need 7 basic variables. The confident

**Tableau 7.5.10**

| $i$ \ $j$ | 1 | 2 | 3 | 4 | 5 | $S_i$ |
|---|---|---|---|---|---|---|
| 1 | | | | | | 25 |
| 2 | | | | | | 45 |
| 3 | | | | | | 50 |
| $D_j$ | 30 | 30 | 20 | 35 | 5 | |

reader should use the blank tableau to attempt the solution before consulting the following discussion or the completed Tableau 7.5.11. We let $x_{11} = \min\{S_1, D_1\} = \min\{25, 30\} = 25$. Since $S_1' = 0$ and $D_1' = 5$, we go to row 2 to satisfy the remainder of the column 1 demand.

We set $x_{21} = \min\{45, 5\} = 5$. Thus $S_2' = 40$ and $D_1' = 0$. Column 1 demand is satisfied, and we continue in row 2 to column 2.

Since $x_{22} = \min\{S_2', D_2\} = \min\{40, 30\} = 30$, we let $S_2' = 40 - 30 = 10$ and let $D_2' = 30 - 30 = 0$. Column 2 demand is satisfied; so we move to column 3, still in row 2.

We let $x_{23} = \min\{S_2', D_3\} = \min\{10, 20\} = 10$ and set $S_2' = 0$ and $D_3' = 10$. Row 2 has been satisfied; so we move on to row 3, still in the unsatisfied column 3.

Since $\min\{S_3', D_3'\} = \min\{50, 10\} = 10$, we let $x_{33} = 10$ and set $S_3' = 40$ and $D_3' = 0$, indicating that column 3 demand has been satisfied.

Moving to column 4, we note that $x_{34} = \min\{40, 35\} = 35$ and let $S_3' = 5$ and $D_4' = 0$, showing that column 4 demand is satisfied.

Finally, $x_{35} = \min\{5, 5\} = 5$, satisfying both the row 3 and column 5 supply and demand constraints. No closed loops exist; so the associated vectors are linearly independent, as desired. The reader should write out the appropriate vectors and verify them by observation of the unit elements. In addition, it is helpful to test the values of the variables in the appropriate row and column equations to verify that each equation is satisfied. ∎

**Tableau 7.5.11**

| $i$ \ $j$ | 1 | 2 | 3 | 4 | 5 | $S_i$ |
|---|---|---|---|---|---|---|
| 1 | (25) ↓ | | | | | ~~25~~ 0 |
| 2 | (5) → | (30) → | (10) ↓ | | | ~~45~~ ~~40~~ ~~10~~ 0 |
| 3 | | | (10) → | (35) → | (5) | ~~50~~ ~~40~~ ~~5~~ 0 |
| $D_j$ | ~~30~~ ~~5~~ 0 | ~~30~~ 0 | ~~20~~ ~~10~~ 0 | ~~35~~ 0 | ~~5~~ 0 | |

## 7.5 Exercises

1. An initial BFS for a TP with $m$ sources and $n$ destinations must contain ________ basic variables, thus ________ cells will contain circles and ________ cells will be blank.

2. (*True* or *false*) Since cost information is not used in obtaining an initial BFS by the Northwest Corner Rule, it should not be expected that the initial feasible solution will be an especially "good" one.

3. If $S_1 = S_2 = 60$ and $D_1 = D_2 = D_3 = 40$ for a TP with 2 sources and 3 destinations, which of the following $X$ vectors would be suitable as a BFS?
   If $X$ is not suitable, say why.
   a. $X^T = [40, 20, 0, 0, 20, 40]$.
   b. $X^T = [20, 20, 20, 20, 20, 20]$.
   c. $X^T = [50, 10, 0, 0, 30, 40]$.
   d. $X^T = [50, 20, -10, -10, 20, 50]$.
   e. $X^T = [40, 0, 20, 0, 40, 20]$.

4. To obtain an initial BFS by the Northwest Corner Rule why is it necessary that the total supply equal the total demand? (Consider the number of constraints satisfied by each choice of a new $x_{ij}$.)

5. Consider Exercise 7.2.1.
   a. Construct a tableau similar to Tableau 7.5.8, and find an initial BFS vector $X$ by the Northwest Corner Rule.
   b. Verify by observation that there are no closed loops connecting any subset of the circle cells. The associated vectors are thus ____________.
   c. The total cost associated with $X$ is ________.

6. Consider Exercise 7.2.7.
   a. Use Tableau 7.5.12 to find an initial BFS by the Northwest Corner Rule.

**Tableau 7.5.12**

| $i$ \ $j$ | 1 | 2 | 3 | $S_i$ |
|---|---|---|---|---|
| 1 | | | | 20 |
| 2 | | | | 15 |
| 3 | | | | 25 |
| $D_j$ | 30 | 15 | 15 | |

   b. The associated objective function value is ________.
   c. Find a second BFS by using a Northwest Corner Rule.

7. Consider Exercise 7.2.9. Construct an appropriate tableau and find an initial BFS.

### Suggested Reading

S. I. Gass. *Linear Programming.* 3d. ed. New York: McGraw-Hill, 1969.

T. H. Naylor, E. T. Byrne, and J. M. Vernon. *Introduction to Linear Programming: Methods and Cases.* Belmont, Ca.: Wadsworth, 1971.

## 7.6 Determination of Optimality and the Entering Basis Vector

The initial BFS that we obtained for Problem $T_2$ is shown in Tableau 7.6.1. The cost coefficients have been added so that the objective function value of any BFS can be calculated and so that the question "Is the current BFS optimal?" can be answered. If it is, of course, the algorithm terminates. If not, then we must decide which non-basic variable should become basic in order to achieve a cheaper solution. Since the non-basic variables associated with the initial BFS are zero, the initial objective function value is

$$2(20) + 3(20) + 6(10) + 7(20) + 6(20) + 4(10) = 460,$$

as we indicate in the lower right corner of the tableau. Additional symbols, $u_i$ and $v_j$, have been added to the tableau. Their purpose will be explained shortly.

**Tableau 7.6.1**

| $i$ \ $j$ | 1 | 2 | 3 | 4 | $S_i$ | $u_i$ |
|---|---|---|---|---|---|---|
| 1 | [2] (20) | [3] | [4] | [5] | 20 | $u_1$ |
| 2 | [3] (20) | [6] (10) | [8] | [3] | 30 | $u_2$ |
| 3 | [4] | [7] (20) | [6] (20) | [4] (10) | 50 | $u_3$ |
| $D_j$ | 40 | 30 | 20 | 10 | $f = 460$ | |
| $v_j$ | $v_1$ | $v_2$ | $v_3$ | $v_4$ | | |

In executing the standard simplex algorithm we answered the optimality question by computing $c_j - z_j$ for each non-basic $A_j$ and checked its algebraic sign, terminating in a minimization problem if $c_j - z_j$ was non-negative in each case. In that algorithm, we computed $z_j = c_B Y_j = c_B B^{-1} A_j$. In the present situation, however, we do not have $Y_j$ directly available; so an alternate means is required. The method to be explained, known as the "*u,v* method," is simple to learn and quick to apply, compared with other alternatives.* The relevant formula for expressing $z_{ij}$ in this case is

$$z_{ij} = u_i + v_j$$

where $u_i$ and $v_j$ are found by solving a simple and readily available set of equations, each of which depends upon the fact that $z_{ij} = c_{ij}$ if $x_{ij}$ is a basic variable, that is, $c_{ij} - z_{ij} = 0$. As we shall see, the structure of the system of equations is so simple that we will be able to solve the equations right on the original tableau.

In accordance with tradition, then, we designate 7 new variables, one for each of the 3 row constraints and one for each of the 4 column constraints. The 3 row variables are denoted $u_1$, $u_2$, $u_3$, and the 4 column variables

* The method is derived from the study of an LP problem intimately related to the given TP. The LP problem is known as the *dual* of the original problem. The $u_i$ and $v_j$ used in this section are the variables in the dual problem. The theory of duality is very rich in practical results. For a detailed discussion, the interested reader should consult one of the references at the end of this chapter.

$v_1$, $v_2$, $v_3$, and $v_4$. If we could find $u_2$ and $v_3$, for example, then the general formula above tells us that $z_{23} = u_2 + v_3$.

Observing Tableau 7.6.1, we see that each basic cell is associated with exactly one $u_i$ and one $v_j$, that is, the ones in its row and column. Proceeding through the *basic* cells from left to right and top to bottom, we use the fact that $z_{ij} = c_{ij}$ to generate the 6 equations (in 7 unknowns), each of the form $u_i + v_j = c_{ij}$, as follows:

$$\begin{array}{llllllll}
u_1 & & & + v_1 & & & & = 2 \\
& u_2 & & + v_1 & & & & = 3 \\
& u_2 & & & + v_2 & & & = 6 \\
& & u_3 & & + v_2 & & & = 7 \\
& & u_3 & & & + v_3 & & = 6 \\
& & u_3 & & & & + v_4 & = 4.
\end{array}$$

Since there are fewer equations ($m + n - 1$, in general) than unknowns ($m + n$, in general), there are an infinite number of potential solutions. For our purposes, however, we need only one solution; and it is readily available. Observing the last equation, we treat $v_4$ as a free variable and for convenience set $v_4 = 0$.* Knowing that $v_4 = 0$ allows us to compute $u_3 = 4$ in the last equation. If $u_3 = 4$, we can use the fifth equation to compute $v_3 = 2$ and the fourth equation to compute $v_2 = 3$. Since $v_2 = 3$, we can use the third equation to compute $u_2 = 3$. This can then be used in the second equation to compute $v_1 = 0$, which, in the first equation, gives $u_1 = 2$. The special format of these derived equations permits us to solve very quickly for a solution which in this case is

$$[u_1, u_2, u_3, v_1, v_2, v_3, v_4] = [2, 3, 4, 0, 3, 2, 0].$$

Before proceeding to use the $u_i$, $v_j$ values to find $c_{ij} - z_{ij}$, we first demonstrate how the above values for the $u_i$, $v_j$ can be found directly from the tableau without explicitly writing down the equations. For this purpose we consider Tableau 7.6.2. We first recognize that there is an equation of the form $u_i + v_j = c_{ij}$ for each of the 6 basic cells. We also know that there is

**Tableau 7.6.2**

| $i$ \ $j$ | 1 | 2 | 3 | 4 | $u_i$ |
|---|---|---|---|---|---|
| 1 | [2] (20) | | | | $u_1 = 2$ |
| 2 | [3] (20) | [6] (10) | | | $u_2 = 3$ |
| 3 | | [7] (20) | [6] (20) | [4] (10) | $u_3 = 4$ |
| $v_j$ | $v_1 = 0$ | $v_2 = 3$ | $v_3 = 2$ | $v_4 = 0$ | |

* Another choice of a free variable to set equal to zero would lead to different values for $u_i$ and $v_j$, but the various sums $u_i + v_j$ would remain the same.

one free variable. We choose arbitrarily to let $v_4 = 0$ and proceed with the computations.

Since $x_{34}$ is a basic variable, $u_3 + v_4 = c_{34} = 4$. Since $v_4 = 0$, $u_3 = 4$; and we record this in the column to the right of the tableau. But $u_3$ is involved with $v_3$, since $x_{33}$ is basic, and with $v_2$, since $x_{32}$ is basic (the $x_{33}$ and $x_{32}$ cells are occupied). Knowing $u_3 = 4$, we see that $v_3 = 2$ and $v_2 = 3$; and we record their values. Whereas the value of $v_3$ gives us no further help (Why?), the knowledge that $v_2 = 3$ tells us that $u_2 = 3$, since $x_{22}$ is basic and $u_2 + v_2 = 6$. We record $u_2 = 3$ and then consider $v_1$ since $x_{21}$ is basic. Since $u_2 + v_1 = 3$, $v_1 = 0$, which we record. Since $x_{11}$ is basic, that is, $u_1 + v_1 = 2$, we have $u_1 = 2$; thus with this equation we conclude the exercise.

Once the $u_i$ and $v_j$ are calculated, it is a simple matter to compute $c_{ij} - z_{ij}$ for each non-basic variable $x_{ij}$, since $z_{ij} = u_i + v_j$. Thus for example, $z_{12} = u_1 + v_2 = 5$, namely the sum of the appropriate row element ($u_1$) and the appropriate column element ($v_2$). Similarly, $z_{23} = u_2 + v_3 = 3 + 2 = 5$.

We write down our calculated $c_{ij} - z_{ij}$ values for non-basic $x_{ij}$, in a compact format such as that of Tableau 7.6.3. The shaded circles indicate the basic cells. To calculate $c_{12} - z_{12}$ we compute $c_{12} - (u_1 + v_2) = 3 - (2 + 3) = -2$ and record the value in the (1, 2) cell. Similarly, we calculate and record

$$\begin{aligned}
c_{13} - z_{13} &= c_{13} - (u_1 + v_3) = 4 - (2 + 2) = 0 \\
c_{14} - z_{14} &= c_{14} - (u_1 + v_4) = 5 - (2 + 0) = 3 \\
c_{23} - z_{23} &= c_{23} - (u_2 + v_3) = 8 - (3 + 2) = 3 \\
c_{24} - z_{24} &= c_{24} - (u_2 + v_4) = 3 - (3 + 0) = 0 \\
c_{31} - z_{31} &= c_{31} - (u_3 + v_1) = 4 - (4 + 0) = 0.
\end{aligned}$$

Since Tableau 7.6.3 has 1 non-basic cell with negative $c_{ij} - z_{ij}$, the present tableau is not optimal; and we would move on to Step 3 of the simplex algorithm to decide which non-basic variable should become basic. The selection criterion is unchanged; we select the $x_{ij}$ with the *most negative* value of $c_{ij} - z_{ij}$.* In this tableau, $x_{12}$ would thus be chosen.

**Tableau 7.6.3**

| $i \backslash j$ | 1 | 2 | 3 | 4 | $u_i$ |
|---|---|---|---|---|---|
| 1 | [ ] ● | [3] −2 | [4] 0 | [5] 3 | $u_1 = 2$ |
| 2 | [ ] ● | [ ] ● | [8] 3 | [3] 0 | $u_2 = 3$ |
| 3 | [4] 0 | [ ] ● | [ ] ● | [ ] ● | $u_3 = 4$ |
| $v_j$ | $v_1 = 0$ | $v_2 = 3$ | $v_3 = 2$ | $v_4 = 0$ | |

* If this results in a contest between equally most negative variables, any of the *tied variables* may be selected. If we go from top to bottom (and left to right) in calculating $c_{ij} - z_{ij}$, we will normally pick the first variable in the *tied* set.

Example 7.6.1 We shall continue the solution of Problem $T_1$, finding the $c_{ij} - z_{ij}$ values for the initial BFS constructed in Example 7.5.1 and reproduced as Tableau 7.6.4. The calculations are recorded as Tableau 7.6.5.

**Tableau 7.6.4**

| $i$ \ $j$ | 1 | 2 | 3 | $S_i$ |
|---|---|---|---|---|
| 1 | [4] (20) | [9] (10) | [8] | 30 |
| 2 | [3] | [5] (10) | [7] (20) | 30 |
| $D_j$ | 20 | 20 | 20 | |

To solve sequentially for the $u_i$, $v_j$, we let $v_3 = 0$. In turn, we find $u_2 = 7$, $v_2 = -2$, $u_1 = 11$, and $v_1 = -7$. For the 2 non-basic cells, we then compute

$$c_{13} - z_{13} = c_{13} - (u_1 + v_3) = 8 - (11 + 0) = -3,$$
$$c_{21} - z_{21} = c_{21} - (u_2 + v_1) = 3 - (7 - 7) = 3,$$

and record the calculated values in the appropriate cells.* Since $c_{13} - z_{13} < 0$, the present tableau is not optimal; and $A_{13}$ would enter the basis. The present value of the objective function $f$ is

$$4(20) + 9(10) + 5(10) + 7(20) = 380.$$ ∎

**Tableau 7.6.5**

| $i$ \ $j$ | 1 | 2 | 3 | $S_i$ | $u_i$ |
|---|---|---|---|---|---|
| 1 | [4] (20) | [9] (10) | [8] −3 | 30 | 11 |
| 2 | [3] 3 | [5] (10) | [7] (20) | 30 | 7 |
| $D_j$ | 20 | 20 | 20 | $f = 380$ | |
| $v_j$ | −7 | −2 | 0 | | |

Example 7.6.2 We now consider the initial BFS generated in Example 7.5.2, as shown in Tableau 7.6.6. The $c_{ij}$ have been added to complete the problem formulation. You should attempt to go through the calculation of the various $u_i$ and $v_j$ and then determine $c_{ij} - z_{ij}$ for the 8 non-basic cells. The completed tableau (Tableau 7.6.7) is shown on the next page.

* The fact that these values are not circled means that they are $c_{ij} - z_{ij}$ values for non-basic variables, and not $x_{ij}$ values for a basic variable.

Tableau 7.6.6

| i \ j | 1 | 2 | 3 | 4 | 5 | $S_i$ | $u_i$ |
|---|---|---|---|---|---|---|---|
| 1 | [4] (25) | [5] | [8] | [3] | [7] | 25 | |
| 2 | [3] (5) | [6] (30) | [5] (10) | [5] | [4] | 45 | |
| 3 | [6] | [4] | [3] (10) | [7] (35) | [6] (5) | 50 | |
| $D_j$ | 30 | 30 | 20 | 35 | 5 | | |
| $v_j$ | | | | | | | |

For those who need assistance, the various steps are outlined below, arbitrarily setting $v_5 = 0$.

With $v_5 = 0$, we compute in sequence $u_3 = 6, v_4 = 1, v_3 = -3, u_2 = 8, v_2 = -2, v_1 = -5$, and $u_1 = 9$. Starting with the first row, we compute $c_{12} - z_{12} = c_{12} - (u_1 + v_2) = 5 - (9 - 2) = -2$, and $c_{13} - z_{13} = 8 - 6 = 2$, $c_{14} - z_{14} = 3 - 10 = -7$, $c_{15} - z_{15} = 7 - 9 = -2$, and so forth. Since at least one of the $c_{ij} - z_{ij}$ is negative, the present tableau is not optimal; and $x_{14}$ would be chosen to become a basic variable. The present value of the objective function, as you should verify, is 650. ∎

Tableau 7.6.7

| i \ j | 1 | 2 | 3 | 4 | 5 | $S_i$ | $u_i$ |
|---|---|---|---|---|---|---|---|
| 1 | [4] (25) | [5] −2 | [8] 2 | [3] −7 | [7] −2 | 25 | 9 |
| 2 | [3] (5) | [6] (30) | [5] (10) | [5] −4 | [4] −4 | 45 | 8 |
| 3 | [6] 5 | [4] 0 | [3] (10) | [7] (35) | [6] (5) | 50 | 6 |
| $D_j$ | 30 | 30 | 20 | 35 | 5 | $f = 650$ | |
| $v_j$ | −5 | −2 | −3 | 1 | 0 | | |

## 7.6 Exercises

1. a. (*True* or *false*) For any basic variable $x_j$ in any *LP problem*, $c_j - z_j = 0$, or $c_j = z_j$.
   b. In the double subscript notation of a TP, the above equation translates into $c_{ij} =$ ________.
   c. (*True* or *false*) If $z_{ij} = u_i + v_j$, then for any basic variable $x_{ij}$, $u_i + v_j = c_{ij}$.

2. In the $u$, $v$ method of determining optimality there is one $u_i$ for each ________ constraint and one $v_j$ for each ________ constraint. For a TP with $m$ sources and

$n$ destinations there are ________ $u_i$ variables and ________ $v_j$ variables. Since one linear equation involving $u_i$ and $v_j$ is obtained from each basic cell, we obtain ________ equations in the ________ unknowns, $u_i$ and $v_j$. There is thus always ________ free variable which may be set equal to 0 to start the sequential solution process.

3. Consider the BFS for Exercise 7.2.1 shown in Tableau 7.6.8.

**Tableau 7.6.8**

| | i \ j | 1 ($v_1$) | 2 ($v_2$) | 3 ($v_3$) | $S_i$ |
|---|---|---|---|---|---|
| $u_1$ | 1 | [2] | [5] (15) | [6] (25) | 40 |
| $u_2$ | 2 | [3] (20) | [4] (15) | [7] | 35 |
| | $D_j$ | 20 | 30 | 25 | |

a. Letting the $u_i$ and $v_j$ be associated with the rows and columns indicated in the tableau, use the fact that $z_{ij} = c_{ij}$ for basic variables to write down a valid set of equations involving the $u_i$ and $v_j$. We have ________ equations in ________ unknowns.

b. In the equations of a. set $v_3 = 0$ and solve sequentially for $u_1$, $v_2$, $u_2$ and $v_1$ in that order.

c. Use the results of b. to compute $z_{ij}$ values for non-basic $x_{ij}$ as follows: $z_{11} = u_1 +$ ________ $=$ ________; $z_{23} =$ ________ $+$ ________ $=$ ________.

d. $c_{11} - z_{11} =$ ________; $c_{23} - z_{23} =$ ________. Thus the present tableau (*is, is not*) optimal. If not, then ________ should be the new basic variable.

e. Suppose that you had set $u_1 = 0$ in b. above, rather than $v_3$. Solve for $v_2$, $u_2$, $v_1$ and $u_3$ in that order. Are the values the same as before?

f. Compute $z_{11}$ and $z_{23}$ using the values of $u_i$ and $v_j$ from e. Are they the same as in c.

4. (*True* or *false*) The special structure of the $u_i$, $v_j$ equations allows them to be solved without explicitly writing them out.

5. Consider Exercise 7.2.3.

a. Using Tableau 7.6.9 below, find an initial BFS using the Northwest Corner Rule.

**Tableau 7.6.9**

| i \ j | 1 | 2 | 3 | $S_i$ | $u_i$ |
|---|---|---|---|---|---|
| 1 | [5] | [8] | [6] | 60 | |
| 2 | [4] | [9] | [6] | 40 | |
| $D_j$ | 35 | 30 | 35 | | |
| $v_j$ | | | | | |

b. Determine the $u_i$ and $v_j$ by setting $v_3 = 0$.
c. Find $c_{13} - z_{13}$ and $c_{21} - z_{21}$. What is $c_{11} - z_{11}$?
d. The present tableau (*is*, *is not*) optimal. If not, then ________ should be the new basic variable.

6. Consider Exercise 7.5.6. Use Tableau 7.5.12 with an extra $u_i$ row and a $v_j$ column to determine if the Northwest Corner Rule BFS is optimal. If not, which vector should enter the basis? (See Exercise 7.2.7.)

7. Consider Exercise 7.5.7. Use Tableau 7.6.10, and determine if the indicated BFS is optimal. If not, which non-basic variable should become basic? (See Exercise 7.2.9.)

**Tableau 7.6.10**

| $i$ \ $j$ | 1 | 2 | $S_i$ | $u_i$ |
|---|---|---|---|---|
| 1 | [10] (500) | [5] | 500 | |
| 2 | [4] (400) | [9] | 400 | |
| 3 | [15] (100) | [7] (700) | 800 | |
| 4 | [6] | [2] (650) | 650 | |
| $D_j$ | 1000 | 1350 | | |
| $v_j$ | | | | |

## 7.7 Selecting a Vector to Leave the Basis and Completing the New Tableau

After a particular tableau is discovered to be non-optimal ($c_{ij} - z_{ij} < 0$ for at least 1 non-basic $x_{ij}$) and the appropriate vector $A_{ij}$ is chosen to enter the basis, we must as before determine the vector $A_{kl}$ which is to leave the basis. As we shall see, the choice is again dependent upon feasibility considerations, as in the standard simplex algorithm.

In order to illustrate the procedure, we continue with our discussion of Problem $T_2$. The present situation, after $A_{12}$ has been chosen to enter the basis, is shown in Tableau 7.7.1, which is a version of Tableau 7.6.3 with the now required *supply* and *demand* data included. Since the $c_{ij}$ are not needed for this selection process, they are omitted. The basic variables in Tableau 7.7.1 are represented by circles while the location of the incoming variable $x_{12}$ is indicated by the hexagon.

Let us recall that in the standard simplex algorithm it was not difficult to find the coordinates of any non-basic vector $A_j$ with respect to the present

**Tableau 7.7.1**

| $i \backslash j$ | 1 | 2 | 3 | 4 | $S_i$ |
|---|---|---|---|---|---|
| 1 | ○ | ⬡ | | | 20 |
| 2 | ○ | ○ | | | 30 |
| 3 | | ○ | ○ | ○ | 50 |
| $D_j$ | 40 | 30 | 20 | 10 | $f = 460$ |

basis $\{B_1, B_2, \ldots, B_m\}$. These coordinates were contained in the vector $Y_j = B^{-1}A_j$, so that

$$A_j = y_{1j}B_1 + y_{2j}B_2 + \cdots + y_{mj}B_m.$$

We determined which vector should leave the basis to preserve feasibility by computing the ratios $x_{Bi}/y_{ij}$ for all $y_{ij} > 0$ and chose that vector to leave which yielded the minimum value of the computed ratios. Essentially the same method will be implemented for the TP but in a rather disguised fashion. The procedure is illustrated in the examples which follow. A general rule appears in Exercise 7.7.1.

*The first problem is to determine the coordinates of the incoming vector $A_{12}$ in terms of the present basis* so that we will know which variables change, as $x_{12}$ is increased, and how they change. This was the type of problem investigated in Section 7.4. Whereas the 6 circles in Tableau 7.7.1 represent the largest possible linearly independent set of vectors, the new set which includes $A_{12}$ is clearly dependent. Furthermore, the closed loop in the upper left corner reveals the coordinate representation $A_{12} = (1)A_{11} + (-1)A_{21} + (1)A_{22}$. These $+1$ and $-1$ coefficients of the coordinate representation of $A_{12}$ are recorded in Tableau 7.7.2.

**Tableau 7.7.2**

| $i \backslash j$ | 1 | 2 | 3 | 4 | $S_i$ |
|---|---|---|---|---|---|
| 1 | +1 (20) | | | | 20 |
| 2 | −1 (20) | +1 (10) | | | 30 |
| 3 | | (20) | (20) | (10) | 50 |
| $D_j$ | 40 | 30 | 20 | 10 | $f = 460$ |

The three associated variables, $x_{11}$, $x_{21}$, and $x_{22}$, are the *only* basic variables which will change if $x_{12}$ is increased from 0. Since $A_{21}$ has a $-1$ coefficient, $x_{21}$ will *increase* one unit for each unit increase of $x_{12}$, and thus not affect feasibility. Both $x_{11}$ and $x_{22}$ will *decrease, and at the same rate*, namely one unit of decrease for each unit increase in $x_{12}$. We, therefore, see that the basic variable which has the smallest value to start with would get to 0 first; and this variable must become non-basic. Since $\min\{x_{11}, x_{22}\} = \min\{20, 10\} = 10$, it is clear that the vector which must leave the basis is $A_{22}$, that is, $x_{22}$ becomes non-basic.

As will always be the case in a TP, the pivot element is $+1$. Thus no division is required in going from one tableau to the next—only addition and subtraction.* Those cells which have a $+1$ are decreased by 10 units. Those cells with a $-1$ are increased by 10 units. The values of the remaining basic $x_{ij}$ remain unchanged. The circle in the $x_{22}$ cell is deleted since $x_{22}$ is no longer basic, and the hexagon is replaced by a circle. The new Tableau 7.7.3 is now complete and ready for Step 2 of the simplex algorithm. (See Exercise 7.7.5.) You should note that all row and column constraints are still satisfied. The value of the new objective function is 440.

**Tableau 7.7.3**

| $i$ \ $j$ | 1 | 2 | 3 | 4 | $S_i$ |
|---|---|---|---|---|---|
| 1 | [2] (10) | [3] (10) | [4] | [5] | 20 |
| 2 | [3] (30) | [6] | [8] | [3] | 30 |
| 3 | [4] | [7] (20) | [6] (20) | [4] (10) | 50 |
| $D_j$ | 40 | 30 | 20 | 10 | $f = 440$ |

Example 7.7.1 In Tableau 7.6.5 we found that for Problem $T_1$ $A_{13}$ should enter the basis. The result is shown in Tableau 7.7.4. The closed loop involving $A_{13}$ is obvious and we may write

$$A_{13} = A_{12} - A_{22} + A_{23}.$$

The unit coefficients are indicated in the tableau. If $x_{13}$ increases, then $x_{12}$ and $x_{23}$ will decrease. Since $\min\{10, 20\} = 10$, the new value of

**Tableau 7.7.4**

| $i$ \ $j$ | 1 | 2 | 3 | $S_i$ |
|---|---|---|---|---|
| 1 | [4] (20) | [9] +1 (10) | [8] ⬡ | 30 |
| 2 | [3] | [5] −1 (10) | [7] +1 (20) | 30 |
| $D_j$ | 20 | 20 | 20 | $f = 360$ |

* As we have previously mentioned, if the initial $R$ vector is all-integer, then subsequent $X_B$ vectors will also be all-integer since only addition and subtraction are performed. No fractions will ever be introduced into any tableau.

$x_{13}$ will be 10 and will be achieved when $x_{12}$ reaches 0. The new value for $x_{23}$ is 10 while $x_{22}$ is increased to 20. Tableau 7.7.5 contains the updated information which may now again be investigated for optimal conditions (see Exercise 7.7.6). ∎

**Tableau 7.7.5**

| $i$ \ $j$ | 1 | 2 | 3 | $S_i$ |
|---|---|---|---|---|
| 1 | [4] (20) | [9] | [8] (10) | 30 |
| 2 | [3] | [5] (20) | [7] (10) | 30 |
| $D_j$ | 20 | 20 | 20 | $f = 330$ |

Example 7.7.2 We have found from Tableau 7.6.7 that $A_{14}$ should enter the basis. The closed loop for the representation of $A_{14}$ is shown in Tableau 7.7.6 and indicated by the $+1$ and $-1$ elements of the coordinate representation. Thus, $x_{11}$, $x_{23}$, and $x_{34}$ will decrease equally as $x_{14}$ is increased. Since $\min\{25, 10, 35\} = 10$, $x_{23}$ will become non-basic. Similarly, $x_{21}$ and $x_{33}$ will increase by 10 units. Tableau 7.7.7 contains the values of the new basic variables. The new value of the objective function should be $7(10) = 70$ less than the original value of 650, namely 580. You should use the cost data of Tableau 7.6.7 to verify this value. ∎

**Tableau 7.7.6**

| $i$ \ $j$ | 1 | 2 | 3 | 4 | 5 | $S_i$ |
|---|---|---|---|---|---|---|
| 1 | +1 (25) | | | ⬡ | | 25 |
| 2 | −1 (5) | (30) | +1 (10) | | | 45 |
| 3 | | | −1 (10) | +1 (35) | (5) | 50 |
| $D_j$ | 30 | 30 | 20 | 35 | 5 | $f = 650$ |

**Tableau 7.7.7**

| $i$ \ $j$ | 1 | 2 | 3 | 4 | 5 | $S_i$ |
|---|---|---|---|---|---|---|
| 1 | (15) | | | (10) | | 25 |
| 2 | (15) | (30) | | | | 45 |
| 3 | | | (20) | (25) | (5) | 50 |
| $D_j$ | 30 | 30 | 20 | 35 | 5 | |

Now we have illustrated each of the steps of the simplex algorithm as they apply to the TP. In the next section we shall put them together in a completely worked example.

## 7.7 Exercises

1. Since every transportation problem is a special LP problem, the choice of vector $A_{kl}$ to leave the basis must be based on ________ considerations. We examine all basic variables which will ________ as the new basic variable $x_{ij}$ increases. Since each of these basic variables decreases at the same rate, the one which will reach ________ first is the one which was the ________ to start with. In case of a tie, any one of the *tied* variables may become non-basic. The other *tied* variable will remain basic but will necessarily have value ________ in the next tableau.

   After determining $A_{kl}$, we find the new value of $x_{ij}$ by setting $x_{ij}$ equal to the old value of $x_{kl}$, say $a$. Each remaining basic variable either ________ by $a$, ________ by $a$, or ________, depending upon whether there is a $-1$, $+1$, or a 0 for that vector in the coordinate representation of $A_{ij}$.

2. (*True* or *false*)
   a. It is possible in a TP to explicitly solve a set of linear equations by Gauss-Jordan reduction to find $Y_{ij}$.
   b. It is necessary in a TP to explicitly solve a set of linear equations by Gauss-Jordan reduction to find $Y_{ij}$.

3. Suppose that $A_{23}$ has been chosen to enter the basis and $A_{23} = A_{11} - A_{12} + A_{22} - A_{31} + A_{33}$, while $X^T = [20, 20, 0, 0, 0, 10, 0, 30, 10, 0, 30, 0]$, $m = 3$, and $n = 4$.
   a. Which variables decrease as $x_{23}$ increases?
   b. Which vector should leave the basis?
   c. What will be the new value of $x_{23}$?
   d. What will be the new value of $x_{12}, x_{33}, x_{24}$?

4. If the new value of an incoming basic variable $x_{ij}$ is $a$ and $c_{ij} - z_{ij} = b$, then the decrease in the objective function will be ________.

**Tableau 7.7.8**

| $i$ \ $j$ | 1 | 2 | 3 | 4 | $S_i$ | $u_i$ |
|---|---|---|---|---|---|---|
| 1 | [2] (10) | [3] (10) | [4] | [5] | 20 | |
| 2 | [3] (30) | [6] | [8] | [3] | 30 | |
| 3 | [4] | [7] (20) | [6] (20) | [4] (10) | 50 | |
| $D_j$ | 40 | 30 | 20 | 10 | $f = 440$ | |
| $v_j$ | | | | | | |

5. Consider the conclusion of Problem $T_2$.
   a. Use the $u$, $v$ method to determine if Tableau 7.7.8 is optimal for Problem $T_2$.
   b. Let $A_{31}$ enter the basis. $A_{31}$ = ________ $- A_{12} +$ ________. Therefore, ________ decrease(s) as $x_{31}$ increases, and ________ must become non-basic. The new value of $x_{31}$ = ________. Since $c_{31} - z_{31}$ = ________, the change in $f$ will be ________.
   c. Fill in the new Tableau 7.7.9 and use the $u$, $v$ method to show that it is optimal. The optimal value of $f$ is ________.

**Tableau 7.7.9**

| $i$ \ $j$ | 1 | 2 | 3 | 4 | $S_i$ | $u_i$ |
|---|---|---|---|---|---|---|
| 1 | [2] | [3] | [4] | [5] | 20 | |
| 2 | [3] | [6] | [8] | [3] | 30 | |
| 3 | [4] | [7] | [6] | [4] | 50 | |
| $D_j$ | 40 | 30 | 20 | 10 | $f =$ | |
| $v_j$ | | | | | | |

6. Consider the conclusion of Problem $T_1$.
   a. Use the $u$, $v$ method to determine if Tableau 7.7.10 is optimal for problem $T_1$.

**Tableau 7.7.10**

| $i$ \ $j$ | 1 | 2 | 3 | $S_i$ | $u_i$ |
|---|---|---|---|---|---|
| 1 | [4] (20) | [9] | [8] (10) | 30 | |
| 2 | [3] | [5] (20) | [7] (10) | 30 | |
| $D_j$ | 20 | 20 | 20 | $f = 330$ | |
| $v_j$ | | | | | |

   b. Show what happens to the objective function if $A_{21}$ is now allowed to enter the basis.
   c. The optimal solution $X$ in a. is $X^T = [\qquad\qquad]$.

7. a. Use Tableau 7.7.11 to find an initial BFS by the Northwest Corner Rule.
   b. Test the solution obtained in a. for optimality by the $u$, $v$ method.
   c. Let $x_{13}$ be a new basic variable. $A_{13} = A_{23} -$ ________ $+$ ________.
   d. Since min{20, ________} = 20, ________ should leave the basis.

**Tableau 7.7.11**

| $i$ \ $j$ | 1 | 2 | 3 | $S_i$ | $u_i$ |
|---|---|---|---|---|---|
| 1 | [2] | [5] | [6] | 40 | |
| 2 | [3] | [4] | [7] | 35 | |
| $D_j$ | 20 | 30 | 25 | $f =$ | |
| $v_j$ | | | | | |

e. Use the above results to fill in the new Tableau 7.7.12

f. Test the solution in e. for optimal conditions.

**Tableau 7.7.12**

| $i$ \ $j$ | 1 | 2 | 3 | $S_i$ | $u_i$ |
|---|---|---|---|---|---|
| 1 | [2] | [5] | [6] | 40 | |
| 2 | [3] | [4] | [7] | 35 | |
| $D_j$ | 20 | 30 | 25 | $f =$ | |
| $v_j$ | | | | | |

## 7.8 The Transportation Problem—an Example

Tableau 7.8.1 contains the initial information for a TP with 3 sources and 4 destinations. Any BFS will contain $3 + 4 - 1 = 6$ basic variables. We shall review the special algorithm developed for the TP by going through the various steps involved. To conserve space, all of the work involved in a particular iteration will be performed on a single tableau.

**Tableau 7.8.1**

| $i$ \ $j$ | 1 | 2 | 3 | 4 | $S_i$ |
|---|---|---|---|---|---|
| 1 | [6] | [4] | [3] | [2] | 30 |
| 2 | [3] | [5] | [4] | [8] | 30 |
| 3 | [7] | [6] | [5] | [7] | 30 |
| $D_j$ | 25 | 25 | 20 | 20 | |

**Iteration 1**

**Step 1** Find an initial BFS.
Using the Northwest Corner Rule, an initial BFS is created as shown in Tableau 7.8.2. The initial value of $f$ is 500.

**Tableau 7.8.2**

| $i$ \ $j$ | 1 | 2 | 3 | 4 | $S_i$ | $u_i$ |
|---|---|---|---|---|---|---|
| 1 | [6] +1 (25) | [4] −1 (5) | [3] 0 | [2] −3 | ~~30~~ ~~5~~ 0 | 5 |
| 2 | [3] ⬡ −4 | [5] +1 (20) | [4] (10) | [8] 2 | ~~30~~ ~~10~~ 0 | 6 |
| 3 | [7] −1 | [6] 0 | [5] (10) | [7] (20) | ~~30~~ ~~20~~ 0 | 7 |
| $D_j$ | ~~25~~ 0 | ~~25~~ ~~20~~ 0 | ~~20~~ ~~10~~ 0 | ~~20~~ 0 | $f = 500$ | |
| $v_j$ | 1 | −1 | −2 | 0 | | |

**Step 2** Is the present BFS optimal?
The appropriate $u_i$, $v_j$ values are generated, starting with $v_4 = 0$. These are used to calculate $c_{ij} - z_{ij} = c_{ij} - (u_i + v_j)$ for each non-basic variable $x_{ij}$. The results are inserted in the non-basic cells of Tableau 7.8.2. The present tableau is not optimal.

**Step 3** $A_{21}$ should enter the basis. To indicate this, a hexagon is placed in the (2, 1) cell.

**Step 4** Which vector should leave the basis?
A closed loop involving $A_{21}$ is evident. The appropriate $+1$ and $-1$ elements have been inserted in the basic cells of the tableau that are used to generate $A_{21}$. Since $\min\{25, 20\} = 20$, $A_{22}$ should leave the basis.

**Step 5** Tableau 7.8.3 contains the new BFS. The new value of $f$ is 420. We

**Tableau 7.8.3**

| $i$ \ $j$ | 1 | 2 | 3 | 4 | $S_i$ |
|---|---|---|---|---|---|
| 1 | [6] (5) | [4] (25) | [3] | [2] | 30 |
| 2 | [3] (20) | [5] | [4] (10) | [8] | 30 |
| 3 | [7] | [6] | [5] (10) | [7] (20) | 30 |
| $D_j$ | 25 | 25 | 20 | 20 | $f = 420$ |

would have expected a reduction of 80 since $c_{21} - z_{21} = -4$, and $x_{21}$ is increased by 20.

## Iteration 2

**Step 2** Is the present BFS optimal?
In Tableau 7.8.4 the $u_i$, $v_j$ are calculated and used to find the $c_{ij} - z_{ij}$ elements shown in the non-basic cells. The tableau is still not optimal.

**Step 3** $A_{14}$ should enter the basis.

**Step 4** Which vector should leave the basis?
The positive and negative units indicate the closed loop containing $A_{14}$. All of the basic vectors except $A_{12}$ are needed. Since $\min\{x_{11}, x_{23}, x_{34}\} = \min\{5, 10, 20\} = 5$, the new value of $x_{14} = 5$; and $x_{11}$ becomes non-basic.

**Tableau 7.8.4**

| i \ j | 1 | 2 | 3 | 4 | $S_i$ | $u_i$ |
|---|---|---|---|---|---|---|
| 1 | [6] +1 (5) | [4] (25) | [3] −4 | [2] ○ −7 | 30 | 9 |
| 2 | [3] −1 (20) | [5] 4 | [4] +1 (10) | [8] 2 | 30 | 6 |
| 3 | [7] 3 | [6] 4 | [5] −1 (10) | [7] +1 (20) | 30 | 7 |
| $D_j$ | 25 | 25 | 20 | 20 | $f = 420$ | |
| $v_j$ | −3 | −5 | −2 | 0 | | |

**Step 5** Tableau 7.8.5 contains the next candidate BFS. The new value of $f$ is 385, representing a reduction of $7 \cdot 5 = 35$, as appropriate.

**Tableau 7.8.5**

| i \ j | 1 | 2 | 3 | 4 | $S_i$ |
|---|---|---|---|---|---|
| 1 | [6] | [4] (25) | [3] | [2] (5) | 30 |
| 2 | [3] (25) | [5] | [4] (5) | [8] | 30 |
| 3 | [7] | [6] | [5] (15) | [7] (15) | 30 |
| $D_j$ | 25 | 25 | 20 | 20 | $f = 385$ |

## Iteration 3

**Step 2** Regarding Tableau 7.8.6, which is not optimal, note that setting $v_4 = 0$ gives values for both $u_3$ and $u_1$.

**Step 3** There is a tie. For convenience, $A_{22}$ is chosen to enter.

**Step 4** The closed path is indicated by the unit elements in the basic cells. Since $\min\{x_{12}, x_{23}, x_{34}\} = \min\{25, 5, 15\} = 5$, $x_{23}$ becomes non-basic, and the new value of $x_{22}$ is 5. We would thus expect a decrease in $f$ of $3(5) = 15$.

**Tableau 7.8.6**

| $i \backslash j$ | 1 | 2 | 3 | 4 | $S_i$ | $u_i$ |
|---|---|---|---|---|---|---|
| 1 | [6] (7) | [4] +1 (25) | [3] 3 | [2] −1 (5) | 30 | 2 |
| 2 | [3] (25) | [5] ⬡ −3 | [4] +1 (5) | [8] 2 | 30 | 6 |
| 3 | [7] 3 | [6] −3 | [5] −1 (15) | [7] +1 (15) | 30 | 7 |
| $D_j$ | 25 | 25 | 20 | 20 | $f = 385$ | |
| $v_j$ | −3 | 2 | −2 | 0 | | |

**Step 5** Tableau 7.8.7 is the new tableau. Now $f = 370$.

**Tableau 7.8.7**

| $i \backslash j$ | 1 | 2 | 3 | 4 | $S_i$ | $u_i$ |
|---|---|---|---|---|---|---|
| 1 | [6] 4 | [4] +1 (20) | [3] 3 | [2] −1 (10) | 30 | 2 |
| 2 | [3] (25) | [5] (5) | [4] 3 | [8] 5 | 30 | 3 |
| 3 | [7] 0 | [6] ⬡ −3 | [ ] (20) | [7] +1 (10) | 30 | 7 |
| $D_j$ | 25 | 25 | 20 | 20 | $f = 370$ | |
| $v_j$ | 0 | 2 | −2 | 0 | | |

**Iteration 4**

**Step 2** See Tableau 7.8.7. The tableau is not optimal.

**Step 3** $A_{32}$ should enter the basis.

**Step 4** The closed path includes $A_{12}$, $A_{14}$, and $A_{34}$. The unit elements are as shown. Since $\min\{x_{12}, x_{34}\} = \min\{20, 10\} = 10$, $x_{34}$ becomes non-basic; and the new value for $x_{32}$ is 10. We would expect a decrease in $f$ of $3(10) = 30$.

**Step 5** See Tableau 7.8.8. Now $f = 340$.

**Tableau 7.8.8**

| $i$ \ $j$ | 1 | 2 | 3 | 4 | $S_i$ | $u_i$ |
|---|---|---|---|---|---|---|
| 1 | [6] 4 | [4] (10) | [3] 0 | [2] (20) | 30 | 2 |
| 2 | [3] (25) | [5] (5) | [4] 0 | [8] 5 | 30 | 3 |
| 3 | [7] 3 | [6] (10) | [5] (20) | [7] 3 | 30 | 4 |
| $D_j$ | 25 | 25 | 20 | 20 | $f = 340$ | |
| $v_j$ | 0 | 2 | 1 | 0 | | |

**Iteration 5**

**Step 2** See Tableau 7.8.8. The tableau is optimal. However, there are alternate optimal solutions as indicated by the fact that $c_{13} - z_{13} = c_{23} - z_{23} = 0$.

Thus the optimal BFS is $X^T = [0, 10, 0, 20, 25, 5, 0, 0, 0, 10, 20, 0]$, with the 90 units being shipped accordingly.

## 7.8 Exercises

1. Solve the TP of Exercise 7.2.3. (See Exercise 7.6.5.)
2. See Tableau 7.8.8. If $A_{24}$ were allowed to enter the basis, $A_{22}$ would leave; and $f$ would ________ by ________ units.
3. Solve the TP of Exercise 7.2.7. (See Exercise 7.6.6.)
4. Solve the TP of Exercise 7.2.5.
5. Solve the TP of Exercise 7.2.9. (See Exercise 7.6.7.)
6. Solve the TP of Exercise 7.2.8 with the dummy source. Interpret your solution.
7. After solving Exercise 6 above, how would you recommend handling a situation in which the total supply exceeds the total demand?

# 8 Probability

## 8.1 Probability and Decision

The topic of this chapter, probability, is one which has attracted wide interest in recent years as its set of applications has continued to increase. Many of the basic concepts of this body of knowledge have been studied for hundreds of years, mostly in association with actuarial insurance problems or with investigations of so-called "games of chance," such as roulette, craps, coin-flipping, etc. More recently, businessmen, medical and military professionals, scientists such as meteorologists and agronomists, and other decision makers in various occupations have turned to these same basic concepts for assistance.

The reason for this unified interest is easily explained. In each of the above-mentioned occupations there is an ever-recurring and laborious decision-making function. These people must choose between alternative courses of action in real-life situations where the eventual results of a given choice are not known in advance. Consider, for example, General Motors' decisions concerning how many cars of various sizes to produce in the face of uncertain demand, or two decision problems of physicians concerning correct diagnosis of a disease and the appropriate treatment for it. We can even include a dilemma faced by President Ford in 1975 which involved making a decision about which of several possible energy policies to recommend that would circumvent future oil embargos.*

An entirely new discipline, known as **decision theory**, has evolved from the combined use of the ideas of the probability of a prescribed possible outcome and various economic/psychological concepts concerned with how

---

* Although many present applications of probabilistic ideas are widely recognized as proper, there is a wide divergence of opinion and a lively philosophical debate about others. For an introduction to some interesting aspects of this controversy, see Fellner in Suggested Reading.

the decision maker "feels" about the consequences of various alternative/outcome combinations. To illustrate one of the latter economic ideas that affects the psychology of decision making, we merely note that whereas a business loss of $100 may not ruin Jack Nicklaus, it might wreck your plans to visit Florida during spring vacation. Although sophisticated decision models will not be covered in detail here, some perspective will be given on a few of the ideas just mentioned. Throughout the example which follows, we shall use the term *probability* in an intuitive fashion, and then attempt to define it more precisely in Section 8.2.

Example 8.1.1 Let us suppose that you have purchased a new bike for $100. The dealer has offered to sell you, at a cost of $15, a warranty policy that will "cover" all necessary repair charges for 1 year. In decision-theoretic terms, you have two alternatives:

$A_1$: "buy the warranty"
$A_2$: "don't buy the warranty."

After considering the dealer's offer, you intuitively conclude that an important factor in your decision process concerns the likelihood that your new bike will require repairs totalling more than $15 during the coming year. As we shall see, this is a very perceptive observation.

Next let us imagine that a consumer magazine that you (being a prudent buyer) consulted reported that they had tested 10 bikes like yours for a year and all ten bikes had required a particular part costing $20. Such information could well induce you to choose the warranty (Alternative $A_1$). On the other hand, if none of the tested bikes had required any repairs, then you would lean toward $A_2$.* To make things more difficult, and more realistic, we now suppose that such conclusive information is not available; but your product research did pay some dividends. The magazine did report that your particular bicycle model had only one part, part $A$, that was subject to failure during the first year, and that the "parts and labor" repair cost for this part was $25 when it failed. Table 8.1.1 contains a **decision matrix** for this set of

**Table 8.1.1 A Sample Decision Matrix**

| | $O_1$ "no failure" $P(O_1) = 0.7$ | $O_2$ "part $A$ fails" $P(O_2) = 0.3$ |
|---|---|---|
| $A_1$ "buy the warranty" | $15 | $15 |
| $A_2$ "don't buy the warranty" | $0 | $25 |

* You undoubtedly would feel better if the magazine had reported a test of 100 bikes, or even 1000; but this, while giving you more security, would also have cost more. The study of such interesting notions as how many bikes to test, and the associated benefits and costs, belongs to the field of statistics—where the ideas of probability provide the theoretical foundation.

circumstances, with the possible alternatives ($A_i$) as rows, and the possible outcomes ($O_j$) as columns. To explain the use of the matrix information we note that either $O_1$ or $O_2$ will occur, but the cost to you will be the $15 warranty fee, regardless. If, however, you decide to chance it and select $A_2$, then the actual cost to you will either be $0 (if $O_1$ occurs) or $25 (if $O_2$ occurs). You now recall that your consumer magazine contained further information—namely, that part $A$ had failed on 30% of the bikes tested. For purposes of making your decision, you choose to treat this fraction, 3/10, as the probability that part $A$ on your own bike will fail. Letting $P(O_2)$ denote the probability that $O_2$ will occur, we set $P(O_2) = 0.3$ and, since 70% of the bikes have no failures, $P(O_1) = 0.7$, as indicated in Table 8.1.1.

With the decision matrix completed, the decision theorist evaluates the alternatives as follows. The first alternative, $A_1$, costs him $15, regardless of the eventual outcome. In considering $A_2$, the ultimate outcome is significant and unknown. When $O_2$ occurs the cost will be $25. $O_2$ will occur "approximately" 30% of the time, that is, on about 30% of the bikes sold. Thus the buyer could expect to pay about 0.3 ($25) or $7.50 in repair costs.* On this basis he might well feel that $15 is too much to pay for the warranty and thus choose $A_2$. ∎

In the following sections we shall consider the basic concepts and mathematical formulas of probability theory that would be used in the solution of problems such as the above and in many other applications. For pedagogical purposes, we shall introduce these concepts in the context of familiar or easily understood "games of chance." In the final sections of this chapter, we will present an introductory discussion of matrices whose elements are probabilities, and illustrate their use in the study of Markov chains, a particular type of probabilistic phenomenon which has had wide practical application. The chapter concludes with a possible application of Markovian ideas in the insurance industry. In Chapter 9 similar methods are used to solve typical problems in manpower planning.

### Suggested Reading

J. C. G. Boot and E. B. Cox. *Statistical Analysis for Managerial Decision.* New York: McGraw-Hill, 1970.

H. Bierman, C. P. Bonini, and W. H. Hausman. *Quantitative Analysis for Business Decisions.* Homewood, Ill.: Irwin, 1973.

W. J. Fellner. *Probability and Profit.* Homewood, Ill.: Irwin, 1965.

G. Hadley and M.C. Kemp. *Finite Mathematics in Business and Economics.* New York: American Elsevier, 1972.

A. N. Halter and G. W. Dean. *Decisions Under Uncertainty.* Cincinnati, Ohio: South-Western Pub., 1971.

W. T. Morris. *Management Science, A Bayesian Introduction.* Englewood Cliffs, N.J.: Prentice-Hall, 1968.

* If 100 such bike buyers formed a pool, each contributing $7.50, and 30 bikes failed, then they would have just enough to meet the repair expenses, namely 30($25) = $750. Such pooling of funds is the basis of all insurance plans.

**New Terms**

decision theory, 311 decision matrix, 312

## 8.2 Random Experiments

From our high school days we are all familiar with experiments of a general nature. Many such experiments, of the type conducted in chemistry and physics classes, for example, have the same outcome each time they are performed; and the instructor could tell you ahead of time what it will be. Assuming that the experiment is properly conducted, the instructor is never "surprised" by the eventual outcome.

The experiments which we now have in mind are conceptually different from those above in that the same outcome does not always occur each time the experiment is performed. The situation is not completely chaotic, however, because we do assume that the set of all possible experimental outcomes is known and that exactly one of them will occur each time the experiment is conducted. Because of the non-certain or random aspects of such experiments, they have been called **random experiments**. Table 8.2.1 contains examples of the types of random experiments we shall use to illustrate the basic rules of probability. In $E_1$, we shall flip a single coin and consider two possible outcomes, "head" or "tail". Although there may indeed be a remote chance that our particular coin will land and remain on its edge, or even be blown away by a sudden gust of wind, we arbitrarily choose to disregard such contingencies in our model of the situation.

**Table 8.2.1 Random Experiments**

| *Experiment* ($E_i$) | *Number of Possible Outcomes* ($N_i$) | *Set of Possible Outcomes* ($S_i$) |
|---|---|---|
| Flip a coin ($E_1$) | $N_1 = 2$ | $S_1 = \{\text{Head, Tail}\} = \{H, T\}$ |
| Flip 2 coins ($E_2$) | $N_2 = 4$ | $S_2 = \{(H, H), (H, T), (T, H), (T, T)\}$ |
| Toss a die ($E_3$) | $N_3 = 6$ | $S_3 = \{1, 2, 3, 4, 5, 6\}$ |
| Toss 2 dice, blue and gold ($E_4$) | $N_4 = 36$ | $S_4 = \{(1, 1), (1, 2), (1, 3), (1, 4), (1, 5), (1, 6),$ $(2,1), (2, 2), (2, 3), (2, 4), (2, 5), (2, 6),$ $(3, 1), (3, 2), (3, 3), (3, 4), (3, 5), (3, 6),$ $(4, 1), (4, 2), (4, 3), (4, 4), (4, 5), (4, 6),$ $(5, 1), (5, 2), (5, 3), (5, 4), (5, 5), (5, 6),$ $(6, 1), (6, 2), (6, 3), (6, 4), (6, 5), (6, 6)\}$ |
| | | $S'_4 = \{2, 3, 4, 5, 6, 7, 8, 9, 10, 11, 12\}$ |
| Draw a card ($E_5$) | $N_5 = 52$ | $S_5 = \{A_c, 2_c, 3_c, \ldots, J_c, Q_c, K_c,$ $A_d, 2_d, 3_d, \ldots, J_d, Q_d, K_d,$ $A_h, 2_h, 3_h, \ldots, J_h, Q_h, K_h,$ $A_s, 2_s, 3_s, \ldots, J_s, Q_s, K_s,\}$ |

In $E_2$ we simultaneously flip two coins, say a nickel and a dime, and record the respective outcomes in the ordered pairs as shown. Thus the outcome $(H, T)$ denotes a "head" with the nickel and a "tail" with the dime. As we shall see later, this same set of outcomes could be used to describe an experiment in which a single coin is flipped twice. Such sequential experiments as the latter are called **multi-stage random experiments**, as distinguished from the single-stage experiments which we describe presently.

Experiments $E_3$ and $E_4$ involve the tossing of a single die, and the simultaneous tossing of two different colored dice, respectively. $S_4$ is the set of all possible experimental outcomes that may occur when the two dice are tossed in $E_4$. In Table 8.2.1, we first list the outcome of the blue die, and then the outcome of the gold die, in each of the 36 possible outcome pairs. As is the case in many random experiments, there is an alternative way of describing the set of possible outcomes. If we are really more interested in the sum of the digits on the individual dice, as we might be in playing Monopoly, for example, we could list the 11 potential outcomes 2–12 shown in $S'_4$. We shall use both $S_4$ and $S'_4$ as we proceed.

Experiment $E_5$ describes the situation when we wish to draw a single card from a standard deck of 52 playing cards. $S_5$ contains the possible outcomes, with $A_c$ denoting the ace of clubs, $A_d$ the ace of diamonds, etc.

It should be re-emphasized that in each of the experiments described above there is some uncertainty about the outcome of a given experimental trial. The uncertainty, however, is of a limited nature, in that we are supposing that exactly one of the relatively limited set of possibilities must occur. Although we are unable to specify in advance which outcome will occur, we have at least narrowed the field.

In the following sections we shall develop a quantitative probability measure to be associated with the elements of $S_i$ that will provide some measure of their relative chance of occurrence. This quantitative measure deals with the problem of reducing the uncertainty associated with random experiments even further, that is, by enabling one to calculate the *likelihood* or *chances* that any particular outcome or subset of outcomes will occur on any given experimental trial or set of trials. For simplicity however, the random experiments which we shall consider in this text will be limited to those with a *finite* number of experimental outcomes. It should be pointed out, however, that there do exist random experiments with an infinite number of possible outcomes. Consider the experiment that consists of tossing a coin (repeatedly, perhaps) until a "head" appears, and recording the number of the flip on which the experiment ends. The set of potential outcomes is $S = \{1, 2, 3, \ldots\}$, an infinite set. Detailed descriptions of such experiments may be found in most standard probability textbooks.

## 8.2 Exercises

1. Consider an experiment $E_6$, in which a nickel, a dime, and a quarter are tossed simultaneously. If each coin may come up "head" or "tail," find
   a. $S_6$, the set of all possible outcomes (let $H_N$ denote a head on the nickel; $H_D$ a

head on the dime, and $H_Q$ a head on the quarter, with $T_N$, $T_D$, and $T_Q$ denoting the corresponding tail outcomes).

b. $N_6$, the number of different possible outcomes.

2. Consider an experiment $E_7$, in which 3 dice are tossed simultaneously—one blue die, one gold, and one red. Let the triple (1, 2, 3), for example, denote a 1 on the blue die, a 2 on the gold die, and a 3 on the red die. Then find
   a. $S_7$, the set of all possible outcomes.
   b. $N_7$, the number of different possible outcomes.
   c. $S'_7$, the set of all possible sums that can result from the $N_7$ different outcomes.

**New Terms**

random experiment, 314 multi-stage random experiment, 315

## 8.3 Events

The possible individual outcomes listed in the various outcome sets, $S_i$, in Table 8.2.1 are called **elementary events**. The set of all elementary events, that is, $S_i$, is called the **outcome space** or **sample space** for the particular random experiment. Thus, when we toss a single die in $E_3$, the elementary events are $\{1\}, \{2\}, \{3\}, \{4\}, \{5\}, \{6\}$; and the sample space is $S_3 = \{1, 2, 3, 4, 5, 6\}$. The sample space is nothing more than the set union of all of the elementary events. Letting $e_1, e_2, \ldots, e_N$ denote the elementary events of an arbitrary random experiment, and $S$ its sample space, then

$$S = e_1 \cup e_2 \cup \cdots \cup e_N.$$

It is obvious from our description and examples of elementary events that once the sample space of a given random experiment has been formulated, there are many other subsets of $S$ which could be created and that conceivably could be of interest. Since $S_3$ above has 6 elements, there are $2^6$ or 64 possible subsets.* Each such subset of a given sample space will be called an **event**. Clearly the original elementary events are also events under this definition, and in $E_3$ they comprise 6 of the 64 total possible events. The event $A = \{1, 3, 5\}$ of $E_3$ could alternatively be described verbally as "the outcome is odd," while the event $B = \{1, 2, 3\}$ could be described as "the outcome is less than 4." Thus an event can be described verbally as well as in set terminology.

Given any two events, such as $A$ and $B$ above, the standard set operations may be used to form new events. Thus, for $A$ and $B$ above, we could form their set intersection to create a new event $C = A \cap B = \{1, 3\}$. The Venn diagrams we described briefly in Chapter 1 are helpful in showing pictorially a given sample space and various events. The Venn diagram of Figure 8.3.1 depicts events $A$ and $B$ with a shaded portion which portrays the new event $C$, another of the 64 possible events.

* Recall that a set with $n$ elements has $2^n$ subsets.

Figure 8.3.1

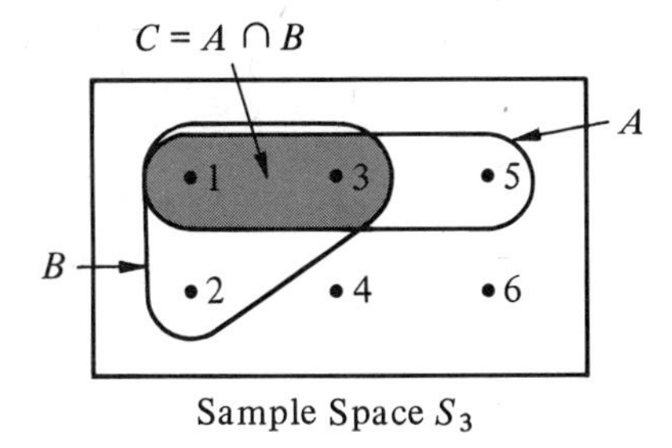


Similarly, we could also form $D = A \cup B = \{1, 3, 5\} \cup \{1, 2, 3\} = \{1, 2, 3, 5\}$, as shaded in Figure 8.3.2. Finally, for any event $A$, we can also consider the event $\tilde{A}$ known as the **complement** of event $A$ formed by taking its set complement in the sample space. $\tilde{A}$ *thus is the event that occurs when* $A$ *does not occur.* Figure 8.3.3 depicts $\tilde{A}$ for the set $A$ considered above. In the notation of Chapter 1, $\tilde{A} = S - A$. It is common to refer to events

Figure 8.3.2

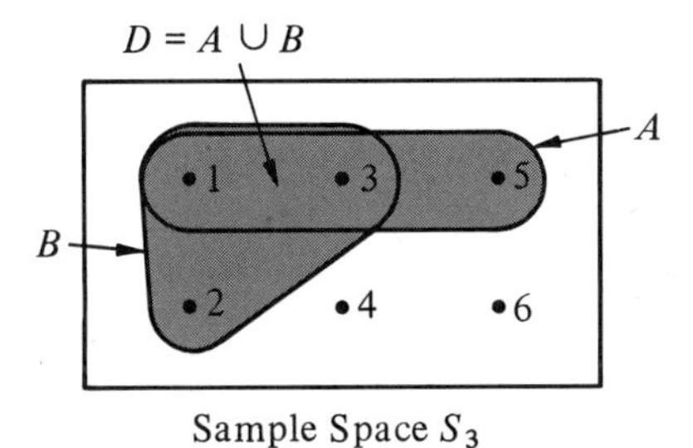


Figure 8.3.3

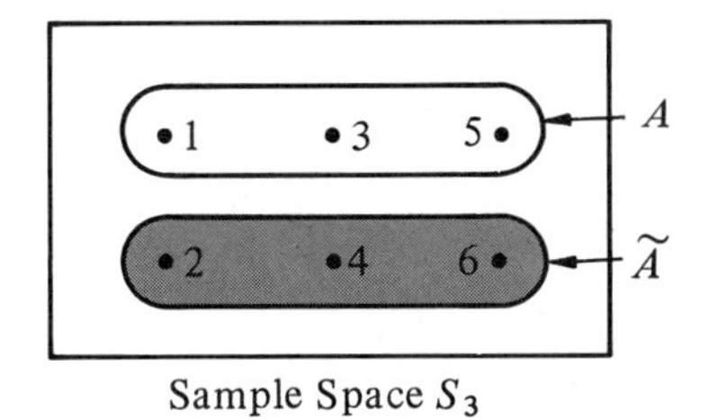


such as $C$, $D$ and $\tilde{A}$ which we have "compounded" from elementary events as **compound events** or **composite events**.

Example 8.3.1 Let us consider experiment $E_2$ of Table 8.2.1 in which two coins are flipped to form $S_2 = \{(H, H), (H, T), (T, H), (T, T)\} = \{e_1, e_2, e_3, e_4\}$. $S_2$ has 4 elements, so there are $2^4 = 16$ possible events if the empty set, $e_5$, is included. The remaining 11 events are all composite:

$$e_6 = \{(H, H), (H, T)\} \qquad e_{12} = \{(H, H), (H, T), (T, H)\}$$
$$e_7 = \{(H, H), (T, H)\} \qquad e_{13} = \{(H, H), (H, T), (T, T)\}$$
$$e_8 = \{(H, H), (T, T)\} \qquad e_{14} = \{(H, H), (T, H), (T, T)\}$$
$$e_9 = \{(H, T), (T, H)\} \qquad e_{15} = \{(H, T), (T, H), (T, T)\}$$
$$e_{10} = \{(H, T), (T, T)\} \qquad e_{16} = \{(H, H), (H, T), (T, H), (T, T)\}$$
$$e_{11} = \{(T, H), (T, T)\}$$

Many of these composite events have convenient verbal descriptions, as shown in Table 8.3.1,* and as is pointed out in the notes, some of the events are representable as various set combinations of the others. ∎

**Table 8.3.1**

| *Event* | *Verbal Description* | *Note* |
|---|---|---|
| $e_6$ | a head on the nickel | |
| $e_7$ | a head on the dime | $e_7 \cap e_6 = e_1$ |
| $e_8$ | both coins have the same outcome | |
| $e_9$ | each coin has a different outcome | $e_8 \cup e_9 = S, \quad e_8 \cap e_9 = \varnothing$ |
| $e_{10}$ | a tail on the dime | |
| $e_{11}$ | a tail on the nickel | |
| $e_{12}$ | at least one head* | $e_{12} = e_6 \cup e_7$ |
| $e_{13}$ | a head on the nickel, or a tail on the dime | $e_{13} = e_6 \cup e_{10}$ |
| $e_{14}$ | a tail on the nickel, or a head on the dime | $e_{14} = e_{11} \cup e_7, \quad e_{14} \cap e_{13} = e_8$ |
| $e_{15}$ | at least one tail† | $e_{15} = e_{10} \cup e_{11}$ |

* $e_{12}$ could be equivalently described as "a head on the nickel, or a head on the dime."
† $e_{15}$ could be equivalently described as "a tail on the nickel, or a tail on the dime."

In any given random experiment there are events that cannot occur in the same experimental trial. In experiment $E_2$, for example, it is clear that only one of the elementary events, $e_1 - e_4$ can occur on a given trial. Furthermore, in Table 8.3.1, we note that $e_8$ and $e_9$ cannot occur at the same time. Since $e_8 \cap e_9 = \varnothing$ it is impossible to have both coins have the same outcome and different outcomes simultaneously. In a given random experiment, events $A$ and $B$ such that $A \cap B = \varnothing$ are called **mutually exclusive events**. Such events are important in the following sections since the fact that they cannot both occur facilitates calculating the probability that at least one of them will occur, as we shall see. We emphasize now, if you failed to notice earlier, that elementary events for a given experiment are necessarily mutually exclusive.

Example 8.3.2

Consider $E_3$ in which we toss a single die with $S_3 = \{1, 2, 3, 4, 5, 6\}$ and in which the events $A = \{1, 3, 5\}$, $B = \{1, 2, 3\}$, $\tilde{A} = \{2, 4, 6\}$, and $F = \{4, 6\}$ have been defined. In this case $A$ and $B$ are not mutually exclusive, since if either 1 or 3 occurs, they both occur. The events $A$ and $\tilde{A}$ are mutually exclusive, however, as would always be the case for any given set and its complement. Since $B \cap F = \varnothing$, then $B$ and $F$ are also mutually exclusive. ∎

* There are in all 1 "0-element" event, 4 "1-element" events, 6 "2-element" events, 4 "3-element" events, and 1 "4-element" event. If you have studied *combinations* or binomial coefficients, you will recognize 1, 4, 6, 4, 1 as $\binom{4}{0}, \binom{4}{1}, \binom{4}{2}, \binom{4}{3}, \binom{4}{4}$, respectively.

## 8.3 Exercises

1. Any ______ of a given sample space is called an event.

2. Consider experiment $E_3$ in which a single die is tossed.
   a. What are the elementary events?
   b. What is the sample space?
   c. If $C$ is the event $\{1, 2\}$, what is $\tilde{C}$?
   d. Are the events $\{1, 2, 4\}$ and $\{5, 3, 2\}$ mutually exclusive?
   e. Draw a Venn diagram to illustrate the situation in d. above.

3. Consider the coin flipping experiment $E_1$ in which a single coin is tossed.
   a. What are the elementary events?
   b. What is the sample space?
   c. How many subsets does this sample space have? List them.
   d. If $A$ is the event $\{H\}$, what is $\tilde{A}$?

4. Consider the events $e_6 - e_{15}$ for experiment $E_2$ as listed in Table 8.3.1. Which of the following pairs of events are mutually exclusive?
   a. $e_7, e_6$ b. $e_7, e_{10}$ c. $e_{10}, e_{11}$ d. $e_{13}, e_{14}$

5. Consider the experiment $E_6$ described in Exercise 8.2.1.
   a. List the elementary events for $E_6$.
   b. How many conceivable events exist for $E_6$?
   c. Which elementary event(s) would satisfy the verbal description "a head on the nickel"?
   d. Which elementary event(s) would correspond to the phrase "at least one head"?
   e. Which elementary event(s) would satisfy the verbal description "all coins have the same outcomes"?
   f. Are any pairs of the 3 verbally described events in c.–e. above mutually exclusive?
   g. Name 2 mutually exclusive events that exist for $E_6$ that are not elementary events.

### New Terms

elementary events, 316
outcome space, 316
sample space, 316
event, 316
complement, 317
compound event, 317
composite event, 317
mutually exclusive events, 318

## 8.4 The Probability of an Event

The most philosophically comfortable basis for the definition of the probability of an event $A$ is the concept of a *long-run relative frequency*. To illustrate this concept we consider experiment $E_1$ in which a single coin is to be flipped. We want to define the phrase "probability of a head" or $P(H)$. If we had our given coin and an automatic coin-flipper, we could perform an experiment and record a running ratio of "heads" to "flips" as shown in tabular form in Table 8.4.1 and, graphically, in Figure 8.4.1. In the first

**Table 8.4.1** **A Relative Frequency Table**

| | | | | | | | | | |
|---|---|---|---|---|---|---|---|---|---|
| Number of heads ($N_H$) | 4 | 9 | 18 | 27 | 56 | 111 | 166 | 214 | 268 |
| Flips ($N$) | 5 | 10 | 25 | 50 | 100 | 200 | 300 | 400 | 500 |
| Relative frequency ($N_H/N$) | 0.8 | 0.9 | 0.72 | 0.54 | 0.56 | 0.555 | 0.5533 | 0.535 | 0.536 |

**Figure 8.4.1**

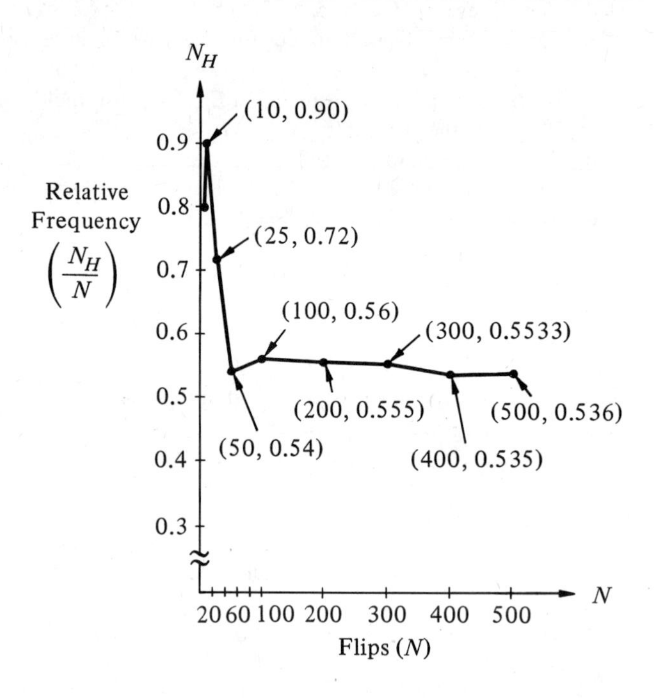


10 flips, for example, there were 9 "heads," so that the relative frequency of "heads" was initially 0.9. In this instance, however, we see that the relative frequency of heads, $N_H/N$, stabilized reasonably quickly in the neighborhood of 0.54 and remained there as $N$ increased. Such a record, including the large early fluctuations in $N_H/N$ and then a gradual stabilization, is a typical occurrence in experimental trials of the type in which we are interested. If now we were asked to estimate $P(H)$ for this particular coin, we assert that it could be reasonable to use $P(H) = 0.536$. If we were fortunate enough to have on hand $N_H/1000$ or $N_H/10{,}000$, then we would, of course, use these data. However, it is characteristic of such data not to show a substantial change from the $N_H/500$ value unless something changed in the coin-flipping process. What we would mean by $P(H)$, so estimated, is then clear. It is the *limiting value* that we suppose $N_H/N$ would approach as $N$ becomes very large. With this particular case as motivation, then, for any event $A$, we say that the **probability of** $A$, or $P(A)$, is the limiting value of $N_A/N$ as $N$ is made very large.* This fraction will serve as the quantitative measure of the likelihood of event A occurring.

* Thus we are supposing that the same limiting value would be approached regardless of how many times the sets of experiments are performed.

In most practical experimental problems, however, it is not economically feasible to conduct an experiment a great number of times in order to obtain relative frequency estimates of the probabilities of the elementary events. Thus it is helpful to form a conceptual model of the experiment and to choose the elementary events in such a way that at least the probabilities of these events can be logically deduced and then be used to calculate the probabilities of other events of interest. This can usually be done in games of chance. If experimental results tend to throw doubt upon the probabilities established by the process of logical deduction, then the logic should be examined.

Example 8.4.1 In considering our experiment $E_1$ we have no particular coin in mind but would like our results to pertain to any "normal" coin. Thus we conceptualize a coin with two sides, "head" and "tail," and after thinking about the situation, decide that there is no good reason why one side should occur more frequently than the other and that their probabilities should therefore be equal. We thus decide that "head" and "tail" are **equally-likely events** and arbitrarily assign $P(H) = P(T) = 1/2$. Thus for this conceptual coin, we are supposing that the relative frequency of heads, $N_H/N$, for a theoretically infinite number of flips, would be 0.5, as would $N_T/N$. Clearly, $N_H/N + N_T/N = 1$. ∎

Thus, in any given random experiment we attempt to specify first the mutually exclusive elementary events, $e_j$, in $S$, and then assign a probability $P(e_j)$, usually on the basis of logical deduction. The probabilities so assigned to the $N$ elementary events must have the following properties:

For any elementary event $e_j$, $P(e_j) > 0$, and

$$P(e_1) + P(e_2) + \cdots + P(e_N) = 1.$$

Once the $e_j$ have been assigned probabilities, $P(e_j)$, the following method can be used to compute the probability of any non-elementary event $A$. We merely express $A$ as the union of the appropriate mutually exclusive elementary events which make it up. *Then $P(A)$ may be written as the sum of the probabilities of the individual elementary events.*

Example 8.4.2 Consider $E_2$ in which a nickel and a dime are flipped. The elementary events are $e_1 = \{(H, H)\}$, $e_2 = \{(H, T)\}$, $e_3 = \{(T, H)\}$, $e_4 = \{(T, T)\}$. Logically seeing no reason why any one of these should occur more frequently than the others,* we assign equal probabilities to each so that $P(e_1) = P(e_2) = P(e_3) = P(e_4) = 1/4$. We note that each $P(e_j) > 0$ and their sum is 1. ∎

Now we shall consider the non-elementary events of $E_2$, namely $e_6 - e_{15}$ of Table 8.3.1 and determine the probability that should be assigned to each. The results are contained in Table 8.4.2. We have not included $e_5$, the empty event *to which we assign the probability* 0, since this, in a sense, is an event which can never occur.

* One must be careful not to apply this strategy where it is not appropriate. Consider the elementary events in $S_4'$ of Table 8.2.1. What probabilities would you assign to the 11 elementary events $e_j'$? We shall consider this particular problem later.

**Table 8.4.2 Non-elementary Events of Experiment $E_2$**

| *Event* | *Verbal Description* | *Elementary Events* | *Probability $P(e_j)$* |
|---|---|---|---|
| $e_6$ | a head on the nickel | $\{(H, H), (H, T)\}$ | $1/2 = 1/4 + 1/4$ |
| $e_7$ | a head on the dime | $\{(H, H), (T, H)\}$ | $1/2 = 1/4 + 1/4$ |
| $e_8$ | both coins have the same outcome | $\{(H, H), (T, T)\}$ | $1/2 = 1/4 + 1/4$ |
| $e_9$ | each coin has a different outcome | $\{H, T), (T, H)\}$ | $1/2 = 1/4 + 1/4$ |
| $e_{10}$ | a tail on the dime | $\{(H, T), (T, T)\}$ | $1/2 = 1/4 + 1/4$ |
| $e_{11}$ | a tail on the nickel | $\{(T, H), (T, T)\}$ | $1/2 = 1/4 + 1/4$ |
| $e_{12}$ | at least one head | $\{(H, H), (H, T), (T, H)\}$ | $3/4 = 1/4 + 1/4 + 1/4$ |
| $e_{13}$ | a head on the nickel or a tail on the dime | $\{(H, H), (H, T), (T, T)\}$ | $3/4 = 1/4 + 1/4 + 1/4$ |
| $e_{14}$ | a tail on the nickel or a head on the dime | $\{(H, H), (T, H), (T, T)\}$ | $3/4 = 1/4 + 1/4 + 1/4$ |
| $e_{15}$ | at least one tail | $\{(H, T), (T, H), (T\ T)\}$ | $3/4 = 1/4 + 1/4 + 1/4$ |
| $e_{16}$ | | $\{(H, H), (H, T), (T, H), (T, T)\}$ | $1 = 1/4 + 1/4 + 1/4 + 1/4$ |

**Example 8.4.3** In $E_3$ we toss a single die. $S_3 = \{1, 2, 3, 4, 5, 6\}$. Our conceptual die, having no shaved corners or mysterious lumpy "loads" inside, is assumed to be such that each of the six elementary events is equally likely. We therefore assign $P(\{1\}) = P(\{2\}) = \cdots = P(\{6\}) = 1/6$.

For $A = \{1, 3, 5\}$,

$$P(A) = P(\{1\}) + P(\{3\}) + P(\{5\}) = 1/6 + 1/6 + 1/6 = 1/2.$$

Similarly,

$$P(B) = P(\{1, 2, 3\}) = 1/6 + 1/6 + 1/6 = 1/2.$$

▮

**Example 8.4.4** In $E_4$ we toss a blue die and a gold die. $S_4$ has 36 elements such as (1, 1), (1, 2), etc., shown in Table 8.2.1. Assuming that these 36 elementary events are equally likely, we assign $P(e_1) = P(e_2) = \cdots = P(e_{36}) = 1/36$. To emphasize the fact that one must be careful of such assignments of probabilities to elementary events, we recall that $S_4' = \{2, 3, \ldots, 12\} = \{e_1', e_2', \ldots, e_{12}'\}$ was an alternate sample space that could have been used to describe $E_4$ in which the sum of the two individual dice was recorded at each toss. In considering $S_4'$ it is important to note that because of the differing numbers of ways in which the various $e_j'$ may occur, elementary events are not equally likely if our previous assignment of probabilities to the elements of $S_4$ was proper. For example, we note that since two of the 36 equally-likely outcomes will form a sum of 3 in $e_2'$, the probability of doing so should be twice as great as the probability of achieving a sum of 2, which can be obtained in only one way, that is, if each die shows a 1. To illustrate this more completely, we depict the relationships between the $e_j$ of $S_4$ and the $e_j'$ of $S_4'$, along with the appropriate probabilities of the $e_j'$ in Table 8.4.3. Thus it turns out

**Table 8.4.3 Equivalent Representation of Sample Space $S_4$**

| *Elementary Event in $S'_4$* | *Equivalent Elementary Event(s) in $S_4$* | *Probability of $e'_j$ in $S'_4$* |
|---|---|---|
| $e'_1 = 2$ | (1, 1) | 1/36 |
| $e'_2 = 3$ | (1, 2), (2, 1) | 2/36 |
| $e'_3 = 4$ | (1, 3), (2, 2), (3, 1) | 3/36 |
| $e'_4 = 5$ | (1, 4), (2, 3), (3, 2), (4, 1) | 4/36 |
| $e'_5 = 6$ | (1, 5), (2, 4), (3, 3), (4, 2), (5, 1) | 5/36 |
| $e'_6 = 7$ | (1, 6), (2, 5), (3, 4), (4, 3), (5, 2), (6, 1) | 6/36 |
| $e'_7 = 8$ | (2, 6), (3, 5), (4, 4), (5, 3), (6, 2) | 5/36 |
| $e'_8 = 9$ | (3, 6), (4, 5), (5, 4), (6, 3) | 4/36 |
| $e'_9 = 10$ | (4, 6), (5, 5), (6, 4) | 3/36 |
| $e'_{10} = 11$ | (5, 6), (6, 5) | 2/36 |
| $e'_{11} = 12$ | (6, 6) | 1/36 |

to be the case that if the sample space does not consist of equally-likely events, then a more careful analysis must take place. Many times the appropriate probability designation may be made even in such more complicated cases by breaking up the various elementary events into other events so that the new events of the new set are equally likely (as in $S_4$). In other cases it might be possible to actually conduct the experiment a large number of times to obtain estimates of the various desired probabilities. ∎

Example 8.4.5 Our final illustrative random experiment was $E_5$, in which a single card is to be drawn from a standard deck of 52 playing cards. In this case, $S_5$ has 52 equally likely possibilities, so we assign 1/52 as $P(e_j)$ for each of the 52 elementary events. In this case we note that $S_5$ has $2^{52}$ subsets. Now to get an idea of how large a number that is, we note that $2^{26} =$ 67,108,864 so that there are over 4.5 quintillion different subsets of $S_5$, or events, that we could consider. The reader will understand that no one has taken the time to calculate the probabilities of each of them, or even to attempt to describe all of them. ∎

For random experiments with sample spaces of large size, it is obvious that we need formulas to help us calculate the probabilities of events in which we are interested from the probabilities of those events which we already know. We consider such formulas in the next section.

## 8.4 Exercises

1. a. If we suppose that the 8 elementary events of experiment $E_6$ of Exercise 8.2.1 are equally likely, what probability would we assign to each?
   b. What probability would we assign to the event "a head on the nickel"?

c. What probability would we assign to the event "at least one head"?
d. What probability would we assign to the event "a head on the nickel, or a tail on the quarter"?
e. What probability would we assign to the event "a head on the nickel and a head on the quarter"?

2. When are events said to be equally likely?

3. Fill in the blanks in the statements below.
   a. For any elementary event $e_j$, $P(e_j)$ ________ 0.
   b. The sum of the probabilities of all the elementary events is ________.
   c. To find the proper probability assignment for an event $A$ which is not an elementary event, we express $A$ as the ________ of the mutually exclusive ________ which make it up.

4. Consider the experiment $E_7$ of Exercise 8.2.2 in which 3 dice are tossed simultaneously.
   a. What probability would be assigned to each elementary event of the type $(1_B, 1_G, 1_R)$?
   b. If a new sample space of the type $S_7' = \{3, 4, \ldots, 18\}$ were formed in which the elements represent the possible sums that can be achieved on one toss of the 3 dice, what probability should be assigned to the events $\{3\}, \{4\}, \ldots, \{18\}$?

**New Terms**

probability of event A, 320 equally-likely events, 321

## 8.5 Formulas for Calculation

We now consider the problem of finding the probability of certain composite events when the probabilities of the individual events are given. The main formula is given in the following theorem.

**THEOREM 8.5.1** *If $A$ and $B$ are events and $P(A)$, $P(B)$, and $P(A \cap B)$ exist, then*

$$P(A \cup B) = P(A) + P(B) - P(A \cap B).$$

Thus the probability that either $A$ or $B$ (or both) occurs can be found by adding the probabilities that $A$ and $B$ occur and then subtracting the probability that they both occur. The situation is depicted in Figure 8.5.1.

**Figure 8.5.1**

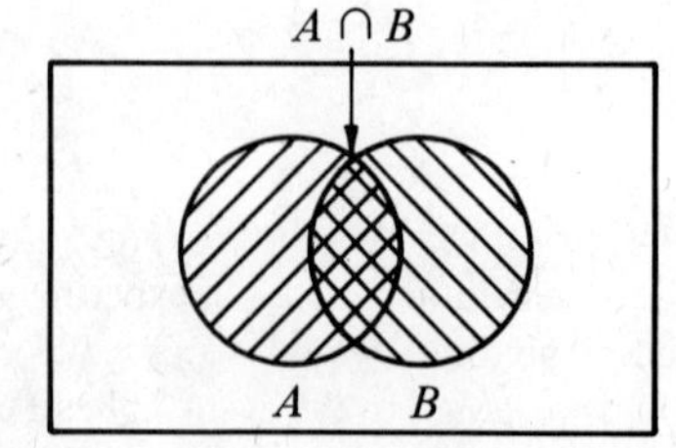

For discussion purposes we suppose that the elementary events of $A$, say $\{a_1, \ldots, a_k\}$, and then the elementary events of $B$, say $\{b_1, \ldots, b_l\}$, are listed and in such a way that the $s$ common elementary events $e_1 = a_1 = b_1$, $e_2 = a_2 = b_2, \ldots, e_s = a_s = b_s$ are listed first. The first $s$ listed events are thus the elementary events of $A \cap B$. If we are to compute $P(A \cup B)$ by adding up the individual probabilities of the elementary events of $A \cup B$ and to do so we take the elementary events of $A$ and of $B$, it is clear that we will have counted certain elementary events twice, namely those in $A \cap B$. In order to correct for this duplication in counting, $P(A \cap B)$ is subtracted from $P(A) + P(B)$ to give the correct sum.

Example 8.5.1

In $E_3$ we have considered the events $A = \{1, 3, 5\}$, $B = \{1, 2, 3\}$, and $A \cap B = \{1, 3\}$, and found that $P(A) = P(B) = 1/2$. We similarly find that

$$P(A \cap B) = P(1) + P(3) = 1/6 + 1/6 = 1/3.$$

We may compute $P(A \cup B)$ directly by the elementary event method to obtain

$$P(A \cup B) = P(\{1, 2, 3, 5\}) = 4/6,$$

or we may use the new formula to obtain the same result, since

$$\begin{aligned} P(A \cup B) &= P(A) + P(B) - P(A \cap B) = 1/2 + 1/2 - 1/3 \\ &= 3/6 + 3/6 - 2/6 = 4/6. \end{aligned}$$ ∎

Having now emphasized through the use of the bracket notation that events are indeed sets, that is, subsets of a sample space, we will now abbreviate our notation for events and discuss, for example, $P(1, 2)$ when we really mean $P(\{1, 2\})$. The new notation is standard in present usage.

Example 8.5.2

In $E_4$ we consider the events

$A$: "a 3 on the blue die"

$B$: "a 5 on the gold die"

$$\begin{aligned} P(A) &= P((3, 1), (3, 2), (3, 3), (3, 4), (3, 5), (3, 6)) = 6/36, \\ P(B) &= P((1, 5), (2, 5), (3, 5), (4, 5), (5, 5), (6, 5)) = 6/36, \\ P(A \cap B) &= P((3, 5)) = 1/36, \end{aligned}$$

so that

$$P(A \cup B) = P(A) + P(B) - P(A \cap B) = 6/36 + 6/36 - 1/36 = 11/36,$$

as you may verify by constructing the event $A \cup B$, to see that the event $\{(3, 5)\}$ is an element of both $A$ and $B$, so that, in reality, $A \cup B$ is composed of 11 equally-likely elementary events. ∎

Example 8.5.3

In $E_4$ we could also consider the events

$A$: "a 3 on the blue die"

$C$: "the sum of the 2 dice is 7."

In this case, we see that 6 of the 36 elementary events have a blue 3 and that 6 of the 36 have sums of 7, so that

$$P(A) = 6/36, \quad P(C) = 6/36, \quad \text{and} \quad P(A \cap C) = P((3, 4)) = 1/36.$$

Thus $P(A \cup C) = 6/36 + 6/36 - 1/36 = 11/36$. Since

$$A \cup C = \{(3,1),(3,2),(3,3),(3,4),(3,5),(3,6),(1,6),(2,5),(4,3),(5,2),(6,1)\}$$

the result is verified. ∎

In an earlier section we discussed the situation for events $A$ and $B$ when $A \cap B = \emptyset$, that is, when $A$ and $B$ are mutually exclusive and cannot occur at the same time. Since $P(\emptyset) = 0$, the above theorem in this case reduces to the following special case:

**THEOREM 8.5.2** *If $A$ and $B$ are mutually exclusive events, then*

$$P(A \cup B) = P(A) + P(B).$$

We have, of course, already been using this principle in the case of mutually exclusive elementary events. The above result also holds for 3 or more pairwise mutually exclusive events, that is, events that have each possible pair mutually exclusive. In this case of 3 such events we have $P(A \cup B \cup C) = P(A) + P(B) + P(C)$.

Example 8.5.4 When a single coin is flipped in $E_1$, what is $P(H \cup T)$? Since $P(H) = 1/2 = P(T)$, and $P(H \cap T) = 0$, then $P(H \cup T) = P(H) + P(T) = 1/2 + 1/2 = 1$, as we would expect, since one or the other must occur. ∎

Example 8.5.5 When two dice are flipped in $E_4$, what is the probability of $\{7\}$ or $\{11\}$? In $S_4'$ the events $\{7\}$ and $\{11\}$ are mutually exclusive; thus

$$P(7, 11) = P(7) + P(11) = 6/36 + 2/36 = 8/36.$$ ∎

Example 8.5.6 Also in $E_4$, what is the probability of $\{2\}$, $\{3\}$, or $\{12\}$? Since the events are mutually exclusive,

$$P(2, 3, 12) = P(2) + P(3) + P(12) = 1/36 + 2/36 + 1/36 = 4/36.$$ ∎

The results of the last two examples are significant in the dice game called "craps" that is played in nearly all gambling casinos. $P(7,11)$ represents the probability of winning on the first roll of the dice, while $P(2, 3, 12)$ represents the probability of losing on the first roll. It can be shown that the roller's probability of winning either on the first or a later roll is approximately 0.4929, so that "craps" is nearly a "fair" game from the roller's standpoint. If your instructor claims any familiarity with this game, you might ask him why it wouldn't pay in the long-run to always bet against the roller at a Las Vegas casino?

We conclude this section with a short discussion of the concept of *expectation.* In experiment $E_2$ in which two coins are flipped, let us suppose that we have a friendly wager in which we receive \$2 if a "head" appears and we lose \$1 if a "tail" appears. Our so-called **expected value** or **expectation** is

$$2P(H) + (-1)P(T) = 2(1/2) + (-1)(1/2) = 1/2,$$

assuming that the coin is fair. In general, we may compute the expectation associated with a particular experimental trial with $N$ possible outcomes by finding the sum

$$EV = W_1P(e_1) + W_2P(e_2) + \cdots + W_NP(e_N)$$

where $W_i$ represents the amount "won" if event $e_i$ occurs.

Example 8.5.7 In $E_4$, as above, let us suppose that we receive \$5 if either $\{7\}$ or $\{11\}$ occur, but that we lose \$4 if either $\{2\}$, $\{3\}$, or $\{12\}$ occur. In this case

$$EV = 5(8/36) + (-4)(4/36) + (0)(24/36) = 2/3.$$

Although on any particular trial we will either win \$5, lose \$4, or break even, we would, *in the long-run*, be ahead about 67¢ per trial. The game could be called "fair" if we were to pay 2/3 of a dollar per trial for the privilege of playing. We would then *expect to break even in the long-run.* ∎

## 8.5 Exercises

1. Consider the events $C = \{1, 2\}$, $D = \{2, 3, 5\}$ and $E = \{3, 4, 6\}$ associated with $E_3$. Find:
   a. $P(C)$  b. $P(D)$  c. $P(E)$
   d. $P(C \cap D)$  e. $P(C \cap E)$  f. $P(D \cap E)$.

   Find the probabilities of the following events both directly by the method of elementary events and also by using Theorems 8.5.1 or Theorem 8.5.2 as appropriate.
   g. $C \cup D$  h. $C \cup E$  i. $D \cup E$

2. In $E_4$, what is the probability that the sum of the 2 dice is either 6 or 8?

3. In $E_2$, what is the probability of either a "head on the nickel" or a "head on the dime"?

4. In $E_3$ compute the probability of tossing either a "number less than 3" or an "even number," given that $P(1, 2) = 1/3$ and $P(2, 4, 6) = 1/2$. Verify your result by determining $\{1, 2\} \cup \{2, 4, 6\} = X$ and finding $P(X)$ from its elementary event properties.

5. Consider $E_7$, as described in Exercise 8.2.2.
   a. Use the results of Exercise 8.4.4 to find the probability that the sum of the three dice is an odd number.
   b. Let $A = \{9, 10, 11, 12\}$ and $B = \{4, 6, \ldots, 18\}$. If $P(A) = 104/216$ and $P(B) = 1/2$, find $P(A \cap B)$ and $P(A \cup B)$.
   c. If $C = \{3, 4, 5, 6, 7, 8\}$ and $P(C) = 7/27$, find $P(A \cup C)$.

6. Consider $E_6$ as described in Exercise 8.2.1.
   a. Find $P(H_N \cap H_D)$.
   b. If $P(H_N) = 4/8$ and $P(H_D) = 4/8$, use Theorem 8.5.1 to find $P(H_N \cup H_D)$.
   c. Why is Theorem 8.5.2 not appropriate in b. above?
   d. Let $X = (H_N, H_D, H_Q)$ and $Y = (T_N, T_D, T_Q)$, so that $P(X) = P(Y) = 1/8$. Find $P(X \cup Y)$.
   e. Let $R$ be the event "at least 1 head" and $S$ the event "no heads." Find $P(R \cup S)$ by 2 methods.

7. Compute the expected value in $E_4$ if one receives
   a. \$5 if an even sum occurs and 0 otherwise.
   b. \$5 if $\{7, 11\}$ occurs, and \$3 if $\{6, 8\}$ occurs, but loses \$4 if any other outcome occurs.

### New Terms

expected value, 327 expectation, 327

## 8.6 Conditional Probability

In this section we shall further consider the probability of an event defined on an original sample space $S$, and see how the probability (that is, relative frequency) of a particular event may be affected when the sample space is reduced in size for some reason—usually because some of its elementary events are not of interest for some special problem.

To motivate the definition of *conditional probability* which follows, we consider experiment $E_3$ in which a single die is tossed. The sample space $S_3 = \{1, 2, 3, 4, 5, 6\}$. We define two events on $S_3$, namely

$A$: "the outcome is less than 4"

$B$: "the outcome is odd."

Thus $A = \{1, 2, 3\}$, $B = \{1, 3, 5\}$, and $A \cap B = \{1, 3\}$, while $P(A) = 3/6$, $P(B) = 3/6$, and $P(A \cap B) = 2/6$. These events are depicted in the Venn diagram of Figure 8.6.1. In the long-run, we would expect $A$ to occur on about 50% of the trials of $E_3$. For our present purpose, however, we are interested in event $A$ in a different sense. We are no longer curious about $A$ every time it occurs, *but only when B occurs and A occurs along with it.* Thus

Figure 8.6.1

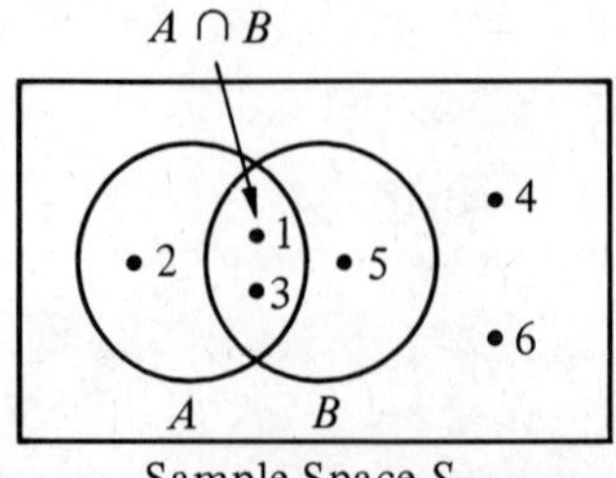

we still focus on the probability of event $A$ but only in the context of its occurrence in the *restricted sample space* associated with event $B$, rather than in its entirety, $S_3$. Phrased still another way, we would now like to know on what proportion of those trials of $E_3$ that $B$ occurs should we expect $A$ also to occur.

To intuitively extablish the correct formula for the desired proportion, we return to relative frequency concepts. If we were to conduct $N$ trials of $E_3$, with events $A$, $B$, and $A \cap B$ occurring $N_A$, $N_B$, and $N_{A \cap B}$ times, respectively, we would expect that $N_A/N$, $N_B/N$, and $N_{A \cap B}/N$ would closely approximate $P(A)$, $P(B)$, and $P(A \cap B)$, in that order.* Since we seek to know the proportion of those $N_B$ trials on which $B$ occurs that $A$ also occurs, we compute the ratio $N_{A \cap B}/N_B$, since $N_{A \cap B}$ represents the number of trials in which $A$ occurs along with $B$. Dividing both the numerator and denominator of the above ratio by $N$, we see that we could compute the desired proportion $N_{A \cap B}/N_B$ as a quotient of two previously established relative frequencies, that is,

$$N_{A \cap B}/N_B = (N_{A \cap B}/N)/(N_B/N)$$

Since $N_{A \cap B}/N$ approximates $P(A \cap B)$ and $N_B/N$ approximates $P(B)$, we are led to define what we mean by the phrase **probability of $A$, given $B$**," or $P(A \mid B)$, by the formula

$$P(A \mid B) = \frac{P(A \cap B)}{P(B)},$$

assuming, of course, that $P(B) \neq 0$. Since the event "A, given B" is *conditional* on the occurrence of $B$, it is called a **conditional probability**. This definition allows us to compute immediately $P(A \mid B)$ when the previously defined probabilities $P(A \cap B)$ and $P(B)$ are known. Thus in the foregoing $E_3$ example,

$$P(A \mid B) = \frac{P(A \cap B)}{P(B)} = \frac{2/6}{3/6} = 2/3,$$

so that we would, in the long run, expect that event $A$ would occur on roughly 66.7% of those trials of $E_3$ on which $B$ occurred. (On what proportion of all the trials of $E_3$ would we expect event $A$ to occur?)

Example 8.6.1

In $E_3$, let $A$ be the event "the outcome $\geqslant 3$," and $B$ be the event "the outcome is even." Thus $A = \{3, 4, 5, 6\}$, $B = \{2, 4, 6\}$, and $A \cap B = \{4, 6\}$. Thus

$$P(A \mid B) = \frac{P(A \cap B)}{P(B)} = \frac{2/6}{3/6} = 2/3.$$

We would thus expect that $A$ would occur on about 2/3 of the occurrences of $B$. It is only coincidental that $A$ will also occur on about 2/3 of the trials of $E_3$. This special case when $P(A \mid B) = P(A)$ will be discussed further in the following section. ∎

---

* The accuracy of the approximation is usually better for large $N$, which we assume here.

Example 8.6.2 In $E_4$, with two dice, let $A$ be the event "the sum of the two dice is 5," and let $B$ be the event "the blue die is a 3." Thus $A \cap B = \{(3, 2)\}$ and $P(A \cap B) = 1/36$, so that

$$P(A \mid B) = \frac{P(A \cap B)}{P(B)} = \frac{1/36}{6/36} = 1/6.$$

We would anticipate that $A$ would occur on about 1/6 of those trials on which $B$ occurs. We recall that we would expect $A$ to occur on about 4/36 of all the trials of $E_4$. In this example, then, $P(A \mid B) \neq P(A)$. ▮

Example 8.6.3 In $E_2$, with two coins flipped, let $B$ be the event "the nickel is a head," and let $A$ be the event "2 heads." Then $A \cap B = \{(H, H)\}$ and

$$P(A \mid B) = \frac{P(A \cap B)}{P(B)} = \frac{1/4}{1/2} = 1/2.$$

How does this compare with $P(A)$? ▮

Before proceeding with a further example of conditional probability, we return to our earlier remarks concerning the *restriction* of the original sample space. We do this in order to show that a conditional probability may be computed in two ways, that is, from the defining formula or from more basic principles applied to the proper **restricted sample space** obtained when certain elementary events are eliminated from consideration. We use the same example in $E_3$, with $A = \{1, 2, 3\}$, $B = \{1, 3, 5\}$, and $A \cap B = \{1, 3\}$, as before. To compute $P(A \mid B)$, we restrict the sample space to $B$, as shown in Figure 8.6.2, and ask for the probability that $A$ occurs *in this restricted setting*, or equivalently, the probability, in the restricted setting, that $A \cap B$ occurs. Since the 3 elementary events are equally likely in the reduced sample space, we compute $P(A \mid B) = P'(1, 3) = 1/3 + 1/3 = 2/3$ directly.* Go back to the beginning of this section and compute the desired conditional probabilities using the *restricted sample space* concept in addition to the defining formula. A further example of this alternate procedure is provided in Example 8.6.4.

**Figure 8.6.2**

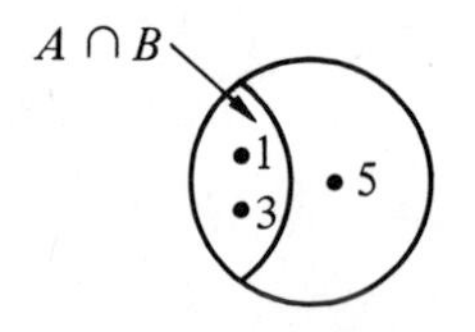


Sample Space $B$

Example 8.6.4 We consider the random experiment of drawing a name from 100 names distributed in two categories as shown in Figure 8.6.3. The four blocks in the upper left corner may be viewed as a special rectangular Venn diagram with 4 mutually exclusive events $B \cap R$, $B \cap D$, $G \cap R$, and $G \cap D$.

* The prime notation is used to denote a probability computed on the reduced sample space.

Figure 8.6.3

| | Republican (R) | Democrat (D) | Total |
|---|---|---|---|
| Boys (B) | 15 | 30 | 45 |
| Girls (G) | 15 | 40 | 55 |
| Total | 30 | 70 | 100 |

a. $P(B) = 45/100$, $P(G) = 55/100$, $P(R) = 30/100$, $P(D) = 70/100$.

b. $P(B \cap R) = 15/100$, $P(B \cap D) = 30/100$, $P(G \cap R) = 15/100$, $P(G \cap D) = 40/100$.

c. $P(D \mid B) = P(D \cap B)/P(B) = (30/100)/(45/100) = 30/45$. To use the restricted sample space concept here, we limit the sample space to the 45 boys in row 1. Then $P(D \mid B) = 30/45$ directly.

d. $P(B \mid D) = P(B \cap D)/(P(D) = (30/100)/(70/100) = 30/70$. The restricted sample space consists of the 70 democrats in column 2 only. We compute $P(B \mid D) = 30/70$ directly.

e. $P(G \mid R) = P(G \cap R)/P(R) = (15/100)/(30/100) = 15/30$.

f. $P(D \mid G) = P(D \cap G)/P(G) = (40/100)/(55/100) = 40/55$.

g. $P(R \mid B) = P(R \cap B)/P(B) = (15/100)/(45/100) = 15/45$. Note that $P(D \mid B) + P(R \mid B) = 1$. ∎

Example 8.6.5

Table 8.6.1 contains life insurance actuarial data representing the proportion of individuals born in the U.S. who live to various ages. The data has been computed from information contained in the Commissioners' 1958 Standard Ordinary Mortality Table as published by the Actuarial Society of America.

Table 8.6.1

| *Age* | *Number Living* | *Number Dying* | *Age* | *Number Living* | *Number Dying* |
|---|---|---|---|---|---|
| 0 | 10,000,000 | | 55 | 8,331,317 | 430,989 |
| 5 | 9,868,375 | 131,625 | 60 | 7,698,698 | 632,619 |
| 10 | 9,805,870 | 62,505 | 65 | 6,800,531 | 898,167 |
| 15 | 9,743,175 | 62,695 | 70 | 5,592,012 | 1,208,519 |
| 20 | 9,664,944 | 78,181 | 75 | 4,129,906 | 1,462,106 |
| 25 | 9,575,636 | 89,358 | 80 | 2,626,372 | 1,503,534 |
| 30 | 9,480,358 | 95,278 | 85 | 1,311,348 | 1,315,024 |
| 35 | 9,373,807 | 106,551 | 90 | 468,174 | 843,174 |
| 40 | 9,241,359 | 132,448 | 95 | 97,165 | 371,009 |
| 45 | 9,048,999 | 192,360 | 100 | 0 | 97,165 |
| 50 | 8,762,306 | 286,693 | | | |

Reading the table, we observe that for every 10,000,000 new births in the U.S., 131,625 infants or young children will die before reaching age 5, leaving 9,868,375 still alive. Of this number, 62,505 will die in the next 5 years, leaving 9,805,870 still alive at age 10. And so it goes, up through age 100, by which time this table assumes that everyone has died. ∎

Such tables provide information for making probability statements. Suppose that an infant is born today. The probability that the child will live to be 5 years old is given by

$$P(5) = \frac{9{,}868{,}375}{10{,}000{,}000} = 0.9868375.$$

The probability that the same child will live to be 50 is 0.8762306, and $P(80) = 0.2626372$. We also observe from the table that today's infant has better than a "50–50" chance, or 0.5 probability, of living to age 70. These statements thus far have all been illustrations of unconditional probabilities.

To consider conditional probability in this context we compute, for example, $P(50 \mid 5)$, the probability of living to age 50, given that a child lives to age 5. Using the defining formula,

$$P(50 \mid 5) = \frac{P(50 \cap 5)}{P(5)} =^{*} \frac{P(50)}{P(5)} = \frac{0.8762306}{0.9868375} = 0.8879178.$$

Surviving to age 5 does not appear to do much to enhance one's chances of living to age 50. Let us look at what living to 50 does to one's chances of living to 80. We recall that originally $P(80) = 0.2626372$. Now we compute that

$$P(80 \mid 50) = \frac{P(80 \cap 50)}{P(50)} = \frac{P(80)}{P(50)} = \frac{0.2626372}{0.8762306} = 0.2997352,$$

still only a relatively small change.

As a person gets older, however, the chances of reaching 80 increase, as would be expected. For example

$$P(80 \mid 70) = \frac{0.2626372}{0.5592012} = 0.4966649.$$

Thus almost half of the people who reach 70 will also reach 80. Mortality tables such as Table 8.6.1 serve as a basis for the computation of all life insurance premiums. The tables, themselves, are based on a certain time period in U.S. history. If that particular period contained a lengthy war or a deadly epidemic, then the table data might well project a very pessimistic outlook for life spans in a peacetime with improved medical care. On what time period is the mortality table for your or your parents' insurance based? Is it representative of the present situation?

## 8.6 Exercises

1. Consider events $C = \{1, 2, 3, 4\}$ and $D = \{3, 4, 5\}$ of $E_3$, in which a single die is tossed.

* Why?

a. Find $P(C|D)$ by two methods. b. Find $P(D|C)$ by 2 methods.
c. Is $P(C|D) = P(C)$? d. Is $P(D|C) = P(D)$?

2. Consider events $C = \{1, 2, 3, 4\}$ and $E = \{4, 5, 6\}$ of $E_3$.
   a. Find $P(C|E)$ by 2 methods. b. Find $P(E|C)$ by 2 methods.
   c. Is $P(C|E) = P(C)$? d. Is $P(E|C) = P(E)$?
   e. Why are the answers to c. and d. in this problem different from those of Exercise 1.c. and 1.d. above?

3. Consider events $A$, $B$ of Example 8.6.1. Find $P(B \mid A)$ by 2 methods.

4. a. Find $P(B \mid A)$ for the events $A$ and $B$ of Example 8.6.2.
   b. Is $P(B \mid A) = P(B)$?

5. a. Find $P(B \mid A)$ for the events $A$, $B$ of Example 8.6.3.
   b. Is $P(B \mid A) = P(B)$?

6. Consider the random experiment of drawing a single name from 200 names distributed as shown in Figure 8.6.4.

**Figure 8.6.4**

| | Protestant ($P$) | Non-Protestant ($NP$) | Total |
|---|---|---|---|
| Boys ($B$) | 48 | 32 | 80 |
| Girls ($G$) | 72 | 48 | 120 |
| Total | 120 | 80 | 200 |

   a. Find $P(B)$, $P(G)$, $P(P)$, and $P(NP)$.
   b. Find $P(B \cap P)$, $P(B \cap NP)$, $P(G \cap P)$, and $P(G \cap NP)$.
   c. Find $P(B \mid P)$ and $P(P \mid B)$. Are they equal? Is $P(B \mid P) = P(B)$?
   d. Find $P(P \mid G)$ using the restricted sample space concept. Show that $P(P \mid G) + P(NP \mid G) = 1$.

7. Using the data of Table 8.6.1, find
   a. $P(85)$ b. $P(85|40)$ c. $P(85|50)$
   d. $P(85|60)$ e. $P(85|70)$ f. $P(85|80)$

8. One million people, each age 20, wish to deposit an equal amount of money into a pool so that those who die at an age between 21 and 49 inclusive will leave a $10,000 estate. Using Table 8.6.1, compute
   a. The number of deaths you would expect during the years in question.
   b. The amount of money that each person would have to put into the pool to insure that a sufficient amount would be present to make the expected pay-offs. (Assume no interest growth of the deposited money.)

### New Terms

probability of $A$, given $B$, 329
conditional probability, 329
restricted sample space, 330

## 8.7 Independent Events

From the previous section we have the conditional probability formula

$$P(A \mid B) = \frac{P(A \cap B)}{P(B)} \quad \text{if } P(B) \neq 0.$$

Multiplying on both sides by $P(B)$, we obtain a formula for $P(A \cap B)$, namely

$$P(A \cap B) = P(A \mid B)P(B).$$

The latter formula could then be used to find $P(A \cap B)$ whenever the probabilities on the right are known. In this section we shall see that *in certain cases* the formula for $P(A \cap B)$ can be simplified even further because $P(A \mid B) = P(A)$. As a matter of fact, we have already seen one such case in Example 8.6.1.

What is it about a particular experimental situation or particular events $A$ and $B$ that would lead to $P(A \mid B) = P(A)$? Example 8.6.2 shows that *it is not always true that* $P(A \mid B) = P(A)$, so that there must be something special that would cause this to be the case. To show how this phenomenon occurs, we start by recalling that $P(A)$ is approximated by the proportion of times that $A$ occurs in $N$ trials of a given experiment, that is, by $N_A/N$. On the other hand, $P(A \mid B)$ is approximated by $N_{A \cap B}/N_B$. If it should turn out that

$$\frac{N_{A \cap B}}{N_B} = \frac{N_A}{N},$$

then in this case outcomes favorable to $A$ would be distributed throughout $B$ (and thus throughout $A \cap B$) in the same proportion that outcomes favorable to $A$ were distributed throughout the entire sample space $S$.

Example 8.7.1 In the die-throwing experiment $E_3$, let $A = \{1, 2, 3\}$ and $B = \{2, 4\}$, as shown in Figure 8.7.1. Then $P(A) = 3/6$, $P(B) = 2/6$, and $P(A \cap B) = P(2) = 1/6$. It follows that

$$P(A \mid B) = \frac{P(A \cap B)}{P(B)} = \frac{1/6}{2/6} = 1/2.$$

**Figure 8.7.1**

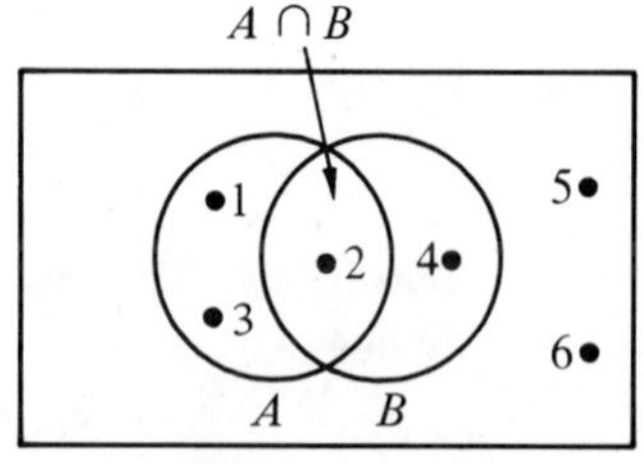


Thus $P(A \mid B) = P(A)$, and the conditional probability of $A$ (on the restricted sample space $B$) is equal to the "full sample space" probability $P(A)$ because (equally likely) outcomes favorable to $A$ are in the same proportion within the restricted sample space $B$ (namely, 1 of 2) as they are within the entire sample space $S_3$ (namely, 3 of 6.) ∎

Events $A$ and $B$ with the property that $P(A \mid B) = P(A)$ are said to be **independent events**. Otherwise, $A$ and $B$ are said to be **dependent events**. If $A$ and $B$ are independent, then we can use the independence property and our earlier formula for $P(A \mid B)$, that is,

$$P(A \cap B) = P(A \mid B)P(B),$$

to obtain a simplified expression for $P(A \cap B)$, a result which we express as a theorem.

**THEOREM 8.7.1** *If A and B are independent events, then*

$$P(A \cap B) = P(A)P(B).$$

Thus, when two events are independent, the probability that they will both occur can be found by simply multiplying their individual probabilities. The proof is left up to you to carry out as an exercise. In addition, you will be asked in the exercises to show that independence is *symmetric* with respect to $A$ and $B$, that is, if $P(A \mid B) = P(A)$, then also $P(B \mid A) = P(B)$.

Example 8.7.2 To illustrate further the concept of independence we use it in another context. We return to our sex-in-politics Example 8.6.4, and look at a slightly different distribution of boys and girls, as shown in Figure 8.7.2.

**Figure 8.7.2**

| | Republicans ($R$) | Democrats ($D$) | Total |
|---|---|---|---|
| Boys ($B$) | 60 | 240 | 300 |
| Girls ($G$) | 40 | 160 | 200 |
| Total | 100 | 400 | 500 |

Here we assert that the events "boy" and "democrat" are independent, and verify this by showing that $P(B \mid D) = P(B)$, as follows:

$$P(B) = \frac{300}{500}; \qquad P(B \mid D) = \frac{P(B \cap D)}{P(D)} = \frac{240/500}{400/500} = \frac{240}{400}.$$

The result is established by reducing the two fractions. As we hinted in our discussion of Theorem 8.7.1, independence is symmetric in this case as well since $P(D \mid B) = P(D) = 0.8$. ∎

Example 8.7.3 Yet another example of independent events is provided by experiment $E_4$, in which two dice are tossed. Let $A$ and $B$ be defined as follows:

$A$: "the sum of the two dice is 7"

$B$: "the outcome of the blue die is 4."

Now, $P(A) = P((1, 6), (2, 5), (3, 4), (4, 3), (5, 2), (6, 1)) = 6/36$, and

$$P(A \mid B) = \frac{P(A \cap B)}{P(B)} = \frac{P((4, 3))}{P((4, 1), (4, 2), (4, 3), (4, 4), (4, 5), (4, 6))}$$
$$= \frac{1/36}{6/36} = 1/6.$$

Thus $P(A \mid B) = P(A)$, and $A$ and $B$ are independent. Even if the two dice were thrown sequentially, with the blue die first, and we knew that the outcome of the blue die was a "4," we would not change the probability that the sum would be 7. ∎

Example 8.7.4 Let us alter the definitions of $A$ and $B$ in $E_4$, so that

$A$: "the sum of the dice is 5"
$B$: "the outcome of the blue die is 4."

$$P(A \mid B) = \frac{P(A \cap B)}{P(B)} = \frac{P((4, 1))}{P((4, 1), \ldots, (4, 6))} = \frac{1/36}{6/36} = 1/6,$$

but $P(A) = 4/36 \neq P(A \mid B)$. Therefore $A$ and $B$ are dependent. ∎

## 8.7 Exercises

1. a. Consider two events $A$ and $B$, such that $P(A \mid B) = 2/3$ and $P(B) = 1/2$. Find $P(A \cap B)$.
   b. If $A$ and $B$ are independent events, find $P(A)$.

2. Show that if $P(A \mid B) = P(A)$, then $P(B \mid A) = P(B)$. (Assume that neither $P(A)$ nor $P(B)$ is 0.)

3. Consider events $A = \{1, 2, 3\}$ and $B = \{2, 4\}$ associated with $E_3$ and discussed in Example 8.7.1. Find $P(B \mid A)$ and show that $P(B \mid A) = P(B)$.

4. Consider events $A$, $B$ of Example 8.7.3.
   a. Find $P(B \mid A)$ and compare it with $P(B)$.
   b. Are the events $A$ and $B$ independent?

5. Consider events $A$, $B$ of Example 8.7.4.
   a. Find $P(B \mid A)$
   b. Are the events $A$ and $B$ independent?

6. Consider the pairs of events of Exercises 8.6.1–8.6.5. Which pairs are independent events?

**New Terms**

independent events, 335 dependent events, 335

## 8.8 Multi-stage Random Experiments

Many random experiments can be thought of as taking place in sequential stages, either because this is the actual case, or because it is merely helpful in some way to consider the experiment in that fashion. To provide a simple example of a 2-stage random experiment in which it is absolutely mandatory that we consider the 2 stages separately, we proceed as follows. Stage 1 consists of $E_3$, in which a single die is tossed. The experiment to be conducted in Stage 2 depends upon the outcome of Stage 1. If the outcome of Stage 1 is in $\{1, 2\}$, we draw a ball from urn 1 as in Figure 8.8.1. If the outcome is in $\{3, 4, 5, 6\}$, we draw from urn 2.

**Figure 8.8.1**

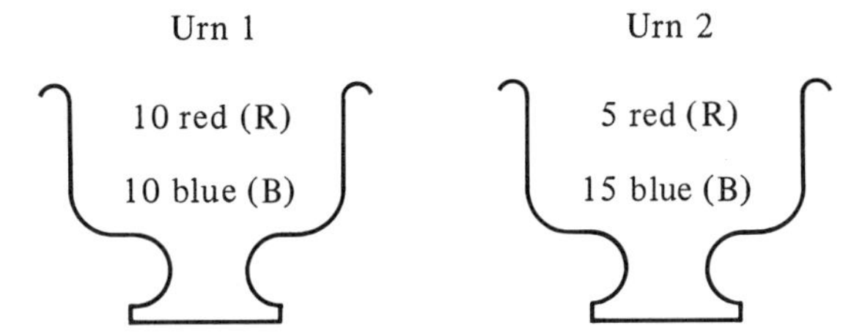


The results for such experiments can be most easily depicted by using a **tree diagram** like that in Figure 8.8.2. The Stage 1 experimental outcomes are shown on the first two branches of the diagram. We see that $P(1, 2) = 2/6$ and that $P(3, 4, 5, 6) = 4/6$. When $\{1, 2\}$ occurs in Stage 1, we proceed along the upper branch and use urn 1 for Stage 2. On this branch we use the composition of urn 1 to compute the probability that a red ball will be drawn given that $\{1, 2\}$ occurred in Stage 1. Thus $P(R \mid 1, 2) = 10/20$. *We emphasize that this is a conditional probability*. Similarly, we compute $P(B \mid 1, 2) = 10/20$. The lower half of the tree diagram shows the case where $C = \{3, 4, 5, 6\}$ occurs in Stage 1. Then urn 2 is used in Stage 2, and we compute $P(R \mid C) = 5/20$ and $P(B \mid C) = 15/20$.

**Figure 8.8.2**

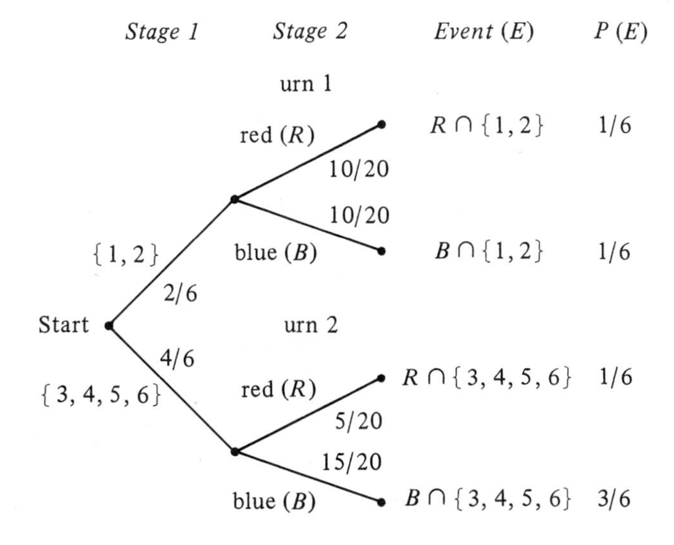

The four possible outcomes of the overall 2-stage experiment are represented by the 4 branch ends at the right side of the tree, along with their respective probabilities. To illustrate the probability computations, we compute

$$P(R \cap \{1, 2\}) = P(R \mid 1, 2)P(1, 2) = (10/20)(2/6) = 1/6$$

and

$$P(B \cap C) = P(B \mid C)P(C) = (15/20)(4/6) = 3/6.$$

You should verify that the two remaining branch probabilities have been correctly calculated. Assuming that this is the case, we could also consider the event "a red ball is drawn" and its probability $P(R)$. Since the event can happen in two mutually exclusive ways, namely $R \cap \{1, 2\}$ and $R \cap C$, we see that $P(R)$ is the sum of the first and third final branch probabilities, or that $P(R) = 1/6 + 1/6 = 2/6$.

Example 8.8.1 A coin is flipped. If $H$ is the outcome, then a ball is drawn from urn 1 of Figure 8.8.3. If $T$ is the outcome, then a ball is drawn from urn 2. We wish to find $P(R \cap H)$, $P(R \cap T)$, and $P(R)$, the probability that a red ball will be drawn in Stage 2.

**Figure 8.8.3**

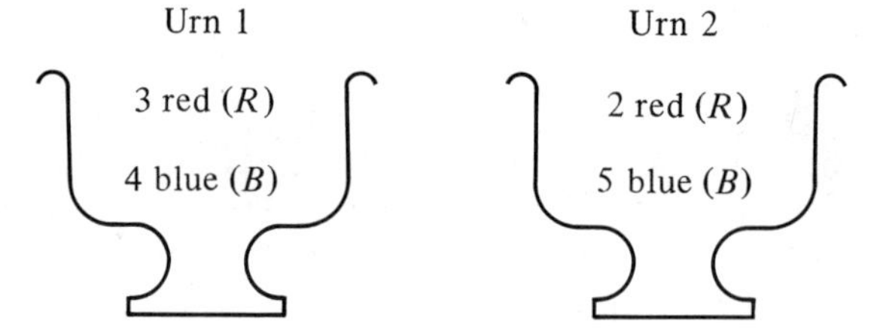


Construct the appropriate tree diagram and verify that

$$P(R \cap H) = P(R \mid H)P(H) = (3/7)(1/2) = 3/14,$$
$$P(R \cap T) = P(R \mid T)P(T) = (2/7)(1/2) = 2/14,$$

while

$$P(R) = P(R \cap H) + P(R \cap T) = 5/14.$$ ∎

Although the multi-stage experiments discussed so far have been couched in the context of games and urns, it is not difficult to imagine serious, real-life instances that fit into the same conceptual category.

Example 8.8.2 Dr. Gentry has a patient with a certain eye disease. The disease may be treated with drug $A$ or drug $B$. Although drug $A$ has a higher cure probability, it also causes dangerous side effects in about 30% of those individuals that take it; and its use must be terminated. In that case, drug $B$, with the lower cure probability, must be used. Drug $B$ has caused no known dangerous side effects. Because of the hazards involved, the treatment program is conducted in 2 stages. During Stage 1 a reduced dosage of drug $A$ is administered for 1 week. If no adverse side effects are noticed, the normal dosage of drug $A$ is then administered. If these side effects become apparent, then drug $B$ is used. The treatment scenario and the appropriate probabilities are shown in Figure 8.8.4. Using the notation and the data given in the figure, we compute the overall probability of a cure as

$$P(C) = P(C \cap N) + P(C \cap S) = 0.63 + 0.15 = 0.78.$$ ∎

Figure 8.8.4

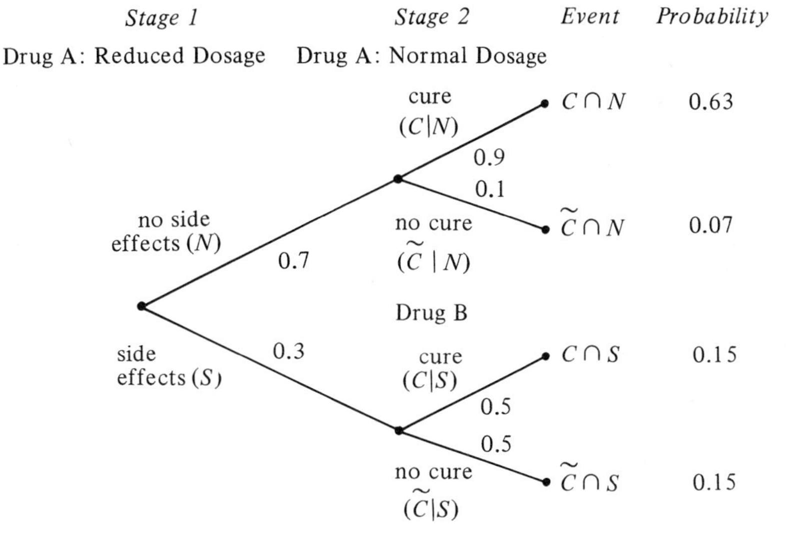


The examples of experimental situations discussed thus far in this section have been cases in which a multi-stage treatment is necessary. There are other cases in which a choice is possible.

In our early example, $E_2$, in which we flipped a nickel and a dime simultaneously, we could have equally well conceptualized the experiment as one in which we first flipped the nickel (Stage 1) and then flipped the dime (Stage 2). The appropriate tree diagram for the multi-stage variation is contained in Figure 8.8.5. From our previous examination of the experiment, we already know that $P(H,H) = P(H,T) = P(T,H) = P(T,T) = 1/4$. This case provides an example in which Stage 2 probabilities are not dependent on Stage 1 outcomes. Using $H_D$ to denote "a head on the dime" and $H_N$ to denote "a head on the nickel" we observe that

$$P(H_D \mid H_N) = 1/2 = P(H_D).$$

Thus the two events are independent in the sense of our earlier definition. The same is true for any pair of events associated with different stages of

Figure 8.8.5

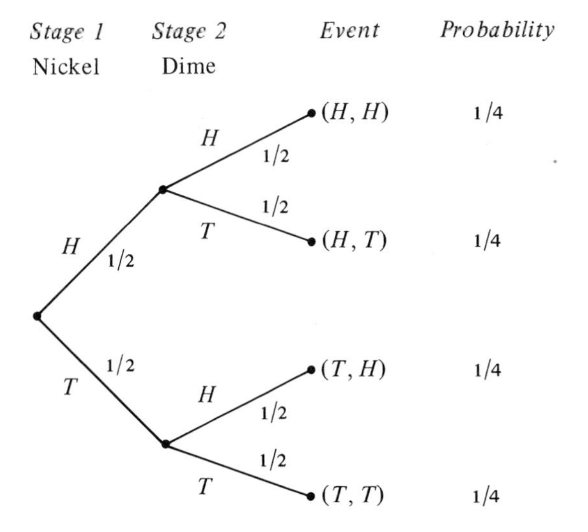

the experiment. It is no surprise, therefore, to find that $P(H_N \cap H_D) = P(H_N, H_D) = P(H_N)P(H_D)$ and that a similar statement holds for all 4 of the original elementary events of $E_2$.

Example 8.8.3 Experiment $E_4$, with two dice, can also be considered as a 2-stage experiment in which the blue die is rolled in Stage 1 and the gold die is rolled in Stage 2. You should construct the tree diagram for this experiment and note, for example, that $P(1, 1)$ can be found by computing the product $P(1_B)P(1_G) = 1/36$. An analogous statement is, of course, true for any one of the 36 possible outcomes of the overall experiment. By the nature of the experiment we would have intuitively thought that any two events individually associated exclusively with different "rolls" should have been "independent," in that the rolls have nothing to do with each other. This is borne out by the fact that $P(A \cap B)$ can be found by computing $P(A)P(B)$ whenever $A$ is associated exclusively with Stage 1 and $B$ is associated exclusively with Stage 2. ∎

As a final illustration of the use of multi-stage experiments we shall examine a real-life situation in which more than one of a particular operational component is placed in a system to provide an increased probability that at least one of the components will work when needed.

Example 8.8.4 Component $A$ is an important part of a new power plant. When $A$ functions properly, electrical output is maintained; and there is no danger. When component $A$ malfunctions, however, there is then the possibility of either a dangerous accident or a long shut-down for repairs, or both. Components $B_1$ and $B_2$ are warning devices, each of which is capable of sensing trouble associated with $A$, and then shutting down the plant without incident—allowing $A$ to be properly repaired. The operation of these safety devices may be portrayed as shown in Figure 8.8.6.

**Figure 8.8.6**

| Stage 1 | Stage 2 | Event | Implication | Probability |
|---|---|---|---|---|
| $B_1$ works $(B_1)$ 0.95 | $B_2$ works $(B_2)$ 0.95 | $(B_1, B_2)$ | proper shutdown | 0.9025 |
| | $B_2$ fails $(\tilde{B}_2)$ 0.05 | $(B_1, \tilde{B}_2)$ | proper shutdown | 0.0475 |
| $B_1$ fails $(\tilde{B}_1)$ 0.05 | $B_2$ works $(B_2)$ 0.95 | $(\tilde{B}_1, B_2)$ | proper shutdown | 0.0475 |
| | $B_2$ fails $(\tilde{B}_2)$ 0.05 | $(\tilde{B}_1, \tilde{B}_2)$ | loss | 0.0025 |

(Tree root: $A$ fails)

If $A$ fails for any reason, the failure should be sensed by both $B_1$ and $B_2$. These units are placed in geographically different locations so that they will not likely be affected and rendered inoperative by the same phenomenon (fire, tornado, etc.). Thus it can be properly said that the sensing units are independent of one another.

The plant will be properly shut down unless both $B_1$ and $B_2$ fail to function, in which case an unknown loss will occur. Using $B_1$ to denote the event that $B_1$ works and $\tilde{B}_1$ to denote the event that $B_1$ fails, we see the probability that both fail is given by

$$P(\tilde{B}_1 \cap \tilde{B}_2) = P(\tilde{B}_1)P(\tilde{B}_2) = (0.05)(0.05) = 0.0025.$$

There are, then, roughly 25 chances in 10,000 that a loss will occur if and when $A$ fails. The use of the redundant system $B_2$ has thus decreased the probability of a loss from 0.05 to 0.0025. How would the addition of a third redundant and independent system affect the same probability? ∎

We conclude this section with a discussion of the terms "conditional probability" and "independent events" as they apply to sequential multi-stage experiments. When we consider a conditional probability of the type $P(A_2 \mid A_1)$, for events $A_2$ associated with Stage 2 and $A_1$ associated with Stage 1, we could, indeed, *know* the result of the Stage 1 experiment before performing the Stage 2 experiment. Thus we could legitimately say that $P(A_2 \mid A_1)$ means *the probability of event $A_2$, given that event $A_1$ has occurred.* This is a different situation from that of the single-stage case of $P(A \mid B)$. If we really knew that event $B$ had occurred, the experiment would already have been performed; and event $A$ would either have already occurred or not. One could then have a philosophical argument about whether the *probability of $A$* had any further meaning for that particular trial.

In sequential situations, the concept of independent events can also be explained intuitively in a slightly different fashion when the events in question are associated exclusively with different stages. In that case we can say that $A_2$ (associated with Stage 2 only) is *independent of $A_1$ if the knowledge that the Stage 1 event has occurred does not alter the probability of occurrence of $A_2$*, that is, $P(A_2 \mid A_1) = P(A_2)$ as before.

## 8.8 Exercises

1. Consider a 2-stage experiment in which Stage 1 consists of rolling a single die. If the outcome is odd, a ball is drawn from an urn containing 5 red balls ($R$) and 3 blue balls ($B$). If the outcome is even, a ball is drawn from an urn containing 4 red balls and 4 blue balls.
   a. Draw a tree diagram to illustrate the 2-stage experiment.
   b. Compute $P(R)$.
   c. Compute $P(B)$ by 2 methods.

2. Consider Experiment $E_6$, in which a nickel, dime, and a quarter are tossed, as a 3-stage sequential experiment in which events associated with different stages are independent.

a. Compute $P(H_N, H_D, H_Q)$ using the probability multiplication property associated with independent events.
b. Compute the probability of at least 1 "head."

3. In the medical problem of Example 8.8.2, suppose that we increase $P(N)$ to 0.9, and change $P(C \mid N)$ to 0.8, while letting $P(C \mid S) = 0.4$. Compute the new values of
a. $P(C \cap N)$
b. $P(\tilde{C} \cap N)$
c. $P(C \cap S)$
d. $P(\tilde{C} \cap S)$.

4. Consider the redundancy problem of Example 8.8.4. Suppose that new components $B_1$ and $B_2$ are avilable such that $P(B_1) = P(B_2) = 0.99$. Compute the revised value for the probability of a loss that would be appropriate for the new equipment.

### New Terms

tree diagram, 337

## 8.9 Stochastic Matrices

Certain types of matrices have been found to be useful in the study of random experiments similar to those which we have been studying. In this and the following section we shall develop the terminology which is used in describing such matrices and experiments and illustrate the use of matrices in the representation and solution of the natural problems that arise. Although we shall confine our examples to experiments which require matrices of small size, it is obviously possible to extend all of our definitions to cover larger and more general situations.

We first introduce the phrase "stochastic vector." The $1 \times 2$ matrix $[p_1, p_2]$ will be called a **stochastic vector** provided that $p_1 + p_2 = 1$ and $p_i \geqslant 0$. We have used the letter $p$ to point out that the elements can, in fact, be interpreted as probabilities.

Example 8.9.1 If we let $p_1$ denote $P(H)$ in the coin-flipping experiment $E_1$, and let $p_2$ denote $P(T)$, then $P = [p_1, p_2] = [1/2, 1/2]$ is a stochastic matrix containing these probabilities. ∎

Example 8.9.2 The vectors $[1/3, 2/3]$, $[0, 1]$, $[1, 0]$, and $[5/6, 1/6]$ are each stochastic vectors. However, the vectors $[-1, 0]$, $[2, 1]$ and $[-1/3, 4/3]$ are not stochastic. Why? ∎

Having seen that appropriate row vectors may be given a probabilistic interpretation, we now proceed to use these row vectors to create a square matrix that may be used, along with the row vectors, to portray random experiments with certain properties. Thus, the $2 \times 2$ matrix $M$ is called a **stochastic matrix**, provided that its rows are stochastic vectors. Such square matrices are also sometimes called **probability matrices**, or **Markov matrices**. It is clear that the elements of a stochastic matrix are non-negative.

**Example 8.9.3** The matrices

$$\begin{bmatrix} 1/2 & 1/2 \\ 1/2 & 1/2 \end{bmatrix}, \quad \begin{bmatrix} 1 & 0 \\ 0 & 1 \end{bmatrix}, \quad \text{and} \quad \begin{bmatrix} 1/3 & 2/3 \\ 1/4 & 3/4 \end{bmatrix}$$

are each stochastic. What about $\begin{bmatrix} 1/2 & 1/3 \\ 1/2 & 2/3 \end{bmatrix}$? ∎

A general $2 \times 2$ stochastic matrix may be easily described. If $M$ is such a non-negative matrix, then it is of the form

$$M = \begin{bmatrix} x & 1-x \\ y & 1-y \end{bmatrix} \quad \text{for some} \quad [x, y] \geqslant 0.$$

We shall use this characterization to help us establish three elementary properties of stochastic matrices. These particular properties have been chosen because they should give you a feeling for what happens when these matrices are manipulated in certain ways.

**Property 1** *If $M$ is a stochastic matrix then $M^2$ is a stochastic matrix.*

**Proof** Let $M = \begin{bmatrix} x & 1-x \\ y & 1-y \end{bmatrix}$ with $[x, y] \geqslant 0$. Then

$$M^2 = \begin{bmatrix} x & 1-x \\ y & 1-y \end{bmatrix}\begin{bmatrix} x & 1-x \\ y & 1-y \end{bmatrix}$$

$$= \begin{bmatrix} x^2 + (1-x)y & x(1-x) + (1-x)(1-y) \\ yx + (1-y)y & y(1-x) + (1-y)(1-y) \end{bmatrix}.$$

From the latter display, we see that each element of $M^2$ is a sum of products of non-negative numbers and is thus non-negative. Furthermore, the sum of the elements in row 1 is

$$\begin{aligned} x^2 + (1-x)y + x(1-x) + (1-x)(1-y) &= x^2 + (1-x)(y + x + 1 - y) \\ &= x^2 + (1-x)(1+x) \\ &= x^2 + 1 - x^2 \\ &= 1. \end{aligned}$$

The reader should verify that the sum of the elements in row 2 is likewise equal to 1, so that $M^2$ is a stochastic matrix. ∎

**Example 8.9.4** If $M = \begin{bmatrix} 1/2 & 1/2 \\ 1/4 & 3/4 \end{bmatrix}$, then

$$M^2 = \begin{bmatrix} 1/2 & 1/2 \\ 1/4 & 3/4 \end{bmatrix}\begin{bmatrix} 1/2 & 1/2 \\ 1/4 & 3/4 \end{bmatrix} = \begin{bmatrix} 3/8 & 5/8 \\ 5/16 & 11/16 \end{bmatrix}$$

which is stochastic. ∎

A more general result is the following:

**Property 2** *If each of $M_1$ and $M_2$ is stochastic, then $M_1M_2$ is stochastic.* We leave the proof of Property 2 to you as an exercise at the end of this section.

Example 8.9.5 Let $M_1 = \begin{bmatrix} 1/2 & 1/2 \\ 1/4 & 3/4 \end{bmatrix}$ and $M_2 = \begin{bmatrix} 1/3 & 2/3 \\ 1/2 & 1/2 \end{bmatrix}$. Then

$$M_1M_2 = \begin{bmatrix} 1/2 & 1/2 \\ 1/4 & 3/4 \end{bmatrix}\begin{bmatrix} 1/3 & 2/3 \\ 1/2 & 1/2 \end{bmatrix} = \begin{bmatrix} 5/12 & 7/12 \\ 11/24 & 13/24 \end{bmatrix}$$

is also stochastic. ∎

Having Property 2 at our disposal enables us to extend the result of Property 1, where we showed that a sufficient reason for $M^2$ to be stochastic was that $M$ was stochastic. Since each of $M$ and $M^2$ is stochastic, we let $M_1 = M$ and $M_2 = M^2$ in Property 2, and see that $M^3 = M_1M_2$ is also stochastic. Continuing in this fashion we could show that $M^n$ is stochastic for any positive integer $n$. An important problem in the study of stochastic matrices is to examine the properties of $M^n$ as $n$ increases. It may be, for example, that $M^n$ tends toward some limiting matrix $M'$ as $n$ increases.

**Property 3** *If $M$ is a stochastic matrix and $P$ is a stochastic vector, then $PM$ is a stochastic vector.*

**Proof** Let $P = [p_1, p_2]$ and $M = \begin{bmatrix} x & 1-x \\ y & 1-y \end{bmatrix}$, $[x, y] \geqslant 0$. Then

$$PM = [p_1, p_2]\begin{bmatrix} x & 1-x \\ y & 1-y \end{bmatrix} = [p_1x + p_2y, p_1(1-x) + p_2(1-y)].$$

The elements of $PM$ are clearly non-negative, and their sum is

$$\begin{aligned} p_1x + p_2y + p_1(1-x) + p_2(1-y) &= p_1(x + 1 - x) + p_2(y + 1 - y) \\ &= p_1 + p_2 \\ &= 1. \end{aligned}$$ ∎

Example 8.9.6 If $M = \begin{bmatrix} 1/2 & 1/2 \\ 3/4 & 1/4 \end{bmatrix}$ and $P = [1/3, 2/3]$, then

$$PM = [1/3, 2/3]\begin{bmatrix} 1/2 & 1/2 \\ 3/4 & 1/4 \end{bmatrix} = [2/3, 1/3].$$ ∎

## 8.9 Exercises

1. Which of the following vectors are stochastic? For each vector which is not, explain why not.

   a. $[1, 2]$  b. $[1/2, 2/3]$  c. $[5/4, -1/4]$

   d. $[2/3, 1/3]$  e. $\left[\frac{\sqrt{2}-1}{\sqrt{2}}, \frac{1}{\sqrt{2}}\right]$

2. Is the sum of two stochastic matrices stochastic? Why?

3. Which of the following matrices are stochastic?

a. $\begin{bmatrix} 1/2 & 1/3 \\ 1/2 & 2/3 \end{bmatrix}$ b. $\begin{bmatrix} 0 & 1 \\ 1 & 0 \end{bmatrix}$ c. $\begin{bmatrix} 1/2 & -1/2 \\ -1/2 & 1/2 \end{bmatrix}$

d. $\begin{bmatrix} 5/6 & 1/6 \\ 2/6 & 4/6 \end{bmatrix}$ e. $\begin{bmatrix} 4/5 & 1/5 \\ 1/5 & 4/5 \end{bmatrix}$ f. $\begin{bmatrix} -1 & 2 \\ 2 & -1 \end{bmatrix}$

g. $\begin{bmatrix} 1/4 & 3/4 \\ 3/4 & 1/4 \end{bmatrix}$

4. A stochastic matrix with the property that its column elements sum to 1 and are non-negative is said to be **doubly stochastic**. Which of the matrices in Exercise 3 are doubly stochastic?

5. Let $M_1, M_2$ be the matrices listed. Show that $M_1M_2$ is stochastic when appropriate.

a. $\begin{bmatrix} 1/3 & 2/3 \\ 1/2 & 1/2 \end{bmatrix}, \begin{bmatrix} 1/2 & 1/2 \\ 1/4 & 3/4 \end{bmatrix}$ b. $\begin{bmatrix} 1/5 & 4/5 \\ 0 & 1 \end{bmatrix}, \begin{bmatrix} 1 & 0 \\ 2/3 & 1/3 \end{bmatrix}$

c. $\begin{bmatrix} -1/3 & 4/3 \\ 0 & 1 \end{bmatrix}, \begin{bmatrix} 1 & 0 \\ 0 & 1 \end{bmatrix}.$

6. If each of $M_1$ and $M_2$ is a stochastic matrix, show that $M_1M_2$ is stochastic.

7. Let $P = [4/5, 1/5]$ and $M = \begin{bmatrix} 2/3 & 1/3 \\ 1/5 & 4/5 \end{bmatrix}$. Show that $PM$ is stochastic.

8. Find two matrices, $A$ and $B$, such that neither $A$ nor $B$ is stochastic but the product $AB$ is stochastic.

### Suggested Reading

J. G. Kemeny, J. L. Snell, and G. L. Thompson. *Introduction to Finite Mathematics.* Englewood Cliffs, N.J.: Prentice-Hall, 1957.

### New Terms

stochastic vector, 342
stochastic matrix, 342
probably matrix, 342
Markov matrix, 342
doubly stochastic matrix, 345

## 8.10 Markov Chains—Known Initial State

We now consider a special type of random process. Suppose that we have a system, which, upon the performance of an experiment, is found to be in one of two **states**, or situations, $S_1$ or $S_2$. The experiment could consist simply of flipping a light switch with possible states $S_1$ (denoting that the light is "on") and $S_2$ (denoting that the light is "off"). We suppose that such an experiment can be performed repeatedly. Thus, we could progress from state to state as the experiments are performed.

As another example of such a system we shall consider a rocket booster engine that is fired periodically, in which case the possible states could be

$S_1$: "the rocket functions properly when fired"

$S_2$: "the rocket malfunctions when fired."

If we know two conditional probabilities, namely,

$p_{11}$: the probability that the rocket will function properly, given that it functioned properly last time, that is, the probability of going from $S_1$ to $S_1$,

$p_{12}$: the probability that the rocket will malfunction this time, given that it functioned properly last time, that is, the probability of going from $S_1$ to $S_2$,

then we could expect that $p_{11} + p_{12} = 1$ and $p_{ij} \geqslant 0$, since the rocket must either function or malfunction this time, regardless of what it did last time. Likewise we may define $p_{21}$ and $p_{22}$, supposing that we are originally in $S_2$ rather than $S_1$. If we are in $S_2$, then the probability of remaining there is $p_{22}$, while the probability of going to $S_1$ is $p_{21}$. These $p_{ij}$ thus represent the conditional probabilities of going from the various $S_i$ to $S_j$ in one particular firing attempt. Since they can be viewed as representing the probability of transition from $S_i$ to $S_j$, as shown in Table 8.10.1, they are also called **transition probabilities**. The matrix $M = [p_{ij}]$ is similarly called a **transition matrix**.

**Table 8.10.1**

| *State* | $S_1$ | $S_2$ |
|---|---|---|
| $S_1$ | $p_{11}$ | $p_{12}$ |
| $S_2$ | $p_{21}$ | $p_{22}$ |

*A stochastic system such as we have described, which moves from state to state in a fashion such that the probability of going from $S_i$ to $S_j$ depends only upon the state we are in*, rather than on a more complete historical description, *is called a 2-state* **Markov chain**.

Example 8.10.1 Suppose that we study repeated firings of the rocket engine of the previous example and that the probability of functioning properly is found to be different, depending upon whether or not the rocket fired correctly last time. Examination of the data shows that a Markov model is appropriate, and that $p_{11} = 0.9$, $p_{12} = 0.1$, while $p_{21} = 0.8$ and $p_{22} = 0.2$. The transition matrix

$$M = \begin{bmatrix} 0.9 & 0.1 \\ 0.8 & 0.2 \end{bmatrix}.$$

If the rocket fired successfully on the previous trial, then we are in $S_1$ and the probability of a successful firing on the next trial is 0.9. ∎

**Figure 8.10.1**

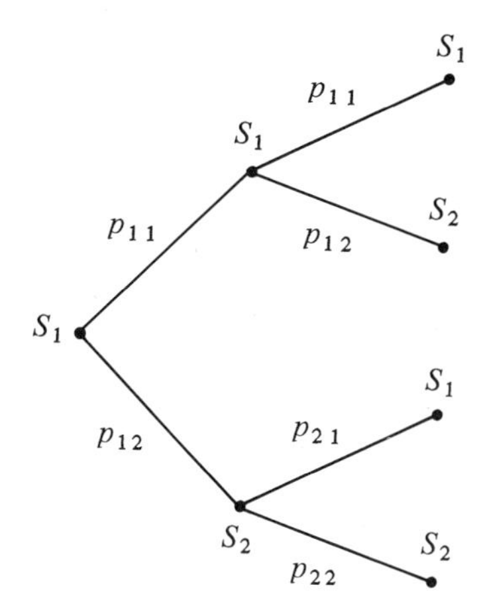


In many such random processes we are interested in knowing the probability of going from $S_1$ to $S_1$ say, in 2, or 3, or even $n$ steps. For $n = 2$ in the rocket case, we would want to know the probability of the rocket firing properly on the second future attempt, given that it functioned properly on the last attempt. In this case we find that we can go from $S_1$ to $S_1$ in 2 steps in two different ways as illustrated in the tree diagram of Figure 8.10.1. We use the symbol $p_{11}^{(2)}$ to denote the probability of going from $S_1$ to $S_1$ in two steps and note that it should be computed as the sum of the probabilities of going from $S_1$ to $S_1$ to $S_1$ and from $S_1$ to $S_2$ to $S_1$. Thus, using the tree as a guide, we obtain

$$p_{11}^{(2)} = p_{11}p_{11} + p_{12}p_{21}.^*$$

Similarly, we define $p_{12}^{(2)}$, $p_{21}^{(2)}$, and $p_{22}^{(2)}$ and obtain

$$\begin{aligned} p_{12}^{(2)} &= p_{11}p_{12} + p_{12}p_{22} \\ p_{21}^{(2)} &= p_{21}p_{11} + p_{22}p_{21} \\ p_{22}^{(2)} &= p_{21}p_{12} + p_{22}p_{22}. \end{aligned}$$

The entire matrix of 2-step transition probabilities, namely,

$$M^{(2)} = [p_{ij}^{(2)}],$$

turns out fortunately to be nothing more than the stochastic matrix $M^2$ since

$$M^2 = \begin{bmatrix} p_{11} & p_{12} \\ p_{21} & p_{22} \end{bmatrix} \begin{bmatrix} p_{11} & p_{12} \\ p_{21} & p_{22} \end{bmatrix} = \begin{bmatrix} p_{11}p_{11} + p_{12}p_{21} & p_{11}p_{12} + p_{12}p_{22} \\ p_{21}p_{11} + p_{22}p_{21} & p_{21}p_{12} + p_{22}p_{22} \end{bmatrix}$$
$$= \begin{bmatrix} p_{11}^{(2)} & p_{12}^{(2)} \\ p_{21}^{(2)} & p_{22}^{(2)} \end{bmatrix}.$$

Example 8.10.2 With $M = \begin{bmatrix} 0.9 & 0.1 \\ 0.8 & 0.2 \end{bmatrix}$ as in Example 8.10.1, we obtain

$$M^{(2)} = M^2 = \begin{bmatrix} 0.9 & 0.1 \\ 0.8 & 0.2 \end{bmatrix} \begin{bmatrix} 0.9 & 0.1 \\ 0.8 & 0.2 \end{bmatrix} = \begin{bmatrix} 0.89 & 0.11 \\ 0.88 & 0.12 \end{bmatrix}.$$

Thus, for example, $p_{12}^{(2)} = 0.11$. ∎

* We suppose that the conditional probabilities are the same for each of the two trials.

It can be shown in general that the **$n$-step transition matrix** for going from $S_i$ to $S_j$ in $n$ steps is such that $M^{(n)} = M^n$.*

## 8.10 Exercises

1. Suppose that you have a transition matrix $M = \begin{bmatrix} 0 & 1 \\ 1/3 & 2/3 \end{bmatrix}$ and you are in $S_1$.
   a. Find the probability of going to $S_2$.
   b. Find the probability of remaining in $S_1$.
   c. Find the probability of going from $S_1$ to $S_2$ to $S_1$.
   d. Use a tree diagram to find $p_{12}^{(2)}$.

2. Use $M^2$ to find $p_{11}^{(2)}$ and $p_{12}^{(2)}$ in Exercise 1 above.

3. a. With $M$ as in Exercise 1 above, find $p_{11}^{(3)}$ and $p_{12}^{(3)}$ and verify that their sum is 1.
   b. Find $p_{21}^{(3)}$ and $p_{22}^{(3)}$, and verify that their sum is 1.

4. In general, the elements of the $n$-step transition matrix $M^{(n)}$ are given by the elements of ________.

5. Suppose that you have a sequence of two identical receiver-transmitter units which are supposed to receive a given message (either "yes" or "no") and transmit it exactly as received. If the probability of sending the message exactly as received is 2/3:
   a. Set up the transition matrix which reflects the situation above, assuming that the unit could have received either message.
   b. Find the 2-step transition matrix.

6. Suppose that a student either rides his bicycle or walks to class. If he rides his bicycle on a given day he is equally likely to walk as ride his bicycle on the next day. He never walks two days in a row. Set up a transition matrix reflecting this situation. Let the states be "$W$" and "$R$," for "walk" or "ride." If he walked on Monday, what is the probability that he will walk on Wednesday? What if he rode his bicycle on Monday? Tuesday, of course, is a class day.

### Suggested Reading

J. G. Kemeny and J. L. Snell. *Finite Markov Chains.* Princeton, N.J.: Von Nostrand Reinhold, 1967.

J. G. Kemeny, J. L. Snell and G. L. Thompson. *Introduction to Finite Mathematics.* Englewood Cliffs, N.J.: Prentice-Hall, 1957.

C. Derman, L. J. Gleser, and I. Olkin. *A Guide to Probability Theory and Application.* New York: Holt, Rinehart and Winston, 1973.

### New Terms

state, 345
transition probability, 346
transition matrix, 346
Markov chain, 346
$n$-step transition matrix, 348

* J. G. Kemeny, *et al.*, *Introduction to Finite Mathematics* (Englewood Cliffs, N.J.: Prentice-Hall, 1957), pp. 218 ff.

## 8.11 Markov Chains—Uncertain Initial State

In the last section we discussed the procedure for describing a random process which moves from $S_i$ to $S_j$ with conditional probability $p_{ij}$. Under the assumption that we knew which state we started in, we obtained the probability of being in state $S_1$ or $S_2$ after 1, 2, . . . , $n$ steps by computing the proper power of $M = [p_{ij}]$.

Here we shall consider the case where we do not know which state we start in, but only have a stochastic vector $P^{(0)} = [a, b]$ expressing our probabilistic estimate of the state we start in. For example, if we are equally likely to be in either state, then $P^{(0)} = [a, b] = [1/2, 1/2]$.

We now compute the total probability $p_1^{(1)}$ of being in $S_1$ after 1 step as follows:

$$p_1^{(1)} = p_{11}a + p_{21}b,$$

where $p_{11}a$ represents the product of the conditional probability $p_{11}$ of going from $S_1$ to $S_1$, and $a$, the probability of starting in $S_1$. The second summand, $p_{21}b$, has an analogous interpretation assuming a start in $S_2$. Similarly, $p_2^{(1)}$, the total probability of arriving in $S_2$ in 1 step is computed as

$$p_2^{(1)} = p_{12}a + p_{22}b.$$

We verify that

$$\begin{aligned} p_1^{(1)} + p_2^{(1)} &= (p_{11}a + p_{21}b) + (p_{12}a + p_{22}b) \\ &= a(p_{11} + p_{12}) + b(p_{21} + p_{22}) \\ &= a + b \\ &= 1. \end{aligned}$$

Example 8.11.1 Let $M = \begin{bmatrix} 1 & 0 \\ 1/2 & 1/2 \end{bmatrix} = \begin{bmatrix} p_{11} & p_{12} \\ p_{21} & p_{22} \end{bmatrix}$ and $P^{(0)} = [a, b] = [1/2, 1/2]$. Then $p_1^{(1)} = 1(1/2) + 1/2(1/2) = 3/4$, and $p_2^{(1)} = 0(1/2) + 1/2(1/2) = 1/4$. Since $p_{11} = 1$ in $M$ shows that we remain in $S_1$ if we are already there, then we would have expected the probabilities to change as they did. Thus, we would think that our probability of being in $S_1$ would increase with the number of trials. Since $a = 1/2$ and $p_1^{(1)} = 3/4$, such is the case. ∎

With $M$ and $P^{(0)}$ defined as above, we see that the matrix

$$P^{(1)} = [p_1^{(1)}, p_2^{(1)}]$$

can also be computed as a matrix product where

$$\begin{aligned} P^{(1)} = P^{(0)}M &= [a, b]\begin{bmatrix} p_{11} & p_{12} \\ p_{21} & p_{22} \end{bmatrix} \\ &= [ap_{11} + bp_{21}, ap_{12} + bp_{22}]. \end{aligned}$$

Example 8.11.2 With $M$ and $P^{(0)}$ as in Example 8.11.1, we have

$$P^{(1)} = P^{(0)}M = [1/2, 1/2]\begin{bmatrix} 1 & 0 \\ 1/2 & 1/2 \end{bmatrix} = [3/4, 1/4].$$

We now let $P^{(2)} = [p_1^{(2)}, p_2^{(2)}]$ where $p_i^{(2)}$ denotes the probability of being in $S_i$ after 2 steps. Replacing our original estimate of state vector $P^{(0)}$ by $P^{(1)}$ we obtain

$$p_1^{(2)} = p_{11}p_1^{(1)} + p_{21}p_2^{(1)}$$

and

$$p_2^{(2)} = p_{12}p_1^{(1)} + p_{22}p_2^{(1)}.$$

In matrix format we have

$$P^{(2)} = [p_1^{(2)}, p_2^{(2)}] = [p_1^{(1)}, p_2^{(1)}]\begin{bmatrix} p_{11} & p_{12} \\ p_{21} & p_{22} \end{bmatrix}$$

$$= P^{(1)}M = (P^{(0)}M)M = P^{(0)}M^2.$$ ∎

As we mentioned in the previous section, it can be shown that, for a general $n$

$$P^{(n)} = P^{(0)}M^n$$

where $P^{(n)} = [p_1^{(n)}, p_2^{(n)}]$ and $p_i^{(n)}$ represents the probability of being in $S_i$ after $n$ steps.*

Example 8.11.3 With $M$ and $P^{(0)}$ as in Example 8.11.2, we compute

$$P^{(2)} = [p_1^{(2)}, p_2^{(2)}] = P^{(0)}M^2 = [1/2, 1/2]\begin{bmatrix} 1 & 0 \\ 3/4 & 1/4 \end{bmatrix}$$

$$= [7/8, 1/8]$$

and

$$P^{(3)} = [p_1^{(3)}, p_2^{(3)}] = P^{(2)}M = [7/8, 1/8]\begin{bmatrix} 1 & 0 \\ 1/2 & 1/2 \end{bmatrix}$$

$$= [15/16, 1/16].$$

We observe that the probability of being in $S_1$ continues to increase with $n$. ∎

## 8.11 Exercises

1. Suppose that the transition matrix for a 2-state Markov chain is $M = \begin{bmatrix} 0 & 1 \\ 1 & 0 \end{bmatrix}$. If you are equally likely to be in $S_1$ or $S_2$ at the beginning of a sequence of trials, find
   a. $p_1^{(1)}$ b. $p_2^{(1)}$ c. $P^{(1)}$ d. $p_1^{(2)}$ e. $p_2^{(2)}$ f. $P^{(2)}$

2. a. Draw a complete tree diagram illustrating the situation for 2 trials of a 2-state Markov chain with $M = \begin{bmatrix} 1/3 & 2/3 \\ 1/2 & 1/2 \end{bmatrix}$.
   b. Compute $M^{(2)}$ by use of the tree diagram.
   c. Compute $P^{(2)}$ by use of the tree diagram, assuming that $P^{(0)} = [1/2, 1/2]$.
   d. Compute $P^{(2)}$ using matrix methods.

---

* J. G. Kemeny, *et al.*, *Introduction to Finite Mathematics* (Englewood Cliffs, N.J.: Prentice-Hall, 1957), pp. 218 ff.

3. In Exercise 8.10.5, let $P^{(0)} = [2/3, 1/3]$ and find $P^{(1)}$, $P^{(2)}$, and $P^{(3)}$.

4. In Exercise 8.10.6, let $P^{(0)} = [1/2, 1/2]$ and find $P^{(1)}$, $P^{(2)}$, and $P^{(3)}$.

### Suggested Reading

C. Derman, L. J. Gleser, and I. Oldin. *A Guide to Probability Theory and Application.* New York: Holt, Rinehart and Winston, 1973.

D. P. Gaver, and G. L. Thompson. *Programming and Probability Models in Operations Research.* Monterey, Ca.: Brooks/Cole, 1973.

J. G. Kemeny, J. L. Snell and G. L. Thompson. *Introduction to Finite Mathematics.* Englewood Cliffs, N.J.: Prentice-Hall, 1957.

J. G. Kemeny and J. L. Snell. *Finite Markov Chains.* Princeton, N.J.: Von Nostrand Reinhold, 1967.

## 8.12 An Insurance Example

We conclude this chapter with a short but significant practical example of the use of stochastic matrices.

Suppose that you are employed by the ABC Automobile Insurance Company. The company, upon completion of an extensive study, has concluded that drivers who have had accidents are more likely to have another accident than are drivers who have had no accidents. Faced with increasing repair costs and rising claims, the company has decided to surcharge the premiums of their drivers depending upon their accident history—that is, a policyholder's annual premium will be increased an amount determined by the number of accidents in which he has been involved and held to be at fault.

For simplicity we shall suppose that ABC policyholders have been divided into three categories or states as follows: a policyholder is placed in state $S_0$ if he has had no accidents, into $S_1$ if he has had exactly one accident, and into $S_2$ if he has had two or more accidents as an ABC policyholder. Surcharges will be added to the premiums of drivers in $S_1$ and $S_2$ in an effort to force these groups to contribute their fair share of the total premium income.

Your job is to build a mathematical model which will help to predict the number of present policyholders who will be in $S_0$, $S_1$, and $S_2$ after $1, 2, \ldots, n$ years. Stochastic matrices and Markov chains are well-fitted for this task. In Table 8.12.1 we list the conditional probabilities of a given

**Table 8.12.1**

| State \ State | 0 | 1 | 2 |
|---|---|---|---|
| 0 | 0.90 | 0.09 | 0.01 |
| 1 | 0 | 0.80 | 0.20 |
| 2 | 0 | 0 | 1.00 |

policyholder's moving from $S_i$ to $S_j$ in a given policy year, all of which we assume coincide with the calendar year. Thus the probability of having no accidents and remaining in $S_0$ is 0.90 for the driver who has had no accidents. The conditional probability of having no accidents and remaining in $S_1$ is 0.80 for the driver who has had 1 accident. Similarly, the probability of having exactly one accident and moving to $S_1$ is 0.09 for the driver in $S_0$. The probability of having one or more accidents is 0.10 (that is, 0.09 + 0.01) for the driver in $S_0$ and 0.20 for the driver in $S_1$. Because of our particular classification scheme, drivers in $S_2$ remain in $S_2$, as indicated by the fact that $p_{22} = 1.00$. In reality, there could be many more states, one of which could represent drivers who were eliminated as policy-holders because of poor driving records. The typical but fictitious probabilities which are listed analytically illustrate our belief that increasingly poor driving records indicate increasingly poor risks on the average.

For our prediction purposes we use the body of Table 8.12.1 to form the stochastic matrix

$$M = \begin{bmatrix} 0.90 & 0.09 & 0.01 \\ 0 & 0.80 & 0.20 \\ 0 & 0 & 1.00 \end{bmatrix}.$$

Let us suppose now that the company had 10,000 policyholders and that management wishes to obtain a probabilistic estimate of the number of these 10,000 policyholders that it can expect to have in each of $S_0, S_1, S_2$ after three years. With this information and knowledge of the various surcharges, one could estimate the total premium income from these policyholders for the fourth year. We let the original state vector $P^{(0)} = [10000, 0, 0]$ indicating that all policyholders start in $S_0$. We then compute

$$P^{(1)} = P^{(0)}M = [10000, 0, 0]\begin{bmatrix} 0.90 & 0.09 & 0.01 \\ 0 & 0.80 & 0.20 \\ 0 & 0 & 1.00 \end{bmatrix} = [9000, 900, 100],$$

$$P^{(2)} = P^{(1)}M = P^{(0)}M^2 = [9000, 900, 100]\begin{bmatrix} 0.90 & 0.09 & 0.01 \\ 0 & 0.80 & 0.20 \\ 0 & 0 & 1.00 \end{bmatrix}$$

$$= [8100, 1530, 370],$$

and that

$$P^{(3)} = P^{(2)}M = P^{(0)}M^3 = [8100, 1530, 370]M = [7290, 1953, 757].$$

Thus, to start the fourth policy year, we would expect that of our original 10,000 policyholders, 7290 would be in $S_0$, 1953 in $S_1$, and 757 in $S_2$. Management may now use this data to help decide what the premium surcharges should be. They may, for example, decide to make the penalty so large that policyholders in $S_2$ will be economically forced to leave the company rather than continue.

In the next chapter we shall see how Markovian ideas may be applied to the important problem of manpower planning for large organizations.

# 9 Manpower Planning

## 9.1 Prediction of Long-Range Structure

One of the more recent and potentially useful applications of Markovian type models has been in the field of manpower planning. See Davies* for an introduction to several such models, one of which provides the background for this section. Brothers† has combined Davies' models with linear programming in an effort to analyze implications of various promotion policies upon the officer corps of the U.S. Air Force.

We shall initially consider a *manpower system* which has $k$ "grade" or "rank" levels. A typical situation is provided by a university faculty system in which there are commonly four teaching levels—instructor, assistant professor, associate professor, and professor—which we shall denote by $G_1$ through $G_4$, respectively. We are interested in the behavior of the rank structure of the faculty system over time, starting at $t = 0$ and extending, say, through $m$ annual periods. Although the models are not restricted to this case, we shall suppose that the faculty size is constant and equal to $s$. The $s$ original faculty members at time 0 (the start of period 1) are distributed in the *proportions* indicated by a row matrix $X(t) = [x_1(t), x_2(t), x_3(t), x_4(t)]$ where, for demonstration purposes, $X(0) = [0.10, 0.45, 0.30, 0.15]$ and the last element 0.15, for instance, indicates that 0.15 or 15% of the faculty are full professors. Suppose that we wish to examine the situation after 2 periods during each of which we hypothesize movement between grades as indicated by the $4 \times 4$ matrices $P(1)$, $P(2)$, where

$$P(1) = P(2) = \begin{bmatrix} 0.5 & 0.4 & 0 & 0 \\ 0 & 0.7 & 0.2 & 0 \\ 0 & 0 & 0.8 & 0.1 \\ 0 & 0 & 0 & 0.9 \end{bmatrix} = P.$$

---

* G. S. Davies, "Structural Control in a Graded Manpower System," *Management Science* 20 (September 1973): 76–84.

† J. N. Brothers, "A Markovian Force Structure Model," Master's Thesis, A.F.I.T., Wright-Patterson Air Force Base, Ohio, 1974.

For our purposes, $P$ is assumed constant over all periods and the element $p_{23}$, for example, represents the proportion of the set of assistant professors ($G_2$) who are promoted to the rank of associate professor ($G_3$) during a particular period. An examination of the elements of each row shows a sum less than 1. Since it happens that each sum is 0.9, we are assuming that in this case 10% of the members of each rank leave the staff for one reason or another during each year. We thus have an (assumed constant) attrition row vector $W = [0.1, 0.1, 0.1, 0.1]$ reflecting this fact. If we did no recruiting to match this attrition, the faculty size would obviously decline. We shall consider this further after observing the 2-period behavior of the system.

We first compute the 1-period distribution matrix $X(1) = X(0)P$, where as you should verify, $X(1) = [0.05, 0.355, 0.33, 0.165]$, indicating that 0.33 $s$ faculty members are associate professors at the end of the first period, while only 0.05 $s$ are instructors. Continuing,

$$X(2) = X(1)P = [X(0)P]P = X(0)P^2 = [0.025, 0.2685, 0.335, 0.1815].$$

Summing the elements of $X(1)$ we see, as we would intuitively expect, that only 90% of the original staff remain after 1 period. A similar calculation for $X(2)$ shows that 81% remain at the end of 2 periods.

Our original plan, however, was to insist that the faculty size remain the same. With attrition allowed, the only way we can recoup the losses is by recruitment. Thus we introduce a recruitment row vector $R(1) = [0.2, 0.4, 0.3, 0.1] = [r_1(1), r_2(1), r_3(1), r_4(1)]$ where $r_i(n)$ denotes the *proportion* of the set of "new recruits" who enter grade $i$ at the start of period $n$ (or the end of period $n - 1$). In some systems we would recruit only into the lowest grade, so that $R'(1) = [1, 0, 0, 0]$. A faculty system is normally more flexible, however, so that each of the $r_i$ may be greater than 0, but not greater than 1.

The question that immediately arises, of course, is: "How many new faculty members are needed at the start of the 2nd period, that is, when $t = 1$? The number who left, certainly! Now the *number* who left during the first period is given by $s[X(0)W^T]$, while the *proportion* who left is given by $X(0)W^T$, where

$$sX(0)W^T = s[0.1, 0.45, 0.3, 0.15]\begin{bmatrix} 0.1 \\ 0.1 \\ 0.1 \\ 0.1 \end{bmatrix} = s(0.1),$$

as expected, and this number must be distributed among the 4 grades as indicated by the recruitment vector $R(1)$. In this case, then, rather than the previously obtained value for $X(1)$, we have

$$\begin{aligned} X(1) &= X(0)P + X(0)W^T R(1) \\ &= [0.05, 0.355, 0.33, 0.165] + [0.02, 0.04, 0.03, 0.01] \\ &= [0.07, 0.395, 0.36, 0.175]. \end{aligned}$$

If we were to assume that $R(2) = R(1) = R$, then, at the end of the next period the proportions would be given by

$$X(2) = X(1)P + X(1)W^T R(2)$$
$$= X(1)[P + W^T R]$$
$$= (X(0)[P + W^T R])(P + W^T R)$$
$$= X(0)(P + W^T R)^2$$

$$= [0.1, 0.45, 0.3, 0.15]\left(\begin{bmatrix} 0.5 & 0.4 & 0 & 0 \\ 0 & 0.7 & 0.2 & 0 \\ 0 & 0 & 0.8 & 0.1 \\ 0 & 0 & 0 & 0.9 \end{bmatrix} + \begin{bmatrix} 0.1 \\ 0.1 \\ 0.1 \\ 0.1 \end{bmatrix}[0.2, 0.4, 0.3, 0.1]\right)^2$$

$$= [0.1, 0.45, 0.3, 0.15]\left(\begin{bmatrix} 0.5 & 0.4 & 0 & 0 \\ 0 & 0.7 & 0.2 & 0 \\ 0 & 0 & 0.8 & 0.1 \\ 0 & 0 & 0 & 0.9 \end{bmatrix} + \begin{bmatrix} 0.02 & 0.04 & 0.03 & 0.01 \\ 0.02 & 0.04 & 0.03 & 0.01 \\ 0.02 & 0.04 & 0.03 & 0.01 \\ 0.02 & 0.04 & 0.03 & 0.01 \end{bmatrix}\right)^2$$

$$= [0.1, 0.45, 0.3, 0.15]\begin{bmatrix} 0.52 & 0.44 & 0.03 & 0.01 \\ 0.02 & 0.74 & 0.23 & 0.01 \\ 0.02 & 0.04 & 0.83 & 0.11 \\ 0.02 & 0.04 & 0.03 & 0.91 \end{bmatrix}^2$$

$$= [0.055, 0.3445, 0.397, 0.2035].$$

We now have the above distribution, then, to start off the next period.

You can use whatever values you wish for $P$, $W$, and $R$; and you will find that, in general,

$$X(n) = X(0)[P + W^T R]^n.$$

You would use the formula to determine the proportions of faculty members in the 4 grades $n$ periods hence.

## 9.1 Exercises

1. Consider a 2-grade manpower system with $X(0) = [0.5, 0.5]$, $W = [0.3, 0.1]$ and $P = \begin{bmatrix} 0.5 & 0.2 \\ 0.1 & 0.8 \end{bmatrix}$.
   a. If the system were allowed to operate without recruitment (that is, the number of members decreases), what would be the resulting distribution between grades after 3 periods?
   b. Assuming that the system in a. started with 1000 members, fill in the blanks on the pictorial representation in Figure 9.1.1. The results of a. will be helpful.

Figure 9.1.1

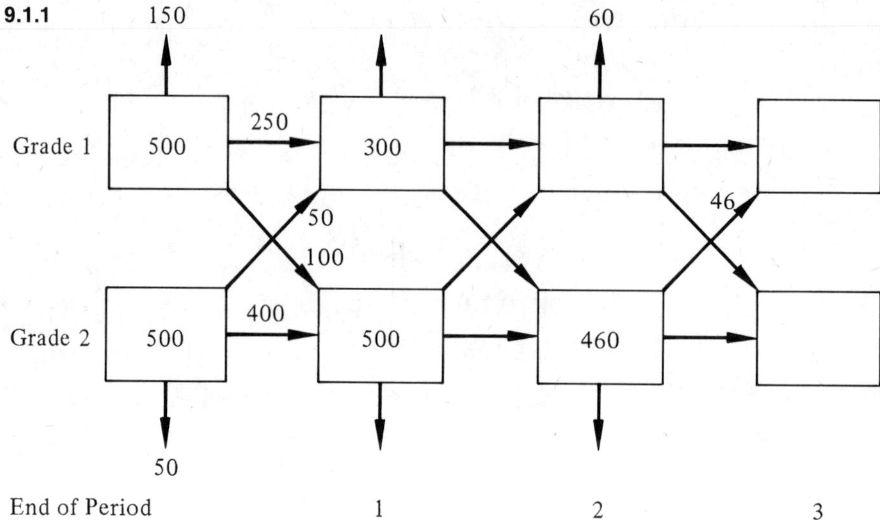


2. a. In Exercise 1, what would be the population distribution after 1 period if a number equal to that of those lost to attrition were allocated to Grade 1 alone? Grade 2 alone?

   b. Considering the results of a. what possible population distributions do you think can be achieved by controlling the recruitment policy vector between the 2 extremes mentioned?

3. Consider the system of Exercise 1 with a constant recruitment vector $R = [0.8, 0.2]$ added to maintain the original population size $s$.

   a. Compute $X(1)$ and $X(2)$ by 2 methods.

   b. Compute $X(0)W^TR$. What does it represent? What does $X(0)W^T$ represent?

   c. What proportion of the original population was lost to attrition during the first period and thus requires replacement?

   d. How is the proportion in b. allocated to Grade 1 and Grade 2 at the end of period 1 and period 2?

   e. Use the above results to help you fill in the missing proportion data in Figure 9.1.2. Each arrow and each box needs a number.

Figure 9.1.2

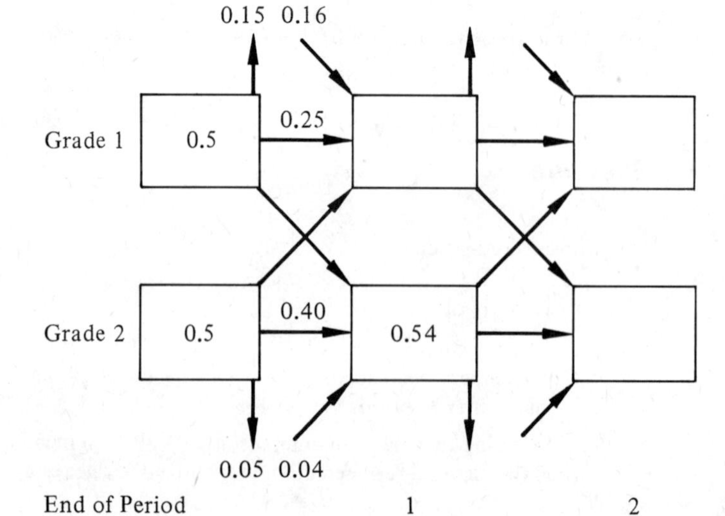

f. If we had started with an initial population of 10,000, how would they be distributed after 2 periods? What is the total attrition over the two periods? The total number recruited?

4. a. In Exercise 3, what would be the proportion distribution after 1 period if a proportion equal to that lost to attrition were allocated to Grade 1 alone? Grade 2 alone? What $R$ vectors would accomplish these allocations?

   b. Compute $X(0)[P + W^T R]$ for $R_1 = [1, 0]$ and $R_2 = [0, 1]$. Compare with a.

   c. Do you think that you could find an $R$ vector such that $X(1) = [0.4, 0.6]$? Such that $X(1) = [0.6, 0.4]$?

## 9.2 The Recruitment Vector as a Control Device

Perhaps a more useful (although distasteful to most college deans) purpose could be served by taking $X(n)$ as a given vector to start with, along with $X(0)$, $P$, and $W$, and inquiring as to whether there exists a sequence of recruitment vectors $R(1), R(2), \ldots, R(n)$, such that $X(n)$ can be reached in exactly $n$ steps. It could well be the case that a given state university faculty, with $X(0) = [0, 0.1, 0.4, 0.5]$, for example, could be deemed by its state legislature to have an overabundance of full professors. The legislators could ask, perhaps, that a recruiting strategy be devised, if one exists, which could keep faculty size constant but reduce the dollar expenditures by creating an $X(3) = [0.2, 0.3, 0.3, 0.2]$.

Such questions can be attacked by the use of Davies' models. Since it is helpful to illustrate the geometry of the situation, we shall consider a smaller dimensional problem in a case where $P$ is $2 \times 2$ and there are only 2 grades or ranks in the system. For this we turn to an example developed by Brothers.* Let $P = \begin{bmatrix} 0.5 & 0.2 \\ 0 & 0.8 \end{bmatrix}$, let the attrition vector $W = [0.3, 0.2]$, and let the initial structure be given by $X(0) = [0.8, 0.2]$. Suppose that we are required to attain the structure $X(2) = [0.2, 0.8]$ by controlling the recruitment vector alone. (We will see shortly that it is impossible to attain the desired structure in one period by controlling the structure with $R(1)$ alone.) We first compute

$$
\begin{aligned}
X(1) &= X(0)[P + W^T R(1)] \\
&= [0.8, 0.2]\left(\begin{bmatrix} 0.5 & 0.2 \\ 0 & 0.8 \end{bmatrix} + \begin{bmatrix} 0.3 \\ 0.2 \end{bmatrix}[r_1(1), r_2(1)]\right) \\
&= [0.8, 0.2]\begin{bmatrix} 0.5 + 0.3r_1(1) & 0.2 + 0.3r_2(1) \\ 0 \quad + 0.2r_1(1) & 0.8 + 0.2r_2(1) \end{bmatrix} \\
&= [0.4 + 0.24r_1(1) + 0.04r_1(1), 0.32 + 0.24r_2(1) + 0.04r_2(1)] \\
&= [0.4 + 0.28r_1(1), 0.32 + 0.28r_2(1)].
\end{aligned}
$$

Now if we were to let $r_1(1) = 1$, so that $r_2(1) = 0$, implying that everyone we recruit would enter $G_1$, the resulting vector would be $[0.68, 0.32]$. On the other hand, if we were to let $r_1(1) = 0$ and $r_2(1) = 1$, we would have $[0.4,$

* J. N. Brothers, "A Markovian Force Structure Model," Master's Thesis, A.F.I.T., Wright-Patterson Air Force Base, Ohio, 1974.

Figure 9.2.1

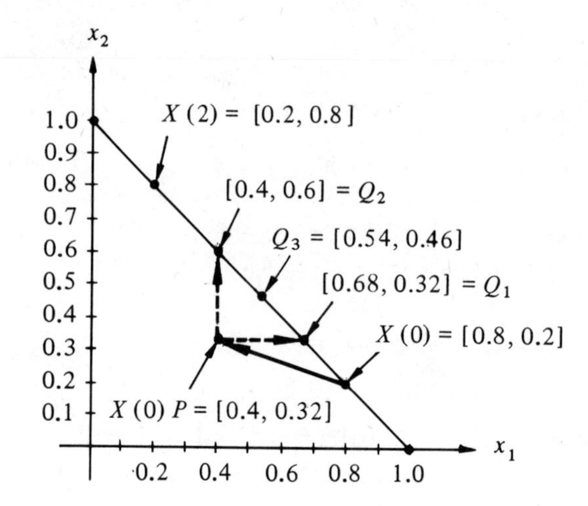


0.6]. Figure 9.2.1 helps to clarify the situation. We see that we started at $X(0)$, and after applying the "promotion" matrix $P$,* we find ourselves at the point $[0.4, 0.32]$ which shows that 28% of our original faculty has been lost. To replace this group we can proportion recruits between $G_1$ and $G_2$ in any way we care to. If we replace all losses by placing them in $G_1$, we proceed horizontally to the point $Q_1 = [0.68, 0.32]$. If we place everyone in $G_2$, we proceed vertically to $Q_2 = [0.4, 0.6]$. If we choose any other assignment scheme, we will end up on the original line with equation $x_1 + x_2 = 1$ somewhere between $Q_1$ and $Q_2$. Thus the only points which are attainable from $X(0)$ in one period are those between $Q_1$ and $Q_2$. As one might intuitively expect, and you should verify this, if we let $r_1(1) = r_2(1) = 0.5$, we land at $X(1) = [0.54, 0.46]$, halfway between $Q_1$ and $Q_2$, denoted by $Q_3$ in the figure. Notably, however, our desired vector $[0.2, 0.8]$ is not a convex combination of $Q_1$ and $Q_2$ (that is, it is not on the line segment $Q_1Q_2$) and is therefore not attainable from $X(0)$ in 1 step. We are thus interested in the set of points that we can reach from $X(0)$ in 2 steps, knowing that we can get to any point between $Q_1$ and $Q_2$ in 1 step. This new set of points consists of the union of two sets, namely the set of points $S_1$ that we can reach in 1 step from $Q_1$ and the set of points $S_2$ that we can reach from $Q_2$ in 1 step.

Now, as Figure 9.2.2 illustrates, $S_1 = Q_1(P + W^TR(2))$ is the set of all points on the line lying between the two points $Q_1' = Q_1(P + W^T[1, 0])$ and $Q_1'' = Q_1(P + W^T[0, 1])$, where

$$Q_1' = Q_1(P + W^T[1, 0]) = [0.68, 0.32]\left(\begin{bmatrix} 0.5 & 0.2 \\ 0.0 & 0.8 \end{bmatrix} + \begin{bmatrix} 0.3 \\ 0.2 \end{bmatrix}[1, 0]\right)$$

$$= [0.68, 0.32]\left(\begin{bmatrix} 0.5 & 0.2 \\ 0 & 0.8 \end{bmatrix} + \begin{bmatrix} 0.3 & 0 \\ 0.2 & 0 \end{bmatrix}\right)$$

$$= [0.68, 0.32]\begin{bmatrix} 0.8 & 0.2 \\ 0.2 & 0.8 \end{bmatrix}$$

$$= [0.608, 0.392],$$

and $$Q_1'' = Q_1(P + W^T[0, 1]) = [0.34, 0.66].$$

* Why, in this case, can we legitimately refer to $P$ as a promotion matrix?

Figure 9.2.2

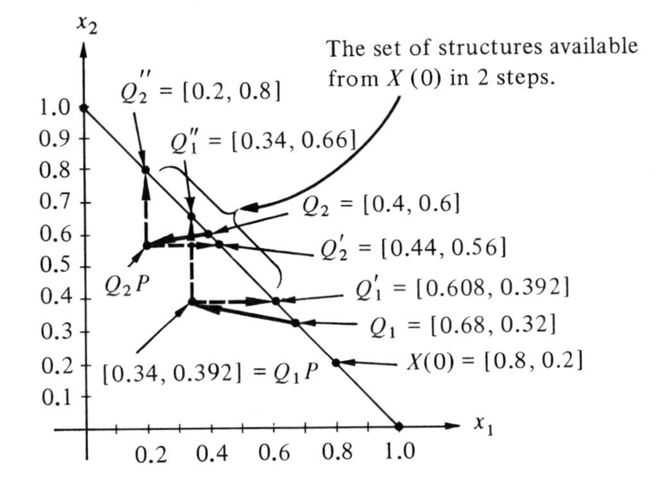


Similarly, $S_2 = Q_2(P + W^T R(2))$ is the set of all convex combinations of the points $Q_2' = Q_2(P + W^T[1, 0]) = [0.44, 0.56]$ and $Q_2'' = Q_2(P + W^T[0, 1]) = [0.2, 0.8]$. Since our desired target is in $S_1 \cup S_2$, we know that it can be reached in 2 steps. (What combination of $R(1)$ and $R(2)$ will achieve it?)

## 9.2 Exercises

1. Consider the manpower system of Exercise 9.1.1 with $X(0) = [0.5, 0.5]$, $W = [0.3, 0.1]$, and $P = \begin{bmatrix} 0.5 & 0.2 \\ 0.1 & 0.8 \end{bmatrix}$.
   a. What is the set of points that can be reached in 1 step from $X(0)$? Illustrate the set graphically as in Figure 9.2.1.
   b. By looking at your graph in a., first guess and then compute the recruitment vector $R = [r_1(1), r_2(1)]$ that will achieve $X(1) = [0.4, 0.6]$.
   c. What is the set of points which can be reached in 2 steps from $X(0)$? Illustrate the set graphically as in Figure 9.2.2.
   d. Is $[0.25, 0.75]$ in the set described in c.?
   e. Find a sequence $R_1 = [r_1(1), r_2(1)]$ and $R_2 = [r_1(2), r_2(2)]$ of recruitment vectors that will achieve $X(1) = [0.35, 0.65]$ and then $X(2) = [0.25, 0.75]$.
   f. Regardless of the original distribution, what is the maximum proportion that you could ever achieve in Grade 1 in one step?

## 9.3 The Promotion and Attrition Matrices as Control Devices

In this section we will discuss two of the three possible control matrices $P$, $W$ and $R$. Both the promotion matrix $P$ and the attrition vector $W$ will be considered as variables to be chosen in order to get from an initial structure

$X(0)$ to a desired final structure $X(1)$ in 1 step. We will not delve into the case when one allows all three matrices to vary since the equations generated would be non-linear.

Let $X(0) = [0.2, 0.8]$ where we have 80% of our group in the higher grade and we wish to lower the expenditure of funds by creating a faculty with structure $X(1) = [0.3, 0.7]$. We will suppose that policy restricts recruitment into the lower grade only, so that $R = [1, 0]$. As we mentioned above,

$$P = \begin{bmatrix} p_{11} & p_{12} \\ p_{21} & p_{22} \end{bmatrix}$$

will be allowed to vary, except that $p_{21} = 0$; no demotions will be permitted. $W = [w_1, w_2]$ will also be considered variable and subject to control of the manpower managers, within the restrictions of the relationships between the $p_{ij}$ and the $w_i$, as we shall see.

From the general formula we see that

$$X(1) = X(0)(P + W^T R)$$

implies the following matrix equation and an equivalent set of linear equations:

$$\begin{aligned}
[0.3, 0.7] &= [0.2, 0.8]\left(\begin{bmatrix} p_{11} & p_{12} \\ 0 & p_{22} \end{bmatrix} + \begin{bmatrix} w_1 \\ w_2 \end{bmatrix}[1, 0]\right) \\
&= [0.2, 0.8]\left(\begin{bmatrix} p_{11} & p_{12} \\ 0 & p_{22} \end{bmatrix} + \begin{bmatrix} w_1 & 0 \\ w_2 & 0 \end{bmatrix}\right) \\
&= [0.2, 0.8]\begin{bmatrix} p_{11} + w_1 & p_{12} \\ w_2 & p_{22} \end{bmatrix}.
\end{aligned}$$

The associated linear equations are thus:

$$\begin{aligned}
0.3 &= 0.2p_{11} + 0.2w_1 + 0.8w_2 \\
0.7 &= 0.2p_{12} + 0.8p_{22}.
\end{aligned}$$

From the nature of the problem it is clear that

$$1 = p_{11} + p_{12} + w_1$$

since everyone in grade 1 must either stay in grade 1, go to grade 2, or leave the system. Similarly,

$$1 = p_{22} + w_2,$$

so that we have the following set of equations:

$$\begin{aligned}
&(1) \quad p_{11} + p_{12} + w_1 = 1 \\
&(2) \quad p_{22} + w_2 = 1 \\
&(3) \quad 0.2p_{11} + 0.2w_1 + 0.8w_2 = 0.3 \\
&(4) \quad 0.2p_{12} + 0.8p_{22} = 0.7
\end{aligned}$$

to be solved for the non-negative values of the $p_{ij}$ and $w_i$.

If, assuming that there is more than one way of achieving the desired structure, we wish to choose one which we prefer in some way, linear programming can sometimes be of assistance. Suppose in the case at hand, for example, that we wish to perform the reduction in average grade level in a way such that many of those employees who remain will be benefitted. We could try, for example, to promote as many as possible of those who remain, that is, make $p_{12}$ as large as possible under the earlier set of restrictions.

We could state the problem as one to—*find* $p_{ij} \geqslant 0$ *and* $w_i \geqslant 0$ *such that equations* (1)–(4) *are satisfied and* $p_{12}$ *is maximized*—a classical linear programming problem such as we considered in Chapter 6. The simplex technique in Chapter 6 leads to the solution

$$p_{11} = 0, \qquad p_{12} = 1, \qquad p_{22} = 0.625, \qquad w_1 = 0, \qquad w_2 = 0.375$$

which, as you should verify, satisfies the equations and certainly maximizes $p_{12}$, in that all who remain in the system are promoted.* In contrast, however, we note that 37.5% of those who were originally in the higher grade must be forced or otherwise enticed to leave. If we were to assume that we started with 1000 members distributed as shown on the left in Figure 9.3.1, they would be finally distributed as shown with 300 of the original 800 $G_2$ members being replaced in the system with 300 members of the recruitment pool. An internal shift of 200 takes place from $G_1$ to $G_2$ to make room for the required new members.

Figure 9.3.1

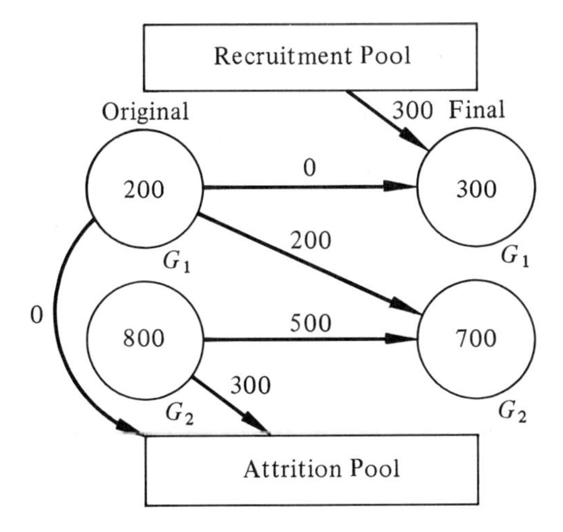


If, rather than maximize the number of promotions, we wished to minimize the attrition required to attain the new structure, we would merely choose to minimize $w_1 + w_2$. This would lead to the LP solution

$$p_{11} = 1, \qquad p_{12} = 0, \qquad p_{22} = 0.875, \qquad w_1 = 0, \qquad w_2 = 0.125,$$

whereby our 1000 original members would redistribute themselves as indicated in Figure 9.3.2. In this case we would have decided that it was better to have preserved the jobs of the 200 original $G_2$ employees than to promote the 200 $G_1$ employees and rehire an additional 200 new employees to fill their positions.

* The LP solution procedures point out that one of the four equations is redundant.

Figure 9.3.2

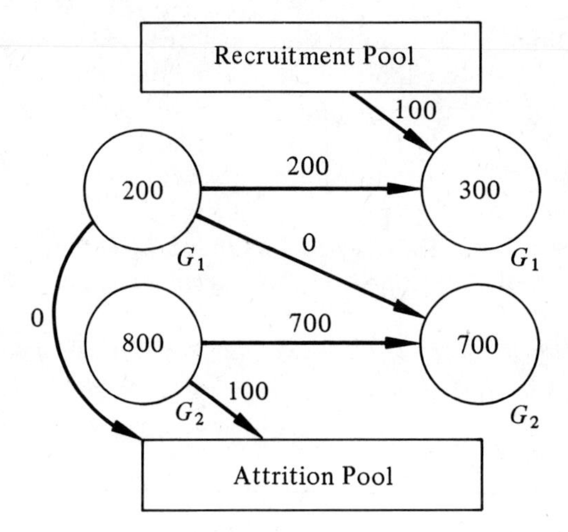


## 9.3 Exercises

1. Our intention is to go from an initial distribution of $X(0) = [0.1, 0.9]$ to a final distribution of $X(1) = [0.3, 0.7]$ in 1 step by proper choice of $P$ and $W$. $R = [0.5, 0.5]$. Promotions ($p_{12}$) are considered good, while demotions ($p_{21}$) and attrition are considered equally bad. Thus, an objective function $f = p_{12} - p_{21} - w_1 - w_2$ is considered appropriate.

   a. Construct an LP problem with 4 equations which will provide the desired solution.

   b. Solve the LP problem. Four artificial vectors are required.

   c. In your final tableau, change the cost coefficient of $p_{21}$ from $-1$ to $-3$, reflecting the viewpoint that demotions are 3 times as bad as attrition. Is the tableau still optimal? If not, find the new solution. Does the change satisfy your intuitive idea about what should happen?

2. The objective is to go from $X(0) = [0.2, 0.8]$ to $X(1) = [0.6, 0.4]$ in 1 step. $R = [0.5, 0.5]$. Let $f = 4p_{12} - 2p_{21} - w_1 - 3w_2$, reflecting the obvious biases about the relative merits of promotions, demotions, and attrition.

   a. Formulate an LP problem which will provide the desired solution—with the additional constraint that there must be at least 10% attrition from Grade 2.

   b. Solve the LP problem.

   c. Suppose that the cost coefficient of $p_{21}$ is changed from $-2$ to $-8$ in your final tableau. It should no longer be optimal. Find the new optimal solution. Does it satisfy your intuition?

# 10 Applications in the Social and Behavioral Sciences—"Two" Least Squares Problems

## 10.1 The Regression Line

Several practical present-day problems have been tackled by techniques that have been referred to as "least squares" techniques. We will make clear the reasons behind these terms in this chapter. We shall consider the application of the *least squares* concept to two seemingly different problems, the first of which may well be familiar to you, especially if you have studied elementary statistics, while the second, in the context described, is probably not.

Today's world is full of situations requiring us to predict outcomes and make decisions based upon those predictions, perhaps weighing the consequences of various types of errors in so doing. An elementary predictive approach is sometimes very helpful, especially in the early stages of such examinations. One of the first uses of a technique known variously as "least squares regression," or "curve fitting," was made by the British statistician Galton (1822–1911). He was interested in the problem of determining the mathematical relationship involved and then of predicting the height of sons, given the height of their fathers. Given data for many father-son combinations, a **scattergram** of height pairs $(f, s)$ can be constructed as shown in Figure 10.1.1.

Suppose now that we have a new father of height $f_1$ who is interested in speculating about the adult height $s_1$ of his newborn son. The method which follows involves using the given data to help discover the hidden, but assumed, linear relationship between the so-called dependent variable $s$ and the single independent variable $f$, and then constructing a straight line representing that relationship as depicted in Figure 10.1.2. For any particular value of $f$, say $f_1$, it would be a simple matter to read the corresponding value of $s_1$ from the graph.

Interestingly, Galton found that sons of tall fathers tended to be, on the average, shorter than their fathers, but that the reverse was true for sons of

Figure 10.1.1

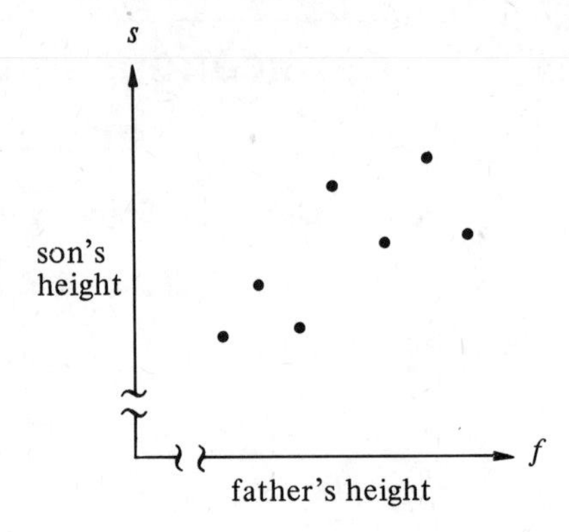


Figure 10.1.2

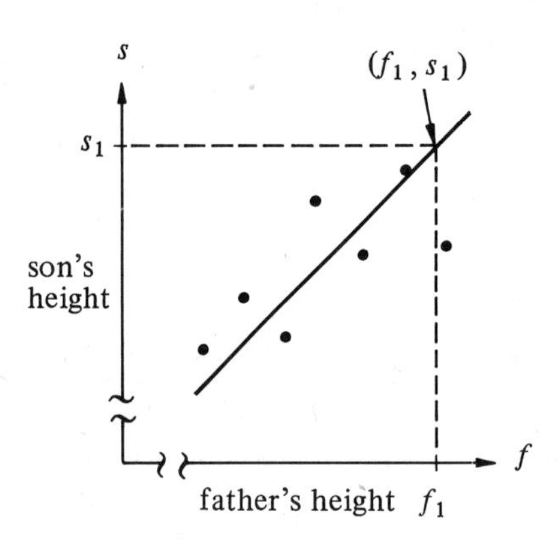


short fathers. Thus, with every new generation he found a tendency to *regress toward an average* height between that of the tall and the short fathers. Thus the constructed line was termed a **regression line**. Today a more sophisticated, but widely used, system of prediction utilized by econometricians and statisticians is known as "regression analysis." In relatively recent years the theory of probability has been brought into the prediction and subsequent decision problems in such a way that one can systematically consider possible errors that may result because of "unlucky" choices of such lines.

Digressing to the problem at hand, however, let us consider an abstract numerical situation in which we are given 4 data points, and go through the details of constructing, *out of the many straight lines one could choose*, a least squares prediction line. This is in lieu, for example, of simply constructing a horizontal predictor line that goes through the average of the ordinate ($s$) values. We could do that in the father-son case, for example, and use that single ordinate value (that is, the average) as a predictor of every son's adult height, regardless of the height of the particular father being considered.

Suppose then that we are given the 4 individually circled data points in Figure 10.1.3. Our task is to choose the scalars $a$ and $b$ so that the straight line with equation

$$x_2 = ax_1 + b$$

will pass *as close as possible* to each of the 4 points. Since it is obviously

Figure 10.1.3

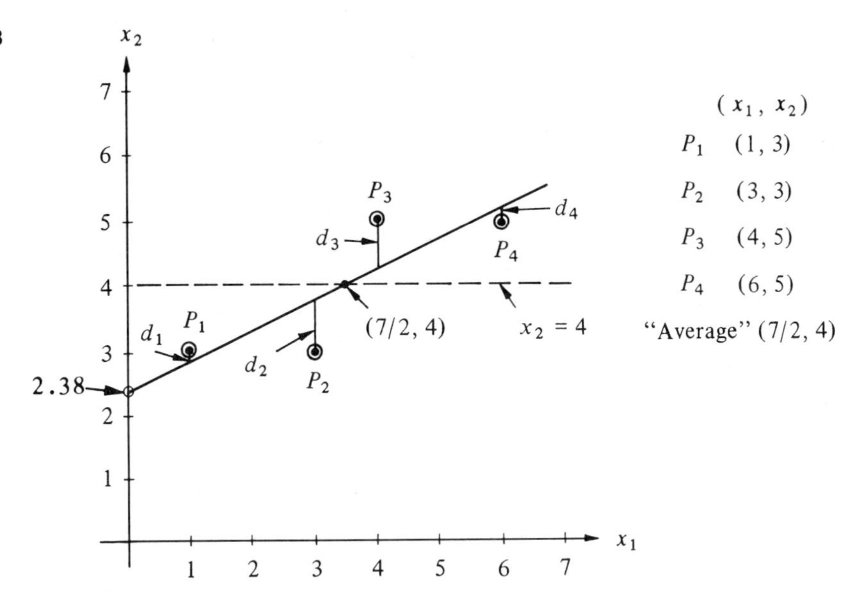


impossible to find a line which will, in general, pass through more than 2 such points, we shall have to be satisfied with less than perfection. Thus, at each data point $P_i$, there will be a vertical deviation between the $x_2$ value on the line and the corresponding $x_2$ value at $P_i$. We denote these vertical deviations by $d_i$ as shown in the figure. Depending upon the order of subtraction, $d_i$ could be positive or negative, but in all cases $d_i^2 \geqslant 0$. A satisfactory definition of what comprises the *best* line would be *that line for which the sum of squares of these deviations is a minimum.**

If, as statisticians in various fields usually do, we number the coordinates of the points with double subscripts so that $P_1(1, 3) = (x_{11}, x_{12})$, $P_2(3, 3) = (x_{21}, x_{22})$, $P_3(4, 5) = (x_{31}, x_{32})$, $P_4(6, 5) = (x_{41}, x_{42})$, we see that we could write the given information in the form of two vectors in 4-space, namely

$$X_1 = \begin{bmatrix} x_{11} \\ x_{21} \\ x_{31} \\ x_{41} \end{bmatrix} = \begin{bmatrix} 1 \\ 3 \\ 4 \\ 6 \end{bmatrix} \quad \text{and} \quad X_2 = \begin{bmatrix} x_{12} \\ x_{22} \\ x_{32} \\ x_{42} \end{bmatrix} = \begin{bmatrix} 3 \\ 3 \\ 5 \\ 5 \end{bmatrix},$$

and describe our problem as one in which we wish to find a predicted vector $\hat{X}_2 = [\hat{x}_{12}, \hat{x}_{22}, \hat{x}_{32}, \hat{x}_{42}]^T$ in such a way that $(\hat{x}_{12} - x_{12})^2 + (\hat{x}_{22} - x_{22})^2 + (\hat{x}_{32} - x_{32})^2 + (\hat{x}_{42} - x_{42})^2$ is minimized.

Our problem can also be rephrased in 4-space terminology. We wish to find $a$ and $b$ in such a way that the predicted vector

$$\hat{X}_2 = a\begin{bmatrix} 1 \\ 3 \\ 4 \\ 6 \end{bmatrix} + b\begin{bmatrix} 1 \\ 1 \\ 1 \\ 1 \end{bmatrix} = aX_1 + bU$$

* Thus perhaps we should refer to a *least sum of squares* method.

is *as close to* the given vector $X_2 = [3, 3, 5, 5]^T$ as possible, in the sense that the square of the distance, or equivalently the distance, between the two vectors is minimized. We recall from Chapter 4 that the distance between two vectors $A$, $B$ is given by the length of $(B - A)$ and that the square of the distance may be found by computing $(B - A)^T(B - A)$. Letting $A = X_2$ and $B = \hat{X}_2$ we find that

$$\begin{aligned}\hat{X}_2 - X_2 &= a[1, 3, 4, 6] + b[1, 1, 1, 1] - [3, 3, 5, 5] \\ &= [a + b - 3, 3a + b - 3, 4a + b - 5, 6a + b - 5],\end{aligned}$$

so that

$$(\hat{X}_2 - X_2)^T(\hat{X}_2 - X_2)$$

$$= [a + b - 3, 3a + b - 3, 4a + b - 5, 6a + b - 5]\begin{bmatrix} a + b - 3 \\ 3a + b - 3 \\ 4a + b - 5 \\ 6a + b - 5 \end{bmatrix}.$$

We have thus formed a function of $a$, $b$, namely, $L(a, b)$ which represents the square of the distance between $\hat{X}_2$ and $X_2$. Performing the indicated matrix multiplication, we find that

$$L(a, b) = (a + b - 3)^2 + (3a + b - 3)^2 + (4a + b - 5)^2 + (6a + b - 5)^2$$

is indeed a sum of squares. Values of $a$ and $b$ which minimize this function of two variables are given by the solution to the following set of linear equations, known as the **normal equations**.*

$$\begin{aligned} 12a + 28b &= 124 \\ 28a + 8b &= 32. \end{aligned}$$

The source of the numerical coefficients will be discussed shortly. Using Gauss-Jordan reduction we solve the two equations to obtain $a = 6/13$ and $b = 31/13$. The desired "least squares" line is thus

$$\hat{x}_2 = (6/13)x_1 + 31/13,$$

which happens to be the original solid prediction line shown in Figure 10.1.3. You may verify that this line does pass through the point (7/2, 4) which happens to be the point which represents the average of the given $x_1$-coordinates and the average of the $x_2$-coordinates.

Now, of course, if we are given any other value of $x_1$, we can readily predict an $\hat{x}_2$ corresponding to it. An important part of statistics and econometrics is concerned with the study of such predictions when probabilistic deviations in $x_2$ are considered.

Although it might be tedious, we can describe a situation with 4 given points without much difficulty. On the other hand, if we were to consider describing a situation with 100 points, the space required would be prohibitive. Matrices enable us to describe the general situation rather succinctly

* For those students who have studied a little calculus, these normal equations are found by differentiating $L(a, b)$ partially, first with respect to $a$ and then with respect to $b$, and setting the results equal to zero.

and to represent in a compact format the so-called "normal equations" from which $a$ and $b$ can be quickly obtained. To develop the general equations, we let $X_1$ and $X_2$ represent, respectively, the given $n$-space vectors of $x_1$ and $x_2$ coordinates of $n$ prescribed points in the plane. Thus, our predicted vector in $n$-space is

$$\hat{X}_2 = aX_1 + b\begin{bmatrix}1\\1\\\vdots\\1\end{bmatrix} = aX_1 + bU.$$

We select $a, b$ so as to minimize the length of the vector

$$\hat{X}_2 - X_2 = aX_1 + bU - X_2.$$

To do so, we note that $X_i^T X_j = X_j^T X_i$ and consider the square of the length of $\hat{X}_2 - X_2$ as given by

$$\begin{aligned} L(a, b) &= (\hat{X}_2 - X_2)^T(\hat{X}_2 - X_2) = (aX_1^T + bU^T - X_2^T)(aX_1 + bU - X_2) \\ &= a^2X_1^TX_1 + abX_1^TU - aX_1^TX_2 + abU^TX_1 + b^2U^TU - bU^TX_2 \\ &\quad - aX_2^TX_1 - bX_2^TU + X_2^TX_2 \\ &= a^2X_1^TX_1 - 2aX_1^TX_2 + 2abX_1^TU - 2bU^TX_2 + b^2U^TU + X_2^TX_2. \end{aligned}$$

You should verify that this matrix equation does lead to $L(a, b)$ as formed in the previous example. The following normal equations can be derived from $L(a, b)$,* namely,

$$\begin{aligned} a(X_1^TX_1) + b(X_1^TU) &= X_1^TX_2 \\ a(X_1^TU) + b(U^TU) &= U^TX_2, \end{aligned}$$

the solution to which is quickly available by Gauss-Jordan reduction. You should also verify that these matrix formulas provide the same equations as those previously presented.

**Example 10.1.1** Consider the 8 points (1, 1), (3, 2), (4, 4), (6, 4), (8, 5), (9, 7), (11, 8), and (14, 9). To find a linear "least squares fit" we let

$$X_1 = [1, 3, 4, 6, 8, 9, 11, 14]^T \quad \text{and} \quad X_2 = [1, 2, 4, 4, 5, 7, 8, 9]^T,$$

while $U$ is an 8-dimensional column vector of ones, and compute

$$X_1^TX_1 = 524, \quad X_1^TU = 56, \quad X_1^TX_2 = 364, \quad U^TX_2 = 40 \text{ and } U^TU = 8.$$

Substituting into the normal equations we obtain

$$\begin{aligned} 524a + 56b &= 364 \\ 56a + \phantom{5}8b &= \phantom{3}40, \end{aligned}$$

---

* Realizing that the various matrix products are all constants, the equations are derived by calculating

$$\frac{\partial f}{\partial a} = 2aX_1^TX_1 - 2X_1^TX_2 + 2bX_1^TU \quad \text{and} \quad \frac{\partial f}{\partial b} = 2aX_1^TU - 2U^TX_2 + 2bU^TU$$

and setting the partial derivatives equal to zero.

from which we compute $a = 7/11$ and $b = 6/11$. For a given $x_1$ value of 5, then, we would associate a predicted value of $x_2$ of $(7/11)(5) + 6/11 = 41/11$. ▮

## 10.1 Exercises

1. a. Construct a scattergram for the points (1, 3), (2, 1), (3, 6), (4, 4), (5, 9), and (6, 7). Draw in your own estimate of a regression line for these 6 points.

   b. To find the actual regression line for these points, we wish to find $a$ and $b$ such that $\hat{X}_2 = aX_1 + bU$ where $X_1^T = [\qquad]$, $U^T = [\qquad]$, and the length of $\hat{X}_2 - X_2$ is minimized, where $X_2^T = [\qquad]$.

   c. The desired vector $[a, b]$ is given by the solution to the so-called ______ equations

$$a \underline{\qquad} + b \underline{\qquad} = X_1^T X_2$$
$$a \underline{\qquad} + b \underline{\qquad} = U^T X_2,$$

   which in this particular case are

$$\underline{\qquad} a + \underline{\qquad} b = \underline{\qquad}$$
$$\underline{\qquad} a + \underline{\qquad} b = \underline{\qquad}.$$

   By Gauss-Jordan reduction we find that

$$a = \underline{\qquad}$$
$$b = \underline{\qquad}.$$

   Plot the regression line $\hat{x}_2 = ax_1 + b$.

   d. Indicate the vertical deviations $\hat{x}_2 - x_2$ on your graph of the regression line. Compute the sum of the squares of the vertical deviations.

   e. Plot the line $\hat{x}_2 = (3/2)x_1 - 1/4$, and compute the sum of the squares of the vertical deviations for this line. Compare the sum with your answer in d.

   f. Plot the line $\hat{x}_2 = (5/4)x_1 + 1$, and repeat the computation and comparison in e.

2. For a certain group of men the following data on years of education and annual income have been collected.

| *Years* | *Income* (in 1000s) |
|---|---|
| 6 | 8 |
| 8 | 9 |
| 12 | 12 |
| 13 | 14 |
| 14 | 18 |

   a. Plot the scattergram for income ($x_2$) vs. years of education ($x_1$).

   b. Find the appropriate normal equations and solve them by Cramer's Rule.

   c. The regression line is ______.

d. For this sample the average education level is ________ years, and the average income is ________. Does the regression line go through the average point?

e. For a person with 11 years education, what income would you predict?

3. Construct a scattergram and plot the regression line appropriate for the given set of points.
   a. (1, 1), (5, 5), (9, 1).
   b. (2, 1), (4, 1.5), (5, 4), (6, 1.5), (7, 4), (8, 3), (9, 6).
   c. (1, 4), (1, 5.5), (2, 3), (2.5, 4), (3, 5.5), (3.5, 4), (4, 2), (4.5, 3).
   d. Find the value that you would predict in each of the above for $x_1 = 6$ and $x_1 = 4.5$.

### Suggested Reading

C. Almon, Jr. *Matrix Methods in Economics.* Reading, Mass.: Addison-Wesley, 1967.

J. C. G. Boot and E. B. Cox. *Statistical Analysis for Managerial Decisions.* New York: McGraw-Hill, 1970.

### New Terms



## 10.2 A "Solution" for Systems of Equations without Any

The second type of least squares problem which we shall examine is of significance in optimization theory—for example, in studying the maximization of a quadratic function of two variables under the condition that a set of linear constraints must also be satisfied, as in linear programming. Problems of this nature are called quadratic programming problems when the solution variables are required to be non-negative.*

The specific problem has to do with a set of linear equations $AX = R$ which has no solution in the usual sense in that there is no vector $X$ such that $AX = R$. In certain applications it has been found helpful to redefine the concept of a solution matrix to allow the inclusion of matrices $X$ such that $AX \neq R$ but, $AX$ "is closest to $R$" in some meaningful way. Let

$$A = \begin{bmatrix} 1 & 1 \\ 1 & -1 \\ 2 & 0 \end{bmatrix} = [A_1, A_2], \quad Y = \begin{bmatrix} y_1 \\ y_2 \end{bmatrix}, \quad \text{and} \quad R = \begin{bmatrix} 1 \\ 0 \\ 2 \end{bmatrix},$$

and solve $AY = R$ by Gauss-Jordan reduction. We obtain a final augmented matrix

$$\left[\begin{array}{cc|c} 1 & 0 & 1/2 \\ 0 & 1 & 1/2 \\ 0 & 0 & 1 \end{array}\right],$$

* See, for example, D. G. Shankland, "Quadratic Programming Using Generalized Inverses," A.F.I.T. Technical Report 75-2, Wright-Patterson Air Force Base, Ohio, May, 1975.

the last row of which indicates that no solution exists for the given set of equations. Considering the vectors $A_1$, $A_2$, and $R$, we have a case in which $R$ does not lie in the plane spanned by $A_1$ and $A_2$. The plane determined by $A_1$ and $A_2$ is shown in Figure 10.2.1 with the vector $R$ projecting out of the plane.

Figure 10.2.1

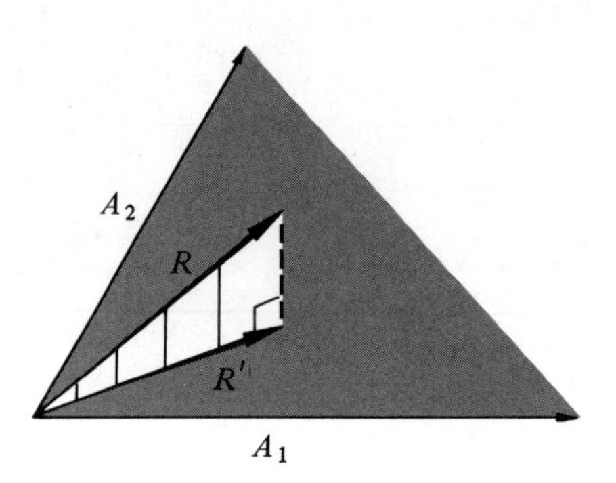


We shall see that when we expand our concept of a solution, our newly defined *solution values* for $y_1$ and $y_2$ will turn out to be the same as they would have been for a problem $AT = R'$ where $R'$ is the vector projection of $R$ onto the plane of $A_1$ and $A_2$ as shown in the figure.

We know that any vector in the $A_1$, $A_2$-plane is represented by

$$y_1A_1 + y_2A_2$$

for the proper choices of $y_1$, $y_2$. It has been found useful to call $\bar{Y}$ the *least squares* solution to the problem provided that $\bar{Y}$ is chosen so that the length of $R - A\bar{Y}$ is minimum over all possible choices for $Y$. Since it is easier to minimize the square of the length of $(R - AY)$, we consider

$$\begin{aligned} f(Y) &= (R - AY)^T(R - AY) = (R^T - Y^TA^T)(R - AY) \\ &= R^TR - Y^TA^TR - R^TAY + Y^TA^TAY, \end{aligned}$$

where each of the final matrix products is a $1 \times 1$ matrix and thus equal to its transpose. Since $(Y^TA^TR)^T = R^TAY$, we have the matrix function

$$\begin{aligned} f(Y) &= R^TR - 2R^TAY + Y^TA^TAY \\ &= [1, 0, 2]\begin{bmatrix}1\\0\\2\end{bmatrix} - 2[1, 0, 2]\begin{bmatrix}1 & 1\\1 & -1\\2 & 0\end{bmatrix}\begin{bmatrix}y_1\\y_2\end{bmatrix} \\ &\quad + [y_1, y_2]\begin{bmatrix}1 & 1 & 2\\1 & -1 & 0\end{bmatrix}\begin{bmatrix}1 & 1\\1 & -1\\2 & 0\end{bmatrix}\begin{bmatrix}y_1\\y_2\end{bmatrix} \\ &= [5] - 2[5, 1]\begin{bmatrix}y_1\\y_2\end{bmatrix} + [y_1, y_2]\begin{bmatrix}6 & 0\\0 & 2\end{bmatrix}\begin{bmatrix}y_1\\y_2\end{bmatrix} \\ &= 5 - 10y_1 - 2y_2 + 6y_1 + 2y_2^2. \end{aligned}$$

The value of $Y$ which will minimize $f(Y)$ is provided by the solution to the matrix equation

$$-2A^TR + 2A^TAY = 0.$$

We see that if $\bar{Y}$ is our choice then

$$A^TA\bar{Y} = A^TR.$$

Since in this example* $A^TA$ is non-singular and hence $(A^TA)^{-1}$ exists, we have our *least squares* solution

$$\bar{Y} = (A^TA)^{-1}(A^TR)$$

$$= \begin{bmatrix} 1/6 & 0 \\ 0 & 1/2 \end{bmatrix} \begin{bmatrix} 1 & 1 & 2 \\ 1 & -1 & 0 \end{bmatrix} \begin{bmatrix} 1 \\ 0 \\ 2 \end{bmatrix}$$

$$= \begin{bmatrix} 1/6 & 0 \\ 0 & 1/2 \end{bmatrix} \begin{bmatrix} 5 \\ 1 \end{bmatrix} = \begin{bmatrix} 5/6 \\ 1/2 \end{bmatrix}.$$

In recapitulation, we started with

$$A_1 = \begin{bmatrix} 1 \\ 1 \\ 2 \end{bmatrix}, \quad A_2 = \begin{bmatrix} 1 \\ -1 \\ 0 \end{bmatrix}, \quad R = \begin{bmatrix} 1 \\ 0 \\ 2 \end{bmatrix}$$

and found no solution in the classical sense for the problem

$$y_1A_1 + y_2A_2 = R.$$

We did find a matrix $\bar{Y}$, however, such that

$$\bar{y}_1A_1 + \bar{y}_2A_2 \cong R, \quad \text{where } \bar{y}_1A_1 + \bar{y}_2A_2 = (5/6)\begin{bmatrix} 1 \\ 1 \\ 2 \end{bmatrix} + (1/2)\begin{bmatrix} 1 \\ -1 \\ 0 \end{bmatrix} = \begin{bmatrix} 4/3 \\ 1/3 \\ 5/3 \end{bmatrix}.$$

The approximation is in the sense that the vector $R - \bar{y}_1A_1 - \bar{y}_2A_2$ has minimum length. In this case, it must be that the vector $R' = \bar{y}_1A_1 + \bar{y}_2A_2$ is the vector projection of the vector $R$ onto the plane of $A_1$, $A_2$. If that is the case, then $R$ and $(R - R')$ should be orthogonal. To verify this, we check that the appropriate dot product is 0.

$$R - R' = \begin{bmatrix} 1 \\ 0 \\ 2 \end{bmatrix} - \begin{bmatrix} 4/3 \\ 1/3 \\ 5/3 \end{bmatrix} = \begin{bmatrix} -1/3 \\ -1/3 \\ 1/3 \end{bmatrix}.$$

$$(R - R') \cdot R' = [-1/3, -1/3, 1/3]\begin{bmatrix} 4/3 \\ 1/3 \\ 5/3 \end{bmatrix}$$

$$= -4/9 - 1/9 + 5/9 = 0.$$

* In general, the rank of $A^TA$ is always the same as the rank of $A$. If $A$ is $m \times n$, $n < m$, and of rank $n$, then $A^TA$ is $n \times n$ and of rank $n$ and hence is invertible.

Example 10.2.1 Consider the set of equations $AY = R$ in which

$$A = \begin{bmatrix} 3 & 2 \\ 1 & 1 \\ 1 & 1 \end{bmatrix} \quad \text{and} \quad R = \begin{bmatrix} 12 \\ 4 \\ 6 \end{bmatrix}.$$

The last two equations indicate that there is no solution in the classical sense. (If you were to pick any point in the plane to be called a *solution* to this set of equations, what point would it be? Draw a graph of the equations and see!)

The least squares solution, as we have seen, is given by

$$\begin{aligned} Y &= (A^T A)^{-1} A^T R \\ &= \begin{bmatrix} 11 & 8 \\ 8 & 6 \end{bmatrix}^{-1} \begin{bmatrix} 3 & 1 & 1 \\ 2 & 1 & 1 \end{bmatrix} \begin{bmatrix} 12 \\ 4 \\ 6 \end{bmatrix} \\ &= \begin{bmatrix} 3 & -4 \\ -4 & 11/2 \end{bmatrix} \begin{bmatrix} 46 \\ 34 \end{bmatrix} \\ &= \begin{bmatrix} 2 \\ 3 \end{bmatrix}. \end{aligned}$$

How does this choice of a solution compare with your intuitively chosen point above? ∎

## 10.2 Exercises

1. Find the least squares solution for the following sets of equations. Graph the lines and the solution

   a. $y_1 = y_2$
   $y_1 = -y_2 + 10$
   $0 = y_2 - 1$

   b. $y_2 = y_1 - 2$
   $y_1 + 3y_2 = 6$
   $y_1 - 3y_2 = 6$
   $y_1 - 3y_2 = -6$

2. Find the regression line that would be constructed for the points of intersection of the equations in Exercise 1b. Does it contain the least squares solution point?

3. Find the least squares solution for the following sets of equations. Graph the lines and the solution.

   a. $8x_1 + 5x_2 = 40$
   $x_1 + 2x_2 = 8$
   $-x_1 + x_2 = 1$

   b. $3x_1 + 2x_2 = 6$
   $-2x_1 - 4x_2 = 1$
   $-2x_1 + 3x_2 = 9$

   c. The 3 equations in part a. together with the equation $x_1 + x_2 = 2$.

4. Let $R' = A\bar{Y}$ in Exercises 1a and 1b above. Show that $R'$ and $R - R'$ are orthogonal, or that $R'$ is the projection of $R$ onto the $A_1, A_2$-plane.

## 10.3 A Comparison of the Two Problems

The question now arises as to how *different* the two problems which we have examined really are. In the first we were given a set of 4 points in 2-space and asked to find a *best linear fit*. In the latter case we were given a set of 3 linear equations in 2 unknowns and asked to find a *best* solution.

In the first case we were given known vectors $X_1, X_2$ and asked to find a vector $\begin{bmatrix} a \\ b \end{bmatrix}$ such that

$$aX_1 + b\begin{bmatrix} 1 \\ 1 \\ 1 \\ 1 \end{bmatrix} = X_2.$$

If we set $a = y_1, b = y_2, X_1 = A, U = A_2$, and $X_2 = R$, then the problem could be expressed in terms of finding $y_1, y_2$, such that $y_1A_1 + y_2A_2 = R$, which is the same as $AY = R$, the original notation for the second problem. Here we know in advance that, except in an unusual case, no solution exists.

Thus it would appear that the former problem is just a special case of a more general problem illustrated in the second situation. If indeed that is the case, it should be true that we can specialize the formula derived in the general case to handle the first situation. Thus we reconsider the formula for $\bar{Y}$, namely,

$$(A^TA)\bar{Y} = (A^TR)$$

in the case where $A = [X_1, U], R = X_2$ and $\bar{Y} = \begin{bmatrix} a \\ b \end{bmatrix}$.

Since

$$A^T = \begin{bmatrix} X_1^T \\ U^T \end{bmatrix}, \quad A^TA = \begin{bmatrix} X_1^T \\ U^T \end{bmatrix}[X_1, U] = \begin{bmatrix} X_1^TX_1 & X_1^TU \\ U^TX_1 & U^TU \end{bmatrix}$$

and $$A^TR = \begin{bmatrix} X_1^T \\ U^T \end{bmatrix}X_2 = \begin{bmatrix} X_1^TX_2 \\ U^TX_2 \end{bmatrix},$$

we have the corresponding matrix equation

$$\begin{bmatrix} X_1^TX & X_1^TU \\ U^TX_1 & U^TU \end{bmatrix}\begin{bmatrix} a \\ b \end{bmatrix} = \begin{bmatrix} X_1^TX_2 \\ U^TX_2 \end{bmatrix},$$

which we recognize as the matrix form of the normal equations presented at the end of Section 10.1. If we reconsider the regression line problem, we can also think of it as requiring us to solve a number of equations in 2 unknowns. In Example 10.1.1 for instance, we wished to solve the equations

$$\begin{aligned} a + b &= 3 \\ 3a + b &= 3 \\ 4a + b &= 5 \\ 6a + b &= 5. \end{aligned}$$

It is now evident that in this context we had merely a special case of a more general situation.

## 10.3 Exercises

1. For $A = \begin{bmatrix} 1 & 1 \\ 3 & 1 \\ 4 & 1 \\ 6 & 1 \end{bmatrix}$, $R = \begin{bmatrix} 3 \\ 3 \\ 5 \\ 5 \end{bmatrix}$ as they would be in the notation of the *second* type of *least squares* problem, verify that the solution obtained by using $\bar{Y} = (A^T A)^{-1} A^T R$ is the same as that obtained earlier.

2. Solve Exercise 10.1.1 by using the formula in Exercise 1 and correctly interpreting $A$ and $R$.

# Completeness Property of the Real Numbers

In order to fully describe the real number system we need to make several definitions which were not included in Section 1.3. First, we say that a subset $S$ of $\mathbb{R}$ is **bounded below** if there is a real number $L$ such that for each $s$ in $S$, $L \leqslant s$, and $L$ is called a **lower bound** for $S$. Similarly, $S$ is said to be **bounded above** if there is a number $U$ such that for each $s$ in $S$, $s \leqslant U$, and $U$ is then called an **upper bound** for $S$. Clearly, if a set $S$ has a lower bound $L$, then it has many lower bounds; for any number $k \leqslant L$ is also a lower bound for $S$. By way of example, consider

$$S = \{x \mid x \text{ is in } \mathbb{R} \text{ and } 5 \leqslant x \leqslant 10\}.$$

Then $S$ is bounded below (5 is a lower bound) and bounded above (10 is an upper bound). The set

$$T = \{x \mid x \leqslant 10\}$$

is bounded above but is not bounded below, while

$$V = \{x \mid 5 \leqslant x\}$$

is bounded below but not above. And, moreover, the set $\mathbb{R}$ of all real numbers is neither bounded below nor above. Thus with respect to upper and lower bounds, there are examples of all types of subsets to be found in $\mathbb{R}$.

We say that a number $u$ is a **least upper bound** for a set $S$ provided $u$ is, first, an upper bound for $S$ and, second, if $v$ is another upper bound for $S$, then $u < v$. (A similar definition may be made for **greatest lower bound**.)

Our final property of the real numbers may now be described by saying that the real numbers are *complete* in the sense that every set $S$ which has an upper bound has a least upper bound. This **completeness property** is the one with which the reader is probably least familiar—and for good reasons. First, it is not a property which everyone finds intuitively obvious (to say the least), and, second, it is not linked to the familiar properties A1–A4, M1–M4, $D_1$ and $D_2$. Indeed, the rational numbers $\mathbb{Q}$, which form an ordered field, do not have the completeness property. For consider the set

$$S = \{x \mid x \text{ is in } \mathbb{Q} \text{ and } x^2 \leqslant 2\}.$$

Now $S$ has plenty of upper bounds—17 is one—but has no least upper bound, since if $u$ is an upper bound for $S$, then $2 \leqslant u^2$; and there are many rational numbers $v$ such that $2 \leqslant v^2 \leqslant u^2$. Any such $v$ is also an upper bound for $S$ and is less than $u$. Hence $S$ has no least upper bound. The "number" that *ought* to be the least upper bound for $S$ is $\sqrt{2}$, but it, of course, is not rational and thus does not enter our discussion. On the other hand,

$$T = \{x \mid x \text{ is in } R \text{ and } x^2 \leqslant 2\}$$

has $\sqrt{2}$ as least upper bound. The difference here is that $S$ is a subset of the rationals, while $T$ is a subset of the reals.

We have now described our familiar set $\mathbb{R}$ of real numbers as a **complete ordered field**. And it is the completeness property which distinguishes $\mathbb{R}$ from all other ordered fields. This fact, which is proved in several of the references at the end of Chapter 1, may be loosely stated as: there is one and only one complete ordered field—namely, the real numbers.

# Sigma Notation and More Matrix Algebra

# B

## B.1 Sigma Notation

In our general notation $A = [a_{ij}]$ for matrices, the elements are designated as doubly subscripted $a$'s, the subscripts denoting respectively the row and the column in which the element lies. In order to present proofs of certain algebraic properties of matrices, it is convenient to have a shorthand notation for addition of the elements of a matrix. We use the Greek capital letter $\sum$ (pronounced "sigma") as our summation symbol, and use it in the following manner.

To indicate the sum of the elements of the third row of the $4 \times 7$ matrix $[a_{ij}]$, instead of the usual

$$a_{31} + a_{32} + a_{33} + a_{34} + a_{35} + a_{36} + a_{37},$$

we write

$$\sum_{j=1}^{7} a_{3j}.$$

That is, the symbol stands for the sum (indicated by $\sum$) of all terms of the form $a_{3j}$, where $j$ (called the index of summation) takes on all integer values from 1 to 7 *inclusive*. In a similar manner

$$\sum_{i=1}^{4} a_{i3}$$

is just an abbreviation for

$$a_{13} + a_{23} + a_{33} + a_{43},$$

which is the sum of the elements of the third column of our aforementioned matrix $[a_{ij}]$.

Consider now the problem of finding the sum of all the elements of the $4 \times 7$ matrix $[a_{ij}]$. In longhand, we might write this, summing by rows, as

$$a_{11} + a_{12} + a_{13} + a_{14} + a_{15} + a_{16} + a_{17} +$$
$$a_{21} + a_{22} + a_{23} + a_{24} + a_{25} + a_{26} + a_{27} +$$
$$a_{31} + a_{32} + a_{33} + a_{34} + a_{35} + a_{36} + a_{37} +$$
$$a_{41} + a_{42} + a_{43} + a_{44} + a_{45} + a_{46} + a_{47}.$$

Here the individual row sums are given by

$$\sum_{j=1}^{7} a_{1j}, \quad \sum_{j=1}^{7} a_{2j}, \quad \sum_{j=1}^{7} a_{3j}, \quad \sum_{j=1}^{7} a_{4j}.$$

Thus our sum might be written as

$$\sum_{j=1}^{7} a_{1j} + \sum_{j=1}^{7} a_{2j} + \sum_{j=1}^{7} a_{3j} + \sum_{j=1}^{7} a_{4j}.$$

But even this is cumbersome, so we write more compactly

$$\sum_{i=1}^{4} \left( \sum_{j=1}^{7} a_{ij} \right).$$

The first part $\sum_{i=1}^{4}$ of the above symbol indicates that we are summing four quantities, and each quantity in the sum is of the form $\sum_{j=1}^{7} a_{ij}$. Note that we also could have formed the original sum by columns,

$$\sum_{i=1}^{4} a_{i1} + \sum_{i=1}^{4} a_{i2} + \sum_{i=1}^{4} a_{i3} + \sum_{i=1}^{4} a_{i4} + \sum_{i=1}^{4} a_{i5} + \sum_{i=1}^{4} a_{i6} + \sum_{i=1}^{4} a_{i7}$$

and that this could have also been more briefly written as

$$\sum_{j=1}^{7} \left( \sum_{i=1}^{4} a_{ij} \right).$$

We thus see that in this case the two sums are equal so that

$$\sum_{i=1}^{4} \left( \sum_{j=1}^{7} a_{ij} \right) = \sum_{j=1}^{7} \left( \sum_{i=1}^{4} a_{ij} \right).$$

More generally, for any finite double summation it is true that

$$\sum_{i=1}^{m} \left( \sum_{j=1}^{n} a_{ij} \right) = \sum_{j=1}^{n} \left( \sum_{i=1}^{m} a_{ij} \right).$$

Thus the order of the summation is immaterial, and we may write the previous sum (without the parentheses) as

$$\sum_{i=1}^{4} \sum_{j=1}^{7} a_{ij} \quad \text{or as} \quad \sum_{j=1}^{7} \sum_{i=1}^{4} a_{ij}.$$

Let us apply the notation in a simple situation. Suppose that

$$A = \begin{bmatrix} a_{11} & a_{12} & a_{13} & a_{14} \\ a_{21} & a_{22} & a_{23} & a_{24} \\ a_{31} & a_{32} & a_{33} & a_{34} \end{bmatrix} \quad \text{and} \quad B = \begin{bmatrix} b_{11} & b_{12} \\ b_{21} & b_{22} \\ b_{31} & b_{32} \\ b_{41} & b_{42} \end{bmatrix}.$$

The element in the second row, first column of the product $AB$ is

$$a_{21}b_{11} + a_{22}b_{21} + a_{23}b_{31} + a_{24}b_{41}.$$

Notice that since our element is in the second row, then the row subscript 2 remains fixed on the *a*'s. And since our element is in the first column of the product, then the column subscript 1 remains fixed for the *b*'s. It is the column subscript on the *a*'s and the row subscript on the *b*'s which change from term to term in the sum, so that the sum may be written as

$$\sum_{k=1}^{4} a_{2k}b_{k1}.$$

We conclude this section by noting that the summation index $k$ is completely arbitrary. Thus, for example, the above sum may be written as

$$\sum_{k=1}^{4} a_{2k}b_{k1} = \sum_{s=1}^{4} a_{2s}b_{s1} = \sum_{t=1}^{4} a_{2t}b_{t1}.$$

## B.2 Matrix Multiplication

In Chapter 2 we gave a rather labored definition of matrix multiplication. In this short section we define matrix multiplication using the $\sum$ notation and give an example. In the following section we prove those properties of matrix algebra which were mentioned in Chapter 2 but not proved there.

Let $A = [a_{ij}]$ and $B = [b_{ij}]$ be respectively an $m \times n$ and an $n \times q$ matrix. Then the product $C = AB$ is defined in that order and $AB = [c_{ij}] = C$ where

$$c_{ij} = \sum_{k=1}^{n} a_{ik}b_{kj}.$$

Example B.2.1 Let

$$A = \begin{bmatrix} a_{11} & a_{12} & a_{13} \\ a_{21} & a_{22} & a_{23} \\ a_{31} & a_{32} & a_{33} \end{bmatrix} \quad \text{and} \quad B = \begin{bmatrix} b_{11} & b_{12} \\ b_{21} & b_{22} \\ b_{31} & b_{32} \end{bmatrix}.$$

Then

$$C = AB = \begin{bmatrix} \sum_{k=1}^{3} a_{1k}b_{k1} & \sum_{k=1}^{3} a_{1k}b_{k2} \\ \sum_{k=1}^{3} a_{2k}b_{k1} & \sum_{k=1}^{3} a_{2k}b_{k2} \\ \sum_{k=1}^{3} a_{3k}b_{k1} & \sum_{k=1}^{3} a_{3k}b_{k2} \end{bmatrix}$$

$$= \left[ \sum_{k=1}^{3} a_{ik}b_{kj} \right] = [c_{ij}].$$

∎

## B.3 More Matrix Algebra

In this section we present in rather terse form the algebraic theorems which were mentioned but not proved in Chapter 2. The last two theorems are linked results, the final one taking a little of the bitterness out of the "divisors of zero" notion mentioned in Section 2.6.

**THEOREM B.1** *Matrix multiplication is associative.*

**Proof** Let $A = [a_{ij}]$, $B = [b_{ij}]$, and $C = [c_{ij}]$ be three matrices which are conformable for multiplication in the order $ABC$. Let $A$ be $m \times n$, $B$ be $n \times q$, and $C$ be $q \times p$. Then each of the products $AB$, $(AB)C$, $BC$, and $A(BC)$ is defined; and we wish to prove that $A(BC) = (AB)C$. We first note that $A$ is $m \times n$ while $BC$ is $n \times p$, so that $A(BC)$ is $m \times p$. Moreover, $AB$ is $m \times q$ while $C$ is $q \times p$ so that $(AB)C$ is $m \times p$ also. Thus the products $A(BC)$ and $(AB)C$ at least have the same dimensions. To show that they are equal, we show that a typical element, say in the $i$th row, $j$th column of $A(BC)$ is the same as the corresponding element in $(AB)C$. Consider the product $BC$ whose $j$th column is

$$\begin{bmatrix} \sum_{k=1}^{q} b_{1k}c_{kj} \\ \sum_{k=1}^{q} b_{2k}c_{kj} \\ \vdots \\ \sum_{k=1}^{q} b_{nk}c_{kj} \end{bmatrix}.$$

Then the element in the $i$th row, $j$th column of $A(BC)$ is

$$\sum_{s=1}^{n} a_{is}\left(\sum_{k=1}^{q} b_{sk}c_{kj}\right),$$

which may be rearranged as

$$\sum_{s=1}^{n} \sum_{k=1}^{q} a_{is}b_{sk}c_{kj}. \qquad \text{(Why?)}$$

Now we calculate the element in the $i$th row, $j$th column of $(AB)C$. To do this, first note that the $i$th row of $AB$ is

$$\left[\sum_{s=1}^{n} a_{is}b_{s1}, \sum_{s=1}^{n} a_{is}b_{s2}, \ldots, \sum_{s=1}^{n} a_{is}b_{sq}\right].$$

Thus the $i$th row, $j$th column element of $(AB)C$ is computed by forming the product of the $i$th row of $AB$ by the $j$th column of $C$. We thus obtain

$$\sum_{k=1}^{q}\left(\sum_{s=1}^{n} a_{is}b_{sk}\right)c_{kj} = \sum_{k=1}^{q}\left(\sum_{s=1}^{n} a_{is}b_{sk}c_{kj}\right).$$

But, as we have seen, the latter may be written as

$$\sum_{k=1}^{q} \sum_{s=1}^{n} a_{is}b_{sk}c_{kj}.$$

Since the order of summation may also be interchanged, this may be expressed as

$$\sum_{s=1}^{n} \sum_{k=1}^{q} a_{is}b_{sk}c_{kj},$$

which is our result for the $i$, $j$ element in $A(BC)$. Thus corresponding elements of $A(BC)$ and $(AB)C$ are equal and hence $A(BC) = (AB)C$, which is what we wished to prove. ∎

**THEOREM B.2** *If $A$, $B$, and $C$ are conformable for the product $A(B + C)$ to be defined, then (left distributive property)*

$$A(B + C) = AB + AC.$$

**Proof** Suppose $A$ is $m \times n$, and each of $B$ and $C$ is $n \times q$. With $A = [a_{ij}]$, $B = [b_{ij}]$, and $C = [c_{ij}]$ we find that the $j$th column of $B + C$ is

$$\begin{bmatrix} b_{1j} + c_{1j} \\ b_{2j} + c_{2j} \\ \vdots \\ b_{nj} + c_{nj} \end{bmatrix}.$$

Thus the element in the $i$th row, $j$th column of $A(B + C)$ is

$$\sum_{k=1}^{n} a_{ik}(b_{kj} + c_{kj}) = \sum_{k=1}^{n} a_{ik}b_{kj} + \sum_{k=1}^{n} a_{ik}c_{kj}.$$

Now we merely observe that the right-hand side of this equation is the element in the $i$th row, $j$th column of the sum $AB + AC$. Thus $A(B + C) = AB + AC$. ∎

**THEOREM B.3** *Matrix multiplication is right distributive; that is, for conforming $A$, $B$, $C$, we have $(A + B)C = AC + BC$.*

**Proof** You should supply your own proof as an exercise.

**THEOREM B.4** *If the product $AB$ is defined, then $(AB)^T = B^T A^T$; that is, the transpose of a product is the product of the transposes in reverse order.*

**Proof** Suppose $A$ is $m \times n$ and $B$ is $n \times q$. Then the element in the $i$th row, $j$th column of $(AB)^T$ is the element in the $j$th row, $i$th column of $AB$, and thus is

$$\sum_{k=1}^{n} a_{jk}b_{ki}.$$

Now, the element in the $i$th row, $j$th column of $B^T A^T$ is formed by multiplying the $i$th row of $B^T$ by the $j$th column of $A^T$. But since the $i$th row of $B^T$ is the $i$th column of $B$, while the $j$th column of $A^T$ is the $j$th row of $A$, we then have

$$\sum_{k=1}^{n} b_{ki}a_{jk}.$$

But since multiplication of real numbers is commutative, this is the same as

$$\sum_{k=1}^{n} a_{jk}b_{ki},$$

which proves our theorem. ∎

We note that for the transpose of the product of three matrices, we have

$$(ABC)^T = ((AB)C)^T = C^T(AB)^T = C^T(B^TA^T) = C^TB^TA^T.$$

The theorem may be generalized to any (finite) number of matrices.

In Section 2.6 we observed all sorts of "bad" things about matrix multiplication. First, it is non-commutative (which is bad enough). Secondly, there are many examples of **divisors of zero**, that is, matrices $A$ and $B$, neither of which is zero but for which $AB = 0$. Now we know that some matrices have inverses and some do not. If $A$ has an inverse, then $AB = AC$ implies $B = C$, since we merely left-multiply the first equation by $A^{-1}$ to obtain the second. Thus in case $A$ has an inverse, we have a left cancellation law. But if $A$ has no inverse, then what are we to conclude? Is $B = C$ or $B \neq C$? The general answer to this question is more elaborate than we want to consider in this text, but the next two theorems (really the second one) provide a partial answer in a special setting. Specifically, we will show that if we are given

$$A^T(AB) = A^T(AC)$$

then we may left-cancel the factor $A^T$ to obtain

$$AB = AC.$$

Note that in this result nothing is assumed about the matrix $A$. It does not have to have an inverse—it does not even have to be square!

First we need a definition. By the **trace of a square matrix** $B$ we mean the sum $\sum_{i=1}^{n} b_{ii}$ of its diagonal elements. Thus the trace of $\begin{bmatrix} 1 & 2 \\ 3 & 4 \end{bmatrix}$ is $1 + 4 = 5$ and the trace of $\begin{bmatrix} -1 & 2 \\ 3 & 1 \end{bmatrix}$ is $-1 + 1 = 0$. Now, by way of introducing our next theorem, notice that while the matrix $A = \begin{bmatrix} 1 & 2 & 3 \\ 4 & 5 & 6 \end{bmatrix}$ has no trace (since it is not square), the matrix $A^TA$, which is calculated as

$$\begin{bmatrix} 1 & 4 \\ 2 & 5 \\ 3 & 6 \end{bmatrix} \begin{bmatrix} 1 & 2 & 3 \\ 4 & 5 & 6 \end{bmatrix} = \begin{bmatrix} 17 & 22 & 27 \\ 22 & 29 & 36 \\ 27 & 36 & 45 \end{bmatrix},$$

is square and its trace is $17 + 29 + 45 = 91$.

**THEOREM B.5** *If $A$ is any matrix such that the trace of $A^TA$ is 0 then $A$ must consist entirely of zeros.*

**Proof** Let $A = [a_{ij}]$ be an $m \times n$ matrix such that the trace of $A^TA$ is 0. (Note that $A^T$ is $n \times m$ and that $A^TA$ is thus a square $n \times n$ matrix.) The first diagonal element of $A^TA$ is computed by multiplying the first row of $A^T$ by the first column of $A$, but the first row of $A^T$ is the same as the first column of $A$, so that the first diagonal element of $A^TA$ is

$$\sum_{i=1}^{m} a_{i1}^2.$$

Similarly, the second diagonal element of $A^T A$ is

$$\sum_{i=1}^{m} a_{i2}^2,$$

and in general the $k$th diagonal element of $A^T A$ is

$$\sum_{i=1}^{m} a_{ik}^2.$$

Thus the trace of $A^T A$, which is assumed to be zero, is

$$\sum_{k=1}^{n} \sum_{i=1}^{m} a_{ik}^2.$$

However, this is precisely the sum of the squares of all the elements in $A$, and thus can be zero only if each element in $A$ is zero, which proves our theorem. ∎

We now apply this result to obtain the previously mentioned

**THEOREM B.6** *If $A$, $B$ and $C$ are matrices such that $A^T AB = A^T AC$, then $AB = AC$.*

**Proof** We prove this theorem by showing that under the hypothesis the trace of $(AB - AC)^T(AB - AC)$ is zero, so that it will follow from Theorem B.5 that $AB - AC = 0$, which is the result we want. Now consider the product $(AB - AC)^T(AB - AC)$. We unsparingly use Theorems B.2, B.3, and B.4, together with the fact that the transpose of a sum is the sum of the transposes, to obtain

$$\begin{aligned}(AB - AC)^T(AB - AC) &= (B^T A^T - C^T A^T)(AB - AC)\\ &= B^T A^T AB - B^T A^T AC - C^T A^T AB + C^T A^T AC\\ &= B^T A^T A(B - C) - C^T A^T A(B - C)\\ &= (B^T A^T A - C^T A^T A)(B - C)\\ &= (A^T AB - A^T AC)^T(B - C).\end{aligned}$$

But by hypothesis, $A^T AB - A^T AC = 0$, so that its transpose, which appears as a factor in the last line of the above display, is 0. Thus our original product $(AB - AC)^T(AB - AC) = 0$ and certainly has a 0 trace. Hence, by Theorem B.5, $AB - AC = 0$, so that $AB = AC$. ∎

# C Vector Spaces

In this appendix we give the general definition of vector space, and several examples of vector spaces which are somewhat different from the ordered $n$-tuple spaces of the text.

Let $F$ be a field (see Chapter 1) and let $\mathscr{V}$ be a non-empty set with a binary operation + defined on $\mathscr{V}$ such that, for all $a$, $b$, $c$ in $\mathscr{V}$,

| | | |
|---|---|---|
| 1. | $(a + b) + c = a + (b + c)$ | + is *associative* |
| 2. | $a + b = b + a$ | + is *commutative* |
| 3. | there is an element 0 in $\mathscr{V}$ such that $0 + a = a + 0 = a$ | existence of an *additive identity* |
| 4. | for each $a$ in $\mathscr{V}$ there is an element $-a$ in $\mathscr{V}$ such that $a + (-a) = 0 = (-a) + a$ | each element has an *additive inverse* |

Further let there be defined an operation of scalar multiplication between elements of the field $F$ and elements of the set $\mathscr{V}$ such that for each $x$, $y$ in $F$ and $a$, $b$ in $\mathscr{V}$,

| | | |
|---|---|---|
| 5. | $xa$ is an element of $\mathscr{V}$ | |
| 6. | $x(ya) = (xy)a$ | *associative* |
| 7. | $(x + y)a = xa + ya$ | scalar multiplication distributes over scalar addition |
| 8. | $x(a + b) = xa + xb$ | scalar multiplication distributes over vector addition |
| 9. | $1a = a$ | |

Then the set $\mathscr{V}$ with its binary operation together with the field $F$ and the scalar multiplication is called a *vector space*. We most frequently say that $\mathscr{V}$ is a *vector space over the field F*.

Now the fields most familiar to you are probably the *real numbers*, the *rational numbers*, and the *complex numbers* (those of you who have studied modular arithmetic will recall that the integers modulo 5, or modulo any prime, also form a field). Let us now consider some examples of vector spaces over the familiar fields, and note that in each case we specify the set $\mathscr{V}$ (vectors) and its binary operation (vector addition), and we also specify the field $F$ and its scalar multiplication with the member of $\mathscr{V}$. Of course, the example of ordered pairs, or ordered triples, or ordered $n$-tuples over the reals was *the* example of vector spaces used in the text. We shall not repeat that example here, and only mention in passing that the set of all ordered $n$-tuples from any field $F$, with the usual addition and scalar multiplication is an example of a vector space. Moreover, such a space has dimension $n$, since with 1 as the multiplicative identity of $F$, the set of $n$-tuples

$$\{(1, 0, \ldots, 0), (0, 1, \ldots, 0), \ldots, (0, 0, \ldots, 0, 1)\}$$

is easily seen to be a basis.

Example C.1

Let the field $F$ be the field $\mathbb{Q}$ of rational numbers, and let $\mathscr{V}$ be the set of all polynomials of degree $\leqslant 3$ in a single variable $t$ with rational coefficients. Then addition in $\mathscr{V}$ is simply addition of polynomials, that is, to obtain the coefficients of the sum of two polynomials we add coefficients of like powers of $t$. Scalar multiplication by a member of $F$ is accomplished by multiplying each coefficient of the polynomial by that scalar. For instance, consider the two members

$$v_1 = 1/2 + 5t \quad \text{and} \quad v_2 = -2 + 1/2t^2 - t^3$$

of $\mathscr{V}$. Their sum is

$$-3/2 + 5t + 1/2t^2 - t^3$$

and if we multiply $v_2$ by $-3/2$ (a scalar in $\mathbb{Q}$) then the result is

$$3 - 3/4t^2 + 3/2t^3.$$

∎

With these operations in mind, you should now go through the list of defining properties for a vector space and assure yourself that this set of rational polynomials over the rational field is indeed a vector space. Finally, a word about the dimension of this space. If you have studied Chapter 4, you will find it easy to see that the set of polynomials

$$\{1, t, t^2, t^3\}$$

is not only linearly independent in this space, but also spans this space and hence forms a basis. Thus the space of all rational polynomials of degree $\leqslant 3$ is a space of dimension 4 (4 being the number of coefficients in such a polynomial). Similarly, for any non-negative integer $n$, the set of all rational polynomials of degree $\leqslant n$ forms a vector space of polynomials of dimension $n + 1$. Moreover, there being nothing special about the rational field $\mathbb{Q}$, we can form in this way vector spaces of polynomials over any field. Thus we may consider the space of all polynomials with complex coefficients of degree $\leqslant 17$ over the complex field.

Example C.2 As an extension of the previous example, let $F$ be any field and this time let $\mathscr{V}$ be the set of *all* polynomials (no restriction on the degrees) with coefficients from the field $F$. Again the vector addition is the usual addition of polynomials and the scalar multiplication is defined as before. You may readily verify that $\mathscr{V}$ is a vector space over $F$. Now in this case the matter of dimension needs a word or two of explanation. Consider the (infinite) set

$$\mathscr{B} = \{1, t, t^2, \ldots, t^n, \ldots\}.$$

We say that $\mathscr{B}$ *spans* $\mathscr{V}$ since each element of $\mathscr{V}$ is a linear combination of a *finite* number of elements of $\mathscr{B}$, and we further say that $\mathscr{B}$ is a *linearly independent* set in that any finite subset of it is linearly independent in our usual sense. Thus we are led to say that $\mathscr{B}$ is a basis for $\mathscr{V}$, and, since $\mathscr{B}$ has infinitely many members, we say that $\mathscr{V}$ is an *infinite dimensional vector space over F*. ∎

Example C.3 Let $F$ now be the real number field $\mathbb{R}$, and let $\mathscr{V}$ be the set of all real valued functions defined on the closed interval $[0, 1]$. In $\mathscr{V}$ we define the addition of two functions $f$ and $g$ to be the function $f + g$ whose value at each $x$ in $[0, 1]$ is given by

$$(f + g)(x) = f(x) + g(x).$$

Our scalar multiplication is also the usual one for functions: namely, for any real number $r$ and any function $f$ in $\mathscr{V}$, define $rf$ to be the function whose value at each $x$ in $[0, 1]$ is given by

$$(rf)(x) = rf(x).$$

Then we may easily verify that $\mathscr{V}$ is a vector space over the reals (but we sidestep the question of dimension, because there is not a simple explanation available). To students who have studied the calculus, it will be evident that among the many subspaces of this example are the space of all continuous functions on $[0, 1]$, the space of all differentiable functions on $[0, 1]$, and the space of all integrable functions on $[0, 1]$. The study of such "function spaces" is vital in many areas of both applied and pure mathematics. ∎

Example C.4 In this final example, we should like to point up the fact that the set $\mathscr{V}$ alone does not determine the algebraic character of a vector space. The field $F$ also plays an important role, and to retain $\mathscr{V}$ but change $F$ may make a real difference. To exhibit all this, consider first that the set of complex numbers $C$ (which is a field) may be considered as a vector space over itself; that is, we may think of $C$ as a set of ordered 1-tuples over the field $C$, with ordinary addition of complex numbers as the vector addition and multiplication of complex numbers as the scalar multiplication. Thus $C$ is a vector space over itself and, since the complex number $1 = 1 + 0i$ is both linearly independent and spans $C$, then $C$ is a 1-dimensional space over $C$. Now let us keep $C$ as the vector set $\mathscr{V}$, but take the real numbers $\mathbb{R}$ as the scalar field (same vector addition; scalar multiplication is simply the multiplication of complex numbers by real numbers). Then $C$ is a vector space over $\mathbb{R}$. But, and this is the point of our example, $C$ is now 2-dimensional over the real field, since over the reals the set

$$\{1, i\}$$

is both a linearly independent and a spanning set for $C$. Thus by keeping the same vector set and changing the scalar field we have changed the vector space from one of dimension 1 to one of dimension 2. ∎

# Answers to Odd-Numbered Exercises

## 1.1 Exercises (page 7)

1. a. The set of elements $x$ such that $x - 5$ is a positive integer; $\{6, 7, 8, \ldots\}$.
   b. The set of elements $y$ such that $y$ is a positive integer; $\{1, 2, 3, \ldots\}$.
   c. The set of elements of the form $2x - 1$ such that $x$ is a positive integer; $\{1, 3, 5, \ldots\}$.

3. a. Yes, no  b. Yes

5. None; 1; no, since $\{\varnothing\}$ contains an element, namely $\varnothing$, while $\varnothing$ contains no elements.

7. $A \cup B = B$; $A \cap B = A$.

9. To show that $A \cap B \subseteq A \cap C$ we need to show that if $x$ is in $A \cap B$ then $x$ is in $A \cap C$. If $x$ is in $A \cap B$ then $x$ is in $A$ and $x$ is in $B$. Since $B \subseteq C$ and $x$ is in $B$, $x$ is in $C$ also. Thus $x$ is in $A$ and $x$ is in $C$, so that $x$ is in $A \cap C$.

11. The subsets of $\{1\}$ are $\varnothing$ and $\{1\}$; thus there are $2 = 2^1$ subsets. The subsets of $\{1, 2\}$ are $\varnothing$, $\{1\}$, $\{2\}$, and $\{1, 2\}$; there are $4 = 2^2$ subsets. The subsets of $\{1, 2, 3\}$ are $\varnothing$, $\{1\}$, $\{2\}$, $\{3\}$, $\{1, 2\}$, $\{1, 3\}$, $\{2, 3\}$, and $\{1, 2, 3\}$; there are $8 = 2^3$ subsets. A set with $n$ elements has $2^n$ subsets.

13. a. No  b. Yes  c. Yes  d. No  e. No

15. $f, h$

17.

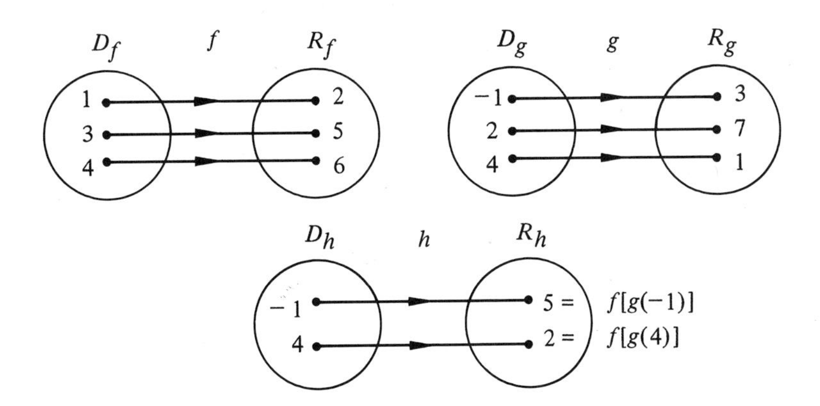


a. $g \circ f = \{(1, 7)\}$

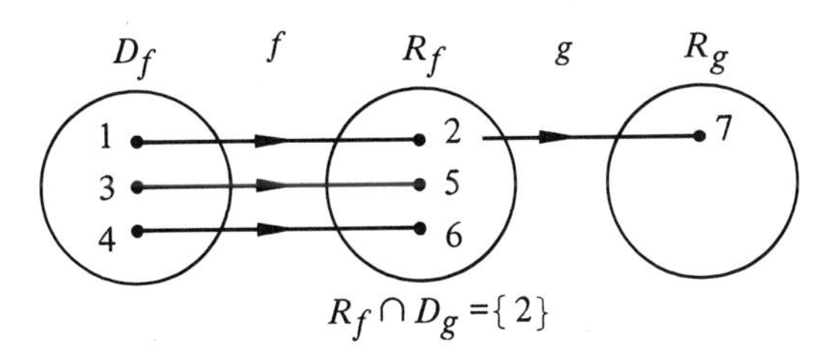


b. $h \circ f = \varnothing$

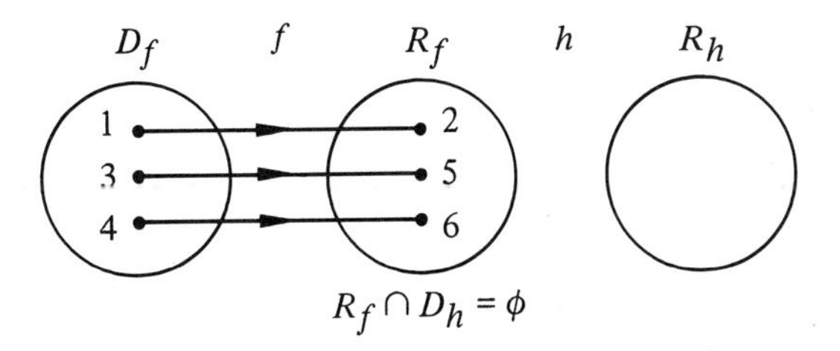


c. $g \circ h = \{(4, 7)\}$

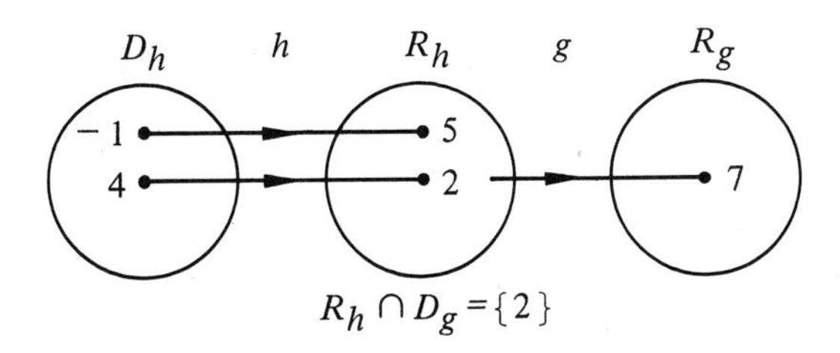

## 1.2 Exercises (page 13)

1. b. $a$
   c. Yes, the table of sums is symmetric with respect to its main diagonal.
   d. $(-a) = a, (-b) = c, (-c) = b$.
   e. Must show that $(x + y) + z = x + (y + z)$ for all possible choices of $x$, $y$, $z$ in $S$.

3. a. 0 b. $(-5)$ c. 5

5. c. $b$ e. $c^{-1} = c$; $a^{-1}$ does not exist.

7. For example, $5 - 3 \neq 3 - 5$, and $5 - (3 - 1) \neq (5 - 3) - 1$.

## 1.3 Exercises (page 16)

1. A matrix is a rectangular array of real numbers. In many applications this definition is extended to allow the elements to be complex numbers.

3. Second row, third column.

5. $b$

7. A–E are each square while $F$ is not square; $B$ has zeros below the left-right diagonal, the only non-zero elements of $C$ and $D$ are on the main diagonal; the elements of $E$ are symmetric with respect to the left-right diagonal.

## 1.4 Exercises (page 19)

1. $b$ 3. $a, c, d, e, f, g, h, i$ 5. $a$ 7. $b, h$ 9. $a, f$

11. a. $\begin{bmatrix} a & b \\ 0 & c \end{bmatrix}, b \neq 0$ b. $\begin{bmatrix} a & 0 \\ 0 & b \end{bmatrix}, a \neq b$ c. none exists d. $\begin{bmatrix} a & b \\ b & d \end{bmatrix}$

13. a. $\begin{bmatrix} 1 & -1 \\ 3 & 2 \end{bmatrix}$

## 1.5 Exercises (page 23)

1. $\begin{bmatrix} -1 & -2 & -3 \\ -4 & -5 & -6 \end{bmatrix}$

3. $[6, 4, 4, 7]$

5 $\begin{bmatrix} 165{,}500 & 191{,}000 & 170{,}000 \\ 6520 & 8700 & 7250 \end{bmatrix}$

7. $\begin{bmatrix} 1 & -1 \\ 2 & -3 \\ 3 & -4 \end{bmatrix}$

9. Yes; we may find it by solving the six equations $a_{11} + x_{11} = b_{11}, a_{12} + x_{12} = b_{12}$, etc. to form $\begin{bmatrix} 1 & 2 & 1 \\ 7 & 3 & -2 \end{bmatrix}$.

11. a. $A^T = \begin{bmatrix} 1 & 4 \\ 2 & 5 \\ 3 & 6 \end{bmatrix}, B^T = \begin{bmatrix} -2 & -8 \\ -4 & -10 \\ -6 & -12 \end{bmatrix}$.

13. $[0, 0, 0], [0, 0, 0, 0]^T, \begin{bmatrix} 0 & 0 & 0 & 0 & 0 & 0 \\ 0 & 0 & 0 & 0 & 0 & 0 \\ 0 & 0 & 0 & 0 & 0 & 0 \end{bmatrix}$.

15. Suppose that both $(-A)$ and $B$ are additive inverses of $A$. Then, since $A + B = 0$, $(-A) + (A + B) = (-A) + 0 = (-A)$ and $((-A) + A) + B = 0 + B = B = -A$. Thus there was really only one.

17. Show that $B + (-A) = B - A$ is a solution and that if $Y$ is a solution then $Y$ must be $B - A$. See Section 1.2 for the analogous equation for real numbers. Yes.

## 1.6 Exercises (page 25, 26 & 32)

1. $[32]$ 3. Undefined 5. $[2a + 3b + 4c]$

7. $[2j + 6k - 12k - 24j - 5k] = [-22j - 11k]$ 9. $[b]$

11. $[c]$ 13. Undefined

15. $[a, b, c]$ 17. $[70, 3, 19, 10, -3]$ 19. $\begin{bmatrix} a & b & c \\ 0 & 0 & 0 \\ 0 & 0 & 0 \end{bmatrix}$

21. $\begin{bmatrix} 0 & 0 & 0 \\ 0 & 0 & 0 \\ g & h & i \end{bmatrix}$ 23. $\begin{bmatrix} 0 & b & 0 \\ 0 & e & 0 \\ 0 & h & 0 \end{bmatrix}$ 25. $\begin{bmatrix} a + d & b + e & c + f \\ d & e & f \\ g & h & i \end{bmatrix}$

27. $\begin{bmatrix} ax + dy & bx + ey & cx + fy \\ d & e & f \\ g & h & i \end{bmatrix}$ 29. $\begin{bmatrix} 2 & 4 & 6 & 11 \\ 3 & 1 & 10 & 10 \end{bmatrix}$

31. $\begin{bmatrix} 20 & 11 \\ 18 & 4 \\ 1 & 1 \\ 2 & 2 \end{bmatrix}$ 33. Yes, both are $\begin{bmatrix} 14 & -8 \\ 17 & -10 \end{bmatrix}$.

## 1.7 Exercises (page 36)

1. $[7] = [2] + [5]$. 3. $[5] = [4] + [1]$.

5. $$\begin{bmatrix} 5 & 5 \\ 2 & 3 \end{bmatrix} = \begin{bmatrix} 7 & 7 \\ 2 & 4 \end{bmatrix} + \begin{bmatrix} -2 & -2 \\ 0 & -1 \end{bmatrix}.$$

7. $$\begin{bmatrix} 4 & 2 & 10 & 6 \\ 0 & 0 & 0 & 0 \\ 0 & 0 & 0 & 0 \\ 2 & 1 & 5 & 3 \end{bmatrix} = \begin{bmatrix} 2 & 1 & 5 & 3 \\ -2 & -1 & -5 & -3 \\ -2 & -1 & -5 & -3 \\ 4 & 2 & 10 & 6 \end{bmatrix} + \begin{bmatrix} 2 & 1 & 5 & 3 \\ 2 & 1 & 5 & 3 \\ 2 & 1 & 5 & 3 \\ -2 & -1 & -5 & -3 \end{bmatrix}.$$

## 1.8 Exercises (page 39, 41 & 43)

9. $\begin{bmatrix} a_{11} & a_{12} & a_{13} & a_{14} \\ a_{21} & a_{22} & a_{23} & a_{24} \end{bmatrix}$ 11. $\begin{bmatrix} 1/3 & 0 & 0 \\ 0 & 1/3 & 0 \\ 0 & 0 & 1/3 \end{bmatrix}$ 13. $\begin{bmatrix} 0 & 1/3 \\ 1/2 & 0 \end{bmatrix}$

15. No inverse exists.

17. $(A + B)^2 = A^2 + BA + AB + B^2 \neq A^2 + 2AB + B^2$ unless $AB = BA$. The answer is "no" in each case.

19. $A^2 = \begin{bmatrix} 0 & 0 & 1 \\ 0 & 0 & 0 \\ 0 & 0 & 0 \end{bmatrix}$, $A^3 = 0$.

23. There are an infinite number of solutions; one is $X = \begin{bmatrix} 1 & -3 \\ 1 & 2 \end{bmatrix}$.

## 2.1 Exercises (page 48)

1. a. $A = \begin{bmatrix} 1 & -4 \\ -2 & 2 \end{bmatrix}$, $B = \begin{bmatrix} -5 \\ 4 \end{bmatrix}$, $X = \begin{bmatrix} x_1 \\ x_2 \end{bmatrix}$.

   b. $A_1 = \begin{bmatrix} 1 \\ -2 \end{bmatrix}$, $A_2 = \begin{bmatrix} -4 \\ 2 \end{bmatrix}$, $B = \begin{bmatrix} -5 \\ 4 \end{bmatrix}$.

   c. No d. $\begin{bmatrix} 1 & -4 \\ -2 & 2 \end{bmatrix}\begin{bmatrix} -1 \\ 1 \end{bmatrix} = \begin{bmatrix} -5 \\ 4 \end{bmatrix}$.

3. a. $A = \begin{bmatrix} 1 & 2 & -1 \\ 2 & -1 & 3 \end{bmatrix}$, $X = \begin{bmatrix} x_1 \\ x_2 \\ x_3 \end{bmatrix}$, $B = \begin{bmatrix} 0 \\ 0 \end{bmatrix}$.

   b. Yes

c. $X = \begin{bmatrix} 0 \\ 0 \\ 0 \end{bmatrix}$.

5. a. $A(X_1 + X_2) = AX_1 + AX_2 = 0 + 0 = 0.$
   b. $A(cX_1) = c(AX_1) = c\,0 = 0.$
   c. $A(c_1X_1 + c_2X_2) = A(c_1X_1) + A(c_2X_2) = c_1(AX_1) + c_2(AX_2)$
   $= c_10 + c_20 = 0.$

7. $x_1 + x_2 = 40$ (hours worked).
   $3x_1 + 2x_2 = 90$ (dollars paid).

## 2.2 Exercises (page 53)

1. Trivial    3. It is the *only* solution.

5. a. The solution set consists of the points on a straight line in the plane.
   b. Two straight lines in the plane.
   c. A plane through the origin in 3-dimensional space.
   d. Two planes in 3-dimensional space.
   e. Three planes through the origin in 3-dimensional space.

7. All points not satisfying $x_1 - x_2 = 0$; that is, all points not on the line whose equation is $x_1 - x_2 = 0$. All points not on the line whose equation is $2x_1 + x_2 = 0$. All points except the origin, or $\begin{bmatrix} 0 \\ 0 \end{bmatrix}$.

9. $x_1 = 10, x_2 = 30.$

## 2.3 Exercises (page 57)

1. a. $\left[\begin{array}{cccc|c} 1 & -1 & 2 & 4 & 1 \\ 0 & 2 & 3 & 7 & 2 \\ 6 & 3 & 4 & 5 & 3 \end{array}\right]$    b. Nonsense

   c. $\left[\begin{array}{cc|c} 2 & 1 & 1 \\ 3 & 3 & 2 \\ 4 & 7 & 3 \end{array}\right]$    d. $\left[\begin{array}{cc|c} 1 & 2 & 3 \\ 3 & 4 & 7 \end{array}\right]$

3. a. $A^{-1} = [X \mid Y]$.
   b. $\begin{bmatrix} 0 & -1 \\ -1 & 0 \end{bmatrix}\begin{bmatrix} x_1 \\ x_2 \end{bmatrix} = \begin{bmatrix} 1 \\ 0 \end{bmatrix}$ has solution $X = \begin{bmatrix} 0 \\ -1 \end{bmatrix}$.

   $\begin{bmatrix} 0 & -1 \\ -1 & 0 \end{bmatrix}\begin{bmatrix} y_1 \\ y_2 \end{bmatrix} = \begin{bmatrix} 0 \\ 1 \end{bmatrix}$ has solution $Y = \begin{bmatrix} -1 \\ 0 \end{bmatrix}$.

   Thus $A^{-1} = [X \mid Y] = \begin{bmatrix} 0 & -1 \\ -1 & 0 \end{bmatrix}$.

## 2.4 Exercises (page 59)

1. a. $\begin{bmatrix} 1 & -1 \\ -1 & 2 \end{bmatrix}\begin{bmatrix} 2 & 1 \\ 1 & 1 \end{bmatrix} = \begin{bmatrix} 1 & 0 \\ 0 & 1 \end{bmatrix}.$

   b. $X = A^{-1}B = \begin{bmatrix} 2 \\ 3 \end{bmatrix}$; $AX = \begin{bmatrix} 1 & -1 \\ -1 & 2 \end{bmatrix}\begin{bmatrix} 2 \\ 3 \end{bmatrix} = \begin{bmatrix} -1 \\ 4 \end{bmatrix} = B.$

   c. $A(A^{-1}Y) = (AA^{-1})Y = I_2Y = Y = A(0) = 0$. Thus $Y = 0$ is the unique solution.

3. a. $AA^{-1} = \begin{bmatrix} 1 & -1 \\ 0 & 1 \end{bmatrix}\begin{bmatrix} 1 & 1 \\ 0 & 1 \end{bmatrix} = \begin{bmatrix} 1 & 0 \\ 0 & 1 \end{bmatrix}.$

   b. $X = A^{-1}B = \begin{bmatrix} 1 \\ 1 \end{bmatrix}$; $AX = \begin{bmatrix} 1 & -1 \\ 0 & 1 \end{bmatrix}\begin{bmatrix} 1 \\ 1 \end{bmatrix} = \begin{bmatrix} 0 \\ 1 \end{bmatrix} = B.$

   c. $A(A^{-1}Y) = (AA^{-1})Y = I_2Y = Y = AB = \begin{bmatrix} -1 \\ 1 \end{bmatrix}.$

## 2.5 Exercises (page 65)

3. a. $\left[\begin{array}{cc|c} 1 & 1 & -1 \\ -1 & 2 & 4 \end{array}\right]$ b. $\left[\begin{array}{cc|c} 1 & 1 & -1 \\ 0 & 3 & 3 \end{array}\right]$ c. $\left[\begin{array}{cc|c} 1 & 1 & -1 \\ 0 & 1 & 1 \end{array}\right]$

   d. $\left[\begin{array}{cc|c} 1 & 0 & -2 \\ 0 & 1 & 1 \end{array}\right]$ e. $X = \begin{bmatrix} -2 \\ 1 \end{bmatrix}.$

5. From 4c, $A\begin{bmatrix} 1/2 \\ 1/2 \end{bmatrix} = \begin{bmatrix} 1 \\ 0 \end{bmatrix}$ and from 4d, $A\begin{bmatrix} -1/2 \\ 1/2 \end{bmatrix} = \begin{bmatrix} 0 \\ 1 \end{bmatrix}$. Thus $A^{-1} = \begin{bmatrix} 1/2 & -1/2 \\ 1/2 & 1/2 \end{bmatrix}.$
   (See Exercise 3 in Section 2.3.)

7. $X = [1, 2, 1, 2]^T$.

9. $X = [120, 100, 60, 50]^T$.

11. a. $X = [-0.6845, 3.5802]^T$.

    b. $X = [4.186197466, 3.157774233, -1.343971699]^T$.

## 2.6 Exercises (page 68)

1. a. $\begin{bmatrix} -1/3 & 2/3 \\ 2/3 & -1/3 \end{bmatrix}$ b. $\begin{bmatrix} 1 & 1 \\ 0 & 1 \end{bmatrix}$ c. $\begin{bmatrix} -5/2 & 3/2 \\ 4/2 & -2/2 \end{bmatrix}$

3. a. $\begin{bmatrix} 1 & -1 & -1 \\ 0 & 1 & 0 \\ 0 & 0 & 1 \end{bmatrix}$ b. $\begin{bmatrix} 1 & 0 & -1 \\ 1 & 1 & -1 \\ 1 & 1 & 0 \end{bmatrix}$ c. $\begin{bmatrix} -4 & 7 & -2 \\ -3 & 5 & -1 \\ 2 & -3 & 1 \end{bmatrix}$

5. $A^{-1} = \begin{bmatrix} 0.2089078504 & -0.171848316 \\ 0.1537058065 & 0.2414405163 \end{bmatrix}$

## 2.7 Exercises (page 72)

1. a. $\left[\begin{array}{rr|r} 1 & 1 & 2 \\ -2 & -2 & -4 \end{array}\right] \sim \left[\begin{array}{rr|r} 1 & 1 & 2 \\ 0 & 0 & 0 \end{array}\right].$

$$X = \begin{bmatrix} 2-a \\ a \end{bmatrix} = \begin{bmatrix} 2-a \\ 0+a \end{bmatrix} = \begin{bmatrix} 2 \\ 0 \end{bmatrix} + a\begin{bmatrix} -1 \\ 1 \end{bmatrix} = X_P + X_H.$$

b. $\left[\begin{array}{rrr|r} 1 & -1 & 2 & 3 \\ 2 & -1 & 1 & 3 \\ -3 & 2 & -3 & -6 \end{array}\right] \sim \left[\begin{array}{rrr|r} 1 & 0 & -1 & 0 \\ 0 & 1 & -3 & -3 \\ 0 & 0 & 0 & 0 \end{array}\right].$

$$X = \begin{bmatrix} 0+a \\ -3+3a \\ 0+a \end{bmatrix} = \begin{bmatrix} 0 \\ -3 \\ 0 \end{bmatrix} + a\begin{bmatrix} 1 \\ 3 \\ 1 \end{bmatrix} = X_P + X_H.$$

c. $\left[\begin{array}{rr|r} 1 & 1 & 0 \end{array}\right]$. $X = \begin{bmatrix} -a \\ a \end{bmatrix} = a\begin{bmatrix} -1 \\ a \end{bmatrix}.$

d. $\left[\begin{array}{rrrr|r} 1 & 1 & 1 & -1 & 1 \\ -1 & 0 & 3 & 1 & -1 \\ 1 & 2 & 5 & -1 & 1 \\ 1 & -1 & -4 & 2 & 1 \end{array}\right] \sim \left[\begin{array}{rrrr|r} 1 & 0 & 0 & 2 & 1 \\ 0 & 1 & 0 & -4 & 0 \\ 0 & 0 & 1 & 1 & 0 \\ 0 & 0 & 0 & 0 & 0 \end{array}\right].$

$$X = \begin{bmatrix} 1-2a \\ 4a \\ -a \\ a \end{bmatrix} = \begin{bmatrix} 1-2a \\ 0+4a \\ 0-a \\ 0+a \end{bmatrix} = \begin{bmatrix} 1 \\ 0 \\ 0 \\ 0 \end{bmatrix} + a\begin{bmatrix} -2 \\ 4 \\ -1 \\ 1 \end{bmatrix}.$$

3. a. $a\begin{bmatrix} -19 \\ -11 \\ 1 \end{bmatrix}$ b. $\begin{bmatrix} -19 \\ -11 \\ 1 \end{bmatrix}, \begin{bmatrix} 38 \\ 22 \\ -2 \end{bmatrix}, \begin{bmatrix} -57 \\ -33 \\ 3 \end{bmatrix}$

5. a. 2 redundant equations.

$$X = \begin{bmatrix} 3+b-2a \\ 4-2b-3a \\ b \\ a \end{bmatrix} = \begin{bmatrix} 3 \\ 4 \\ 0 \\ 0 \end{bmatrix} + a\begin{bmatrix} -2 \\ -3 \\ 0 \\ 1 \end{bmatrix} + b\begin{bmatrix} 1 \\ -2 \\ 1 \\ 0 \end{bmatrix} = X_P + X_H.$$

For $a = b = 1$, $X = \begin{bmatrix} 2 \\ -1 \\ 1 \\ 1 \end{bmatrix}.$

b. 2 redundant equations. No free variables. $X = \begin{bmatrix} 4 \\ -1 \\ 0 \end{bmatrix}.$

c. 2 redundant equations. 1 free variable, $x_3$. $X = \begin{bmatrix} -1 \\ 3 \\ 0 \end{bmatrix} + a\begin{bmatrix} -2 \\ 0 \\ 1 \end{bmatrix}$. For $a = 1$,

$$X = \begin{bmatrix} -3 \\ 3 \\ 1 \end{bmatrix}.$$

5. d. 1 redundant equation. $x_4, x_5, x_6$ arbitrary. Letting $x_4 = c, x_5 = b, x_6 = a$,

$$X = \begin{bmatrix} 2 \\ 3 \\ -1 \\ 0 \\ 0 \\ 0 \end{bmatrix} + a\begin{bmatrix} 1 \\ -3 \\ -1 \\ 0 \\ 0 \\ 1 \end{bmatrix} + b\begin{bmatrix} -5 \\ -4 \\ 0 \\ 0 \\ 1 \\ 0 \end{bmatrix} + c\begin{bmatrix} 0 \\ -1 \\ 0 \\ 1 \\ 0 \\ 0 \end{bmatrix}.$$

When $a = b = c = 1, X = [-2, -5, -2, 1, 1, 1]^T$.

## 2.8 Exercises (page 78)

1. a. $X = a\begin{bmatrix} 2 \\ 0 \\ -1 \\ 1 \end{bmatrix} + b\begin{bmatrix} 1 \\ 1 \\ 0 \\ 0 \end{bmatrix}$. b. $\begin{bmatrix} 3 \\ 1 \\ -1 \\ 1 \end{bmatrix}$ c. $\begin{bmatrix} -9 \\ -3 \\ 3 \\ -3 \end{bmatrix}$

3. a. $X = a\begin{bmatrix} -2 \\ 1 \\ 0 \end{bmatrix}$. b. $2, 3, 3 - 2 = 1, 1$. c. $X = \begin{bmatrix} -4 \\ 2 \\ 0 \end{bmatrix}$; yes.

5. a. $5 - 2 = 3, 2, 2$. b. $X = \begin{bmatrix} 2 \\ 2b \\ b \\ -3a \\ a \end{bmatrix}$.

c. $X = \begin{bmatrix} 2 \\ 0 \\ 0 \\ 0 \\ 0 \end{bmatrix} + a\begin{bmatrix} 0 \\ 0 \\ 0 \\ -3 \\ 1 \end{bmatrix} + b\begin{bmatrix} 0 \\ 2 \\ 1 \\ 0 \\ 0 \end{bmatrix} = X_P + X_H$.

## 2.9 Exercises (page 84)

1. a. $2; X = \begin{bmatrix} 1 \\ 7/3 \\ -2 \\ 0 \\ 0 \end{bmatrix} + a\begin{bmatrix} 0 \\ -7/3 \\ 2 \\ 0 \\ 1 \end{bmatrix} + b\begin{bmatrix} 0 \\ 10/3 \\ -3 \\ 1 \\ 0 \end{bmatrix}$.

b. $3;\ X = [-1/2, 6, 0, 0, 0]^T + a[1/2, -4, 0, 0, 1]^T + b[1/2, 0, 0, 1, 0]^T + c[0, -2, 1, 0, 0]^T$.

c. $0; X = [1, 3, 2, 2]^T$.

d. $2; X = [13/5, 0, 1/5, 0]^T + a[-1/5, 0, 11/10, 1]^T + b[1, 1, 0, 0]^T$.

e. Since $n - m = 3 - 4 < 0$, there will be no solution at all unless at least one of the equations is redundant; $X = [1, 1, -1]^T$.

f. 0; no solution exists.

3. a. $E_1 = \begin{bmatrix} 1/3 & 0 \\ 0 & 1 \end{bmatrix}, E_2 = \begin{bmatrix} 1 & 0 \\ -5 & 1 \end{bmatrix}, E_3 = \begin{bmatrix} 1 & 0 \\ 0 & 3 \end{bmatrix}$

$E_4 = \begin{bmatrix} 1 & -4/3 \\ 0 & 1 \end{bmatrix}.$

b. $E_4E_3E_2E_1 = \begin{bmatrix} 7 & -4 \\ -5 & 3 \end{bmatrix}.$ c. $I_2$ d. $E_4E_3E_2E_1 = A^{-1}.$

5. $n - (m - k)$; we choose those variables which are not "fixed" by having one of the "first 1's" in their respective columns.

7. $x_1 = 1, x_2 = -2.$

9. $B = A_1 - A_2 + A_3$, as we see when we solve $AX = B$ and obtain $X = [1, -1, 1]^T$.

## 3.1 Exercises (page 90)

1. a. $A = \begin{bmatrix} 0.2 & 0.3 \\ 0.1 & 0.4 \end{bmatrix}.$

b. $40 of $I_1$ product and $45 of $I_2$ product; compute $AX$.

c. $10 of $I_1$ product and $55 of $I_2$ product

d. $X^T = [100, 200]$.

e. $D = (I - A)^{-1} - I = \begin{bmatrix} 1/3 & 2/3 \\ 2/9 & 7/9 \end{bmatrix}.$

3. a. Since $a_{21} = 0$, for example, there is not any requirement for Product 2 in making Product 1.

b. $(I - A)^{-1} = \begin{bmatrix} 1/8 & 1/114 & 3/76 \\ 0 & 7/57 & 1/19 \\ 0 & 2/57 & 3/19 \end{bmatrix}$. This computation should convince you of the value of a computer program to obtain inverses.

c. $X^T = [7900, 8000, 8800]$.

d. $AX = X - R = [3340, 3440, 4240]^T$.

## 3.2 Exercises (page 96)

1. a. $(I - A)^{-1} \approx I + A + A^2 + A^3 + A^4 + A^5$

$$= \begin{bmatrix} 1 & 0 \\ 0 & 1 \end{bmatrix} + \begin{bmatrix} 0.2 & 0.3 \\ 0.1 & 0.4 \end{bmatrix} + \begin{bmatrix} 0.07 & 0.18 \\ 0.06 & 0.19 \end{bmatrix} + \begin{bmatrix} 0.032 & 0.093 \\ 0.031 & 0.094 \end{bmatrix}$$

$$+ \begin{bmatrix} 0.0157 & 0.0468 \\ 0.0156 & 0.0469 \end{bmatrix} + \begin{bmatrix} 0.00782 & 0.02343 \\ 0.00157 & 0.00624 \end{bmatrix}$$

$$= \begin{bmatrix} 1.32552 & 0.64323 \\ 0.21441 & 1.75434 \end{bmatrix}.$$

b. $X \approx R + AR + A^2R + A^3R + A^4R + A^5R$

$$= \begin{bmatrix} 20 \\ 110 \end{bmatrix} + \begin{bmatrix} 37 \\ 46 \end{bmatrix} + \begin{bmatrix} 21.2 \\ 22.1 \end{bmatrix} + \begin{bmatrix} 10.87 \\ 10.96 \end{bmatrix} + \begin{bmatrix} 5.462 \\ 5.471 \end{bmatrix} + \begin{bmatrix} 2.7337 \\ 2.7346 \end{bmatrix}$$

$$= \begin{bmatrix} 97.2657 \\ 197.2656 \end{bmatrix}.$$

3. a. $(I - A)^{-1} \approx \begin{bmatrix} 1.2496 & 0.0764 & 0.3738 \\ 0 & 1.2155 & 0.5034 \\ 0 & 0.3356 & 1.5551 \end{bmatrix}$

   b. $X \approx [7751.088, 7838.184, 8621.592]^T$.

5. a. $A^6R = [220.8, 220.8, 220.5]^T$.

   b. $X \approx [10450.1, 10450.1, 15175.3]^T$.

## 4.1 Exercises (page 103)

1. a. $\begin{bmatrix} 2 \\ 0 \end{bmatrix} = 2E_1 + 0E_2$. b. $\begin{bmatrix} 3 \\ 5 \end{bmatrix} = 3E_1 + 5E_2$.

   c. $\begin{bmatrix} \pi \\ e \end{bmatrix} = \pi E_1 + eE_2$. d. $\begin{bmatrix} 1/2 \\ 3/8 \end{bmatrix} = (1/2)E_1 + (3/8)E_2$.

3. The augmented solution matrix is $\left[\begin{array}{cc|cccc} 1 & 0 & 1 & 3 & 6/5 & -2 \\ 0 & 1 & 1 & -1 & -4/5 & 1 \end{array}\right]$, where, for example, $\begin{bmatrix} 1 & -1 \\ 2 & 3 \end{bmatrix}\begin{bmatrix} 3 \\ -1 \end{bmatrix} = AX_2 = \begin{bmatrix} 4 \\ 3 \end{bmatrix} = B_2$.

5. a. Yes

   b. $A = \begin{bmatrix} 0 \\ 1 \\ 1 \end{bmatrix}$ and $B = \begin{bmatrix} 1 \\ 0 \\ 0 \end{bmatrix}$ will suffice, although any 2 linear combinations of these would serve as well.

   c. $A + B = 1A + 1B$.

   d. Let $T = \left\{ Y \mid Y = x_1 \begin{bmatrix} 0 \\ 1 \\ 1 \end{bmatrix} \right\}$ for example.

7. Since $0 + 0 = 0$ and $c(0) = 0$ for all $c$, $\{0\}$ is a vector space.

9. Since $\begin{bmatrix} 1 \\ 1 \end{bmatrix} = \begin{bmatrix} 1 \\ 0 \end{bmatrix} + \begin{bmatrix} 0 \\ 1 \end{bmatrix}$ is not in $S$, the set is not closed under addition. However, if $Y$ is in $S$, then $Y$ is of the form $\begin{bmatrix} a \\ 0 \end{bmatrix}$ or $\begin{bmatrix} 0 \\ b \end{bmatrix}$. Any scalar multiple of either of these is of the same form and is in $S$.

11. Let $S_1, S_2$ be vector spaces with the non-empty intersection $S_1 \cap S_2$. If $Y_1$ and $Y_2$ are in $S_1 \cap S_2$, then $Y_1$ and $Y_2$ are each in $S_1$ and $S_2$. Since $S_1$ is a vector space,

$Y_1 + Y_2$ is in $S_1$. Since $S_2$ is a vector space $Y_1 + Y_2$ is in $S_2$. Thus $Y_1 + Y_2$ is in $S_1 \cap S_2$ and thus $S_1 \cap S_2$ is closed under addition. A similar argument holds for scalar multiplication.

## 4.2 Exercises (page 109)

1. Any vector in $S$ may be expressed as a linear combination of the vectors in $T$; that is, if $T = \{A_1, A_2, \ldots, A_p\}$ and $Y$ is in $S$, then $Y = c_1A_1 + c_2A_2 + \cdots + c_pA_p$ for some set of scalars $c_i$.

3. Not all 0; $c_1A_1 + c_2A_2 + c_3A_3 = 0$. ... the only scalars $c_i$ such that $c_1A_1 + c_2A_2 + c_3A_3 = 0$ are $c_i = c_2 = c_3 = 0$.

5. a. Linearly independent  b. $-1A_1 + (1/2)A_2 + 1A_3 = 0$.
   c. $-3A_1 + 2A_2 + 5A_3 = 0$  d. Linearly independent

7. a. $\left[\begin{array}{cccc|c} 1 & 1 & 1 & -1 & 0 \\ 1 & 0 & 0 & -1 & 0 \\ 1 & 1 & 3 & 1 & 0 \\ 1 & 0 & 1 & 0 & 0 \end{array}\right] \sim \left[\begin{array}{cccc|c} 1 & 0 & 0 & -1 & 0 \\ 0 & 1 & 0 & -1 & 0 \\ 0 & 0 & 1 & 1 & 0 \\ 0 & 0 & 0 & 0 & 0 \end{array}\right]$, so that the equation $AC = 0$ has solution $a[1, 1, -1, 1]^T = [1, 1, -1, 1]^T$ for $a = 1$. Thus, $1A_1 + 1A_2 + (-1)A_3 + 1A_4 = 0$, and $T_1$ is linearly dependent. Any one of the vectors may be expressed as a linear combination of the others since each has a non-zero coefficient.

   b. Linearly independent
   c. Linearly independent
   d. Linearly independent
   e. $2A_1 + 2A_2 - 2A_3 - A_4 = 0$. Thus the set is linearly dependent. $A_4 = 2A_1 + 2A_2 - 2A_3$.

9. Show that each of the vectors in $T_1$ can be expressed as a linear combination of the vectors in $T_2$, and vice-versa.

11. There exist $c_i$, not all zero, such that $c_1A_1 + c_2A_2 + c_3A_3 = 0$. Suppose that $c_2 \neq 0$. Then $c_2A_2 = -c_1A_1 - c_3A_3$ and $A_2 = (-c_1/c_2)A_1 - (c_3/c_2)A_3$.

13. We wish to show that if $A$ is linearly dependent, then so is $B$. Use the scalars showing dependence of $A$ and assign 0 as a scalar for each element of $B$ not in $A$. We then have 0 expressed as a linear combination of the elements of $B$ with at least one non-zero scalar; no.

15. Suppose that you have a linear combination of the vectors in the latter set which yields the zero vector. Rewrite the equation as a linear combination of $A_1$, $A_2$, $A_3$ and see that each of the coefficients must be 0 because of the linear independence of $\{A_1, A_2, A_3\}$.

## 4.3 Exercises (page 114)

1. Linearly independent; spans $S$.

3. a. True  b. False  c. True

5. a. $T_1$ is neither linearly independent, nor does it span $S$.
   b. $T_2$ spans $S$ but is not linearly independent.

7. $\begin{bmatrix} 2 \\ 1 \end{bmatrix}$ could be replaced (see Exercise 10).

9. $Y = -A_1 + 3A_3 + 2A_4$.

11. Since $\left[\begin{array}{cccc|c} 1 & -1 & 2 & 2 & 0 \\ 0 & -3 & 3 & 0 & 0 \\ 1 & 1 & 0 & 2 & 0 \\ 1 & 0 & 1 & 2 & 0 \end{array}\right] \sim \left[\begin{array}{cccc|c} 1 & 0 & 1 & 2 & 0 \\ 0 & 1 & -1 & 0 & 0 \\ 0 & 0 & 0 & 0 & 0 \\ 0 & 0 & 0 & 0 & 0 \end{array}\right]$, we conclude that $A_3 = A_1 - A_2$ and $A_4 = 2A_1$. Thus, the dimension is 2 and $\{A_1, A_2\}$ could serve as a basis, as could any two of the vectors except $A_1$ and $A_4$.

## 4.4 Exercises (page 119)

1. (1, −1), (3, 4). We could express these as matrices $\begin{bmatrix} 1 \\ -1 \end{bmatrix}$ and $\begin{bmatrix} 3 \\ 4 \end{bmatrix}$ and use the matrix to represent both the vector and its coordinates.

3. (1, 0), (1, 1), (1, −2).

5. a. Any pair
   b. The coordinates are the same as the matrix elements.
   c. $\begin{bmatrix} 2 \\ 3 \end{bmatrix} = 0A_3 + 1A_1, \begin{bmatrix} 3 \\ 2 \end{bmatrix} = (5/3)A_3 + (2/3)A_1, \begin{bmatrix} 1 \\ 0 \end{bmatrix} = 1A_3 + 0A_1,$
   $\begin{bmatrix} 0 \\ 1 \end{bmatrix} = (-2/3)A_3 + (1/3)A_1.$
   d. $\begin{bmatrix} 1 \\ 0 \end{bmatrix} = (3/5)A_2 - (2/5)A_1, \begin{bmatrix} 0 \\ 1 \end{bmatrix} = -(2/5)A_2 + (3/5)A_1,$
   $\begin{bmatrix} 3 \\ 2 \end{bmatrix} = 1A_2 + 0A_1, \begin{bmatrix} 2 \\ 3 \end{bmatrix} = 0A_2 + 1A_1.$
   e. $\begin{bmatrix} 6 \\ 6 \end{bmatrix} = (6/5)A_2 + (6/5)A_1.$

7. a. $U = \begin{bmatrix} 1 & 4 \\ 4 & 2 \end{bmatrix}$. b. $X_U = \begin{bmatrix} 2 \\ -1 \end{bmatrix}$. c. $X_U = \begin{bmatrix} -1/7 \\ 2/7 \end{bmatrix}$. d. $X = \begin{bmatrix} 5 \\ 6 \end{bmatrix}$.

## 4.5 Exercises (page 123)

1. a. True b. False c. True d. True e. True f. True
   g. False h. False i. True j. True k. True

3. a. 3 b. 0 c. 3 d. 5 e. 1.

## 4.6 Exercises (page 125)

3. a. No b. No c. Yes d. No e. Yes f. No g. Yes
   h. Yes

## 4.8 Exercises (page 135)

1. a. 0 b. 4 c. 0 d. 1 e. 10 f. −15

3. [3/5, 4/5]

5. a. 0 b. $2\sqrt{5}/5$ c. 0 d. $\sqrt{50}/50$ e. 1 f. −1

7. [−4/5, 3/5] or [4/5, −3/5]

## 5.1 Exercises (page 142)

1. −2 3. 0 5. 56 7. 0 9. −24

## 5.2 Exercises (page 145, 146 & 150)

1. $A_{11} = -3$, $A_{12} = -1$, $A_{13} = 2$, $A_{21} = 3$, $A_{22} = 1$, $A_{23} = -2$, $A_{31} = 3$, $A_{32} = 1$, $A_{33} = -2$.

3. $A_{21} = 2$, $A_{22} = 6$, $A_{23} = 2$, $A_{24} = -4$.

5. $A_{12} = -1$, $A_{22} = 7$, $A_{32} = 1$, $A_{42} = -3$.

7. $A_{11} = -4$, $A_{12} = 10$, $A_{13} = -1$.

9. −5. 11. 7. 17. $-2 = |A| = |B|$; $|AB| = 4$.

19. $|A| = -18$, $|B| = 1$, $|AB| = -18$. 21. $|A| = -2$, $|B| = -21$, $|AB| = 42$.

23. $|A| = 4$ 25. $|A| = 0$. 27. $|A| = -4$.

## 5.3 Exercises (page 153 & 156)

1. $\text{adj } A = \begin{bmatrix} 2 & -3 \\ -1 & 1 \end{bmatrix}$.

3. $\text{adj } A = \begin{bmatrix} 5 & 5 & -5 \\ -14 & -14 & 14 \\ -3 & -3 & 3 \end{bmatrix}$.

5. $A^{-1} = (1/2)\begin{bmatrix} -4 & 2 \\ 3 & -1 \end{bmatrix}$.

7. $A^{-1} = (1/2)\begin{bmatrix} 11 & -7 & -1 \\ -2 & 2 & 0 \\ -5 & 3 & 1 \end{bmatrix}$.

9. $A^{-1} = \begin{bmatrix} 0 & 3 & -1 & -1 \\ 1 & -3 & 0 & 1 \\ 0 & 7 & -3 & -2 \\ -1 & -6 & 4 & 2 \end{bmatrix}.$ 15. $X = \begin{bmatrix} 11/5 \\ 2/5 \end{bmatrix}.$

17. $X = \begin{bmatrix} 1 \\ -1 \\ 2 \end{bmatrix}.$ 19. $X = \begin{bmatrix} 1 \\ 1 \\ -1 \\ -1 \\ 0 \\ 0 \end{bmatrix}.$

## 5.4 Exercises (page 157, 160 & 161)

1. 1 3. 3 5. 2 7. $r(A) = 3.$ 9. $r(A) = 2.$

11. $r(A) = 2.$ 13. $r(A) = 3.$ 15. 2; full rank.

17. 3; full rank. 19. 3

21. $(AA^T)^{-1} = \begin{bmatrix} 2/3 & -1/3 \\ -1/3 & 1/3 \end{bmatrix}$. The matrix $(A^TA)$ is a $4 \times 4$ matrix of rank 2 and hence has no inverse by Theorem 5.4.4.

## 5.5 Exercises (page 165)

1. a. $X_U, X_V; X_U, X_V, X_U, V, X_V$ b. $\begin{bmatrix} 5 \\ 5 \end{bmatrix}$ c. $\begin{bmatrix} 13/5 \\ 1/5 \end{bmatrix}.$

   d. $U_1 = V_1 - V_2, U_2 = 2V_1 + 3V_2; 2U_1 + U_2 = 4V_1 + V_2.$

3. a. $A = \begin{bmatrix} 4 & 7 \\ 3 & 5 \end{bmatrix}.$ b. $X_U = \begin{bmatrix} 1 \\ 1 \end{bmatrix}, X_V = \begin{bmatrix} 11 \\ 8 \end{bmatrix}.$

   c. $X_V = \begin{bmatrix} 1 \\ 1 \end{bmatrix}, X_U = \begin{bmatrix} 2 \\ -1 \end{bmatrix}.$

5. a. $\begin{bmatrix} 1 & 2 & 1 \\ 0 & 1 & 1 \\ 1 & 1 & 2 \end{bmatrix}.$ b. $X = \begin{bmatrix} 4 \\ 2 \\ 4 \end{bmatrix}.$ c. $X_V = \begin{bmatrix} 4 \\ 2 \\ 4 \end{bmatrix}.$

## 5.6 Exercises (page 171)

1. Linear 3. Non-linear 5. Non-linear

9. $T$ is linear since $a_1T(X) + a_2T(Y) = T(a_1X + a_2Y).$

11. $\begin{bmatrix} -a \\ 2a \\ a \end{bmatrix}$

## 5.7 Exercises (page 176)

1. $A = I_2; \begin{bmatrix} 2 \\ 3 \end{bmatrix}$.

3. $A = \begin{bmatrix} 2 & 1 \\ 1 & 2 \end{bmatrix}; \begin{bmatrix} 7 \\ 8 \end{bmatrix}$.

5. $A = \begin{bmatrix} 1 & 2 \\ -2 & 0 \end{bmatrix}; \begin{bmatrix} 8 \\ -4 \end{bmatrix}$.

7. $A = I_3; \begin{bmatrix} -1 \\ 2 \\ 3 \end{bmatrix}$.

9. $A = \begin{bmatrix} 1 & 0 & 0 \\ 0 & 1 & 0 \\ 1 & 1 & 1 \end{bmatrix}; \begin{bmatrix} -1 \\ 2 \\ 4 \end{bmatrix}$.

11. $A = \begin{bmatrix} 1 & 1 & 1 \\ 1 & -1 & -1 \\ 1 & 1 & -1 \end{bmatrix}; \begin{bmatrix} 4 \\ -6 \\ -2 \end{bmatrix}$.

13. a. $T\left(\begin{bmatrix} 1 \\ 2 \end{bmatrix}\right) = (17/3)U - (1/3)V; T\left(\begin{bmatrix} 2 \\ 1 \end{bmatrix}\right) = (16/3)U - (2/3)V$.

b. $B = \begin{bmatrix} 17/3 & 16/3 \\ -1/3 & -2/3 \end{bmatrix}$. c. $\begin{bmatrix} 5/3 \\ 2/3 \end{bmatrix}$. d. $\begin{bmatrix} 13 \\ -1 \end{bmatrix}$. e. $\begin{bmatrix} 11 \\ 25 \end{bmatrix}$.

15. a. $A = \begin{bmatrix} -3 & -4 \\ 4 & 4 \end{bmatrix}$. b. $\begin{bmatrix} 1 \\ 1 \end{bmatrix}$. c. $\begin{bmatrix} -7 \\ 8 \end{bmatrix}$. d. $\begin{bmatrix} 9 \\ -6 \end{bmatrix}$.

## 5.8 Exercises (page 182)

1. $A = \begin{bmatrix} 2 & 0 \\ 0 & 2 \end{bmatrix}$.

3. a. $A = \begin{bmatrix} 1 & 0 \\ 0 & 0 \end{bmatrix}, P\left(\begin{bmatrix} a \\ b \end{bmatrix}\right) = \begin{bmatrix} a \\ 0 \end{bmatrix}$.

b. $\begin{bmatrix} 0 & 0 \\ 0 & 1 \end{bmatrix}, \begin{bmatrix} 0 \\ b \end{bmatrix}$.

c. $\begin{bmatrix} 1/2 & -1/2 \\ -1/2 & 1/2 \end{bmatrix}, \begin{bmatrix} (a-b)/2 \\ (-a+b)/2 \end{bmatrix}$.

d. $\begin{bmatrix} 1/2 & 1/2 \\ 1/2 & 1/2 \end{bmatrix}, \begin{bmatrix} (a+b)/2 \\ (a+b)/2 \end{bmatrix}$.

e. $\begin{bmatrix} 1/5 & 2/5 \\ 2/5 & 4/5 \end{bmatrix}, \begin{bmatrix} (a+2b)/5 \\ (2a+4b)/5 \end{bmatrix}$.

f. $\begin{bmatrix} 4/5 & 2/5 \\ 2/5 & 1/5 \end{bmatrix}, \begin{bmatrix} (4a+2b)/5 \\ (2a+b)/5 \end{bmatrix}$.

5. a. $(1/2)\begin{bmatrix} \sqrt{3} & -1 \\ 1 & \sqrt{3} \end{bmatrix}, \begin{bmatrix} (\sqrt{3}-1)/2 \\ (1+\sqrt{3})/2 \end{bmatrix}$.

b. $(1/\sqrt{2})\begin{bmatrix} 1 & -1 \\ 1 & 1 \end{bmatrix}, \begin{bmatrix} 0 \\ \sqrt{2} \end{bmatrix}$.

c. $(1/2)\begin{bmatrix} -1 & \sqrt{3} \\ -\sqrt{3} & -1 \end{bmatrix}, \begin{bmatrix} (-1+\sqrt{3})/2 \\ (-1-\sqrt{3})/2 \end{bmatrix}$.

d. $(1/2)\begin{bmatrix} 1 & -\sqrt{3} \\ \sqrt{3} & 1 \end{bmatrix}, \begin{bmatrix} (1-\sqrt{3})/2 \\ (\sqrt{3}+1)/2 \end{bmatrix}$.

e. $(1/\sqrt{2})\begin{bmatrix} -1 & 1 \\ -1 & -1 \end{bmatrix}, \begin{bmatrix} 0 \\ -\sqrt{2} \end{bmatrix}$.

f. $\begin{bmatrix} 1 & 0 \\ 0 & 1 \end{bmatrix}, \begin{bmatrix} 1 \\ 1 \end{bmatrix}$.

7. a. $A = \begin{bmatrix} -1 & 0 \\ 0 & 1 \end{bmatrix}, \begin{bmatrix} -a \\ b \end{bmatrix}, \begin{bmatrix} -1 \\ 1 \end{bmatrix}.$ b. $\begin{bmatrix} 0 & -1 \\ -1 & 0 \end{bmatrix}, \begin{bmatrix} -b \\ -a \end{bmatrix}, \begin{bmatrix} -1 \\ -1 \end{bmatrix}.$

c. $A = \begin{bmatrix} 0 & -1 \\ -1 & 0 \end{bmatrix}, \begin{bmatrix} -b \\ -a \end{bmatrix}, \begin{bmatrix} -1 \\ -1 \end{bmatrix}.$

d. $(1/5)\begin{bmatrix} -3 & 4 \\ 4 & 3 \end{bmatrix}, (1/5)\begin{bmatrix} -3a + 4b \\ 4a + 3b \end{bmatrix}, \begin{bmatrix} 1/5 \\ 7/5 \end{bmatrix}.$

e. $(1/5)\begin{bmatrix} 4 & 3 \\ 3 & -4 \end{bmatrix}, (1/5)\begin{bmatrix} 4a + 3b \\ 3a - 4b \end{bmatrix}, \begin{bmatrix} 7/5 \\ -1/5 \end{bmatrix}.$ f. $\begin{bmatrix} 0 & 1 \\ 1 & 0 \end{bmatrix}, \begin{bmatrix} b \\ a \end{bmatrix}, \begin{bmatrix} 1 \\ 1 \end{bmatrix}.$

## 5.9 Exercises (page 188)

1. a. Eigenvalues are 0, 1; corresponding to 0 the general form is $a\begin{bmatrix} 0 \\ 1 \end{bmatrix}$; corresponding to 1, the general form is $a\begin{bmatrix} 1 \\ 0 \end{bmatrix}$.

b. Eigenvalues are 0, 3; corresponding to 0 the general form is $a\begin{bmatrix} -1 \\ 1 \end{bmatrix}$; corresponding to 3, $a\begin{bmatrix} 1/2 \\ 1 \end{bmatrix}$.

c. Eigenvalue is 2; general form is $a\begin{bmatrix} 0 \\ 1 \end{bmatrix} + b\begin{bmatrix} 1 \\ 0 \end{bmatrix}$ (all of $\mathcal{M}_{2,1}$)

d. Eigenvalues: 1, $-2$; $a\begin{bmatrix} 4 \\ 1 \end{bmatrix}$ corresponds to 1; $a\begin{bmatrix} 1 \\ 1 \end{bmatrix}$ corresponds to $-2$.

5. a. $c^2 - 4c + 3; A^{-1} = \begin{bmatrix} 1 & -2/3 \\ 0 & 1/3 \end{bmatrix}.$

b. $-c^3 + 8c^2 - 14c + 2; A^{-1} = (1/2)\begin{bmatrix} 11 & -7 & -1 \\ -2 & 2 & 0 \\ -5 & 3 & 1 \end{bmatrix}.$

c. $-c^3 + 3c^2 + 6c + 4; A^{-1} = (1/4)\begin{bmatrix} -2 & -1 & 2 \\ 4 & -4 & 0 \\ 0 & 2 & 0 \end{bmatrix}.$

## 5.10 Exercises (page 191)

1. a. $X^TAX$ where $A = \begin{bmatrix} 1 & 2 \\ 3 & 5 \end{bmatrix}.$

b. $X^TAX$ where $A = \begin{bmatrix} 2 & 1 \\ 0 & -1 \end{bmatrix}.$

c. $X^TAX$ where $A = \begin{bmatrix} 3 & 1 \\ -1 & 1 \end{bmatrix}$.

5. a. Positive definite  b. Positive semi-definite; (1, −1)
   c. Neither; (0, 1)  d. Neither; (1, 0)  e. Positive semi-definite; (2, 1)

## 6.1 Exercises (page 197)

1. Let $x_i$ denote the amount of $P_i$ to be produced. Find $x_1, x_2$ such that $x_i \geqslant 0$ and

$$\begin{aligned} 5x_1 + 4x_2 &\leqslant 480 \quad (M_1) \\ 10x_1 + 8x_2 &\leqslant 960 \quad (M_2) \\ 3x_1 + 6x_2 &\leqslant 480 \quad (M_3) \end{aligned}$$

and $f(x_1, x_2) = 4x_1 + 3x_2$ is maximized.

Do you notice anything peculiar about the first two constraints?

3. Find $x_{ij} \geqslant 0$ such that

$$\begin{array}{llllllllllll} (M_1) & x_{11} + x_{12} + x_{13} & & & = 1 \\ (M_2) & & x_{21} + x_{22} + x_{23} & & = 1 \\ (M_3) & & & x_{31} + x_{32} + x_{33} & = 1 \\ (J_1) & x_{11} & + x_{21} & + x_{31} & = 1 \\ (J_2) & \quad x_{12} & \quad + x_{22} & \quad + x_{32} & = 1 \\ (J_3) & \qquad x_{13} & \qquad + x_{23} & \qquad + x_{33} & = 1 \end{array}$$

and $f(x_{ij}) = 3x_{11} + 5x_{12} + 4x_{13} + 4x_{23} + \cdots + 5x_{33}$ is minimized.

5. Let $x_i$ represent the number of units of $c_i$ to be produced.
Find $x_i$ such that $x_i \geqslant 0$ and

$$\begin{aligned} 3x_1 + 6x_2 + 7x_3 + 8x_4 &\leqslant 10{,}000 \\ 2x_1 + 5x_2 + 8x_3 + 10x_4 &\leqslant 14{,}000 \\ x_2 &\geqslant 250 \\ x_3 &\geqslant 500 \end{aligned}$$

and $f(x_i) = 3x_1 + 4x_2 + 6x_3 + 8x_4$ is maximized.

7. Find $X = \begin{bmatrix} x_1 \\ x_2 \end{bmatrix} \geqslant 0$ such that $AX \leqslant R$ and

$f(X) = CX$ is maximized, where $A = \begin{bmatrix} 5 & 4 \\ 10 & 8 \\ 3 & 6 \end{bmatrix}$, $R = \begin{bmatrix} 480 \\ 960 \\ 480 \end{bmatrix}$, $C = [4, 3]$.

9. Find $x_{ij}$ such that

i. $x_{ij} \geqslant 0 \quad i = 1, 2; \quad j = 1, 2, 3.$

ii. $$\begin{aligned} x_{11} + x_{12} + x_{13} \quad\quad\quad\quad\quad\quad\quad\quad &= 200 \\ x_{21} + x_{22} + x_{23} &= 300 \\ x_{11} \quad\quad + x_{21} \quad\quad\quad\quad &\leqslant 200 \\ x_{12} \quad\quad + x_{22} \quad\quad &\leqslant 250 \\ x_{13} \quad\quad + x_{23} &\leqslant 150 \end{aligned}$$

and $30x_{11} + 38x_{12} + 40x_{13} + 40x_{21} + 35x_{22} + 36x_{23} = f(x_{ij})$ is maximized.

## 6.2 Exercises (page 206)

1. a–d.

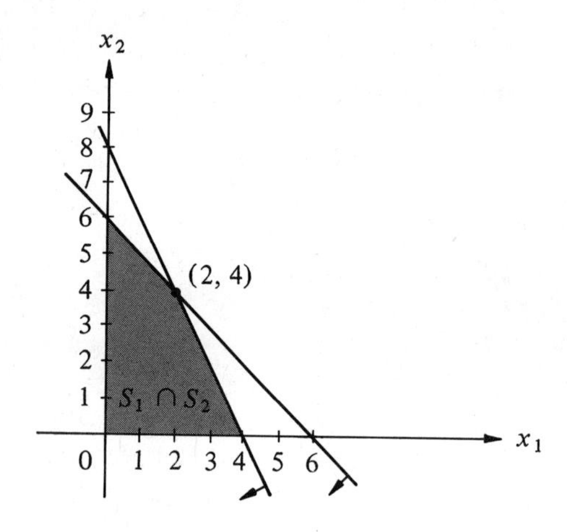


e. (0, 0), (0, 6), (2, 4), (4, 0) f. 0, 12, 10, 4 g. $f(1, 1) = 3$, $f(1, 3) = 7$, $f(3, 1) = 5$, $f(1, 5) = 11$.

3. a–d.

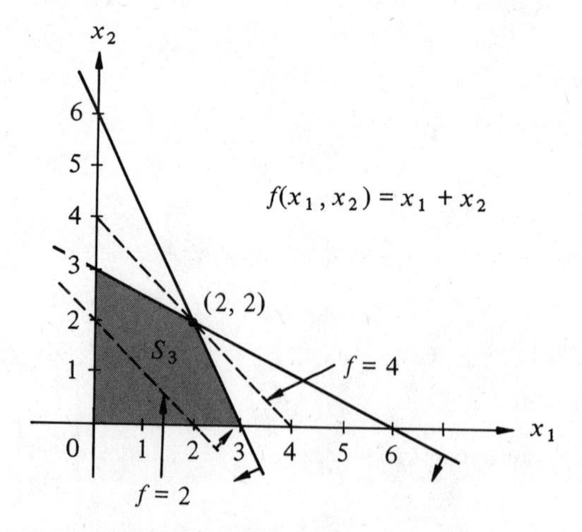


e. (2, 2); $f(2, 2) = 4$.

f. (0, 0); $f(0, 0) = 0$.

g. Any point $(x_1, x_2)$ on the line between (0, 3) and (2, 2).

5.

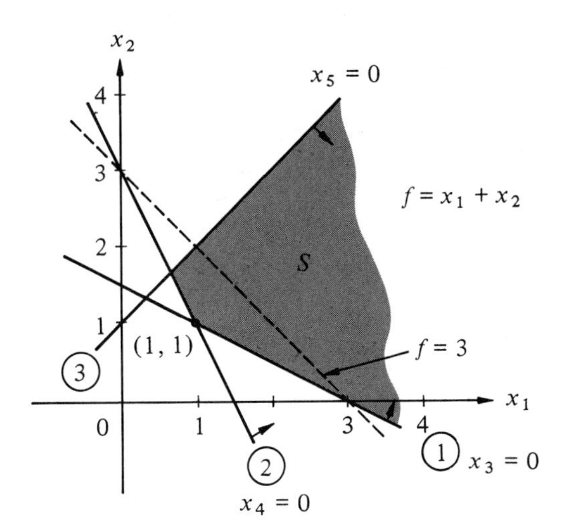


a. See graph.
b. (1, 1), (3, 0), (2/3, 5/3)
c. $f(1, 1) = 2$, $f(3, 0) = 3$, $f(2/3, 5/3) = 7/3$.
d. See graph.
e. (1, 1), $2 = f$.
f. $f$ has no maximum in $S$. $f$ can be made as large as you please by, for example, letting $x_2 = 0$ and increasing $x_1$.
g. The maximum value of $g$ occurs at (1, 1) where $g(1, 1) = -2 = -f(1, 1)$. $g$ has no minimum value over $S$.

7.

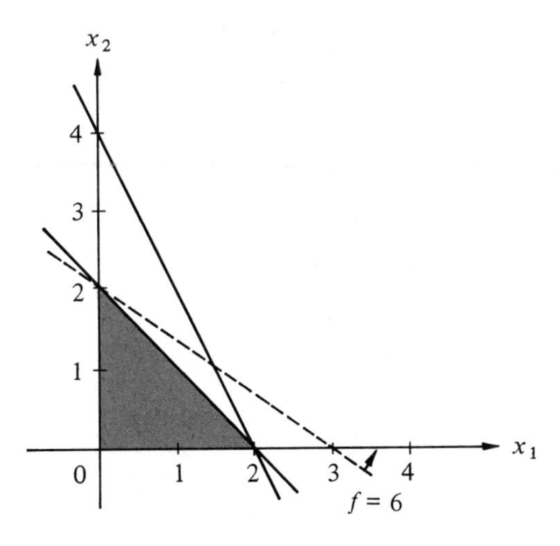


a. The maximum value of $f$ occurs at (0, 2) where $f(0, 2) = 6$.
b. The second constraint was redundant in the sense that it excluded no new points from consideration.

9. Let $x_1, x_2$ denote the amount of $A$ and $B$ to be produced.

a. Find $x_1, x_2$ such that

1) $x_i \geqslant 0 \quad i = 1, 2$

2) $x_1 + 2x_2 \leqslant 480 \quad (M_1)$

$3x_1 + x_2 \leqslant 480 \quad (M_2)$

3) $f(x_1, x_2) = 5x_1 + 4x_2$ is maximized.

b.

| Point | $f$ |
|---|---|
| $O$ | 0 |
| $A$ | 960 |
| $B$ | 1248 |
| $C$ | 800 |

c.

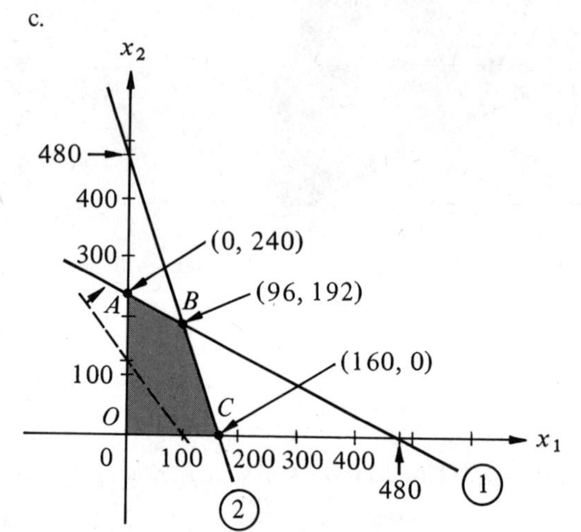


11. $x_1 = 4, \quad x_2 = 3; \quad f = \$23{,}000.$

## 6.3 Exercises (page 215)

1. Negative, $2x_1 + 3x_2 + x_3 = 6$, guard, non-negative, not satisfied, slack

3. $$\begin{aligned} 2x_1 + x_2 + x_3 \qquad &= 6 \\ x_1 + 2x_2 \qquad + x_4 &= 6 \end{aligned}$$

$f(x_1, x_2, x_3, x_4) = x_1 + x_2$ is to be maximized.

5. $$\begin{aligned} x_1 + 2x_2 - x_3 \qquad\qquad &= 3 \\ 2x_1 + x_2 \qquad - x_4 \qquad &= 3 \\ -x_1 + x_2 \qquad\qquad + x_5 &= 1 \end{aligned}$$

$f(x_i) = x_1 + x_2$ is to be minimized.

7. Find $X \geqslant 0$ such that $AX = R$ and $f(X) = CX$ is minimized.

$$A = \begin{bmatrix} 2 & 1 & 1 & 0 \\ 1 & 2 & 0 & 1 \end{bmatrix}, \quad R = \begin{bmatrix} 6 \\ 6 \end{bmatrix}, \quad C = [1, 1, 0, 0].$$

9. Find $X \geqslant 0$ such that $AX = R$ and $f(X) = CX$ is minimized.

$$A = \begin{bmatrix} 1 & 2 & -1 & 0 & 0 \\ 2 & 1 & 0 & -1 & 0 \\ -1 & 1 & 0 & 0 & 1 \end{bmatrix}, \quad R = \begin{bmatrix} 3 \\ 3 \\ 1 \end{bmatrix}, \quad C = [1, 1, 0, 0, 0].$$

11. a. (1, 2, 2), non-negative b. (2, 2, −2), negative, is not
c. (2, 1, 1), (2, 1)

## 6.4 Exercises (page 221)

1. a.
$$\begin{aligned} x_1 + 2x_2 + x_3 \quad &= 3 \\ 2x_1 + x_2 \quad + x_4 &= 3. \end{aligned}$$

b. $A = \begin{bmatrix} 1 & 2 & 1 & 0 \\ 2 & 1 & 0 & 1 \end{bmatrix}, X = \begin{bmatrix} x_1 \\ x_2 \\ x_3 \\ x_4 \end{bmatrix}, R = \begin{bmatrix} 3 \\ 3 \end{bmatrix}.$

c. See Figure 7.4.2. d. Check that $AX_i = R$. e. $X_1, X_3$
f. $X_1, X_2$ g. $X_1$

3. a. Non-basic, basic b. (1/2, 1/2, 3/2, 3/2), yes, no

c.

| *Point* | *2-Space Solution* | *Non-basic Variables* | *Basic Variables* | *4-Space Solution, $X^T$* | *Solution Feasible?* |
|---|---|---|---|---|---|
| *O* | [0, 0] | $x_1, x_2$ | $x_3, x_4$ | [0, 0, 3, 3] | yes |
| *A* | [0, 3/2] | $x_1, x_3$ | $x_2, x_4$ | [0, 3/2, 0, 3/2] | yes |
| *B* | [0, 3] | $x_1, x_4$ | $x_2, x_3$ | [0, 3, −3, 0] | no |
| *C* | [1, 1] | $x_3, x_4$ | $x_1, x_2$ | [1, 1, 0, 0] | yes |
| *D* | [3, 0] | $x_2, x_3$ | $x_1, x_4$ | [3, 0, 0, −3] | no |
| *E* | [3/2, 0] | $x_2, x_4$ | $x_1, x_3$ | [3/2, 0, 3/2, 0] | yes |

5. a. $A = \begin{bmatrix} 1 & 1 & 1 & 0 \\ 2 & 1 & 0 & 1 \end{bmatrix}, R = \begin{bmatrix} 2 \\ 4 \end{bmatrix}, A_1 = \begin{bmatrix} 1 \\ 2 \end{bmatrix}, A_4 = \begin{bmatrix} 0 \\ 1 \end{bmatrix}.$

b.

| *Point* | *2-Space Solution* | *Non-basic Variables* | *Basic Variables* | *4-Space Solution, $X^T$* | *Solution Feasible?* |
|---|---|---|---|---|---|
| *O* | [0, 0] | $x_1, x_2$ | $x_3, x_4$ | [0, 0, 2, 4] | yes |
| *A* | [0, 2] | $x_1, x_3$ | $x_2, x_4$ | [0, 2, 0, 2] | yes |
| *B* | [0, 4] | $x_1, x_4$ | $x_3, x_2$ | [0, 4, −2, 0] | no |
| *C* | [2, 0] | $x_2, x_3$ | $x_1, x_4$ | [2, 0, 0, 0] | yes |
| *C* | [2, 0] | $x_2, x_4$ | $x_3, x_1$ | [2, 0, 0, 0] | yes |
| *C* | [2, 0] | $x_3, x_4$ | $x_1, x_2$ | [2, 0, 0, 0] | yes |

c. [1, 1, 0, 1], basic, feasible

## 6.5 Exercises (page 231)

1. a. $\begin{bmatrix} 5 \\ 29 \end{bmatrix}$ b. 29 c. $\begin{bmatrix} 6 & -1 \\ -5 & 1 \end{bmatrix}$ d. $a = 5, b = 29$.

3. a. $\begin{bmatrix} 3 \\ 1 \end{bmatrix}$ b. $\begin{bmatrix} 4 \\ 4 \end{bmatrix}$, 4 c. $\begin{bmatrix} 4 \\ -8 \end{bmatrix}$, $-8$ d. $\begin{bmatrix} 1 \\ 1 \end{bmatrix}$, 1 e. 4

5. a. $x_1A_1 + x_2A_2 + x_3A_3 + x_4A_4 = R$.

c.

| *Point* | *Basic Variables* | $B$ | $B^{-1}$ | $X_B$ | $X^T$ | $x_{B1}$ |
|---|---|---|---|---|---|---|
| $O$ | $x_3, x_4$ | $[A_3, A_4] = \begin{bmatrix} 1 & 0 \\ 0 & 1 \end{bmatrix}$ | $\begin{bmatrix} 1 & 0 \\ 0 & 1 \end{bmatrix}$ | $\begin{bmatrix} 3 \\ 3 \end{bmatrix}$ | [0, 0, 3, 3] | 3 |
| $A$ | $x_2, x_4$ | $\begin{bmatrix} 2 & 0 \\ 1 & 1 \end{bmatrix}$ | $\begin{bmatrix} 1/2 & 0 \\ -1/2 & 1 \end{bmatrix}$ | $\begin{bmatrix} 3/2 \\ 3/2 \end{bmatrix}$ | [0, 3/2, 0, 3/2] | 3/2 |
| $B$ | $x_2, x_3$ | $\begin{bmatrix} 2 & 1 \\ 1 & 0 \end{bmatrix}$ | $\begin{bmatrix} 0 & 1 \\ 1 & -2 \end{bmatrix}$ | $\begin{bmatrix} 3 \\ -3 \end{bmatrix}$ | [0, 3, −3, 0] | 3 |
| $C$ | $x_1, x_2$ | $\begin{bmatrix} 1 & 2 \\ 2 & 1 \end{bmatrix}$ | $\begin{bmatrix} -1/3 & 2/3 \\ 2/3 & -1/3 \end{bmatrix}$ | $\begin{bmatrix} 1 \\ 1 \end{bmatrix}$ | [1, 1, 0, 0] | 1 |
| $D$ | $x_1, x_4$ | $\begin{bmatrix} 1 & 0 \\ 2 & 1 \end{bmatrix}$ | $\begin{bmatrix} 1 & 0 \\ -2 & 1 \end{bmatrix}$ | $\begin{bmatrix} 3 \\ -3 \end{bmatrix}$ | [3, 0, 0, −3] | 3 |
| $E$ | $x_1, x_3$ | $\begin{bmatrix} 1 & 1 \\ 2 & 0 \end{bmatrix}$ | $\begin{bmatrix} 0 & 1/2 \\ 1 & -1/2 \end{bmatrix}$ | $\begin{bmatrix} 3/2 \\ 3/2 \end{bmatrix}$ | [3/2, 0, 3/2, 0] | 3/2 |

## 6.6 Exercises (page 235)

1. $[1, -1], [1, 0, 2, 0], -1$

3. a. $B_1 = [A_3, A_4], X^T = [0, 0, 4, 4]; B_2 = [A_2, A_4]$, $X^T = [0, 4/3, 0, 8/3]$. $B_3 = [A_2, A_1], X^T = [1, 1, 0, 0]$; $B_4 = [A_1, A_3], X^T = [4/3, 0, 8/3, 0]$
   b. $C_{B_1}X_{B_1} = 0, C_{B_2}X_{B_2} = 8/3, C_{B_3}X_{B_3} = 3, C_{B_4}X_{B_4} = 4/3$.
   c. $[1, 1, 0, 0]^T$ with $f = 3$.
   d. $C_B\ X_B = 0, 0, 1, 4/3; [4/3, 0, 8/3, 0]^T$ with $f = 4/3$.

5. a. At $O$, $C_BX_B = 0$; at $A$, $C_BX_B = -3/2$; at $C$, $C_BX_B = -4$; at $E$, $C_BX_B = -9/2$.
   b. $E$ c. $[3/2, 0, 3/2, 0], f = -9/2$.
7. a. At $O$, 0; at $A$, 12; at $C$, 14; at $E$, 12.
   b. $\begin{bmatrix} 2 \\ 4 \end{bmatrix}$, 14 c. $[2, 4, 0, 0]^T$

## 6.7 Exercises (page 240)

1. a. $x_3 = 6 - 2x_1 - 3x_2, x_4 = 6 - 3x_1 - 2x_2$.
   b. $x_3, x_4, x_4; x_4; x_2, x_4$

c. $x_2$; $x_3$, $x_1$, decrease, $x_3$, $x_3$; $x_2$, $x_1$, $x_2 = 6/5 - (3/5)x_3 + (2/5)x_4$, $x_1 = 6/5 + (2/5)x_3 - (3/5)x_4$.

3. The optimum $BFS$ is $[1, 1, 0, 0]^T$ with $f = 3$.

5. a. positive b. $a$, negative c. $-6/c, 0 - e(6/c), 5 - (6/c)a$
d. $x_4 < 0$.

## 6.8 Exercises (page 250)

1. a.

| | | $C$ | 4 | 3 | 0 | 0 |
|---|---|---|---|---|---|---|
| $B$ | $C_B^T$ | $R$ | $A_1$ | $A_2$ | $A_3$ | $A_4$ |
| $A_3$ | 0 | 6 | 2 | 3 | 1 | 0 |
| $A_4$ | 0 | 6 | (3) | 2 | 0 | 1 |
| $c_j - z_j$ | | 0 | 4 | 3 | 0 | 0 |

b. no, 3, positive c. $A_1$ d. feasibility, $A_4$; 3

e.

| | | $C$ | 4 | 3 | 0 | 0 |
|---|---|---|---|---|---|---|
| $B$ | $C_B^T$ | $X_B$ | $Y_1$ | $Y_2$ | $Y_3$ | $Y_4$ |
| $A_3$ | 0 | 2 | 0 | 5/3 | 1 | $-2/3$ |
| $A_1$ | 4 | 2 | 1 | 2/3 | 0 | 1/3 |
| $c_j - z_j$ | | 8 | 0 | 1/3 | 0 | $-4/3$ |

;

$x_{B1} = 2$, $y_{12} = 5/3$, $y_{24} = 1/3$.

f. 1/3, $A_2$; 6/5, 3; $x_3$, 5/3

g.

| | | $C$ | 4 | 3 | 0 | 0 |
|---|---|---|---|---|---|---|
| $B$ | $C_B^T$ | $X_B$ | $Y_1$ | $Y_2$ | $Y_3$ | $Y_4$ |
| $A_2$ | 3 | 6/5 | 0 | 1 | 3/5 | $-2/5$ |
| $A_1$ | 4 | 6/5 | 1 | 0 | $-2/5$ | 3/5 |
| $c_j - z_j$ | | 42/5 | 0 | 0 | $-1/5$ | $-6/5$ |

4, 0, $-2/5$

3. The optimal tableau is

| | | $C$ | 3 | 1 | 0 | 0 |
|---|---|---|---|---|---|---|
| $B$ | $C_B^T$ | $X_B$ | $Y_1$ | $Y_2$ | $Y_3$ | $Y_4$ |
| $A_3$ | 0 | 3/2 | 0 | 3/2 | 1 | $-1/2$ |
| $A_1$ | 3 | 3/2 | 1 | 3/2 | 1 | 1/2 |
| $c_j - z_j$ | | 9/2 | 0 | $-1/2$ | 0 | $-3/2$ |

## 6.9 Exercises (page 256)

1. $[X_B \mid Y]$, $R$, $A_j$

3. a. The optimal tableau is the third; $X^T = [1, 1, 0, 0]$.

| | | $C$ | 1 | 2 | 0 | 0 |
|---|---|---|---|---|---|---|
| $B$ | $C_B^T$ | $X_B$ | $Y_1$ | $Y_2$ | $Y_3$ | $Y_4$ |
| $A_2$ | 2 | 1 | 0 | 1 | 3/8 | $-1/8$ |
| $A_1$ | 1 | 1 | 1 | 0 | $-1/8$ | 3/8 |
| $C - Z$ | | 3 | 0 | 0 | $-5/8$ | $-1/8$ |

5. a. We would choose the vector with the most negative $(c_j - z_j)$ value.
   b. Since we are only maintaining feasibility with this criterion, it has nothing to do with the type of problem being solved. Therefore it would not change.

7. a. The optimal tableau is the third: $X^T = [96, 192, 0, 0]$.

| | | $C$ | 5 | 4 | 0 | 0 |
|---|---|---|---|---|---|---|
| $B$ | $C_B^T$ | $X_B$ | $Y_1$ | $Y_2$ | $Y_3$ | $Y_4$ |
| $A_2$ | 4 | 192 | 0 | 1 | 3/5 | $-1/5$ |
| $A_1$ | 5 | 96 | 1 | 0 | $-1/5$ | 2/5 |
| $C - Z$ | | 1248 | 0 | 0 | $-7/5$ | $-6/5$ |

9. a. The optimal tableau is the fourth; $X^T = [3, 1, 0, 3, 0]$.

| | | C | 2 | 3 | 0 | 0 | 0 |
|---|---|---|---|---|---|---|---|
| B | $C_B^T$ | $X_B$ | $Y_1$ | $Y_2$ | $Y_3$ | $Y_4$ | $Y_5$ |
| $A_4$ | 0 | 3 | 0 | 0 | 3 | 1 | −2 |
| $A_2$ | 3 | 1 | 0 | 1 | −1 | 0 | 1 |
| $A_1$ | 2 | 3 | 1 | 0 | 2 | 0 | −1 |
| C − Z | | 9 | 0 | 0 | −1 | 0 | −1 |

c. $B = [A_4, A_2, A_1]$. d. 3, −1, 2

## 6.10 Exercises (page 261)

1. 
$$\begin{aligned} x_1 + x_2 + x_3 \qquad\qquad &= 3 \\ -x_1 + 2x_2 \qquad - x_4 + x_5 &= 4 \end{aligned}$$

3. The optimal tableau is the third; $X^T = [4, 3, 0, 0, 0]$.

| | | C | 2 | 5 | 0 | 0 | M |
|---|---|---|---|---|---|---|---|
| B | $C_B^T$ | $X_B$ | $Y_1$ | $Y_2$ | $Y_3$ | $Y_4$ | $Y_5$ |
| $A_2$ | 5 | 3 | 0 | 1 | −3/2 | −1/2 | 3/2 |
| $A_1$ | 2 | 4 | 1 | 0 | 2 | 1 | −2 |
| C − Z | | 23 | 0 | 0 | 7/2 | 1/2 | ∞ |

5. The optimal tableau is the fourth; $X^T = [1, 1, 0, 0, 1, 0, 0]$.

| | | C | 1 | 1 | 0 | 0 | 0 | M | M |
|---|---|---|---|---|---|---|---|---|---|
| B | $C_B^T$ | $X_B$ | $Y_1$ | $Y_2$ | $Y_3$ | $Y_4$ | $Y_5$ | $Y_6$ | $Y_7$ |
| $A_1$ | 1 | 1 | 1 | 0 | 1/3 | −2/3 | 0 | −1/3 | 2/3 |
| $A_5$ | 0 | 1 | 0 | 0 | 1 | −1 | 1 | −1 | 1 |
| $A_2$ | 1 | 1 | 0 | 1 | −2/3 | 1/3 | 0 | 2/3 | −1/3 |
| C − Z | | 2 | 0 | 0 | 1/3 | 1/3 | 0 | ∞ | ∞ |

## 6.11 Exercises (page 266)

3. $CX = c_1x_1 + c_2x_2$; $ZX = z_1x_1 + z_2x_2$.
$CX - ZX = (c_1 - z_1)x_1 + (c_2 - z_2)x_2$. Since each $x_i \geqslant 0$ and each $(c_i - z_i) \leqslant 0$, the entire sum is non-positive; that is, $CX - ZX \leqslant 0$ or $CX \leqslant ZX$.

## 7.2 Exercises (page 271)

1. a. $6; x_{11}, x_{12}, x_{13}, x_{21}, x_{22}, x_{23}$

   b. $x_{11} + x_{12} + x_{13} = 40$

   c. $x_{12} + x_{22} = 30$

   d. Find $x_{ij} \geqslant 0$ such that

$$\begin{aligned} x_{11} + x_{12} + x_{13} \qquad\qquad\qquad\qquad &= 40 \\ x_{21} + x_{22} + x_{23} &= 35 \\ x_{11} \qquad\qquad + x_{21} \qquad\qquad &= 20 \\ x_{12} \qquad\qquad + x_{22} \qquad &= 30 \\ x_{13} \qquad\qquad + x_{23} &= 25 \end{aligned}$$

   and $2x_{11} + 5x_{12} + 6x_{13} + 3x_{21} + 4x_{22} + 7x_{23} = f(x_{ij})$ is minimized.

3. Let $x_{ij}$ equal the amount to be sent from warehouse $i$ to retail outlet $j$.

   a. Find $x_{ij} \geqslant 0$ such that

$$\begin{aligned} x_{11} + x_{12} + x_{13} \qquad\qquad\qquad\qquad &= 60 \\ x_{21} + x_{22} + x_{23} &= 40 \\ x_{11} \qquad\qquad + x_{21} \qquad\qquad &= 35 \\ x_{12} \qquad\qquad + x_{22} \qquad &= 30 \\ x_{13} \qquad\qquad + x_{23} &= 35 \end{aligned}$$

   and $5x_{11} + 8x_{12} + 6x_{13} + 4x_{21} + 9x_{22} + 6x_{23} = f(x_{ij})$ is minimized.

   b.

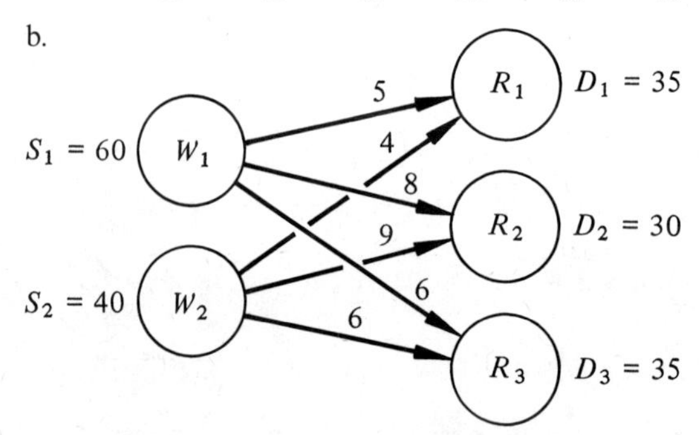


5. a. (See cost coefficients below.)

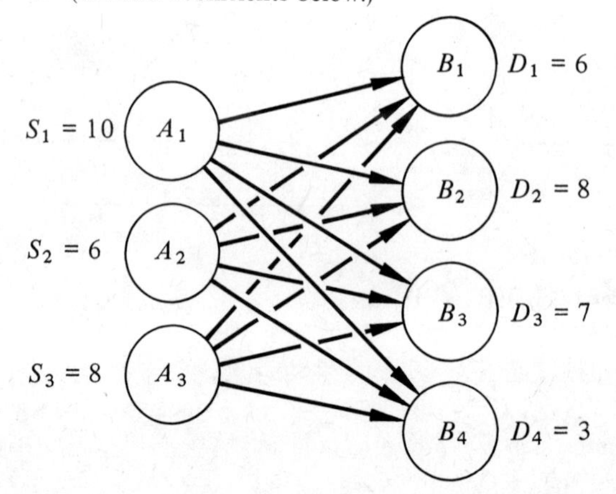

b. Find $X \geqslant 0$ such that $AX = R$ and $CX$ is minimized, where

$$A = \left[\begin{array}{cccccccccccc} 1 & 1 & 1 & 1 & 0 & 0 & 0 & 0 & 0 & 0 & 0 & 0 \\ 0 & 0 & 0 & 0 & 1 & 1 & 1 & 1 & 0 & 0 & 0 & 0 \\ 0 & 0 & 0 & 0 & 0 & 0 & 0 & 0 & 1 & 1 & 1 & 1 \\ \hline 1 & 0 & 0 & 0 & 1 & 0 & 0 & 0 & 1 & 0 & 0 & 0 \\ 0 & 1 & 0 & 0 & 0 & 1 & 0 & 0 & 0 & 1 & 0 & 0 \\ 0 & 0 & 1 & 0 & 0 & 0 & 1 & 0 & 0 & 0 & 1 & 0 \\ 0 & 0 & 0 & 1 & 0 & 0 & 0 & 1 & 0 & 0 & 0 & 1 \end{array}\right],$$

$X = [x_{11}, x_{12}, x_{13}, x_{14}, x_{21}, x_{22}, x_{23}, x_{24}, x_{31}, x_{32}, x_{33}, x_{34}]^T$,

$R = [10, 6, 8, 6, 8, 7, 3]^T$, and

$C = [20, 30, 40, 35, 15, 20, 25, 40, 25, 35, 20, 35]$.

7. Find $X \geqslant 0$ such that $AX = R$ and $CX$ is minimized, where

$$A = \left[\begin{array}{ccccccccc} 1 & 1 & 1 & 0 & 0 & 0 & 0 & 0 & 0 \\ 0 & 0 & 0 & 1 & 1 & 1 & 0 & 0 & 0 \\ 0 & 0 & 0 & 0 & 0 & 0 & 1 & 1 & 1 \\ \hline 1 & 0 & 0 & 1 & 0 & 0 & 1 & 0 & 0 \\ 0 & 1 & 0 & 0 & 1 & 0 & 0 & 1 & 0 \\ 0 & 0 & 1 & 0 & 0 & 1 & 0 & 0 & 1 \end{array}\right], \quad R = [20, 15, 25, 30, 15, 15]^T, \quad \text{and}$$

$C = [8, 6, 4, 5, 4, 5, 6, 7, 9]$.

9. Find $X \geqslant 0$ such that $AX = R$ and $CX$ is minimized, where

$$A = \left[\begin{array}{cccccccc} 1 & 1 & 0 & 0 & 0 & 0 & 0 & 0 \\ 0 & 0 & 1 & 1 & 0 & 0 & 0 & 0 \\ 0 & 0 & 0 & 0 & 1 & 1 & 0 & 0 \\ 0 & 0 & 0 & 0 & 0 & 0 & 1 & 1 \\ \hline 1 & 0 & 1 & 0 & 1 & 0 & 1 & 0 \\ 0 & 1 & 0 & 1 & 0 & 1 & 0 & 1 \end{array}\right], \quad R = [500, 400, 800, 650, 1000, 1350]^T,$$

and $C = [10, 5, 4, 9, 15, 7, 6, 2]$.

## 7.3 Exercises (page 278)

1. a. $A = \begin{bmatrix} 1 & 1 & 1 & 0 & 0 & 0 \\ 0 & 0 & 0 & 1 & 1 & 1 \\ 1 & 0 & 0 & 1 & 0 & 0 \\ 0 & 1 & 0 & 0 & 1 & 0 \\ 0 & 0 & 1 & 0 & 0 & 1 \end{bmatrix}$; $A_{12}^T = [1, 0, 0, 1, 0]$.

   b. $R^T = [40, 35, 20, 30, 25]$; $C = [2, 5, 6, 3, 4, 7]$.

   c. Independent  d. and e. $A_{22} = -A_{11} + A_{12} + A_{21}$.

   f. 4  g. $A_{13} = 0A_{11} + A_{12} - A_{22} + A_{23}$.

3. 2; destination

5. a. See answer for Exercise 7.2.5; $A_{23}^T = [0, 1, 0, 0, 0, 1, 0]$.
   b. See answer for Exercise 7.2.5.
   c. Independent
   d. and e. $A_{13} = 0A_{11} + A_{12} - A_{22} + A_{23}$.
   f. $Y_{14}^T = [0, +1, -1, +1, -1, +1]$.

7. 12; 6; 7; 12

## 7.4 Exercises (page 285)

1. See Example 7.2.1.

3. a. False b. True

5. $A_{33} = -1A_{11} + 0A_{12} + 1A_{13} + 0A_{14} + 0A_{21} + 1A_{31} = -A_{11} + A_{13} + A_{31}$.

7. a. Dependent
   b. $A_{13} = A_{11} - A_{21} + A_{22} - A_{32} + A_{33}$.

## 7.5 Exercises (page 292)

1. $m + n - 1; m + n - 1; (m - 1)(n - 1)$

3. a. Suitable b. Not basic c. Not feasible, i.e., $x_{22} + x_{23} = 70 \neq 60$.
   d. Has negative $x_{ij}$ e. Suitable

5. a. $X^T = [20, 20, 0, 0, 10, 25]$. b. Linearly independent e. 320

7. $X^T = [500, 0, 400, 0, 100, 700, 0, 650]$.

## 7.6 Exercises (page 298)

1. a. True b. $z_{ij}$ c. True

3. a. 4 equations in 5 unknowns

$$\begin{aligned} u_1 \qquad\qquad + v_2 \qquad &= 5; \\ u_1 \qquad\qquad\qquad + v_3 &= 6 \\ u_2 + v_1 \qquad\qquad &= 3 \\ u_2 \quad + v_2 \qquad &= 4 \end{aligned}$$

   b. $u_1 = 6, v_2 = -1, u_2 = 5, v_1 = -2$
   c. $v_1, 4; u_2, v_3, 5$.
   d. $-2; 2$; is not, $x_{11}$.

e. $u_1 = 0, v_2 = 5, u_2 = -1, v_1 = 4, v_3 = 6$; the $u_i, v_j$ values are different.
f. $z_{11} = 4, z_{23} = 5$; the $z_{ij}$ are the same.

5. a. $X^T = [35, 25, 0, 0, 5, 35]$. b. $u_1 = 5, u_2 = 6, v_1 = v_3 = 0, v_2 = 3$.
c. 1, −2, 0 d. is not; $x_{21}$

7. The BFS is not optimal. $x_{41}$ should become the new basic variable.

## 7.7 Exercises (page 304)

1. Feasibility; decrease; 0, smallest; 0; increases, decreases, remains unchanged.

3. a. $x_{11}, x_{22}, x_{33}$ b. Since $\min\{20, 10, 30\} = 10$, $A_{22}$ leaves.
c. 10 d. 30, 20, 30.

5. a. Since $c_{31} - z_{31} = -2$, the tableau is not optimal.
b. $A_{11}, A_{32}$; $x_{11}$ and $x_{32}, x_{11}$; 10; −2, $10(-2) = -20$.
e. 420

7. a. $X_T = [20, 20, 0, 0, 10, 25]$.
b. The tableau is not optimal; $c_{13} - z_{13} = -2$.
c. $A_{22}, A_{12}$ d. 25, $A_{12}$ e. $X = [20, 0, 20, 0, 30, 5]^T$.
f. $c_{12} - z_{12} = 2, c_{21} - z_{21} = 0$; the tableau is optimal.

## 7.8 Exercises (page 310)

1. Three tableaus are required. The optimal $X^T = [0, 30, 30, 35, 0, 5]$, with $f = 590$; $c_{11} - z_{11} = c_{22} - z_{22} = 1$.

3. Five tableaus are required. The optimal $X^T = [0, 5, 15, 5, 10, 0, 25, 0, 0]$, with $f = 305$; $c_{11} - z_{11} - 1, c_{23} - z_{23} = 3, c_{32} - z_{32} = 2, c_{33} - z_{33} = 6$.

5. Three tableaus are required. The optimal $X^T = [0,500, 400,0, 0, 800, 600, 50]$; with $f = 13,400$; $c_{11} - z_{11} = 1, c_{22} - z_{22} = 9, c_{31} - z_{31} = 4$.

7. Create a dummy destination to absorb the leftover supply. Allow units to be shipped from any of the sources. Let all $c_{ij} = 0$ for this destination.

## 8.2 Exercises (page 315)

1. a. $\{H_N H_D H_Q, H_N H_D T_Q, H_N T_D H_Q, H_N T_D T_Q, T_N H_D H_Q, T_N H_D T_Q, T_N T_D H_Q, T_N T_D T_Q\} = S_6$.
b. $N_6 = 8$.

## 8.3 Exercises (page 319)

1. Subset

3. a. $\{H\}, \{T\}$ b. $\{H, T\}$ c. 4; $\{\ \}, \{H\}, \{T\}, \{H, T\}$ d. $\{T\}$

5. a. See Exercise 8.2.1a.
   b. $2^8 = 256$.
   c. $(H_N H_D H_Q), (H_N H_D T_Q), (H_N T_D H_Q), (H_N T_D T_Q)$
   d. All except $(T_N T_D T_Q)$
   e. $(H_N H_D H_Q)$ and $(T_N T_D T_Q)$
   f. No, i.e., no two of the above events are mutually exclusive.
   g. e.g., "exactly one head" and "exactly one tail"

## 8.4 Exercises (page 323)

1. a. 1/8 b. 4/8 c. 7/8 d. 6/8 e. 2/8

3. a. Is greater than
   b. One
   c. Union, elementary events

## 8.5 Exercises (page 327)

1. a. 2/6 b. 3/6 c. 3/6 d. 1/6 e. 0 f. 1/6

3. 3/4

5. a. 108/216 = 1/2
   b. $P(A \cap B) = 52/216$, $P(A \cup B) = 160/216$
   c. 160/216

7. a. 5/2 b. −2/36

## 8.6 Exercises (page 332)

1. a. 2/3 b. 1/2 c. Yes d. Yes

3. 1/2 5. a. 1/2 b. No; $P(B \mid A) = 1$.

7. a. 0.1311348 b. 1311348/9241359 = 0.1418999
   c. 0.1496579 d. 0.1709739 e. 0.2345038 f. 0.4993002

## 8.7 Exercises (page 336)

1. a. 1/3 b. 2/3

3. 1/3

5. a. 1/4 b. No

## 8.8 Exercises (page 341)

1. b. 9/16 c. 7/16

3. a. 0.72 b. 0.18 c. 0.04 d. 0.06

## 8.9 Exercises (page 344)

1. The matrices in d. and e. are stochastic.

3. The matrices in b., d., e., g. are stochastic.

## 8.10 Exercises (page 348)

1. a. 1 b. 0 c. 1/3 d. 2/3

3. a. 2/9, 7/9 b. 7/27, 20/27

5. a. $\begin{bmatrix} 2/3 & 1/3 \\ 1/3 & 2/3 \end{bmatrix}$ b. $\begin{bmatrix} 5/9 & 4/9 \\ 4/9 & 5/9 \end{bmatrix}$

## 8.11 Exercises (page 350)

1. a. 1/2 b. 1/2 c. [1/2, 1/2] d. 1/2 e. 1/2 f. [1/2, 1/2]

3. $P^{(1)} = [5/9, 4/9]$
   $P^{(2)} = [14/27, 13/27]$
   $P^{(3)} = [41/81, 40/81]$

## 9.1 Exercises (page 355)

1. a. [0.146, 0.408]

b.

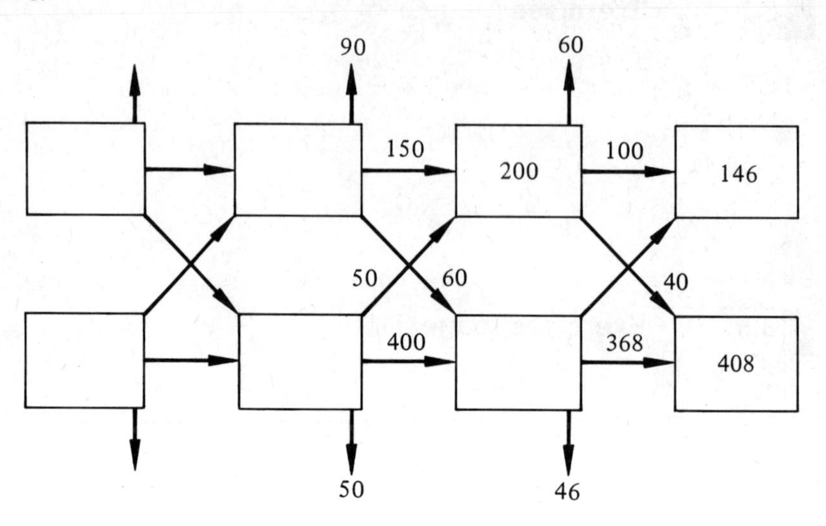


3. a. $X(1) = [0.46, 0.54]$, $X(2) = [0.4376, 0.5624]$.

   b. $[0.16, 0.04]$ is the "new recruit" vector. $X(0)W^T$ is the proportion lost to attrition. It must be replaced to maintain the original population size.

   c. 0.2

   d. Period 1—0.16 to $G_1$, 0.04 to $G_2$; period 2—0.1536 to $G_1$, 0.0384 to $G_2$.

   e.

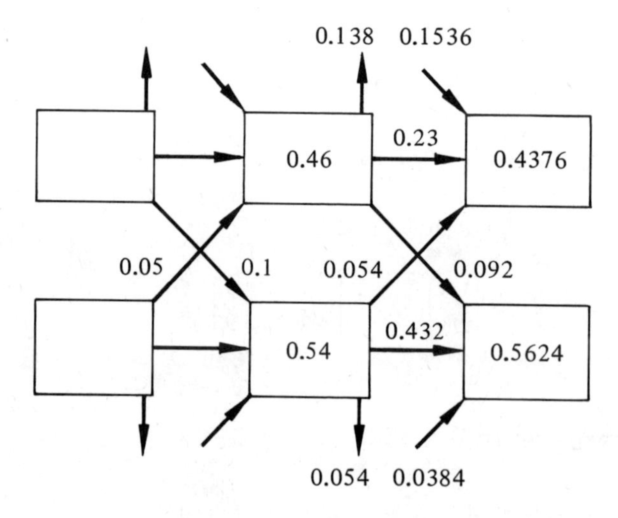


   f. $[4376, 5624]$; total attrition = 3920 = total recruited.

## 9.2 Exercises (page 359)

1. a. All points between $[0.5, 0.5]$ and $[0.3, 0.7]$, since

$$[0.5, 0.5] = X(0)[P + W^T[1, 0]] \quad \text{and} \quad [0.3, 0.7] = X(0)[P + W^T[0, 1]].$$

   b. $R = [0.5, 0.5]$.

c. Let $S_1$ be the set of points between $[0.5, 0.5]$ and $[0.3, 0.7]$, and $S_2$ the set of points between $[0.38, 0.62]$ and $[0.22, 0.78]$. $S_1 \cup S_2$ is the solution.

d. Yes

e. $R_1 = [0.25, 0.75], R_2 = [1/17, 16/17]$.

f. If $X(0) = [1, 0]$ and $R = [1, 0]$, $X(0)[P + W^T R] = [0.8, 0.2]$.

## 9.3 Exercises (page 362)

1. a. Find $p_{ij}, w_i \geqslant 0$ such that

$$\begin{aligned}
p_{11} \quad + 9p_{21} \quad + 0.5w_1 + 4.5w_2 &= 3\\
p_{12} \quad + 9p_{22} + 0.5w_1 + 4.5w_2 &= 7\\
p_{11} + p_{12} \quad + w_1 &= 1\\
p_{21} + p_{22} \quad + w_2 &= 1
\end{aligned}$$

and $p_{12} - p_{21} - w_1 - w_2$ is maximized.

b. $P = \begin{bmatrix} 0 & 1 \\ 1/3 & 2/3 \end{bmatrix}, W = [0, 0]$.

An artificial vector will stay in the solution at value 0. This means that the associated original equation was redundant.

c. $P = \begin{bmatrix} 0 & 1 \\ 0 & 1/3 \end{bmatrix}, W = [0, 2/3]$.

## 10.1 Exercises (page 368)

1. b. $X_1^T = [1, 2, 3, 4, 5, 6], U^T = [1, 1, 1, 1, 1, 1]$,
$X_2^T = [3, 1, 6, 4, 9, 7]$.

c. Normal, $X_1^T X_1, X_1^T U, X_1^T U, U^T U$;

$$\begin{aligned}
91a + 21b &= 126\\
21a + 6b &= 30\\
a = 6/5, b &= 4/5.
\end{aligned}$$

d. The sum of the squared deviations is 16.8.

e. The sum is 18.375.

f. The sum is 17.688.

3. a. The equation of the regression line in $\hat{x}_2 = 7/3$, Thus the mean of the $x_2$ values is used in this case as the best predictor.

b. $\hat{x}_2 = 0.57377x_1 - 0.36056$

c. $\hat{x}_2 = -0.485640x_1 + 5.18016$

d. In a. $\hat{x}_2(4.5) = \hat{x}_2(6) = 7/3$.
In b. $\hat{x}_2(4.5) = 2.22131, \hat{x}_2(6) = 3.08197$
In c. $\hat{x}_2(4.5) = 2.994778, \hat{x}_2(6) = 2.266319$

## 10.2 Exercises (page 372)

1. a. $\overline{Y} = \begin{bmatrix} 5 \\ 11/3 \end{bmatrix}$. b. $\overline{Y} = \begin{bmatrix} 3 \\ 1 \end{bmatrix}$.

3. a. $\overline{X} = \begin{bmatrix} 913/299 \\ 915/299 \end{bmatrix} = \begin{bmatrix} 3.053512 \\ 3.060201 \end{bmatrix}$.

   b. $\overline{X} = \begin{bmatrix} -338/429 \\ 611/429 \end{bmatrix} = \begin{bmatrix} -0.787878 \\ 1.424242 \end{bmatrix}$.

   c. $\overline{X} = \begin{bmatrix} 1001/313 \\ 855/313 \end{bmatrix} = \begin{bmatrix} 3.198083 \\ 2.731629 \end{bmatrix}$.

# Index